Self-Organized Mobile Communication Technologies and Techniques for Network Optimization

Ali Diab
Al-Baath University, Syria

A volume in the Advances in Wireless Technologies and Telecommunication (AWTT) Book Series

Information Science REFERENCE
An Imprint of IGI Global

Published in the United States of America by
Information Science Reference (an imprint of IGI Global)
701 E. Chocolate Avenue
Hershey PA, USA 17033
Tel: 717-533-8845
Fax: 717-533-8661
E-mail: cust@igi-global.com
Web site: http://www.igi-global.com

Library of Congress Cataloging-in-Publication Data

Names: Diab, Ali, 1976- editor.
Title: Self-organized mobile communication technologies and techniques for network optimization / Ali Diab, editor.
Description: Hershey, PA : Information Science Reference, [2016] | Includes bibliographical references and index.
Identifiers: LCCN 2016001046| ISBN 9781522502395 (hardcover) | ISBN 9781522502401 (ebook)
Subjects: LCSH: Mobile communication systems--Automatic control. | Wireless communication systems--Automatic control. | Self-organizing systems. | Cognitive radio networks.
Classification: LCC TK5103.2 .S458 2016 | DDC 621.3845/6--dc23 LC record available at http://lccn.loc.gov/2016001046

This book is published in the IGI Global book series Advances in Wireless Technologies and Telecommunication (AWTT) (ISSN: 2327-3305; eISSN: 2327-3313)

British Cataloguing in Publication Data
A Cataloguing in Publication record for this book is available from the British Library.

For electronic access to this publication, please contact: eresources@igi-global.com.

Advances in Wireless Technologies and Telecommunication (AWTT) Book Series

Xiaoge Xu
The University of Nottingham Ningbo China, China

ISSN: 2327-3305
EISSN: 2327-3313

Mission

The wireless computing industry is constantly evolving, redesigning the ways in which individuals share information. Wireless technology and telecommunication remain one of the most important technologies in business organizations. The utilization of these technologies has enhanced business efficiency by enabling dynamic resources in all aspects of society.

The **Advances in Wireless Technologies and Telecommunication Book Series** aims to provide researchers and academic communities with quality research on the concepts and developments in the wireless technology fields. Developers, engineers, students, research strategists, and IT managers will find this series useful to gain insight into next generation wireless technologies and telecommunication.

Coverage

- Radio Communication
- Mobile Communications
- Mobile Web Services
- Cellular Networks
- Network Management
- Telecommunications
- Wireless Sensor Networks
- Digital Communication
- Mobile Technology
- Broadcasting

IGI Global is currently accepting manuscripts for publication within this series. To submit a proposal for a volume in this series, please contact our Acquisition Editors at Acquisitions@igi-global.com or visit: http://www.igi-global.com/publish/.

The Advances in Wireless Technologies and Telecommunication (AWTT) Book Series (ISSN 2327-3305) is published by IGI Global, 701 E. Chocolate Avenue, Hershey, PA 17033-1240, USA, www.igi-global.com. This series is composed of titles available for purchase individually; each title is edited to be contextually exclusive from any other title within the series. For pricing and ordering information please visit http://www.igi-global.com/book-series/advances-wireless-technologies-telecommunication-awtt/73684. Postmaster: Send all address changes to above address.

Titles in this Series

For a list of additional titles in this series, please visit: www.igi-global.com

Emerging Innovations in Wireless Networks and Broadband Technologies
Naveen Chilamkurti (La Trobe University, Australia)
Information Science Reference • copyright 2016 • 292pp • H/C (ISBN: 9781466699410) • US $165.00 (our price)

Critical Socio-Technical Issues Surrounding Mobile Computing
Norshidah Mohamed (Prince Sultan University, Kingdom of Saudi Arabia and Universiti Teknologi Malaysia, Malaysia) Teddy Mantoro (USBI-Sampoerna University, Indonesia) Media Ayu (USBI-Sampoerna University, Indonesia) and Murni Mahmud (International Islamic University Malaysia, Malaysia)
Information Science Reference • copyright 2016 • 357pp • H/C (ISBN: 9781466694385) • US $210.00 (our price)

Handbook of Research on Next Generation Mobile Communication Systems
Athanasios D. Panagopoulos (National Technical University of Athens, Greece)
Information Science Reference • copyright 2016 • 604pp • H/C (ISBN: 9781466687325) • US $370.00 (our price)

Game Theory Framework Applied to Wireless Communication Networks
Chungang Yang (Xidian University, China) and Jiandong Li (Xidian University, China)
Information Science Reference • copyright 2016 • 502pp • H/C (ISBN: 9781466686427) • US $235.00 (our price)

Enabling Real-Time Mobile Cloud Computing through Emerging Technologies
Tolga Soyata (University of Rochester, USA)
Information Science Reference • copyright 2015 • 399pp • H/C (ISBN: 9781466686625) • US $195.00 (our price)

Emerging Perspectives on the Design, Use, and Evaluation of Mobile and Handheld Devices
Joanna Lumsden (School of Engineering and Applied Science, Aston University, UK)
Information Science Reference • copyright 2015 • 334pp • H/C (ISBN: 9781466685833) • US $205.00 (our price)

Technological Breakthroughs in Modern Wireless Sensor Applications
Hamid Sharif (University of Nebraska – Lincoln, USA) and Yousef S. Kavian (Shahid Chamran University of Ahvaz, Iran)
Information Science Reference • copyright 2015 • 417pp • H/C (ISBN: 9781466682511) • US $200.00 (our price)

Handbook of Research on Software-Defined and Cognitive Radio Technologies for Dynamic Spectrum Management
Naima Kaabouch (University of North Dakota, USA) and Wen-Chen Hu (University of North Dakota, USA)
Information Science Reference • copyright 2015 • 927pp • H/C (ISBN: 9781466665712) • US $505.00 (our price)

IGI GLOBAL
DISSEMINATOR OF KNOWLEDGE
www.igi-global.com

701 E. Chocolate Ave., Hershey, PA 17033
Order online at www.igi-global.com or call 717-533-8845 x100
To place a standing order for titles released in this series, contact: cust@igi-global.com
Mon-Fri 8:00 am - 5:00 pm (est) or fax 24 hours a day 717-533-8661

Table of Contents

Detailed Table of Contents

Conflicts-free auto-configuration of Physical Cell Identities (PCIs) is a main challenge in LTE systems. This challenge is the focus of this chapter, which provides a survey of present PCI assignment as well as allocation schemes and evaluates by means of simulative studies the performance of four well-known distributed PCI assignment approaches. The simulation is achieved applying two types of scenarios, namely static and dynamic. Static scenario stands for an already deployed LTE network that is a subject to a configuration using a specific PCI assignment approach, while the dynamic scenario tries to mimic the situations LTE networks are expected to face in real-life scenarios. Our results show that network dynamics heavily affect the performance of PCI assignment approaches, and even may transit LTE networks to an unstable state. So, it is essential to consider network dynamics by concept in any PCI assignment scheme to be developed.

This chapter discusses network optimization methods for enabling self-organization in current cellular networks such as Long Term Evolution (LTE)/LTE-Advanced (LTE-A), and the upcoming 5G networks. Discrete and continuous optimization models are discussed for developing distributed algorithms for self-configuration and self-optimization. The focus is on Self-Organized Networking (SON) problems, which are relevant to small cell networks. Examples include Physical Cell-ID (PCI) assignment, Primary Component Carrier (PCC) selection, Inter-Cell Interference Coordination (ICIC), and network synchronization. A conflict-graph model is considered for PCI assignment and PCC selection problems, which paves the way for different graph coloring algorithms with self-organizing properties. Algorithms for self-organized ICIC and network synchronization are also developed in a principled manner, through a network utility maximization framework. This systematic approach leads to a variety of algorithms which adhere to self-organization principles, but have varying requirements in terms of inter-cell coordination and computation complexity. Fully distributed self-organizing algorithms do not involve any inter-cell dedicated message-passing, and thus are faster and more scalable than the ones that are distributed but require local coordination via exchange of messages between cells. However, local coordination enables higher network utility and better convergence properties.

During the past few decades wireless technology has seen a tremendous growth. The recent introduction of high-end mobile devices has further increased subscribers' demand for high bandwidth and quality of service. The number of nodes in future cellular systems will be too large to be configured and maintained manually. Further the mobility of users, the varying nature of the wireless channel and variation in user demand systems requires optimization of network parameters and delay in configuration may cause congestion and poor Quality of Service. Due to this increased complexity and the huge scale of wireless systems the network configuration, optimization and maintenance process performed by radio engineers has become inefficient and therefore, lot of research is being done to introduce self-optimizing capabilities within the network, which is expected to permit higher end user Quality of Service (QoS) and less operation cost and maintenance cost for telecom service providers. This chapter details the key aspects related to self optimization of next generation cellular networks.

Self-Organized Networks (SONs) are an advancement of today's communication that arises to overcome the problems in traditional communications in terms of their planning, configuration, optimization, healing and management since the rate of utilizing communication technology is gradually increasing day by day hence an optimum mechanism is needed to cope up with all the changes in the communication era, the result is self-organized networks. The success of SONs depends on how it is effectively utilized without any compromise in its security. However its unique characteristics offer both opportunities and challenges. The aim of this chapter is to begins with the essential concept of SONs such as Mobile Ad Hoc Networks (MANET), Vehicular Ad Hoc Networks (VANET), Wireless Sensor Networks (WSN), Wireless Mesh Networks (WMN), Peer to Peer Networks (P2P), Cognitive Radio Networks, Bio-Inspired Networks and Internet of Things (IoT) and their limitations in different perspectives. As these networks have penetrated into the human life with an anytime anywhere capability, the security of the data being processed and communicated through these networks become vital. This chapter tries to bring out the issues and challenges in providing a trust based solution mechanisms for this type of networks.

elaboration and simulation results as well. Estimating the DOA in this chapter is regarded to the purpose of using Smart antenna system. It is possible to estimate the location of a user by considering the uplink transmission system of the mobile communication system. Estimating the channel and accurate path delay is also an important task which might be done by using 1D Uniform Linear Array (ULA) or 2D Uniform Rectangular (URA) array antenna system. In this chapter, 1D ULA is considered in order to utilize some popular techniques. The performance of a communication system between two ends is substantially determined by the behaviors of the channel characteristics. It determines signal transformation while propagating through the channel between receivers and transmitters. Accurate channel information is crucial for both the transmitter and receiver ends to perform their best services. The ultimate focus of this chapter is to estimate the channel based on 2D parameter estimation. Uniform Rectangular Array (URA) is used to perform the 2D parameter estimation. It is possible to estimate Azimuth and Elevation of a source by using URA model.

Section 3
Network Design, Control, and Performance

Chapter 10
A Key to 5G Networks of Tomorrow 237
Aqeel ur Rehman, Hamdard University, Pakistan
Syed Muhammad Kashif Alam, Hamdard University, Pakistan
Ahmar Murtaza, Hamdard University, Pakistan
Iqbal Uddin Khan, Hamdard University, Pakistan

The emergence of mobile telephony increases the options of its deployment and utilization. As the research goes further, it also delivered Generations of Mobile Communication to us after rapid intervals of time. Each advancement is considers as the generation. Till now Forth Generation (4G) is in its maturity phase and researchers are planning to see 5G commercially around 2020. In upcoming Fifth Generation (5G), vision is to reduce latency typically it will be of less than 1ms, as the latency decrease one more thing will be achieved which can be named as Virtually Zero Distance. As the world is becoming the connected world, the availability of data on global network is going to be huge in size. In addition to that the new emerging concepts like Internet of Things (IoT) are giving boost to the requirement of big data storage and analysis. Internet of Things (IoT) is a concept of providing uniquely identifiable objects connectivity to Internet. This chapter is proposed to highlight the IoT concept and its requirements that are directly linked with the 5G Network technology services.

Chapter 11
Synthetic Modeling of Human Mobility Patterns 259
Ali Diab, Al-Baath University, Syria & Ilmenau University of Technology, Germany
Andreas Mitschele-Thiel, Ilmenau University of Technology, Germany

One of the wide used modeling techniques applied to model human mobility patterns is synthetic mobility modeling. Synthetic models are largely preferred and widely applied to simulate mobile communication networks. They try to capture the patterns of human movements by means of a set of equations. These models are traceable, however, not capable until now of generating realistic mobility models. The chapter highlights the state of art and provides a comprehensive investigation of current research efforts in the field of synthetic mobility modeling. It reviews well-known approaches along with their pros and cons and finalizes with a qualitative comparison between them.

The 5th Generation (5G) of mobile communication networks is being developed to address the demands and business contexts of 2020 and beyond. Its vision is to enable a fully mobile and connected society and also to trigger socio-economic transformations in ways eventually unimagined today. This means that the physical world to be covered with planned 5G networks including communication networks, humans and objects is becoming a type of an information system. So as to improve the experience of individuals, communities, societies, etc. within such systems, a thorough comprehension of intelligence processes responsible of generating, handling and controlling those data is fundamental. One of the major aspects in this context and also the focus of this chapter is the development of novel methods to model human mobility patterns, which have crucial role in forthcoming communication technologies. Human mobility patterns models can be categorized into synthetic, trace-based and community-based models. Synthetic models are largely preferred and widely applied to simulate mobile communication networks. They try to capture the patterns of human movements by means of a set of equations. These models are traceable, however, not capable of generating realistic mobility models. The key idea of trace-based models is the exploitation of available measurements and traces achieved in deployed systems to reproduce synthetic traces characterized by the same statistical properties of real traces. A main drawback of trace-based modeling of human patterns is the tight coupling between the trace-based model and the traces collected, the network topology deployed and even the geographic location, where the traces were collected. This is why the results of various trace-based models deviate clearly from each other. Sure, this prohibits the generalization of trace-based models. When one also considers that the traces themselves are rarely available, one can understand why synthetic models are preferred over trace-based ones. Community-based modeling of human movements depends on the fact stating that mobile devices are usually carried by humans, which implies that movement patterns of such devices are necessarily related to human decisions and socialization behaviors. So, human movement routines heavily affect the overall movement patterns resulting. One of the major contributions in this context is the application of social networks theory to generate more realistic human movement patterns. The chapter highlights the state of art and provides a comprehensive investigation of current research efforts in the field of trace- and social-based modeling of human mobility patterns. It reviews well-known approaches going through their pros and cons. In addition, the chapter studies an aspect that heavily relates to human mobility patterns, namely the prediction of future locations of users.

Section 4
Testbeds, Experiments, Applications, and Services

Preface

Mobile communication technologies are significantly expanding in terms of increased complexity of applications as well as services, required capacities, heterogeneity in Radio Access Technologies (RATs) and involved mobile devices types, etc. The continuous evolution of consumer uptake of mobile broadband, which is expected to experience 33 times increase compared with that of 2010, is one of the main challenges of the upcoming version of mobile communication technologies, namely the 5th Generation (5G). Other challenges include reduced energy consumption, reduced costs, cognitive spectrum support, etc. The increasing complexity of future mobile networks needs to have an adequate wireless networking architecture that supports a high degree of scalability, performance and reliability in a cost-effective manner without compromising security or quality of service. Cost-effectiveness gets more crucial when one also considers the additional cost resulting from the expected affordable operation of overlaid multi-standard networks (2G, 3G, 4G, etc.). Therefore, operators need to keep their Operational Expenses (OPEX) minimized, while simplifying the manageability of their networks and enhancing the performance. Sure, OPEX costs strongly relate to the solutions implemented in the infrastructure. This means that OPEX costs reduction implies implementing adequate solutions that, for sure, will be best if self-organized.

The book investigates important developments, research challenges and limitations of methods, models and new strategies for latest topics of next generation mobile networks. The first part of the book presents a major contribution, namely the self-organization activities in future networks. This part highlights the state of art and provides a comprehensive investigation of current research efforts in the field of SONs in the upcoming 5G mobile communication networks. It goes through SON functionalities and discusses their cyclic behavior. Discrete and continuous optimization models for developing distributed algorithms for self-organization principles are discussed. Furthermore, the coordination between various SON activities to guarantee systems' stability and reliability in addition to performance optimization is a focus of this part. Important present SON use cases and those expected to appear in the 5G networks are handled as well. The second part of the book presents radio aspects and access technologies in 5G wireless networks focusing on radio resource management, MIMIO antennas applied, cognitive radio support, etc. The third part of the book deals with network design, control and performance aspects. This part discusses the principles of Internet-of-Things (IoT) expected to be part of next generation mobile communication networks. An important tpic handeled in this part of the book is the investigation and development of novel methods to model human mobility patterns, which have myriad uses in crucial fields such as mobile communication research, urban planning, ecology, epidemiology, etc. The last part of the book deals with the results of already deployed testbeds and performed experiments. It also covers the new applications and services expected to appear in upcoming 5G networks.

This book will serve as a useful reference for industries, researchers, educators, and governments to accelerate standardization process and promote applications and services for future wireless networks.

Section 1
Self-Organization in Future Mobile Communication Systems

Chapter 1
Impact of Network Dynamics on the Performance of Distributed Physical Cell Identities Assignment Schemes

Ali Diab
Al-Baath University, Syria & Ilmenau University of Technology, Germany

Andreas Mitschele-Thiel
Ilmenau University of Technology, Germany

ABSTRACT

Conflicts-free auto-configuration of Physical Cell Identities (PCIs) is a main challenge in LTE systems. This challenge is the focus of this chapter, which provides a survey of present PCI assignment as well as allocation schemes and evaluates by means of simulative studies the performance of four well-known distributed PCI assignment approaches. The simulation is achieved applying two types of scenarios, namely static and dynamic. Static scenario stands for an already deployed LTE network that is a subject to a configuration using a specific PCI assignment approach, while the dynamic scenario tries to mimic the situations LTE networks are expected to face in real-life scenarios. Our results show that network dynamics heavily affect the performance of PCI assignment approaches, and even may transit LTE networks to an unstable state. So, it is essential to consider network dynamics by concept in any PCI assignment scheme to be developed.

1. INTRODUCTION

Ubiquitous access to information anywhere, anytime and anyhow is a main feature of future communication networks which experience a tremendous development clearly evident from the wide variety of new services and broadband applications way beyond the classical application, i.e. telephony (Diab & Mitschele-Thiel, 2014). The Internet itself emerges towards what is today termed "*the Internet of things*", which assumes that users of future communication networks are objects, not only humans. Each object

DOI: 10.4018/978-1-5225-0239-5.ch001

is embedded with sensors and is able to communicate. In other words, the physical world itself including communication networks, humans and objects is becoming a type of huge information system. To satisfy always increasing demands of users within such communication and information systems, the better support of mobility, the provision of services at low costs, etc. are the drivers standing behind the standardization of 4th Generation (4G) mobile communication networks, termed Long Term Evolution (LTE). Researchers expect that the number of LTE subscribers will reach the base of 3G/UMTS networks (1.087 million subscribers) by the year 2015 (Garza, Ashai, Monturus & Syputa, 2010).

A major challenge in LTE networks is a cost-effective network operation, which necessitates carefully controlling the Operational Expenses (OPEX). Cost-effectiveness gets more crucial when considering the additional costs resulting from the expected affordable operation of overlaid multi-standard networks. 3G mobile communication technologies have shown that network-related OPEX costs contribute by approximately 30% to the total cost, see (Motorola, 2009). Note that OPEX costs strongly depend on the solutions implemented. This means that the reduction of OPEX costs implies the provision of adequate solutions. Self-Organization plays in this context a major role since it significantly contributes in reducing OPEX costs, see (Østerbø & Grøndalen, 2012) and (Diab & Mitschele-Thiel, 2014).

So, the release being standardized by the 3G Project Partnership (3GPP) (3GPP, 2015) with self-organization capabilities is referred to as LTE-Advanced (Dahlman, Parkvall & Skoeld, 2011). This standard aims at minimizing OPEX costs, while optimizing network performance as a whole. To further contribute to Self-Organizing Networks (SONs), the Next Generation Mobile Networks (NGMN) alliance (NGMN, 2015) was also constructed. This alliance has summarized SONs requirements in a number of operation use cases, see (Ramiro & Hamied, 2012) and (Lehser, 2008). One of the use cases is an auto-configuration of evolved NodeBs (eNBs) with conflicts-free Physical Cell Identities (PCIs). This chapter focuses on this use case, addresses the problems faced in its context and highlights the state of art. The major contribution of this chapter is the investigation of the impact of network dynamics on the performance of four well-known distributed PCI assignment schemes, namely the Stable Graph Coloring (SGC) approach, the GC approach, the LTE Standard Distributed (LSD) proposal and the Nokia-Siemens (NS) solution. This investigation is of a great importance since there are no studies until now that tried to analyze this issue, although it is accepted that LTE networks will have topologies of a dynamic nature.

The reminder of the chapter is organized as follows: section 2 addresses the problems faced in the context of PCI assignment and allocation. Present PCI assignment and allocation techniques are overviewed in sections 3 and 4, respectively. These sections finalize with qualitative comparisons between overviewed techniques. Simulation assumptions, scenario and results are provided in section 5 followed by a conclusion and an outlook in section 6.

2. PHYSICAL CELL IDENTITY (PCI) AUTOMATIC CONFIGURATION

Different from 2G and 3G systems, the handling of handovers in LTE systems is delegated to eNBs, which directly signal control messages between them via X2 interfaces. A Neighbor Cell Relation (NCR) list with detected neighbor eNBs is constructed by each eNB. This list contains in principle the eNBs User Equipments (UEs) eventually may hand off in the near future. Neighbors eNBs are typically detected based on reports sent from UEs, which identify cells either based on broadcasted PCIs or Global cell IDentifiers (GIDs).

Reading the PCI of a cell necessitates about 5 ms, while acquiring the GID of the cell requires considerably more time (may reach 160 ms (Lee, Jeong, Saxena & Shin, 2009)). So, acquiring the GID needs long measurement gaps, increases handoffs latencies, consumes the battery power of UEs, etc. Note also that UEs cannot exchange data during measurement gaps without multiple receiver capabilities. For these reasons, the use of GIDs is restricted to when it is absolutely necessary. This implies that UEs should be able to identify cells in measurement reports only via their PCIs. This is possible as long as there are no PCIs conflicts. If identification errors arise, UEs are instructed to do measurements to provide the GID of the detected cell.

The identification of cells only via their PCIs is not that simple. The following discusses this issue in more details and provides the problem statements addressed in this context and the requirements each PCI assignment solution has to satisfy.

2.1. Problem Statements

PCIs are reference signal sequences (comparable to 3G scrambling codes (Bandh, Carle & Sanneck, 2009) used for several purposes, e.g. as regional unique identifiers on the physical level (Diab & Mitschele-Thiel, 2012), time and frequency references (3GPP, 2009), etc. Actually, the synchronization signals (e.g. the Primary Synchronization Signal (PSS) and the Secondary Synchronization Signal (SSS)) and also reference signals (e.g. the Cell-specific Reference Signal (CRS) and the UE-specific Reference Signal (URS)) of any cell are constructed based on sequences with unique mapping to its PCI. Furthermore, the PCI is used to generate the scrambling sequences of most physical channels (Wu, Jiang, Wu & Zhang, 2010), e.g. the Physical Broadcast Channel (PBCH), the Physical Control Format Indicator Channel (PCFICH), etc., see table 1 for more details.

The PCI itself is given by synchronization channels during the cell search procedure and needs less than 5 ms to be decoded by UEs. Different from PCIs, GIDs are globally unique. They are not reference

Table 1. Using the PCI of a cell in initializing the scrambling sequence of its physical channels

Physical Channel	Lower Initialization Value of the Scrambling Sequence	Re-Initialization Period
PBCH	PCI	Fourth radio frame
PCFICH	$PCI + (Subframe - Num + 1) * (2 * PCI + 1) * 2^9$	Sub frame
PHICH	$PCI + (Subframe - Num + 1) * (2 * PCI + 1) * 2^9$	Sub frame
PDCCH	$PCI + Subframe - Num * 2^9$	Sub frame
PDSCH	$PCI + Subframe - Num * 2^9 + \text{Stream}_{ID} * 2^{13} + RNTI * 2^{14}$	Sub frame
PUCCH	$(Subframe - Num + 1) * (2 * PCI + 1) * 2^{16} + RNTI$	Sub frame
PUSCH	$PCI + Subframe - Num * 2^9 + RNTI * 2^{14}$	Sub frame

signals and are longer than PCIs (Golaup, Mustapha & Patanapongpibul, 2009), (Kwak, Lee, Kim, Saxena & Shin, 2008). To obtain a GID of a cell, UEs must read the system information.

As mentioned above, eNBs benefit from NCR lists that contain possible handoff candidate eNBs with connectivity information (IP addresses, PCIs, GIDs, etc.). In fact, each eNB maps between the PCI and the globally unique GID of each of its neighbors. So, this eNB is capable of identifying the eNBs UEs detect in their measurement reports only via their PCIs as long as there are no PCIs conflicts. If identification errors arise, UEs are instructed to do measurements to provide the GID of the detected cell. Therefore, it is essential to carefully distribute PCIs among cells, so that PCIs conflicts disappear. This is, however, challenging due to the limited range of PCIs available. There are only 168 pseudo random sequences termed cell identity groups, each has three orthogonal sequences, which leads to a space of 504 values (3GPP, 2009). Sure, the space of PCIs should be handled carefully, so that PCIs are re-used without causing problems. The challenge addressed gets more crucial when the available range of PCIs is further divided into smaller sub-ranges, which can be done for multiple reasons, e.g. a sub-range of PCIs may be reserved for different types of cells (macro-cells, micro-cells, etc.) (Lee, Jeong, Saxena & Shin, 2009), newly switched on cells (Nokia Siemens Networks and Nokia, 2008), etc. Network operators may also introduce several artificial subgroups of PCIs for specific reasons, e.g. handling planning constraints at network borders or instance. Keep also in mind that several properties of the channels used in any cell depend on its PCI, which further constraints the PCI assignment process, see (3GPP, 2009).

2.2. PCI Automatic Configuration Requirements

The NGMN alliance has considered the auto-configuration of PCIs as a part of the auto-configuration process accomplished in the pre-operational phase. Following requirements are, therefore, essential:

- **Collisions-Free PCI Assignment:** A collision appears when a given eNB has a neighbor eNB assigned the same PCI (Amirijoo, Frenger, Gunnarsson, Kallin, Moe & Zetterberg, 2008). As known, UEs continuously monitor the received signal strength and detect based on monitoring results possible candidate eNBs, to which they may hand off. No handoff decisions are taken as long as the same PCI is reported in UEs measurements reports. In fact, UEs are in the same cell from the serving eNB point of view. So, UEs will move outside the coverage area of the serving eNB without detecting the eNB that serves the new cell. Of course, this means communication disruptions or dropping.
- **Confusions-Free PCI Assignment:** A confusion means that a specific eNB has two neighbor eNBs configured with the same PCI (3GPP, 2010). Of course, one of them will be listed in the NCR list of the given eNB. Such situation appears, for instance, when an eNB was not able to detect all neighbors during the auto-configuration process. The serving eNB gets confused in such cases since handovers are recognized as handovers events to the eNB listed in the NCR list. This in turn results in failing handovers to the eNB not stored in the NCR list, see Figure 1.
- **Adaptive PCI Assignment:** As known, LTE networks will have a high degree of dynamics. They will contain cells structured in a layered manner, i.e. macro-cells, micro-cells, etc. present in the same geographical area. These networks will also continuously be adapted to satisfy operator requirements, e.g. cells' sizes changed (Bandh, Carle & Sanneck, 2009), new cells added (not only at network borders, but also in areas where higher capacity is needed) (Bandh, Carle & Sanneck, 2009), etc. Moreover, the operation of these networks itself creates a continuous dynamicity in

Figure 1. An example scenario illustrating confusions consequences

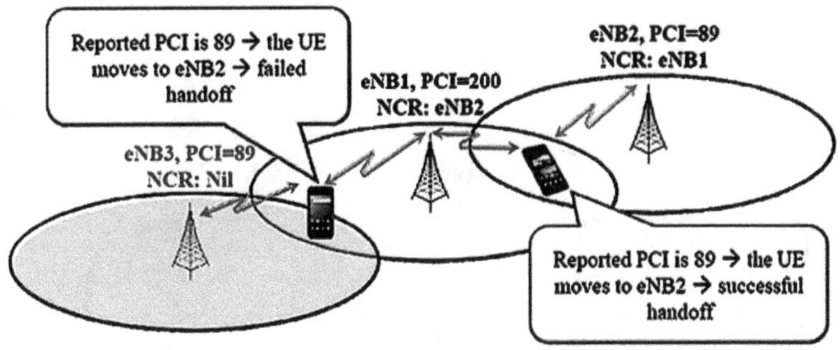

their topologies. The NGMN, for instance, has specified the use case "coverage and capacity optimization" to benefit from self-organization as well, see (3GPP, 2010). This use case dynamically adapts cells' sizes based on current network status, e.g. load, kind of traffic, number of users, time of operation, etc. This will often impact NCR lists of eNBs and eventually may cause conflicts. So, the applied PCI assignment scheme should be capable of automatically doing PCIs re-configuration tasks when necessary. Consider also the utilization of Home eNBs (HeNBs), which further introduces a high degree of dynamic to LTE networks, see (Lee, Jeong, Saxena & Shin, 2009). HeNBs, termed femtocells as well, are low-power cellular base stations operated in licensed spectrum. They provide an excellent experience for indoor scenarios through extended throughput, services, etc. LTE network operators consider the use of HeNBs in their networks to offload traffic from their macro- and/or micro-cells, so that the performance of their networks is optimized (FemtoForum, 2011). In other words, they are considered parts of the LTE networks, they are switched on in their coverage. Note, however, that HeNBs are frequently turned on and off by personal owners, not by network operators and must be integrated with existing LTE networks when switched on. This introduces, for sure, a high degree of dynamics to present LTE networks (Lee, Jeong, Saxena & Shin, 2009). Figure 2 illustrates the issues discussed above.

- **Adequate PCI Allocation Technique:** As mentioned above, the introduction of HeNBs highly affects the performance of deployed LTE networks since they significantly contribute in increasing the dynamics of deployed LTE networks. In fact, this is the half of the story. There are other important issues to be considered, e.g. expected higher operational complexity and management costs, access control, mobility support, interference management, etc., see (Wu, Jiang, Wu & Zhang, 2010). It is worth mentioning that a performance degradation of backhaul networks will be the result if HeNBs are not well managed and properly integrated in these networks. Keep also in mind that HeNBs are expected to be deployed in large amounts.

From the PCI assignment point of view, in addition to assigning conflict-free PCIs to HeNBs, HeNBs should also be recognizable by UEs based on their PCIs. This will simplify facing the above mentioned challenges, especially the access control issue. In detail, HeNBs are configured according to release 9 with different access modes, namely closed, hybrid and open access mode, see [Ref6]. HeNBs configured with closed access mode form Closed Subscriber Groups (CSGs) and allow certain subscribers (CSG members) to benefit from these HeNBs (the cells served by such HeNBs are termed CSG-cells). HeNBs configured with hybrid access mode are accessible by both CSG and non-CSG members. Sure,

Figure 2. An example scenario illustrating consequences of network dynamics

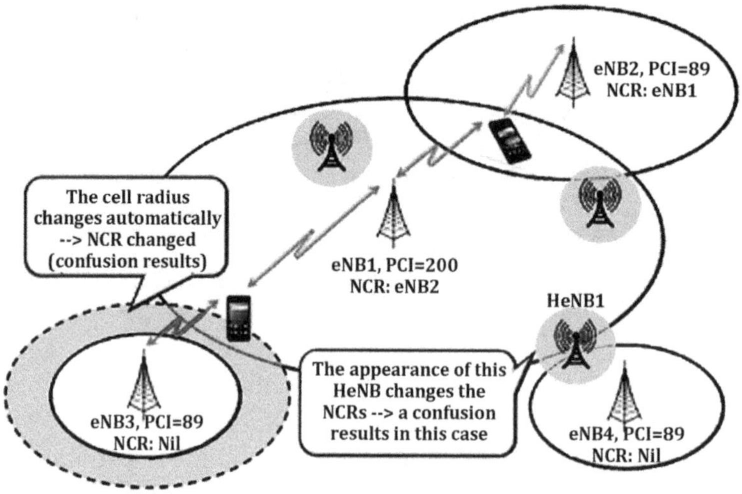

CSG-members obtain different access rights and priorities. In the open access mode, HeNB operates as a normal eNB, see Table 2 for more detailed view. UEs determine the access mode by reading the system information (concrete the System Information Block 1 (SIB1)) using two parameters, namely the CSG-Indication and CSG-Identity, see Table 3.

In LTE, UEs are designed to be aware of the existence of HeNBs. This implies that UEs should search for HeNBs. Of course, this consumes the battery power of UEs, conducts signaling overhead to the core network and affects inbound handovers (handovers from macro-/micro-cells to HeNBs) since long search delays result in communication disruptions. So, to optimize the search procedure, UEs should be able to recognize whether the detected cell is a HeNB. If the UE is not interested in accessing HeNBs, no need to hand off to the detected cell and, thus, no need to complete the search procedure. Sure, UEs will know that the cell detected is served by a HeNB after reading the system information, which takes a long time as mentioned earlier. Therefore, UEs must have capabilities to differentiate HeNBs from eNBs based on their PCIs. Of course, if the access mode is also distinguishable using the PCI, this would be ideal. In

Table 2. HeNBs access modes

HeNB Access Mode	Use Case	UE - CSG Membership	
		Member	Non-Member
Closed (CSG cell)	Private	Access	Prohibited
hybrid	Private/ Public	Access with privilege	Access
Open	Public	Access	Access

Table 3. Parameters required to determine access modes

Definitions in System Information Block SIB1	
CSG-Indication	CSG-Identity
True	Present
False	Present
False	Absent

brief, the solution of the mentioned problem is an adequate PCI allocation scheme that allocates which PCIs are assigned to eNBs and which are reserved for HeNBs.

- **Secure, scalable and quick** PCI assignment process with minimal

2.3. Summary

In brief, the auto-configuration of PCIs cannot be considered as just selecting a suitable PCI during the pre-operational phase. In fact, it is a continuous task done also during the operational phase. Sure, PCIs re-configurations during a commercial operation are not desirable, since they may disrupt ongoing applications and services. Consider that the recover from PCIs conflicts may trigger consecutive conflicts, which may lead to network instability. Keep also in mind that PCIs re-configurations cannot be done at any time. Such actions must be judged against their consequences on ongoing applications and services (3GPP, 2008).

3. PHYSICAL CELL IDENTITY (PCI) ASSIGNMENT SCHEMES

Researchers classify PCI assignment schemes into centralized and distributed schemes (3GPP, 2008). Centralized schemes depend on a central entity with global knowledge, while distributed schemes utilize local knowledge only (3GPP, 2009a). The following provides more details about both categories.

3.1. Centralized Schemes

As mentioned, centralized approaches depends on a central entity with global knowledge. This entity either implements PCI assignment routines or provides all information necessary to assign adequate PCIs (3GPP, 2008). Examples include the LTE Standard Centralized (LSC) scheme (3GPP, 2009a), centralized graph coloring approaches (Bandh, Carle & Sanneck, 2009), (Liu, Li, Zhang & Lu, 2010), (Shahab & Bhatti, 2013) and the Centralized Measurements-based (CM-b) approach (Amirijoo, Frenger, Gunnarsson, Kallin, Moe & Zetterberg, 2008a).

3.1.1 LTE Standard Centralized (LSC) Scheme

LTE considers a centralized scheme as a possible solution to assign PCIs to eNBs. This scheme uses the Operation, Administration and Maintenance center (OAM) as a central entity. This entity selects and signals a certain PCI to the eNB under configuration during the pre-operational phase. The given eNB then has to configure itself using this PCI. No implementation details are described in the specification. This is left open to the provider.

3.1.2 Centralized Graph Coloring Schemes

Centralized GC approaches map the PCI assignment problem to the well-known problem of graph coloring, which has a long history in mobile communication world, see (Eisenblaetter, A., 2003). The proposal presented in (Bandh, Carle & Sanneck, 2009) is implemented in the OAM as well. It considers eNBs as vertices and X2 interfaces as edges. It also considers a graph of 2-hops neighbors to guarantee conflicts freeness. The developers state that the minimal number of colors, termed chromatic number, necessary to color a graph is known to be NP-complete and equals to the degree of the graph+1 in the worst case. So, to approach the graph chromatic number, the eNB under configuration checks first if it is possible to reuse a PCI from those used by its 3-hops neighbors.

The enhanced GC scheme (Liu, Li, Zhang & Lu, 2010) works similarly. Once the new eNB is switched on, it obtains configuration information from the OAM or a specific entity/database. Following that, the OAM constructs a neighbor graph including 1-hop and 2-hops neighbors and selects an adequate PCI by applying the enhanced graph coloring algorithm. The enhancement, in principle, aims at improving the PCI utilization ratio through a better re-use of PCIs. This is done by sorting the 2-hops neighbors of the eNB under configuration in descending order based on the number of their 1-hop neighbors. It selects then the first eNB in the sorted list and tries to reuse a PCI from those assigned to the 1-hop neighbors of the selected eNB. If this fails, the second eNB in the sorted list is checked, and so on. To further improve the PCIs utilization ratio, the authors implement the GC concepts in (Shahab & Bhatti, 2013) based on neural networks principles.

3.1.3 Centralized Measurements-Based (CM-b) Scheme

The CM-b approach operates in the Operation Support System (OSS), which is assumed to have a global knowledge of the network. When a new eNB is switched on, it first contacts a specific configuration server to begin auto-configuration tasks, during which an initial PCI is assigned. After that, the new eNB benefits from UEs measurement reports to update NCR lists and to detect PCI confusions. When the serving eNB receives a measurement report containing a PCI not present in its NCR list, the specific UE is instructed to read and report the GID of the detected cell. After that, the serving eNB communicates with the new eNB, which consequently updates its NCR list. PCI confusions are detected during the NCR update procedure and, subsequently, the OSS is notified. The OSS may initiate a network-wide re-planning of PCIs or restrict the PCI update to one of the detected conflicting cells, termed the selected conflicting cell and chosen based on many factors such as the lowest GID, the shortest NCR, etc. Finally, all NCR lists containing the old PCI of the selected conflicting cell are updated.

The scheme developers state that the detection of collisions cannot be done based on UEs measurement reports since UEs will not report anything when they receive the PCI the serving eNB is currently using. So, to detect collided PCIs, the serving eNB issues transmission gaps at pre-defined times, during which UEs search for the PCI the serving eNB has. The receipt of the mentioned PCI indicates a collision. In this case, the serving eNB notifies the OSS to carry out a re-configuration process to update its PCI (the current cell is the selected conflicting cell).

3.1.4 Pros and Cons

Concerning the pros and cons of the proposals presented above, one notes that the schemes are simple to manage since the OAM/OSS controls the whole PCI assignment process. An optimized network-wide configuration of PCIs is, therefore, possible. This, in turn, reduces the probability of PCIs conflicts. However, the approaches suffer from the single point of failure problem since defects in the OAM/OSS may stop the PCI assignment process. The scalability of the system is known to be problematic due to the dependency on the central entity to achieve all actions related to the PCI assignment process. The enhanced and neural networks-based GC schemes advance other approaches in terms of the PCIs utilization ratio since these schemes invest in re-using PCIs in a planned manner.

The utilization of measurement reports, which exist anyway, to detect and recover from conflicts is an advantage. However, the introduction of transmission gaps to detect the existence of collisions is a drawback since UEs moving into a new cell have no information about the timing of transmission gaps. This will cause, for sure, problems to them and may prohibit the detection of the cell. So, extensive studies are necessary to judge the introduction of transmission gaps.

3.2. Distributed Schemes

These schemes do not require an existence of a central entity since PCI assignment routines are executed based on local knowledge. Example approaches include the LTE Standard Distributed (LSD) scheme (3GPP, 2009a), distributed GC approaches (Ahmed, Tirkkonen, Peltomäki, Koljonen, Yu & Alava, 2010), (Diab & Mitschele-Thiel, 2013), the consultation-based solution (Liu, Li, Zhang & Yu, 2010), the Distributed Measurements-based (DM-b) scheme (Amirijoo, Frenger, Gunnarsson, Kallin, Moe & Zetterberg, 2008), the Nokia-Siemens (NS) proposal (Nokia Siemens Networks and Nokia, 2008) and the synchronization-based approach (Kwon & Lee, 2011).

3.2.1. LTE Standard Distributed (LSD) Scheme

LSD scheme makes use of the OAM as well. However, the task of the OAM is restricted to the transmission of a list of PCIs values, from which the given eNB selects a value different from those obtained from UEs measurement reports, reported by neighbor eNBs over X2 interfaces or acquired through other implementation-dependent methods, e.g. heard over the air via a downlink receiver. Again, no implementation details are provided. This is left open to network operators.

3.2.2. Distributed GC Approaches

Distributed GC algorithms (Ahmed, Tirkkonen, Peltomäki, Koljonen, Yu & Alava, 2010) also use graph coloring principles, however, each eNB has local knowledge only (1-hop, 2-hops and may be 3-hops neighbors). These schemes invest in re-using PCIs as most as possible. The authors in (Ahmed, Tirkkonen, Peltomäki, Koljonen, Yu & Alava, 2010) evaluated two types of distributed GC algorithms, namely those depending on real-valued interference pricing (Neel & Reed, 2006), (Babadi & Tarokh, 2008) and binary interference pricing of conflicts (Zhang, Wang, Xing & Wittenburg, 2005). The evaluation is achieved deploying growing networks, within new eNBs are gradually switched on at network borders. The main result states that in terms of the attempt to find a global optimum, the binary interference pricing algorithms outperform those depending on a real-valued pricing, which converge rapidly to a local optimum.

The Stable GC (SGC) scheme (Diab & Mitschele-Thiel, 2013) improves the PCIs utilization ratio through configuring the eNB under configuration with a PCI selected from the PCIs used by its 3-hops neighbors. The stability comes from prohibiting new switched on eNBs from conflicting already configured ones by reserving a sub-range of PCIs for newly switched on eNBs.

3.2.3. Consultation-Based Solution

The consultation-based proposal operates similar to the LSD scheme. The OAM sends a list of PCIs to the eNB under configuration, which in turn selects an adequate PCI from it. The eNB under configuration filters the received PCI list. Concrete, it filters the PCIs heard over the air interface and those known from the neighbor information exchanged via X2 interfaces. Following that, a PCI is randomly selected from the PCIs remaining in the list. The initiative is the handle of the case where many eNBs are under configuration and have shared neighbor eNB(s). So, when multiple eNBs under configuration exchange neighbor information with a shared neighbor eNB, they will not obtain information of each other since they are still under configuration. This, of course, may cause conflicts. Therefore, the proposal says: when an eNB receives multiple X2 setup requests from many eNBs under configuration, the shared eNB does not respond directly to all requests. Rather the shared eNB queues the requests and responds to the first queued one by sending an X2 setup response. Remaining requests will be respond to by sending X2 setup failures. These messages force the eNBs under configuration to wait for a specific duration. After the eNB under configuration, that has received an X2 setup response, completes the configuration process, the shared eNB responds to the next request in the queue following the same steps presented above. Note that the response to the next request in the queue contains the information about the eNB that has already completed the configuration process. Sure, this will improve the performance of the PCI assignment process.

3.2.4. Distributed Measurements-Based (DM-b) Scheme

The DM-b scheme is the distributed form of the scheme discussed in (Amirijoo, Frenger, Gunnarsson, Kallin, Moe & Zetterberg, 2008a). The developers assume that each eNB under configuration contacts first a certain configuration server, which provides the eNB under configuration with a GID identity, an IP address and other initial configuration parameters. An initial PCI may be optionally assigned to the eNB. Following that, the eNB benefits from UEs measurement reports to update its NCR list and also to detect PCIs' confusions. When a measurement report containing a PCI not present in the NCR list is received, the serving eNB instructs the UE to report the GID of the cell detected. Once the requested GID is reported, the serving eNB obtains connectivity information and signals the detected eNB, either directly via an X2 interface or through the core network, to update its NCR list. PCI confusions are detected during NCR lists update procedure. The eNB that detects the conflict is termed the detecting cell. This eNB chooses which eNB of the conflicting ones has to change its PCI. The chosen eNB is termed the selected conflicting cell and is selected based on specific criteria, e.g. the eNB with the lowest GID, shortest NCR list, etc. The selected conflicting cell randomly selects a PCI different from all PCIs assigned to its 1-hop and 2-hops neighbors. The last step is to notify all neighbors of the change in the PCI. For collisions, they are handled by introducing measurement gaps, during which UEs search for the PCI of the serving eNB, see section 3.1.3.

3.2.5. Nokia-Siemens (NS) Proposal

The basic idea of the NS scheme is to reserve a sub-range of PCIs, termed temporary PCIs, for newly switched on eNBs. Once a new eNB is switched on, a randomly selected PCI is chosen from the temporary PCIs. The eNB under configuration operates with this PCI for a limited time duration. During this duration, the new switched on eNB tries to detect the eNBs locating in its vicinity. Thereafter, the new eNB randomly selects a PCI used neither by its 1-hop nor by its 2-hops neighbors. Following that, the new eNB boots up and notifies its neighbors of the new PCI.

3.2.6. Synchronization-Based Approach

The synchronization-based approach handles the case where the number of 1-hop neighbors is greater than the number of available PCIs. This situation cannot be handled by any method of those presented above. The identification of cells is only possible using their GIDs, which has serious problems as mentioned earlier. The authors propose a virtual extension scheme that enables multiple HeNBs locating within the coverage area of a macro-cell to have the same PCI and to be uniquely identifiable without reading their GIDs. Two assumptions are assumed, namely HeNBs and their closest macro-cell are synchronized and a sub-range of PCIs is reserved for HeNBs. Each HeNB sets its own System Frame Number (SFN) offset in relation to the SFN of the macro-cell. HeNBs are then differentiable within the macro-cell via the combination between their PCIs and their relating SFN offsets. Note that the SFN is 10 bits long, which implies 1024-fold increase in the number of identifiers for HeNBs.

3.2.7. Pros and Cons

When discussing the pros and cons of the presented approaches, one notes that they depend on local knowledge and do not require a central entity. Thus, there is no single point of failure. The local knowledge, however, may not be accurate or up to date. Thus, the assignment of PCIs may not be optimal. A network-wide planning of PCIs is, of course, not possible. For scalability, it is expected that the approaches scale well.

The developers of the NS approach claim that their mechanism is robust as long as the temporary PCIs are sufficient. Note that the introduction of the temporary PCIs improves the performance in terms of reducing the number of conflicts with eNBs that are already correctly configured. The studies provided in (Diab & Mitschele-Thiel, 2012) and (Abdullah, Baba & Ali, 2014) showed that the NS approach outperforms the LSD and GC approaches in studied scenarios (static and growing ones) since the network converges faster to a steady state and experiences fewer collisions and confusions. The deployment scenario heavily affects the performance of PCI assignment approaches. Sure, some approaches are affected more or less than the others. The best deployment scenario for the NS approach is the growing scenario, while the best scenario for the analyzed LSD and GC approaches is the static scenario.

The deal with the temporary PCIs range in NS scheme is, however, not simple, many questions must be answered, such as what is the optimal size of the temporary PCIs range?, should the size of this range be static or dynamic?, etc. There are no studies till now to answer such questions. Note also that the proposed mechanism depends on the NCR lists of the eNB under configuration and its neighbors. These NCR lists should be correctly established during the configuration phase. Otherwise, PCIs conflicts may be the consequence. It is worth mentioning that when an eNB receives a temporal PCI reported from a UE, the eNB considers a new eNB in its vicinity and starts establishing a NCR with it.

The main challenge of the synchronization-based proposal is the synchronization itself. The developers claim that the synchronization can be done based on many technologies, e.g. the Global Positioning System (GPS) (Misra & Enge, 2001), the Precise Time Protocol (PTP) (Dreher & Mohl, 2015), (IEEE standard, 2008), the Network Time Protocol (NTP) (Mills, 1992), etc. Failing in achieving an accurate synchronization results in PCI conflicts. Sure, this complicates the problem of PCI assignment. This is the reason why it is impractical to obtain the 1024-fold increase in the number of identifiers for HeNBs.

3.3. Summary

As seen, there are several efforts done to autonomously solve the problem of PCI assignment. While centralized approaches depend on a central entity with global knowledge, distributed schemes do not and, thus, do not suffer from the single point of failure problem centralized approaches suffer from, see Figure 3.

Centralized approaches are simple to manage and enable a network-wide planning of PCIs. This is not possible in distributed mechanisms, which are also hard to manage. Concerning scalability, distributed schemes scale better than centralized ones. The most important metric to be highlighted is the convergence to a steady state. Any applied PCI assignment approach has to guarantee that the network will reach a steady state after a limited number of configurations iterations. There exist, until now, no study to investigate the impact of network dynamics on the performance of PCI assignment approaches, especially the convergence to a steady state. This chapter handles this issue and focuses on investigating the performance of three distributed PCI assignment approaches, namely the LSD, GC and NS approaches, under a dynamically changing scenario.

4. PHYSICAL CELL IDENTITY (PCI) ALLOCATION SCHEMES

As mentioned in section 2.2, the capability of recognizing HeNBs by UEs based on their PCIs is essential for an optimized performance. For this purpose, researchers propose to reserve sub-ranges of PCIs for an exclusive use by HeNBs. Keep also in mind that 3GPP proposed a PCI reservation mechanism to simplify the management of HeNBs within existing LTE networks, see (Golaup, Mustapha & Pa-

Figure 3. Qualitative comparison between centralized and distributed schemes

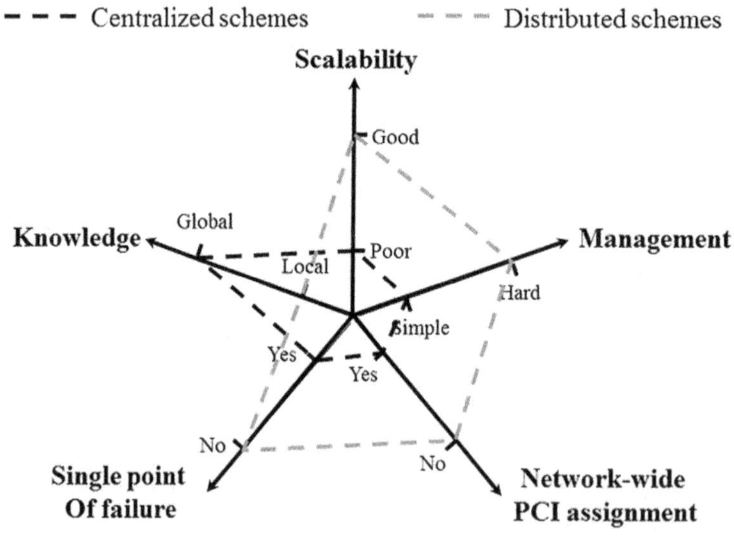

tanapongpibul, 2009) and (Sesia, Toufik & Barker, 2011). The selection of the sub-ranges themselves is left back to network operators. Research questions in this context include how to properly allocate sub-ranges of PCIs while considering the current status of the network?, which types of cells should be considered?, should the allocation scheme be static or dynamic?, etc. The goal is, of course, an optimized performance. Note also that an adequate PCI allocation scheme will, for sure, improve the performance of applied PCI assignment schemes.

Similar to PCI assignment schemes, PCI allocation proposals can be categorized into centralized and distributed ones. The allocation process is controlled by a central entity when applying centralized schemes, while the allocation process is executed locally by eNBs when employing distributed schemes. The following provides more insight into both categories, goes through existing approaches and discusses their pros and cons.

4.1. Centralized Allocation Schemes

As mentioned above, these schemes relay on a central entity to control the allocation process. Examples of such schemes include the framework for an Automation PCI Allocation System (APCIAS) (Wu, Rui, Xiong & Guo, 2010), the Dynamic Group PCI (DG-PCI) allocation scheme (Zhang, Zhou, Xiao, Liu, Zhang & Glasunov, 2011) and the Layered-structured PCI (Ls-PCI) allocation scheme (Wu, Jiang, Wu & Zhang, 2010). The following provides more insight into the mentioned schemes.

4.1.1. Automation PCI Allocation System (APCIAS)

APCIAS framework consists of six components, namely a data collecting/configuration interface, a cell information base, a PCI allocation module, a policy base, a policy management module and a PCI resource base. The data collecting/configuration interface collects cells-related information, e.g. cells states, NCR lists, etc. The cell information base contains cell-related information. The policy base includes the policies that control the PCI allocation process. The framework developers assume that the allocation policy can either be luxury or multiplex, where luxury means that PCIs are selected from the reserved group directly without caring of re-using them. In contrary, multiplex PCI allocation tries to re-use one of the assigned PCIs first before selecting a new one. The policy management module is used to manage the policies applied in the framework. The PCI resource base stores PCIs and their usage conditions. Finally, the PCI allocation module is used to allocate a PCI for an eNB or a HeNB under configuration.

The operation of APCIAS is pretty simple and can be summarized as follows: in the luxury allocation, APCIAS analyzes applied allocation policies and randomly picks up a new PCI from the PCI resource base. Following that, the eNB under configuration is configured with the selected PCI and the PCI resource base as well as the cell information base is updated. Similar actions are undertaken when PCIs are allocated applying the multiplex policies. However, the framework analyzes, first, if one of the PCIs used can be re-used.

4.1.2. Dynamic Group PCI (DG-PCI) Scheme

DG-PCI scheme introduces the term Busy Femtocells (BFemtocells), which is defined as HeNBs more incorporated in inbound handovers than other HeNBs. These HeNBs must obtain a special treatment from the proposal's point of view. The basic idea of the scheme is to group the PCIs in two main groups, one contains the PCIs to be assigned to macro cells, while the other includes those reserved for HeNBs. The group of PCIs reserved for HeNBs is further divided into two sub-groups, namely a CSG and non-CSG sub-group, see Figure 4. Sur, the sub-groups names say that they express the PCIs reserved for CSG and non-CSG HeNBs. The developers go further and divide the CSG and non-CSG sub-groups into unique PCI and reuse PCI sub-groups. The unique PCIs are assigned to BFemtocells. These PCIs are not re-used in the network which guarantees that BFemtocells will experience no PCI conflicts. The re-use PCIs are assigned to remaining HeNBs. Here, efforts should be carried out to guarantee conflicts-free PCI assignment.

The developers assume that the core network is responsible of determining which HeNBs are BFemtocells. Moreover, the core network is also responsible of detecting whether HeNBs are still present or not. This is necessary to release their PCIs in case they are no more available. The main feature of the DG-PCI scheme is the dynamic adaptation of unique and re-use PCIs so that PCIs can be transferred between these both sub-groups based on the current status of the network. This results in flexibility and, for sure, better performance.

4.1.3. Layered-Structured PCI Allocation Scheme

The proposal's developers assume a central entity with global knowledge to control the PCI allocation process. They do not put restrictions on the location of this entity. The layers proposed are, in principle, sub-groups of the whole PCIs range available. The scheme assumes a layer for macro and open access cells, a layer for hybrid access cells and a layer for CSG cells. In other words, the PCIs range is divided into three sub-groups. The PCI assignment procedure itself is similar to that applied using the LSC scheme, see section 3.1.1. The difference lies only in the sub-groups (termed layers in the proposal), from which PCIs are picked up.

4.1.4. Pros and Cons

Let us now discuss the pros and cons of the proposals presented above. One notes that the schemes are simple to manage since a central entity controls the whole PCI allocation process. An optimized network-wide allocation of PCIs is, of course, possible. This, in turn, enhances the performance of the network

Figure 4. Groups of PCIs in DG-PCI

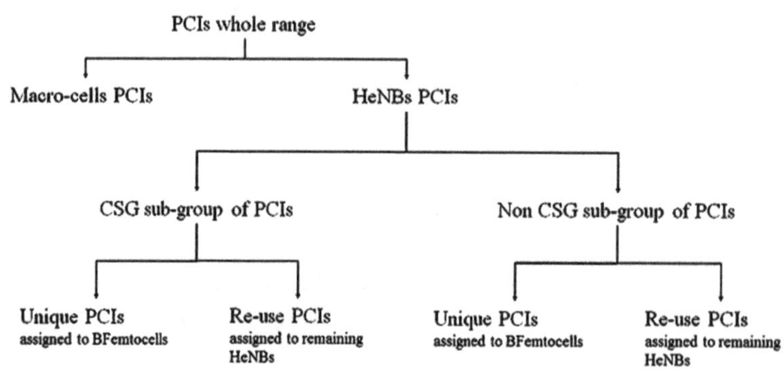

as a whole and reduces the probability of PCIs conflicts. The schemes presented suffer, however, from the single point of failure problem and poor.

When considering the APCIAS framework, one notes that the framework is flexible since new policies can be simply introduced. The extensibility of the framework as well as the interaction with the network management system is considered by concept. The authors claim that new policies can be automatically created. This feature is, however, not proved yet. The reservation of PCIs for HeNBs is also flexible since network operators can simply change the reservation scheme. Questions like how many PCIs must be reserved for eNBs and HeNBs?, how and when to change policies?, etc. are not answered. The focus of APCIAS is to control the allocation process in a flexible manner.

For the DG-PCI scheme, their developers state that because the core network is responsible of determining which HeNBs are BFemtocells and whether they are still present, the management of the allocation scheme is pretty simple. A view of the whole system status is also available. This in turn enables network-wide optimization tasks when required. A major shortcoming of the proposal is that it neglects UEs. DG-PCI simplifies the management of different types of cells. However, it does not provide a solution to enable UEs to distinguish which cells are CSG cells, hybrid cells, etc. The layered-structured PCI allocation scheme is simple and considers all access modes of HeNBs. A main shortcoming of the proposal lies, however, in fixing the sub-groups of PCIs. This prohibits adapting them based on network status.

4.2. Distributed Allocation Schemes

As mentioned earlier, there exist no central entity and no global knowledge of the whole system. So, the allocation process depends on local information collected locally by eNBs. A well-known example of distributed PCI allocation schemes is the PCI Dynamic Reservation (PCI-DR) scheme (Lee, Jeong, Saxena & Shin, 2009).

4.2.1. PCI Dynamic Reservation (PCI-DR) Scheme

PCI-DR has been proposed to enable UEs to identify CSG cells only based on their PCIs. The basic idea of the scheme is the division of the whole range of PCIs into two sub-ranges, one for the CSG cells and the other for remaining types of cells. Each eNB is responsible of determining the number of PCIs that will be reserved for CSG cells. The scheme also defines ($2^n - 1$) reservation types for CSG cells, where n is a positive integer bigger than 2. Of course, each type reserves different number of PCIs for CSG cells. The eNB is capable of switching among reservation types, thus adapting the reserved number of PCIs in response to changes observed in the number of CSG cells present in the network. The authors have also proposed a mechanism to enable UEs to automatically determine the applied reservation type. They proposed to use the front n-bits of the PCI to detect the type. Sure, efforts must be done to distribute PCIs among the mentioned types.

4.2.2. Pros and Cons

When considering the pros and cons of PCI-DR scheme, one notes that the proposal contributes to reducing the delay required for the search procedure, thus, accelerating inbound handovers. This also saves the battery power of UEs and reduces signaling overhead. Moreover, the proposal is flexible in terms of adapting the reservation type based on current network status. So, one expects a noticeable good performance in deployed LTE networks, which will be of a dynamic nature as we mentioned earlier. As with other distributed schemes, no need for a central entity or global knowledge, thus no existence of the single point of failure problem. A network-wide optimization of the performance is, of course, not possible. Other shortcomings of the PCI-DR can be observed in the adaptation of reservation types procedure. This adaptation is not primitive and will not be possible in all cases. It is possible, for instance, that some PCIs, which are parts of a reservation type not applied at the moment, are assigned to macro cells just because they are PCIs for macro cells in the used reservation type. Such situation will prohibit the switch to that reservation type. Other problems appear when neighbor macro cells use different reservation types. In such cases, UEs may recognize a part of CSG cells as macro cells. The problem becomes more crucial if a neighbor eNB is present, but not detected by the serving macro cell yet. It is also worth mentioning that the statement the authors claim, namely CSG cells are recognizable only via their PCIs, is not really true since UEs must know the type of the reservation to judge whether a detected cell is a CSG cell. The reservation type is obtained by reading the system information which induces additional overhead and delay and also implies updating the system information of the network. The update of NCR lists to include the reservation type was also proposed as an alternative. In this case, UEs have to contact the serving cell to judge whether the cell is a CSG cell.

4.3. Summary

As by PCI assignment approaches, centralized PCI allocation schemes suffer from the single point of failure, are simple to manage and enable a network-wide planning of PCIs, which is not possible with distributed mechanisms that do not see the single point of failure and are also hard to be managed. Distributed allocation approaches scale better than centralized ones.

5. PERFORMANCE EVALUATION

5.1. Basic Assumptions

The investigation is accomplished by means of simulative studies done using the simulator LTESim (Diab & Mitschele-Thiel, 2012), (Diab & Mitschele-Thiel, 2013). The simulation is achieved in an area of 100 km2 with three types of cells, namely macro-cells, micro-cells and HeNBs. Each macro-cell has a radius randomly chosen between 0.8 and 2 km. The radius of each micro-cell is randomly selected between 0.4 and 0.6 km. Each HeNB has a radius between 50 and 200 m, see (Diab & Mitschele-Thiel, 2012). The LSD and GC approaches utilize the whole range of PCIs. On the contrary, the temporary PCIs sub-range contains 63 values when employ the NS scheme. Each eNB has an empty NCR list upon switching on. Thereafter, each eNB detects the eNBs locating in its vicinity, updates its NCR list and recovers PCIs conflicts.

Throughout our studies, we do not simulate movements of UEs since we are interested only in studying the impact of the applied scenario, network size and also dynamics on the PCI assignment issue. To ensure reliable simulation results, several measurements were conducted. That is, each simulation was repeated 20 times.

5.2. Simulation Scenarios

So as to comprehensively study the performance of the three selected approaches, two types of scenarios are considered, namely static and dynamic scenarios. For static scenarios, all eNBs are deployed and switched on once. These scenarios contain micro- and macro-cells distributed randomly within the simulation area. They differ from each other in the size of the network in terms of the number of eNBs present. Concrete, there are 14 static scenarios, each with a specific numbers of eNBs (between 50 and 900 eNBs). The number of macro-cells in each scenario is selected randomly. Remaining cells are then micro-cells. Note that a static scenario stands for an already deployed LTE network that is a subject to a configuration using a specific PCI assignment approach.

The dynamic scenario tries to mimic the situations LTE networks are expected to face in real-life scenarios. First, the simulation area is symmetrically covered by 200 micro-cells, each has a radius of 0.5 km. After the simulation starts and all eNBs serving the micro-cells are correctly configured applying the studied PCI assignment scheme, 500 HeNBs start to appear in the scenario at different times and locations. The appearance of HeNBs follows the same behavior of the growing scenario. That is, HeNBs appear one after one at random times and locations, however at network borders. So, the majority of HeNBs reside in the center of the simulation area. This reflexes the situation in cities` centers, where most HeNBs are expected to appear in commercial centers, restaurants, shops, etc. These commercial centers exist mainly in cities centers, as known. Now, in order to investigate how the update of NCR lists at run-time affects the performance of the PCI assignment scheme studied, 100 eNBs configured applying the investigated PCI assignment scheme start to appear and be detected in the scenario. As known, in real-life scenarios not all neighbors are detected directly after switching on, some neighbors may be detected after the eNB gets configured. So, it is interesting to study the impact of such updates of NCR lists.

Another important issue concerning dynamic scenarios is the change of cells` radiuses due to adapting to changes in the network and its environment. For instance, due to the execution of tasks to satisfy the use case "coverage and capacity optimization", cells` sizes change dynamically based on the current state of the network. Such changes do not affect only the mentioned use case. Rather, it may change the NCR lists of some eNBs. This in turn may result in PCI conflicts, affect handoffs, etc. Thus, to consider such changes in our study, 20 cells are selected randomly each specific time interval (400 time steps in our study). The radiuses of these cells are also changed randomly, meaning that the radius of each selected cell will either be incremented or decremented while preserving ranges` restrictions. That is, the upper and lowest range`s limits of the selected cell are maintained. In brief, the dynamic scenario applied in our simulation covers the following issues:

- Appearance of HeNBs in the network.
- NCRs lists updates due to the detection of previously not detected neighbor cells.
- NCR lists updates due to changing cells` radiuses resulting from various reasons, e.g. self-organization use cases, changing network conditions, etc.

Note that the dynamic scenario investigated mimics real-life scenarios already deployed LTE topologies will, for sure, experience.

5.3. Performance Measures

The performance measures we are interested in include the following:

- **The Average Number of Collisions Experienced:** We measure in this context the average number of collisions experienced in the network at each simulation iteration and also scenario.
- **The Average Number of Confusions Experienced:** Again, the average number of confusions the network sees at each simulation scenario and iteration is provided in this concern.
- **The Average Number of Used PCIs:** This figure was neglected in most previous studies in spite of its major importance since it affects the size of the network that can be addressed by a given PCI assignment approach. Moreover, it provides arguments to decide how the available PCI range can be divided. In the context of this performance measure, we also study the PCI utilization ratio.

Throughout our evaluation we study the impact of the network size on the performance of the three investigated PCI assignment approaches with respect to the above mentioned performance measures. Furthermore, we investigate specific selected scenarios to provide an in-depth illustration. Finally, we summarize the results of our studies.

5.4. Impact of Network Size

First, let us take a look at the density of the network when varying the number of eNBs, see Figure 5 which provide the average, minimum and maximum number of 1-hop and also 2-hops neighbor eNBs as a function of network size.

Sure, the number of 1-hop and 2-hops neighbors increases as network size increases in a fixed area. When the number of eNBs in the network starts to exceed 350, the network gets pretty dense. Note that the maximum number 1-hop and 2-hops neighbors exceed 100 and 200, respectively. In networks with 700 eNBs, the average number of 1-hop and 2-hops neighbors approaches 80 and 275, respectively. This implies that conflicts-free PCI assignment in such networks is not possible anymore since the number of 2-hops neighbors exceed 504, the number of PCIs available.

Now, take a look at Figure 6 which shows the average collisions experienced per simulation iteration when employing the four studied PCI assignment approaches as a function of the network size. In general, we see increasing collisions with increasing network size, which is also expected. However, it is clearly seen that network size affects the approaches differently. In other words, some approaches work with dense networks better than others.

The GC scheme is the most one affected by network size. The reason lies in the concepts the studied GC approach follows. In detail, this scheme selects conflicts-free PCIs while considering the reduction as most as possible of the number of PCIs used. In other words, no new PCI will be used until no reusing of an already assigned PCI is possible. In dense networks, this increases the probability of causing conflicts with other eNBs. The LSD proposal sees collisions at all network densities. However, the approach performs clearly better than the distributed GC scheme. The amount of performance improvement also increases when network size goes up. The main reason is that the reuse of PCIs is not a focus of the LSD proposal. So, the probability that the selected PCI will result in collisions with already used PCIs is smaller than that seen by the distributed GC approach. Sure, as network size increases, the impact of ignoring the reuse of PCIs gets more and more clear. This is also shown in the figure.

Figure 5. Average, minimum and maximum number of 1-hop and 2-hops neighbors as a function of network size

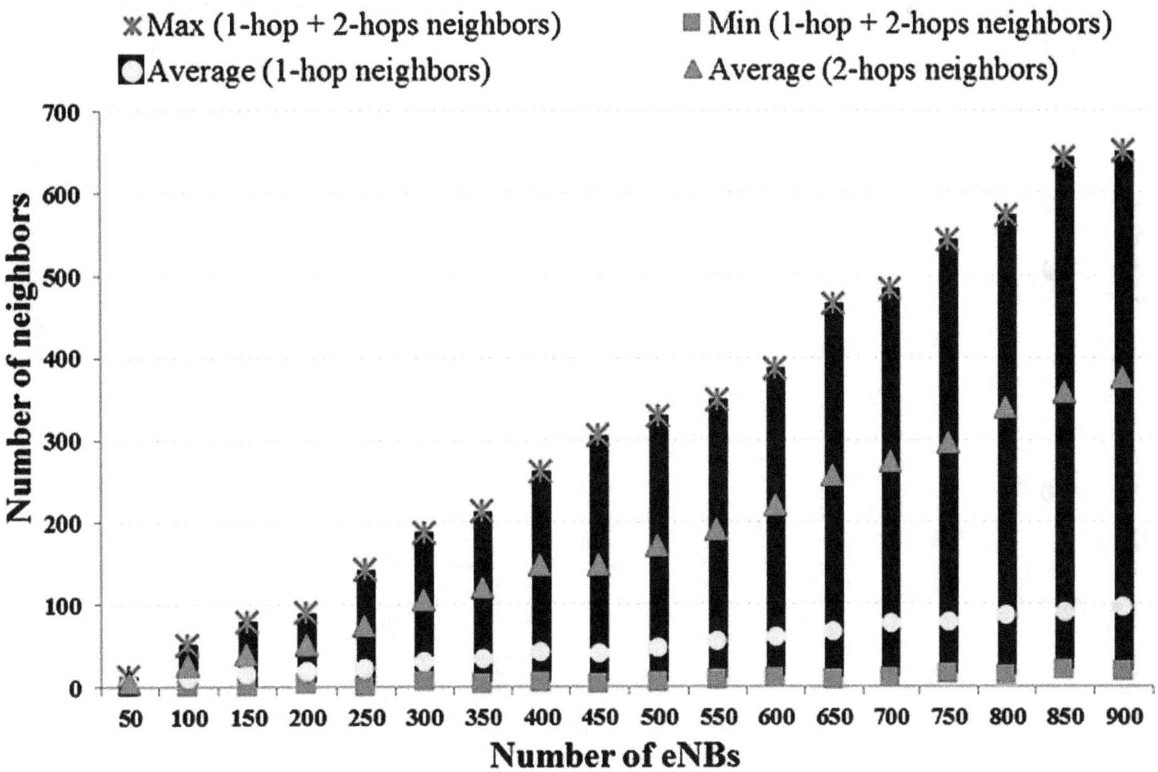

The NS approach is less affected by network size than both the studied GC and LSD schemes. Networks with small to medium sizes (until 350 eNBs in our simulation scenarios) do not see collisions. Networks with high densities experience collisions, however, significantly less than when they operate other both mentioned approaches. The main reason behind this performance improvement is the introduction of a temporal PCIs range. So, fewer collisions than those observed by the other both above mentioned approaches are the result. Keep in mind that there are no collisions between the eNBs configured with permanent PCIs and those assigned temporal PCIs. Note also that collisions among eNBs configured with temporal PCIs are excluded from our study since we are interested in investigating the collisions resulting from the configuration process itself. Keep also in mind that the temporal PCIs lifetime is limited. Sure, in low to middle network sizes, one does not expect to see collisions since the available range of permanent PCIs is big enough to guarantee collisions-free PCIs among 1-hop and 2-hops neighbors. However, as the network size increases, the probability of experiencing collisions also increases.

The performance is considerably improved when employing the SGC scheme as compared to the GC approach and the LSD proposal. The results show that in spite of the efforts done to re-use PCIs when possible, the impact of PCIs re-use is strongly minimized when comparing to the GC approach studied. The reason is the introduction of the temporal PCIs sub-range, which resulted in considerable performance improvements as we have already seen by the NS approach. Let us now compare between the performance of the NS and SGC approaches. Our simulation results show that the NS approach performs slightly better than the SGC in networks with low to middle sizes. When network size starts

Figure 6. The average number of collisions experienced as a function of network size

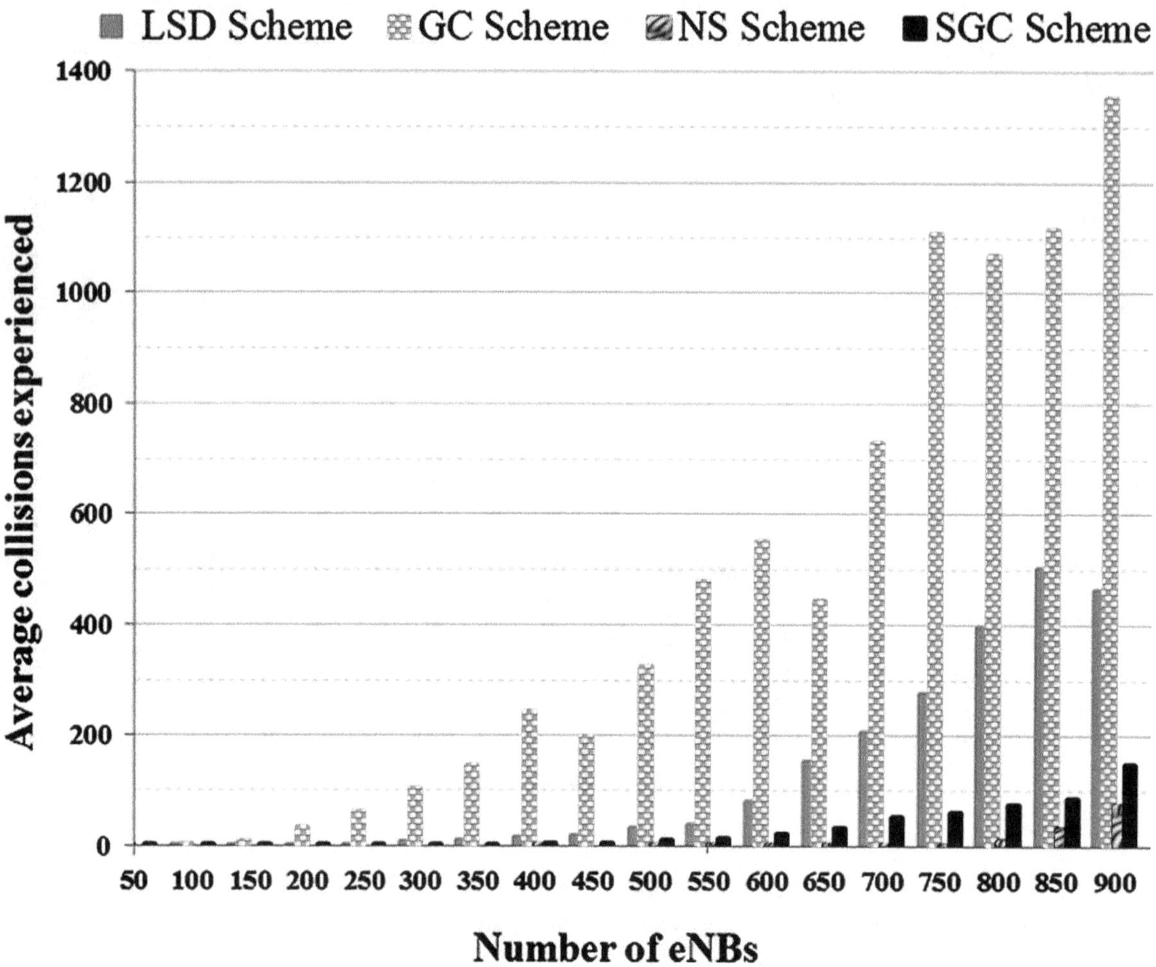

to be high (number of eNBs exceeds 400 in our study), the NS approach starts to show considerably performance improvements compared to the SGC approach. As network size gets very high (networks with more than 800 eNBs), the difference in the performance gets considerably smaller and smaller. The reason for this decrease will be highlighted later on.

Let us now consider the confusions the network experience, see Figure 7. First, it is clear that the number of confusions experienced is significantly greater than the number of collisions, see the figure above. Furthermore, the number of confusions increases approximately in an exponential manner as a function of network size. One also notes that the impact of network size on the performance of PCI assignment schemes is the same when considering the average number of confusions and collisions. Again, NS approach is the best followed by the SGC scheme. LSD proposal also performs better than the GC approach studied.

Now, let us consider the utilization ratio of available PCIs required to obtain an LTE network without PCIs conflicts as a function of network size, see Figure 8. The result we first note is a nonlinear increase in PCIs utilization ratio with increasing network size.

Figure 7. The average number of confusions experienced as a function of network size

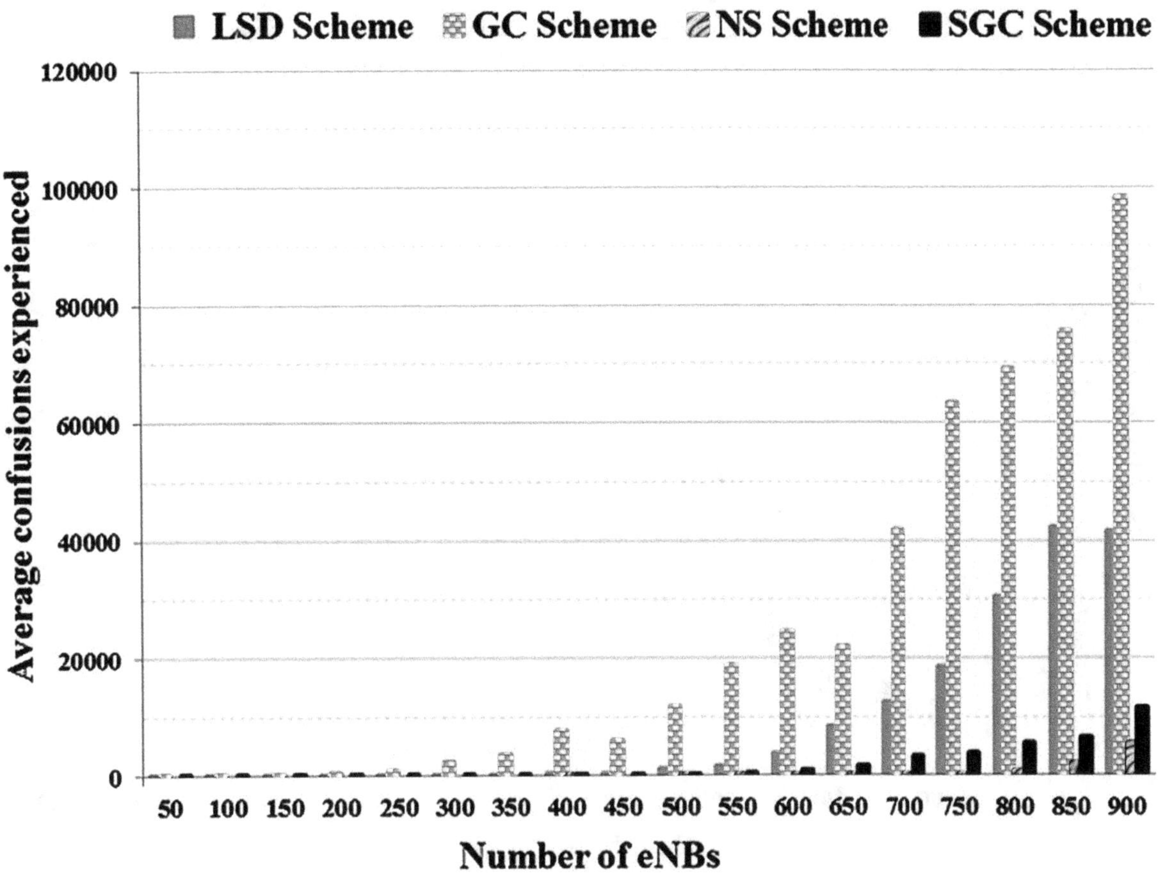

Our results show that the behavior when considering the PCIs utilization ratio is completely different from that we have already seen as we discussed collisions and confusions experienced. The best PCIs utilization ratio is observed with the GC approach. Note that this approach focuses on assigning conflicts-free PCIs while minimizing the number of PCIs used as most as possible. The figure shows that the PCIs utilization ratio does not exceed 65% even for the greatest network size studied (networks with 900 eNBs). For LSD approach and NS solution, one notes that both follow similar behavior with performance improvements for the LSD approach especially seen at big-sized networks. In principle, both approaches follow similar principles when assigning PCIs. This is the reason for the similarity we see in the behavior followed. The LSD proposal consumes less PCIs utilization ratio than the NS solution does because the LSD proposal utilizes the whole range of PCIs available, while the NS solutions assigns PCIs from the permanent sub-group (441 PCIs in our studies).

The SGC approach follows similar behavior to that followed by the GC approach since they assign PCIs in similar way. The utilization ratio is, however, greater when employing the SGC approach since the range of available PCIs is only the permanent sub-range. The PCIs utilization ratio when employing the SGC scheme is smaller than that resulting when employing the LSD proposal and the NS solution. This implies that the SGC approach will be more capable of serving dense networks than the NS and LSD approaches.

Figure 8. PCIs utilization ratio as a function of network size

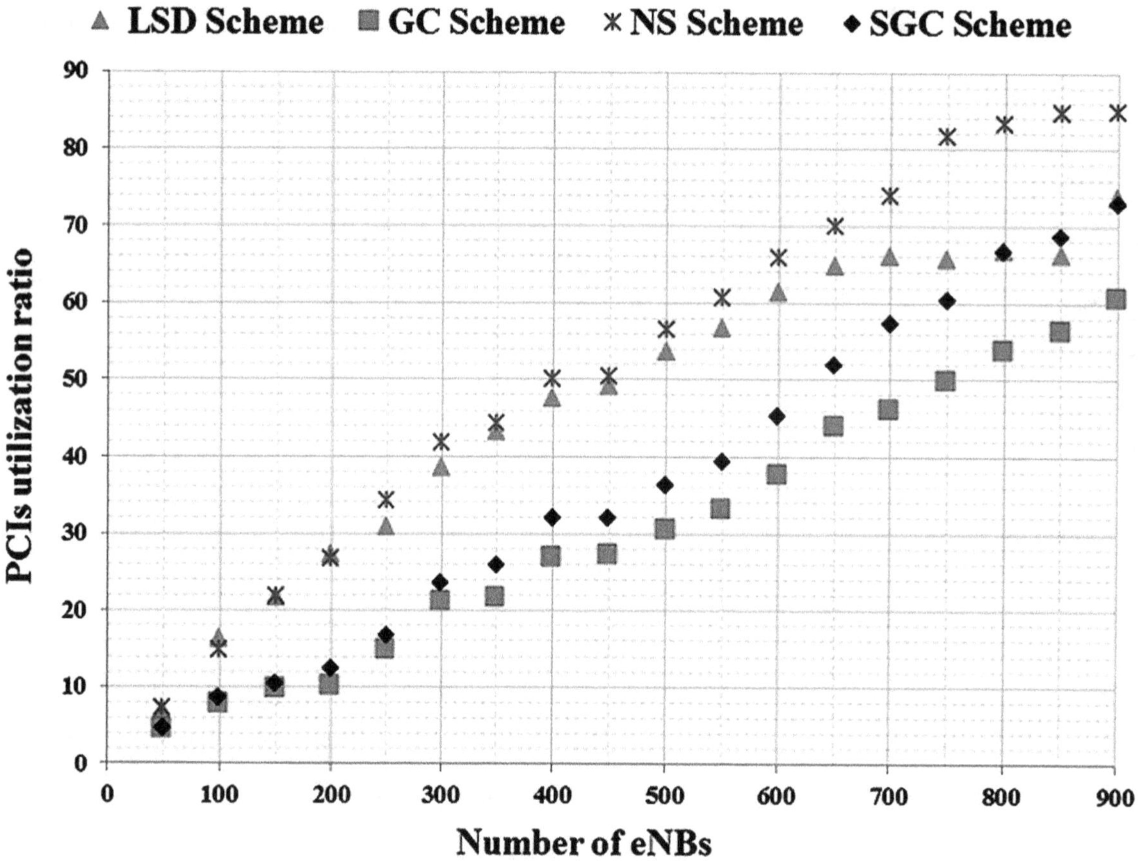

5.5. Sample Static Scenarios

So as to provide an in-depth analysis of the studied schemes, let us consider two sample scenarios, a middle-dense scenario (a network with 550 eNBs) and a high-dense scenario (a network with 850 eNBs). Let us first consider the average number of collisions experienced, see Figure 9 and Figure 10, which provide the average number of raised collisions deploying both mentioned scenarios.

Let us first consider the scenario with 550 eNBs. The first result one notes is that the system converges to a steady state regardless of the employed PCI assignment scheme. The algorithms differ in the time they require transfer the system to a steady state and in the amount of collisions the system experiences. The quickest scheme is the NS mechanism followed by the SGC approach, the GC approach and the LTE standard proposal, respectively. Our results show that the NS scheme converges 90.9%, 94.12% and 99.9% faster than the SGC, GC and the LSD schemes, respectively. The SGC approach in turn converges 35.29% and 80% faster than the GC and the LSD proposals, respectively. The main reason standing behind the fast convergence to a steady state seen with the NS scheme is the introduction of the temporal PCIs range, so that the eNBs not configured yet are prohibited from affecting those already configured. This is also the reason why the SGC scheme shows significant performance improvements compared to the GC

Figure 9. The average number of collisions experienced in a middle-dense scenario (a network with 550 eNBs)

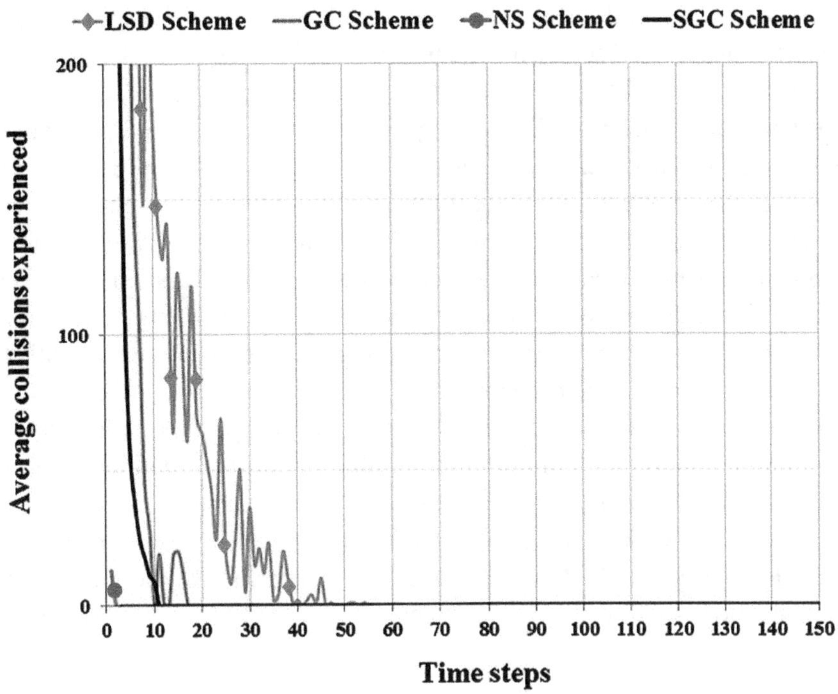

Figure 10. The average number of collisions experienced in a high-dense scenario (a network with 850 eNBs)

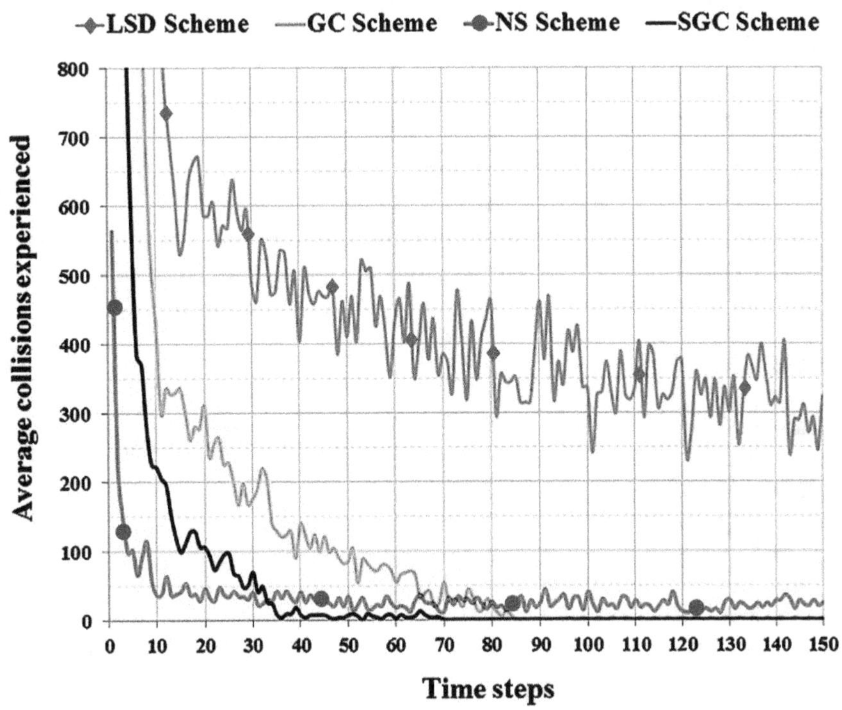

approach, although they follow similar principles in assigning PCIs. The efforts achieved to re-use PCIs are, however, the reason behind the advancements seen in the NS solution compared to the SGC scheme.

Actually, the scenario with 850 eNBs is more interested. The fastest convergence is observed in the network when employing the SGC approach. The network also converges to a steady state when employing the GC approach, however, 17.86% slower. Steady state is, however, not reached when employing both the NS and LSD proposals. This shows the great performance of the SGC against other studied schemes since it also experiences 92.12% fewer collisions than the GC approach. Here, one notes the advantages of PCIs re-usage, which enables properly serving networks of high densities.

Let us now take a look at Figure 11 and Figure 12, which show the average number of confusions experienced deploying both mentioned scenarios for the four studied PCI assignment schemes. Again, the behavior the number of confusions experienced follows is the same when considering the average number of raised collisions. Furthermore, the number of confusions experienced is significantly greater than the number of collisions for all considered schemes. The SGC scheme is outperformed by NS solution in middle-dense scenarios. In high-dense scenarios, however, the SGC scheme shows the best performance since the network quickly converges to a steady state, what we don't see with other studied schemes.

Let us now discuss the PCIs utilization ratio deploying both scenarios, see Figure 13. The figures show that SGC scheme follows similar behavior as the GC approach with more PCIs utilization ratio due to the smaller PCIs range the SGC scheme makes available for use (permanent PCIs sub-range). The PCIs utilization ratio is significantly less than that seen by the LSD and the NS solutions. The most

Figure 11. The average number of confusions experienced in a middle-dense scenario (a network with 550 eNBs)

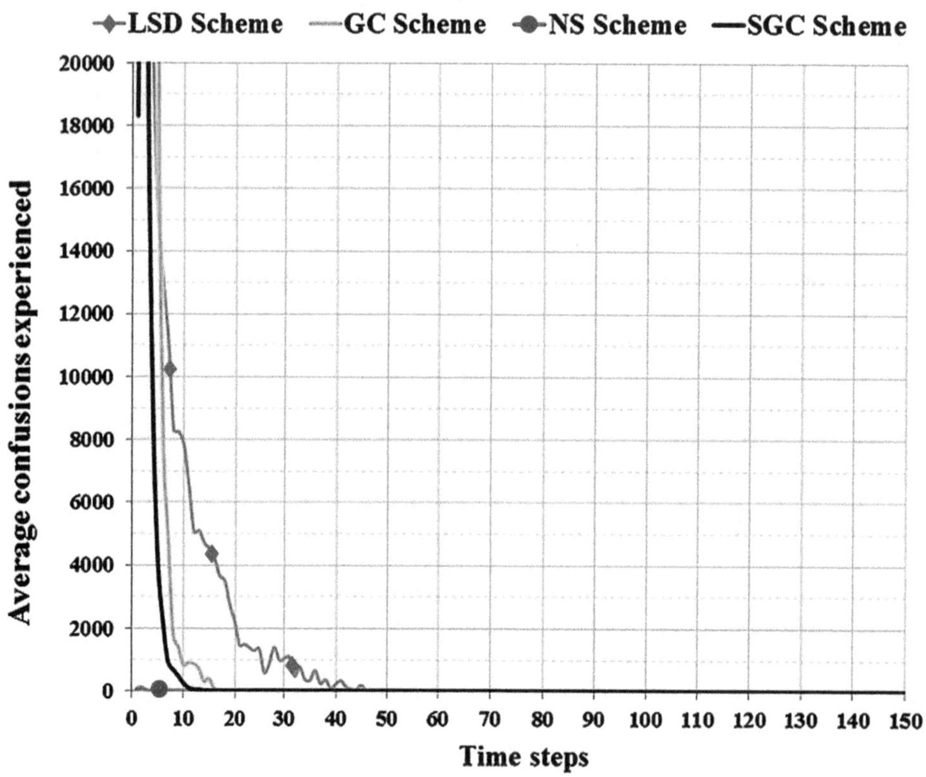

Figure 12. The average number of confusions experienced in a high-dense scenario (a network with 850 eNBs)

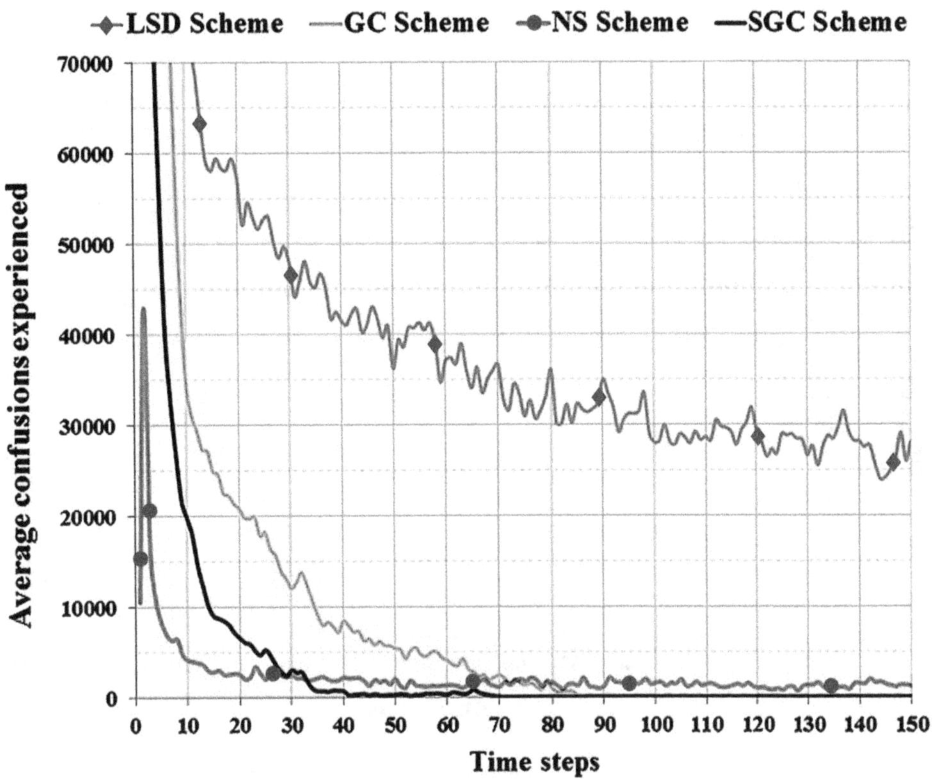

interested result is the behavior of the studied schemes in high-dense network scenarios (a scenario with 850 eNBs in our study). One notes that even with such scenarios the SGC scheme does not consume more than 70% of the PCIs available to guarantee the system's stability. The NS solution approximately uses 85% of the PCIs and does not bring the system into a stable state.

5.6. Impact of Network Dynamics

Figure 14. shows the average number of collisions experienced per time steps. The results shown in the figure show that the NS scheme is the best in terms of collisions experienced. One also notes that the appearance of HeNBs heavily affects LSD and GC schemes. Network instability results until no more HeNBs appear in the network. Both NS and SGC schemes are more robust since the network suffers from few collisions and quickly restores its stability. The strongest impact on the performance of PCI assignment schemes is observed when detecting eNBs previously not detected by their neighbors. All studied approaches are extremely affected and see collisions when updating their NCR lists. This means that the more accurate the NCR lists are established, the fewer the collisions raised. Let us now take a look at the impact of changing cells radiuses. One notes that collisions may appear and may not. The NS scheme is even not affected. The SGC scheme is more affected than the NS solution, however, less affected than the LSD and GC approaches, which experience sometimes collisions. This necessitates PCIs re-configuration to recover from raised collisions.

Figure 13. PCIs utilization ratio experienced in a middle-dense scenario (a network with 550 eNBs)

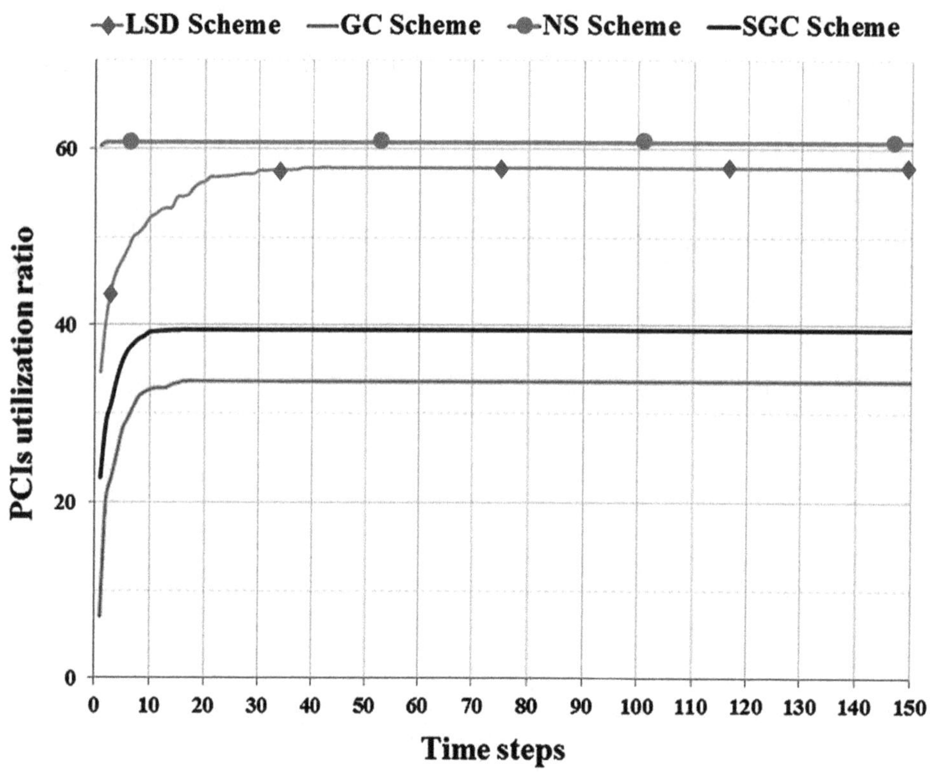

Let us now consider the confusions experienced, see Figure 15. The first result one notes is that extremely more confusions are observed than collisions. The network completely loses its stability when employing both the GC and LSD schemes. The NS scheme is again the best followed by the SGC scheme. The network sees with the NS solution few PCIs confusions when integrating HeNBs with the network and when cells radiuses change in a dynamic manner. There may be even no confusions in such cases. However, considerable amount of confusions is experienced when updating the NCR lists due to detecting neighbor eNBs previously not detected. This also proves that the more accurate the NCR lists are established, the better the performance of the NS scheme. The reason for the great performance of the NS scheme against the other two approaches is the introduction of a temporary PCIs sub-range, which prohibits newly switched on cells from affecting others in the network. The absence of such temporal PCIs is also the main reason why network stability gets lost when considering the LSD and GC approaches, which show similar behavior in terms of the impact of network dynamics on their performance. For the SGC approach, few PCIs confusions also appear when integrating HeNBs with the network. However, considerable amount of confusions appear when updating NCR lists due to the detection of neighbor eNBs previously not detected. The impact of changing cells radiuses in a dynamic manner is clearly obvious since the network completely losses its stability. The reason for this performance is the efforts done to reuse PCIs as most as possible. In detail, one of the PCIs used by 3-hope neighbors of the eNB under configuration is mostly used for this eNB, which implies that increasing the radius of 3-hops neighbors will result in confusions.

Figure 14. PCIs utilization ratio experienced in a high-dense scenario (a network with 850 eNBs)

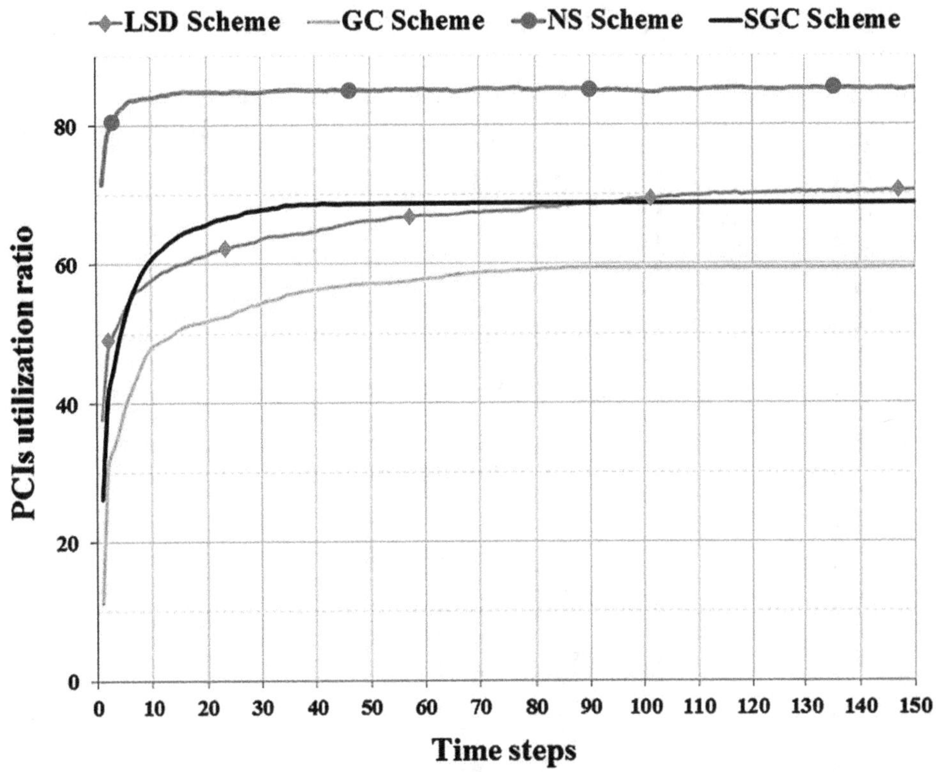

Let us now consider the PCIs required to restore network stability, see Figure 16, which shows the average number of PCIs used per time steps when operating the four studied PCI assignment approaches.

Let us first discuss the appearance of the 500 HeNBs. Sure, it is expected that extra PCIs are required to configure new switched on cells. However, the number of PCIs required strongly depends on the approach employed. The lowest number of PCIs required is observed by the SGC approach, which saves 25.49%, 40.16% and 56.77% PCIs compared to the GC, LSD and NS schemes, respectively. The GC approach comes next and saves 19.68% and 41.97% PCIs compared to the LSD and NS schemes, respectively. The reason is the PCIs reuse principle the studied SGC and GC approaches follow since no new PCI will be assigned until no used one from those assigned to the 3-hops neighbors of the eNB under configuration can be reused again. Going back to the impact of the appearance of these 500 HeNBs on experienced PCIs conflicts shown in the above two figures (Figures 16 and 17), one observes that the save in PCIs is not cost-free. The network has to bear considerable amount of PCIs conflicts. The advancement the SGC approach show is due to the reduced number of PCIs available since only permanent PCIs are available to use with SGC (441 PCIs in our simulation), while the GC approach utilizes the whole range of PCIs (504 values).

The average number of PCIs required by the LSD proposal is greater than that required by both the SGC and GC approaches, however, considerably smaller than that the NS scheme necessitates (27.75% less based on our results). In fact, neither the LSD nor the NS approach invests in reusing PCIs. Both randomly choose a PCI from available ones. We can explain the difference in the number of PCIs required by both the LSD and NS approaches based on the results shown in the two figures above. The LSD proposal

Figure 15. Impact of network dynamics on the average number of collisions experienced

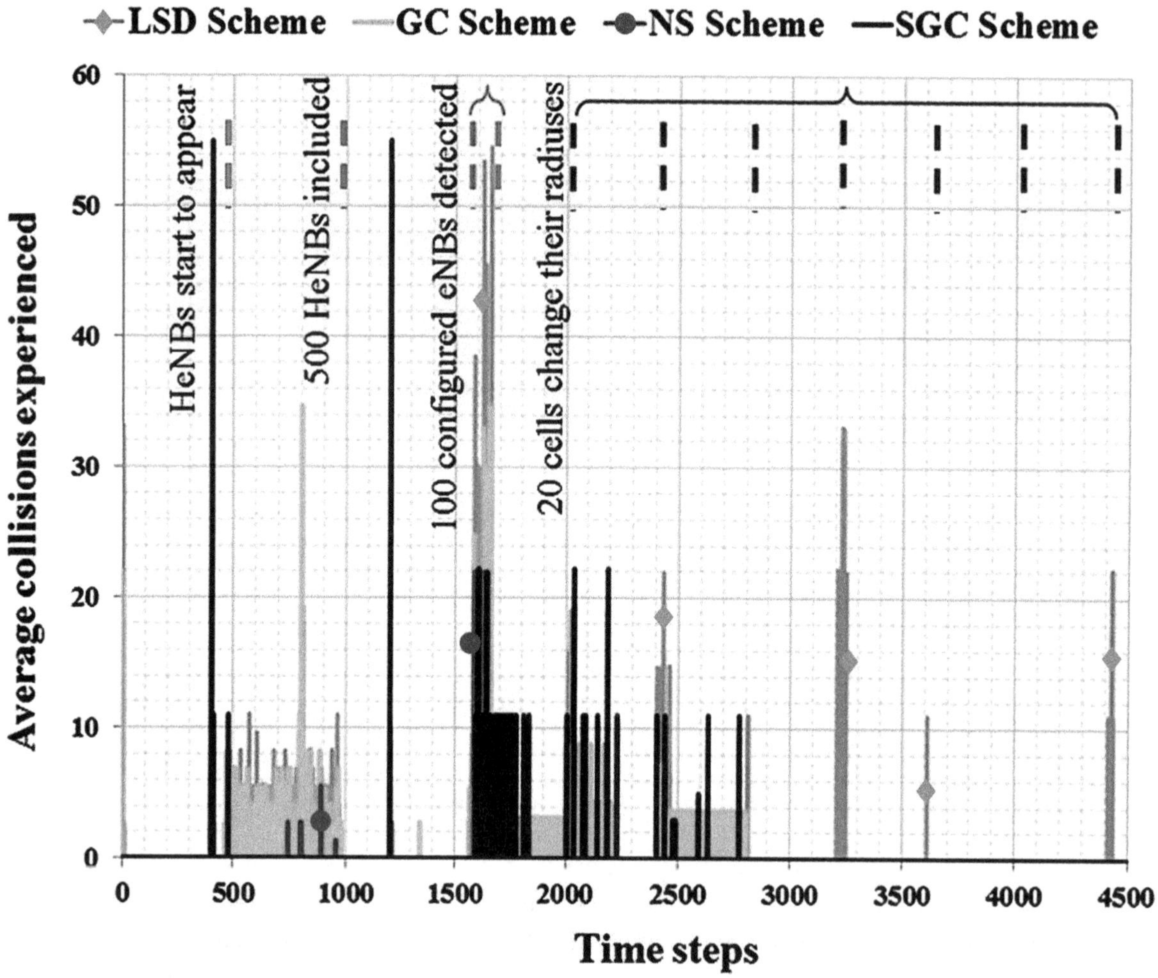

suffers from considerable amount of conflicts. This necessitates subsequent PCIs re-configuration tasks, which in turn has an ancillary effect, namely an increased PCIs reuse factor. Concerning the NS scheme and due to the introduction of the temporary PCIs sub-range, considerably less PCIs re-configuration activities are executed compared to those observed when operating the LSD proposal. This has also an ancillary effect, namely less PCIs reuse factor.

Let us now consider the 100 eNBs to be detected. The figure shows similar behavior to that we already discussed when integrating the 500 HeNBs. The reasons have been already discussed. A main difference to be noted, namely the number of PCIs used by the SGC approach considerably increases to exceed that necessitated by the GC approach. The reason is the less number of PCIs available to used (permanent ones) compared to that used by the GC scheme. This increases the possibilities of conflicts and then the number of PCIs required to eliminate these conflicts.

Now, let us investigate the impact of changing NCR lists due to changing the coverage areas of cells. The figure shows that, in general, the average number of PCIs required decreases when employing all studied PCI assignment approaches. Note however that some PCIs re-configurations may result in an

Figure 16. Impact of network dynamics on the average number of confusions experienced

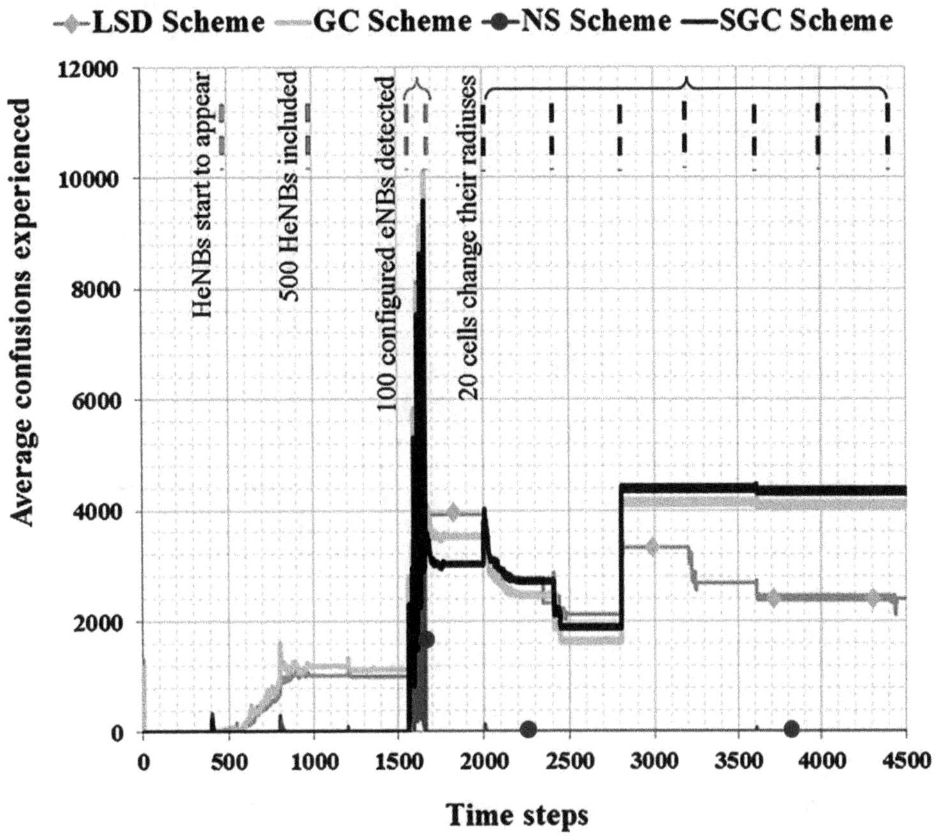

increased number of PCIs compared to the number resulting from previous PCIs re-configuration actions, especially with LSD, GC and SGC approaches.

The SGC approach sees a decrease in the average PCIs number by 35.34%, while the GC approach sees a decrease by 14.23%. This results from the efforts done to reuse PCIs as most as possible. Note that each re-configuration process attempts to reuse PCIs, which will probably result after adequate time duration in a sub-optimized or possibly an optimized PCIs assignment. This is actually the reason, why we do not see any change in the number of PCIs utilized after approximately 2800 time steps, see the figure. For the NS approach, we have mentioned previously that this approach quickly restores network stability at the cost of using more PCIs. In fact, the optimal average number of PCIs required to operate the network without PCIs conflicts is less than what the NS approach required directly after integrating all new cells. This is why we see a continuous decrease in the average number of PCIs utilized after each re-configuration process until the number of PCIs remains constant regardless of the executed PCIs re-configurations (10.64% decrease according to our results). See the impact of network dynamics on PCIs conflicts again and note that NCR lists updates have minimal impact on the performance of NS approach. Re-configuration tasks result also in more optimized PCIs usage.

The LSD scheme experiences a minimal decrease in the number of PCIs required, less than 1% according to our results. In fact, this scheme is heavily affected by NCR lists updates and, thus, considerable amount of PCIs conflicts arise. This implies that re-configuration actions aim primarily at recovering

Figure 17. Impact of network dynamics on the average number of used PCIs

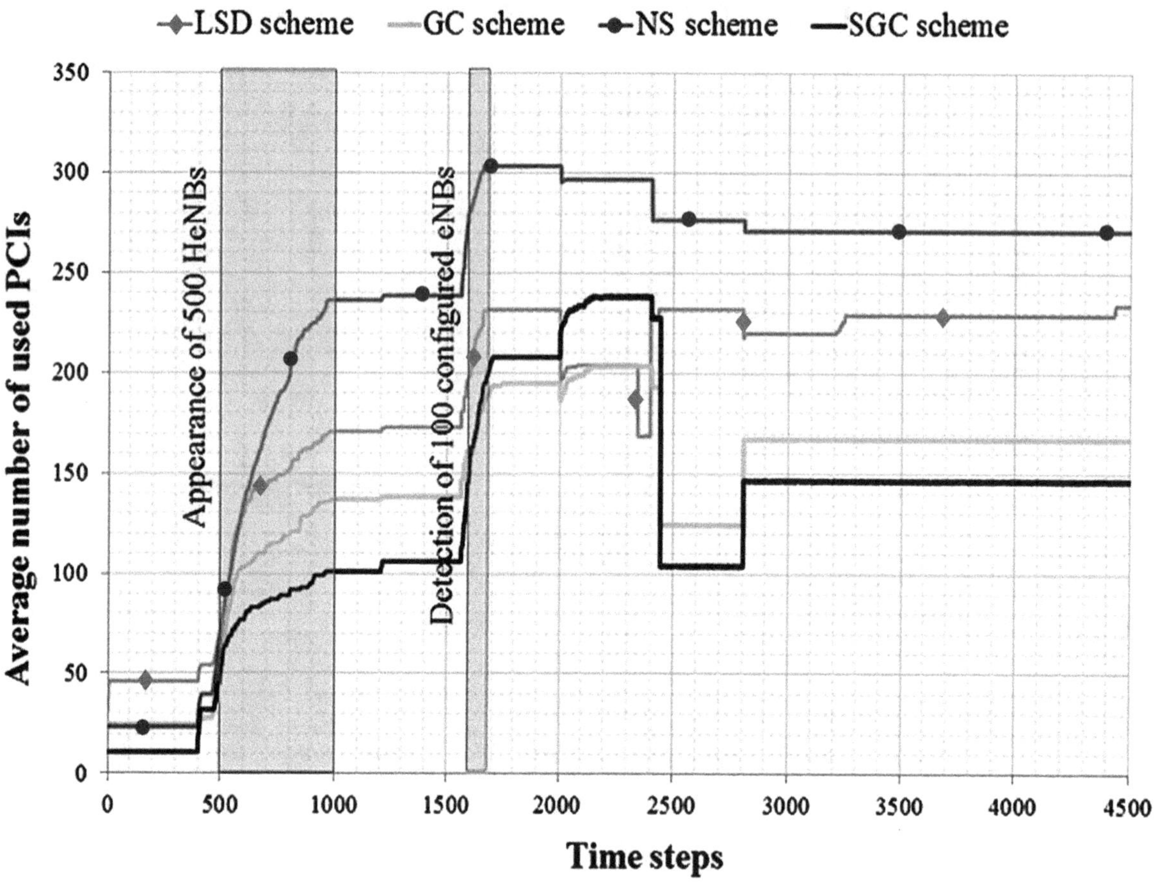

from conflicts not at saving PCIs. This results in increased average number of PCIs sometimes and decreased number sometimes.

6. CONCLUSION

The chapter has focused on the auto-configuration of eNBs with conflicts-free PCIs. It has addressed the problems faced in its context and highlighted the state of art. The chapter has gone through PCI assignment and also allocation schemes. The major contribution of this chapter was the investigation of the impact of network dynamics on the performance of four well-known distributed PCI assignment schemes, namely SGC, GC, LSD and NS schemes. The main results of the chapter can be concluded as follows:

1. PCI assignment and also allocation schemes are grouped into centralized and distributed schemes. Centralized schemes depend on a central entity with global knowledge, while distributed schemes utilize local knowledge only.
2. The number of confusions experienced is significantly greater than the number of collisions.

3. Concerning the impact of network size:
    a. For PCIs conflicts:
        i. The GC scheme is the most one affected by network size when considering the average number of PCIs conflicts experienced. The LSD proposal performs clearly better than the distributed GC scheme. The NS approach is less affected by network size than both the studied GC and LSD schemes. The performance is also considerably improved when employing the SGC scheme as compared to the GC approach and the LSD proposal. The NS approach performs, however, slightly better than the SGC in networks with low to middle sizes. When network size starts to be high (number of eNBs exceeds 400 in our study), the NS approach starts to show considerably performance improvements compared to the SGC approach.
    b. For the PCIs utilization ratio:
        i. The SGC scheme follows similar behavior as the GC approach with more PCIs utilization ratio due to the smaller PCIs range the SGC scheme makes available for use. The PCIs utilization ratio is significantly less than that seen by the LSD and the NS solutions.
4. Concerning the impact of network dynamics:
    a. For PCIs conflicts:
        i. The appearance of new cells heavily affects LSD and GC schemes. Network instability results until no more new cells appear in the network. Both NS and SGC schemes are more robust since the network suffers from few confusions and quickly restores its stability.
        ii. The strongest impact on the performance of PCI assignment schemes is observed when detecting cells previously not detected by their neighbors. All studied approaches are extremely affected and see PCIs conflicts when updating their NCR lists. This means that the more accurate the NCR lists are established, the fewer the PCIs conflicts raised.
        iii. The impact of changing cells radiuses differs based on the scheme employed. The NS scheme is even not affected. The network completely loses its stability when employing the GC, SGC and LSD schemes
    b. For the PCIs utilization ratio:
        i. The lowest number of PCIs required is observed by the SGC approach. The GC approach comes next followed by the LSD and NS schemes.

REFERENCES

Abdullah, L. M., Baba, M. D., & Ali, S. G. A. (2014). Self-Configuration Concept to Solve Physical Cell ID Conflict for SON LTE-based Femtocell Environment. *Int. J. on Recent Trends in Engineering and Technology*, *10*(2).

Ahmed, F., Tirkkonen, O., Peltomäki, M., Koljonen, J.-M., Yu, C.-H., & Alava, M. (2010). Distributed Graph Coloring for Self-Organization in LTE Networks. *Journal of Electrical and Computer Engineering*. doi:10.1155/2010/402831

Amirijoo, M., Frenger, P., Gunnarsson, F., Kallin, H., Moe, J., & Zetterberg, K. (2008). Neighbor Cell Relation List and Measured Cell Identity Management in LTEs. In *IEEE Symposium on Network Operations and Management* (NOMS`08). doi:10.1109/NOMS.2008.4575129

Amirijoo, M., Frenger, P., Gunnarsson, F., Kallin, H., Moe, J., & Zetterberg, K. (2008a). Neighbor Cell Relation List and Physical Cell Identity Self-Organization in LTE. In *Proceeding of IEEE International Conference on Communications Workshops* (ICC'08). doi:10.1109/ICCW.2008.12

Babadi, B., & Tarokh, V. (2008). A Distributed Asynchronous Algorithm for Spectrum Sharing in Wireless Ad Hoc Networks. In *Proceedings of the 42nd Annual Conference on Information Sciences and Systems* (CISS '08). doi:10.1109/CISS.2008.4558635

Bandh, T., Carle, G., & Sanneck, H. (2009). Graph Coloring Based Physical-Cell-ID Assignment for LTE Networks. In *Proceeding of the ACM International Wireless Communications and Mobile Computing Conference* (IWCMC '09). doi:10.1145/1582379.1582406

Dahlman, E., Parkvall, S., & Skoeld, J. (2011). *4G - LTE/LTE-Advanced for Mobile Broadband*. Academic Press.

Diab, A., & Mitschele-Thiel, A. (2012). Comparative Evaluation of Distributed Physical Cell Identity Assignment Schemes for LTE-Advanced Systems. In *Proceeding of the Performance Monitoring, Measurement and Evaluation of Heterogeneous Wireless and Wired Networks Workshop.*

Diab, A., & Mitschele-Thiel, A. (2013). Development of Distributed and Self-Organized Physical Cell Identity Assignment Schemes for LTE-Advanced Systems. In *Proceeding of the 8th ACM workshop on Performance Monitoring and Measurement of Heterogeneous Wireless and Wired Networks* (PM-2HW2N'13).

Diab, A., & Mitschele-Thiel, A. (2014). Self-Organization Activities in LTE-Advanced Networks. In Handbook of Research on Progressive Trends in Wireless Communications and Networking. IGI Global. Doi:10.4018/978-1-4666-5170-8

Dreher, A., & Mohl, D. (2015). *Precision Clock Synchronization*. The Standard IEEE 1588. White paper. Available at https://www.belden.com/docs/upload/Precision_Clock_Synchronization_WP.pdf

Eisenblaetter, A. (2003). *Assigning Frequencies in GSM Networks*. Technical Report.

FemtoForum. (2011). *HeNB (LTE Femto) Network Architecture*. White paper. Retrieved from http://www.smallcellforum.org/

Garza, C., Ashai, B. H., Monturus, E., & Syputa, R. (2010). *LTE Operator Commitments: Deployment Scenarios and Growth Opportunities*. Research report. MARAVEDIS Wireless Market Research & Analysis.

Golaup, A., Mustapha, M., & Patanapongpibul, L. (2009). Femtocell Access Control Strategy in UMTS and LTE. IEEE Communications Magazine, 47(9). doi:10.1109/MCOM.2009.5277464

IEEE Standard. (2008). *IEEE Standard for a Precision Clock Synchronization Protocol for Networked Measurement and Control Systems*. DOI: .10.1109/IEEESTD.2008.4579760

Kwak, H., Lee, P., Kim, Y., Saxena, N., & Shin, J. (2008). Mobility Management Survey for Home-eNB Based 3GPP LTE Systems. Journal of Information Processing Systems, 4(4). doi:10.3745/JIPS.2008.4.4.145

Kwon, S., & Lee, N. H. (2011). Virtual Extension of Cell IDs in a Femtocell Environment. In Proceeding of IEEE Wireless Communications and Networking Conference (WCNC`11). doi:10.1109/WCNC.2011.5779171

Lee, P., Jeong, J., Saxena, N., & Shin, J. (2009). Dynamic Reservation Scheme of Physical Cell Identity for 3GPP LTE Femtocell Systems. Journal of Information Processing Systems, 5(4).

Lehser, F. (Ed.). (2008). *Next generation mobile networks - recommendation on SON and O&M requirements*. Technical Report. NGMN Alliance.

Liu, Y., Li, W., Zhang, H., & Lu, W. (2010). *Graph-Based Automatic Centralized PCI Assignment in LTE. IEEE Symposium on Computers and Communications (ISCC'10)*, Riccione, Italy.

Liu, Y., Li, W., Zhang, H., & Yu, L. (2010). Distributed PCI Assignment in LTE Based on Consultation Mechanism. In *Proceeding of 6th International Conference on Wireless Communications Networking and Mobile Computing* (WiCOM`10). doi:10.1109/WICOM.2010.5601210

Mills, D. L. (1992). *Network Time Protocol (Version 3) Specification, Implementation and Analysis*. IETF RFC 1305.

Misra, P., & Enge, P. (2001). *Global Positioning System: Signals, Measurements, and Performance*. Ganga-Jamuna Press.

Motorola. (2009). *LTE Operations and Maintenance Strategy - Using Self-Organizing Networks to Reduce OPEX*. White paper.

Neel, J. O., & Reed, J. H. (2006). Performance of Distributed Dynamic Frequency Selection Schemes for Interference Reducing Networks. In *Proceedings of the IEEE Military Communications Conference* (MILCOM '06). doi:10.1109/MILCOM.2006.302016

Next Generation Mobile Networks (NGMN) Official Website. (2015). Retrieved May 07, 2015, from https://www.ngmn.org/

Nokia Siemens Networks and Nokia. (2008). *SON Use Case: Cell Phy ID Automated Configuration*. R3-080376, Technical Report.

Østerbø, O., & Grøndalen, O. (2012). Benefits of Self-Organizing Networks (SON) for Mobile Operators. *Computer Networks and Communications, 2012*. doi:10.1155/2012/862527

Ramiro, J., & Hamied, K. (2012). *Self-Organizing Networks (SON) Self-Planning - Self-Optimization and Self-Healing for GSM, UMTS and LTE*. John Wiley & Sons.

3rd Generation Project Partnership (3GPP). (2008). *Automatic Physical Cell ID Assignment*. 3GPP TSG-SA5 (Telecom Management). S5-081185.

3rd Generation Project Partnership (3GPP). (2009). *TS 36.211 v8.6.0: Technical Specification Group Radio Access Network; Evolved Universal Terrestrial Radio Access (E-UTRA); Physical Channels and Modulation (Release 8)*. Technical Report.

3rd Generation Project Partnership (3GPP). (2009a). *TS 36.300v8.9.0: Evolved Universal Terrestrial Radio Access (E-UTRA) and Evolved Universal Terrestrial Radio Access Network (E-UTRAN); Overall Description (release 8).* Technical Specification.

3rd Generation Project Partnership (3GPP). (2010). *Tr 36.902v9.2.0: LTE; Evolved Universal Terrestrial Radio Access Network (E-UTRAN); Self-Configuring and Self-Optimizing Network (SON) Use Cases and Solutions (Release 9).* Technical Specification.

3rd Generation Project Partnership (3GPP) official website. (2015). Retrieved May 07, 2015, from http://www.3gpp.org/

Sesia, S., Toufik, I., & Barker, M. (2011). *LTE, The UMTS Long Term Evolution: From Theory to Practice* (2nd ed.). Wiley. doi:10.1002/9780470978504

Shahab, M. B., & Bhatti, A. A. (2013). Neural Networks Based Physical Cell Identity Assignment for Self Organized 3GPP Long Term Evolution. *International Journal of Advances in Telecommunications, Electrotechnics, Signals and Systems, 2*(3).

Wu, T., Rui, L., Xiong, A., & Guo, S. (2010). An Automation PCI Allocation Method for eNodeB and Home eNodeB Cell. In *Proceeding of the 6th International Conference on Wireless Communications Networking and Mobile Computing* (WiCOM'10). doi:10.1109/WICOM.2010.5600764

Wu, Y., Jiang, H., Wu, Y., & Zhang, D. (2010). Physical Cell Identity Self-Organization for Home eNodeB Deployment in LTE. In *Proceeding of the 6th International Conference on Wireless Communications Networking and Mobile Computing* (WiCOM). doi:10.1109/WICOM.2010.5600778

Zhang, W., Wang, G., Xing, Z., & Wittenburg, L. (2005). Distributed Stochastic Search and Distributed Breakout: Properties, Comparison and Applications to Constraint Optimization Problems in Sensor Networks. Journal of Artificial Intelligence, 161(1-2).

Zhang, X., Zhou, D., Xiao, Z., Liu, E., Zhang, J., & Glasunov, A. A. (2011). Dynamic Group PCI Assignment Scheme. In *Proceeding of the 7th International Conference on Wireless and Mobile Communications.*

Chapter 2
Network Optimization Methods for Self-Organization of Future Cellular Networks:
Models and Algorithms

Furqan Ahmed
Aalto University, Finland

Alexis A. Dowhuszko
Aalto University, Finland

Olav Tirkkonen
Aalto University, Finland

ABSTRACT

This chapter discusses network optimization methods for enabling self-organization in current cellular networks such as Long Term Evolution (LTE)/LTE-Advanced (LTE-A), and the upcoming 5G networks. Discrete and continuous optimization models are discussed for developing distributed algorithms for self-configuration and self-optimization. The focus is on Self-Organized Networking (SON) problems, which are relevant to small cell networks. Examples include Physical Cell-ID (PCI) assignment, Primary Component Carrier (PCC) selection, Inter-Cell Interference Coordination (ICIC), and network synchronization. A conflict-graph model is considered for PCI assignment and PCC selection problems, which paves the way for different graph coloring algorithms with self-organizing properties. Algorithms for self-organized ICIC and network synchronization are also developed in a principled manner, through a network utility maximization framework. This systematic approach leads to a variety of algorithms which adhere to self-organization principles, but have varying requirements in terms of inter-cell coordination and computation complexity. Fully distributed self-organizing algorithms do not involve any inter-cell dedicated message-passing, and thus are faster and more scalable than the ones that are distributed but require local coordination via exchange of messages between cells. However, local coordination enables higher network utility and better convergence properties.

DOI: 10.4018/978-1-5225-0239-5.ch002

INTRODUCTION

Self-organization is a general concept with numerous applications across a wide range of areas related to physics, computer science, and mathematics. Some common examples include distributed systems, multi-agent systems, and complex networks (Haken, 2006). It has gained popularity in the field of mobile cellular communications systems recently, and is considered to be a cornerstone in the success of contemporary cellular networks such as Long Term Evolution (LTE)/LTE-Advanced (LTE-A) (Hämäläinen, Sanneck, & Sartori, 2011). Moreover, it is expected to play a promising role in development of enabling technologies for future cellular networks such as 5G, which will comprise of 10-100 times more devices and much higher data rates compared to LTE/LTE-A. Thus, the proliferation of wireless devices, along with a steady increase in popularity of wireless applications and services, will lead to an unrelenting demand for high data rates and ubiquitous connectivity. Dense deployment of small cell networks (SCNs) will be the key to high capacity and improved coverage. The core principle that enables capacity enhancement in such networks is the reduced cell size compared to traditional macrocell networks, resulting in higher received signal-to-interference-plus-noise power ratios (SINRs), as the distance between transmitter and receiver is decreased. In indoor deployments, transmissions are shielded from macro interference due to penetration losses. Other advantages include significant reduction in the power consumption of mobile stations communicating with their serving base stations, and offloading of traffic to avoid congestion in the macrocellular network.

The deployment of small cells usually complements the macrocell network providing service in a larger coverage area. A dense deployment of small cells in conjunction with macrocells is considered a major enabler for meeting the capacity requirements of future mobile cellular networks. In such networks, the self-organized networking (SON) paradigm will be of key importance due the following factors (Aliu, A. Imran, M. A. Imran, & Evans, 2013; Prehofer & Bettstetter, 2005; Hämäläinen et al., 2011):

1. The network densification due to massive deployments of small cells makes the configuration, optimization, and maintenance of the network difficult. Distributed and scalable solutions based on self-organization are essential for carrying out these tasks in an efficient manner.
2. Small cells are usually classified as microcells, picocells, and femtocells depending on their coverage areas and transmit powers. Among these, the outdoor picocells are deployed by operators, whereas the femtocells are plug-and-play devices that are usually deployed by the end users in an *ad hoc* manner. The resulting heterogeneous cellular network comprises of multiple layers of cells of different sizes and characteristics. Unplanned deployment of femtocells, as well as the disparity in transmit power of cells belonging to different layers, may result in very high cross-layer interference for some users. Effective mitigation of this interference is an important SON issue in dense small cell and heterogeneous networks.
3. In order to maximize spectral-efficiency, it is important to optimize the overall resource allocation in the network over all available degrees of freedom. This is especially true for future mobile networks, which will comprise of billions of devices connected seamlessly, through different types of cells and radio access technologies. The increased complexity will result in more degrees of freedom which can be exploited to improve performance. Resources should be utilized optimally, satisfying the quality of service (QoS) requirements of all users. The use of self-organization schemes for an efficient allocation of resources along the available degrees of freedom, such as time, frequency,

space and users, in a non-interfering way, is an effective way to get close to the capacity limits imposed by the wireless channel.

4. Self-organization is of key importance in reducing capital expenditure (CAPEX) and operational expenditure (OPEX) of the cellular networks, mainly due to the fact that networks with automated features such as self-configuration, self-optimization, and self-healing do not require large skilled workforce for ensuring seamless operation. SON paradigm facilitates automated commissioning, installation, performance evaluation, and monitoring of large-scale networks, thereby reducing the expenditures incurred in carrying out these processes manually.
5. Apart from a reduction in power expenditure, energy efficient design of future cellular networks will lead to the minimization of environmental footprint. Moreover, it can be enabled at a network level through the use of self-organizing energy management techniques.

To sum up, the concept of self-organization is important for cost-effective design of future mobile cellular networks, as it leads to highly efficient network operation, ease of maintenance, and reduced CAPEX/OPEX. In fact, SON has been adopted by 3rd Generation Partnership Project (3GPP) as a requirement for LTE/LTE-A systems, and is expected to play a leading role in the design of enabling technologies for future wireless cellular networks as well.

In the following sections, the background of self-organization and motivation for its application in the field of wireless cellular networks is discussed. This involves a close look at some common SON use-cases that are of current interest from the research and standardization perspective, and a discussion on network optimization methods that can be used for a systematic design of SON algorithms. The generic algorithm design approach considered here underpins a number of SON use-cases. Algorithms discussed in the following sections are based on distributed network optimization over discrete or continuous solution spaces, but the basic idea is essentially the same. By exchanging messages locally, the network nodes are able to make more intelligent and informed decisions, which lead to a significant improvement in performance. The discrete optimization case is modeled as a graph coloring problem, whereas for the continuous optimization network utility maximization (NUM) framework is considered. Different algorithms for the following SON use-cases related to current and future mobile cellular networks are discussed:

- Physical cell-ID (PCI) assignment
- Primary component carrier (PCC) selection
- Inter-cell interference coordination (ICIC)
- Network synchronization

Numerical examples are presented throughout the chapter to support the analysis, and to compare the performance of various algorithms.

BACKGROUND

Definition

Self-organization can be defined as a process in which local interactions among the components of a system lead to an emergence of a global order (Ashby, 1947). This phenomenon can be observed, in some form, in different branches of science such as physics, chemistry, biology, and computer science. The most interesting and inspiring occurrences are found in nature, and especially in biological systems comprising of living organisms. For example, swarming insects, fire-flies firing in synchrony, fishes swimming in a school, ants foraging for food, and flocking birds, to name a few (Camazine, Deneubourg, Franks, Sneyd, Theraulaz, & Bonabeau, 2003). All such systems exhibit emergent behavior, i.e., passive interactions of a large number of individuals contributes in achieving a system-level goal. For example, a self-organization mechanism in bird flocking entails mutual interaction between birds, where each bird adjusts its position in a flock by repelling its neighbors individually while maintaining its collective attraction and alignment with its neighbors. Self-organizing mechanisms are also of significant interest in studying economic systems, human psychology, cybernetics, traffic flow, and population growth.

Motivation

The main feature of a self-organizing system is the lack of a central control, in coordinating the actions of agents in the system. The system is fully decentralized, and each agent in the system interacts only with a limited number of neighbors. These interactions collectively drive the system to an ordered state. Thus, self-organizing systems are inherently scalable, and the underlying principles are well suited for designing highly complex and large-scale dynamic systems. This motivates the engineering of self-organization in practical systems. Principles of self-organization have been successfully applied for solving complex problems in electrical engineering and computer science, particularly in areas such as multi-agent systems, complex networks, and distributed systems. Its popularity stems from the fact that it enables distributed solutions, with limited coordination and computation requirements for system components. Ideally, self-organizing systems should involve implicit coordination among system components, rather than dedicated message-passing. This is the main difference between self-organizing and distributed systems. Another unique feature of self-organizing systems, which distinguishes them from decentralized and distributed systems, is the emergence of global behavior patterns, achieved through local interactions. Fully distributed systems that do not involve dedicated communication between components may thus be termed as self-organizing systems, provided they exhibit the phenomenon of emergence. From a perspective of computer networks, a hierarchy of networking approaches is shown in Figure 1, ranging from centralized to self-organizing networks (Baran, 1964). The level of coordination and system scalability are illustrated along this axis of centralization, emphasizing that engineering of self-organization enhances the scalability of the system. In Figure 2, pictorial representations of networks along this axis are presented. A decentralized network consists of a network of multiple centers. In a distributed network, the nodes are peers, and decisions are made based on bilateral message-passing. In a self-organizing network, no dedicated message-passing happens. Local information exchange is based on passively observing changes in the environment caused by actions of neighbors, or on receiving generic multicast or broadcast signals.

The main features of self-organizing systems can be summarized as follows:

1. **Scalability:** One of the foremost aspects of self-organizing systems is scalability, and it is of utmost importance in large-scale complex networks. It ensures that the system can be scaled up with manageable complexity. Thus, the system can be expanded without an unbounded increase in complexity. Scalability emanates from the self-organizing principle itself. Due to the local nature of interactions between the system components, addition of new components results in limited additional complex-

Figure 1. Local interactions and limited coordination leads to high scalability in self-organizing systems

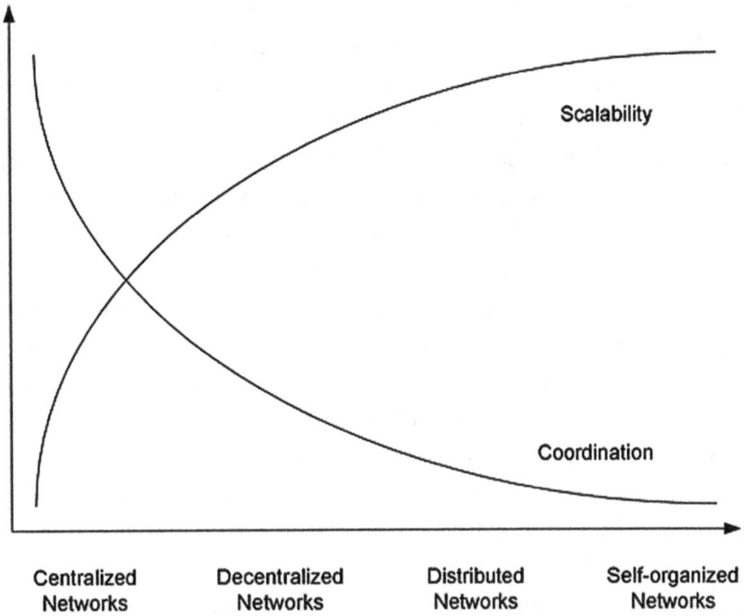

Figure 2. Classification of networks based on the level and type of coordination between the nodes

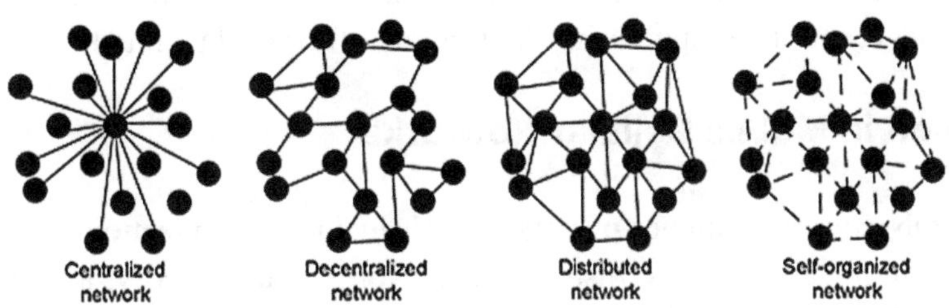

ity. This is especially important for dynamic systems, which evolve over time in terms of size and number of components. A design principle based on local interactions yields scalable solutions for a wide range of problems. Local interaction between system components can be formulated in many different ways, depending on the problem at hand. For ideal self-organization, the local interaction should be entirely passive, and based on mere observations of the local neighborhood. Ideal self-organization with entirely passive local interactions is common in nature, e.g., in bird flocking. However, when dealing with engineering of self-organization, e.g., in wireless networks, the local interactions may also be active and involve an exchange of information or cooperation in the form of message-passing. In such cases, it is desirable to keep the information exchange local, and at a minimum, to reduce the overhead resulting from the addition of new components. The interactions among the system components may be based on cooperation, but they should always be local. Long-ranged cooperation among components jeopardizes the scalability, as it may lead to an unmanageable increase in overhead information as the system grows.

2. **Adaptability:** Due to a high degree of adaptability, self-organizing systems can effectively respond to changes in the environment and adjust operation parameters accordingly to reach equilibrium.

This means that a system can reorganize in order to adapt to changes like the addition/removal of new components in the system, or loss in system performance resulting from both external and internal factors. In this regard, agility is also important as the system should be able to react to the changes in a timely manner to ensure that performance remains unaffected. Therefore, adaptability and agility play an important role in guaranteeing rapid convergence to a desirable state.

3. **Emergent behavior:** A key property of self-organizing systems is that they possess emergent properties, i.e., the local interactions among system components lead to an emergence of desirable properties at the global level. All the system components contribute towards the desirable global behavior, and are not capable of achieving it individually. This collective intelligence leads to the emergence of global behavior, which transcends the capabilities of individual components.
4. **Robustness:** Self-organizing systems are adaptable and decentralized, which make them particularly robust against failures. This is due to the fact that system can adapt to changes in environment, and possesses the capability to reconfigure itself in case of failures. The decentralized architecture is intrinsically robust, and leads to properties such as stability and graceful degradation. In order to enhance the robustness, the self-organizing systems are often designed with self-healing mechanisms to tackle failures in a systematic manner.

It should be noted that self-organization as defined here, and in most of the scientific literature is partially in a different context than the SON paradigms discussed for the mobile cellular networks by 3GPP and other standardization bodies. 3GPP standardization activities focus on automation of network functions, using methods which may be fully self-organizing adhering to the above discussed definition, or based on more or less centralized approaches. This point is further elaborated in the following sections.

Self-Organization in Mobile Cellular Networks

The research on SON covers a number of areas which include optimization, design and analysis of complex systems, multi-agent systems, distributed networks, game theory, and wireless networking (Olfati-Saber, Fax, & Murray, 2007; Hubaux, Gross, Le Boudec, & Vetterli, 2001; Strogatz, 2001). Most practical applications of SON are related to advanced networks that are large-scale, diverse and highly interconnected, and thus can be modeled as complex systems. Future generation wireless networks are predicted to be much more complex than the contemporary ones, comprising of billions of devices. Different problems pertaining to such massive networks can be effectively handled by engineering the system using self-organizing principles.

In wireless networks, SON based mechanisms can enhance performance by adding artificial-intelligence and adaptability to network nodes, allowing them to react to the changes in network and environment state (Hubaux et al., 2001). The adaptive nature of such networks enables the utilization of the situational and local knowledge for achieving optimal end-to-end performance. In particular, enabling the network nodes to do intelligent adaptations at an individual level, based on local information guarantees a significant performance improvement over traditional networking technologies. It provides the ultimate solution to the problem of managing a large number of bandwidth intensive network nodes competing for scarce resources. For wireless networks, the term "self-organization" has been used with different meanings in literature, and the concept has been applied to different types of networks in recent years: it has been used in the context of mobile *ad hoc* networks by Hubaux et al. (2001), peer-to-peer networks by Hales and Arteconi (2006), multi-hop networks by Mitton, Fleury, Lassous, and Tixeuil (2005), and finally in wireless sensor networks (WSNs) by Sohrabi, Gao, Ailawadhi, and Pottie (2000).

In general, SON based methods have been studied extensively for networks that do not involve any central planning or coordination. Most popular examples include WSNs and *ad hoc* networks. Thus, the key benefit of SON is that it leads to a decentralized architecture with scalability properties (Prehofer & Bettstetter, 2005). In contrast, design of cellular networks has traditionally been based on centralized planning. For example, 2G networks such as GSM require central frequency planning and network dimensioning. However, the recent increase in popularity of high data rate wireless connectivity has led to a paradigm shift in the design of cellular networks. Among the emerging technologies, SON based mechanisms have proved to be extremely useful, and are expected to play a key role in meeting demands of the future cellular networks (Hämäläinen et al., 2011).

Therefore, SON represents a broad research interest in the area of wireless networking, which spans not only the well-established areas of distributed and *ad hoc* networking, but also the standardization and implementation of contemporary and future mobile cellular networks. SON features can address the challenges of ensuring high data rates, improved coverage, scalability, efficient radio resource management (RRM), energy efficiency, reduced CAPEX/OPEX, fault tolerance, and quick recovery. The overall objective is network automation through approaches that are capable of adapting to varying channel conditions and QoS requirements of the users, which leads to better network performance, higher scalability, and improved robustness. Consequently, a number of SON use-cases have been considered by the 3GPP in the standardization of LTE and LTE-A. These have been identified in different research projects, and in standardization activities. Notable examples include the use-cases proposed in Next Generation of Mobile Communication (NGMN) report by Lehser et al. (2008), SOCRATES project by Scully et al. (2008), and the 3GPP (3GPP, 2010). The SON use-cases related to network deployment and operation are classified into three main categories: self-configuration, self-optimization, and self-healing. In 3GPP Rel. 8, the following SON use-cases for self-configuration were introduced:

- Automatic software download
- Automatic PCI assignment
- Automatic Inventory
- Automatic neighbor relation (ANR)

This was followed by the addition of the following self-optimization use-cases in 3GPP Rel. 9:

- Mobility load balancing (MLB) optimization
- ICIC
- Handover/ mobility robustness optimization (MHO/MRO)
- Random access channel (RACH) optimization

The 3GPP Rel. 10 focused on enhancements of the existing use-cases, along with the addition of new self-healing and energy saving functionalities:

- Enhanced inter-cell interference coordination (eICIC)
- Cell outage detection and compensation
- Self-healing functions

- Automatic coverage and capacity optimization (CCO)
- Minimization of drive testing (MDT)
- Energy savings (ES)

In addition to CCO, MHO, ES, and MDT, in Release 11 the following enhancement was considered

- SON function coordination

Similar concepts were discussed in SOCRATES project and by NGMN. However, due to substantial overlap, it is possible to identify the main categories as given by Aliu et al. (2013). The aforementioned SON use-cases can be tackled using a variety of approaches, each of them having different requirements on local coordination and information exchange between cells. Distributed approaches may involve direct cooperation among the network nodes through dedicated message-passing. On the other hand, fully distributed approaches that do not involve message-passing are also possible in some cases, and are more attractive from a practical standpoint. Such approaches are closer to the research community definition of ideal self-organization, as they involve only implicit local coordination in a passive way, rather than explicit cooperation through the exchange of dedicated messages. In standardization activities of cellular networks, SON algorithms may be fully distributed algorithms that do not involve message-passing, distributed approaches with message-passing, or even centralized mechanisms of network operation. Distributed algorithms with message-passing are important due to the fact that in the 3GPP LTE/LTE-A radio access network architecture, message-passing between base stations is possible via the X2 interface. The main advantage of message-passing algorithms over fully distributed ones is guaranteed convergence, and often substantially better performance. Thus, according to the 3GPP standardization terminology, apart from ideal self-organizing approaches, distributed and fully distributed approaches also come under the rubric of SON in mobile cellular networks. A fully distributed approach may be called an ideal self-organizing one, provided that it exhibits emergent behavior (i.e., equilibrium exists and it is possible to reach it).

This chapter discusses the application of distributed, fully distributed, and ideally self-organizing algorithms for SON use-cases. Broadly speaking, the focus is on distributed algorithms that can be applied to SON use-cases defined by NGMN, SOCRATES, and 3GPP. The first step towards developing effective SON algorithms is to identify the SON problems from the SON use-cases, and classify them accordingly. A use-case may involve multiple problems, and it is also possible that the same or closely related problems are part of different use-cases. The problems may be modelled as static or dynamic, and the requirements on inter-cell coordination may vary according to the characteristics of the scenario. For example, in multi-vendor and multi-operator networks, complete coordination among all the cells may not be possible. As mentioned before, the most popular classification of SON use-cases is according to their respective role in the network life cycle, i.e., self-configuration, self-optimization, and self-healing. Both centralized and distributed SON approaches have been discussed for cellular networks in 3GPP standardization. Self-configuration entails configuring different network parameters. Therefore, the identification of configurable parameters in a mobile cellular network yields a partial classification of SON problems. Problems related to self-optimization are numerous and more complicated in nature.

In this chapter, SON problems are classified according to the properties that influence the solution methodology. Some techniques can be applied to multiple types of problems. At a high level, one can identify SON problems related to the well-known *mutual synchronization* and *consensus* problems. Both involve reaching an agreement among cells at a global level. On the other hand, there are several problems that require the computation of a local variable. In problems such as PCI assignment and PCC selection, these variables are logical, and it is essential that neighboring cells have different values. Hence, they can be categorized as *neighbor avoidance* problems. In many problems, real and logical-valued variables are controlled to tradeoff performance in the serving cell against the interference caused to other cells. Such problems, e.g. transmit power allocation, fall under the *inter-cell RRM* category. Tracking area formation and coordination area formation are typical examples, and can be identified as instances of a *clustering* problem. Problems such as ANR, mobile station association, MHO, etc. require active cooperation among neighboring cells, and thus can be called *neighbor cooperation* problems. In many cases, it is important to find the best set of cooperation parameters or, otherwise, divide the network to parts. Moreover, these generic SON problems can be regarded as primitives that can be used when realizing SON solutions for complicated use-cases. A summary of generic problems and approaches that can be employed for their solution are listed in Table 1.

In the following sections, different methodologies for enabling SON use-cases are explained along with several examples supported by simulation results. The discussion focuses on how to formulate different use-cases as optimization problems, and what computational methods can be employed to solve them. To this end, concepts from both discrete and continuous optimization theory are used to develop a systematic approach towards the design of efficient SON algorithms. In the discrete or combinatorial optimization model considered here, graph coloring formulation is discussed along with different coloring algorithms that have self-organizing properties. For the continuous optimization case, NUM is considered for maximizing some performance metric of the network with or without message-passing among the cells. The methods discussed here are applicable to a large range of problems in wireless networking, and adhere to the following principles that are important for engineering self-organization:

- Local rules that lead to global behavior
- Implicit coordination
- Limited information exchange
- Adaptability

Table 1. Classification of SON problems and solution approaches

Category	Global Variable	Local Variable	Mathematical Models
Mutual synchronization	Network synchronization		Distributed synchronization algorithms, NUM
Neighbor avoidance	Global cell ID selection	Uplink reference signals PCC/PCI selection RACH optimization	Conflict graphs, constraint satisfaction
Inter-cell-RRM		Transmit power, ICIC, ES	NUM
Clustering		Tracking area/ coordination area formation, CCO, ES	Clustering algorithms, clique algorithms
Neighbor cooperation		MLB optimization, MRO, ANR, CCO, ES	

GRAPH BASED NETWORK MODELS AND CONSTRAINT SATISFACTION

A number of problems in the area of wireless networking can be modelled as discrete or combinatorial optimization problems. The system model comprises of a network cost function, formulated in terms of logical network variables, which may be optimized. Examples include topology control, routing optimization, QoS provisioning, and general resource allocation problems (Cheng, Li, & Du, 2006). Most of the existing literature focuses on these problems in the context of *ad hoc* networks, satellite networks, and WSNs. Recently, the benefits of SON in cellular networks have paved the way for an application of both continuous and discrete optimization methods to problems related to SON use-cases. To address these problems, different approaches have been considered in the literature; see e.g. (Zhu & Ray Liu, 2008). A number of problems such as conflict-free channel selection (or frequency assignment), PCI assignment, PCC selection, optimal user association with base stations, and power allocation can be solved by discrete network optimization methods. Some SON problems specific to cellular networks have never been addressed before in other types of networks, and have not been solved using systematic optimization methods. These include PCI assignment, RACH optimization, and load balancing, to name a few. Among the methods that can be used for solving discrete network optimization problems pertinent to cellular networks, distributed constraint satisfaction and graph coloring are of notable importance.

A constraint satisfaction problem (CSP) arises when multiple autonomous agents controlling logical variables coexist in a system with constraints among their possible actions. In a CSP, values are assigned to a set of variables under a given set of constraints. The aim is to obtain a consistent assignment of values to variables, so that the constraints are met. A common example of such problems is an N - queens problem, where the problem is to place N queens on the chess board in a way that they do not threaten each other. A wide range of problems in artificial intelligence can be formalized as CSPs, with applications in a diverse range of areas such as machine vision, belief maintenance, scheduling, and belief propagation, among others. (Kumar, 1992). A subclass of CSPs can be reduced to graph coloring problems. In graph coloring, the aim is to assign colors to the vertices of a graph in non-conflicting manner, so that no two vertices connected by an edge have the same color. Graph coloring has an old and rich history in wireless systems, which dates back to the design of first generation of cellular systems (Balasundaram & Butenko, 2006). It has been studied extensively for the frequency assignment problem (FAP) in wireless networks. Graph coloring for FAP entails a conflict-graph model of the network, where vertices represent the cells and the edges represent the interference couplings. The aim is to color the graph using a fixed number of colors, where each color represents a frequency. To construct a conflict-graph, the interference statistics from the interfering base stations are required, which can be used to create edges between the vertices. In order to color the graph for FAP, a number of approaches have been discussed in the literature (Eisenblätter, Grötschel, & Koster, 2000). To benefit from graph coloring based solutions for SON in cellular networks, distributed algorithms are required.

As example solutions of distributed CSPs, SON algorithms based on distributed graph coloring are discussed in the remaining part of this section, along with different applications of these algorithms. Multiple algorithms based on local search metaheuristics can be devised to solve CSPs and graph coloring instances (Galinier & Hertz, 2006). Here distributed versions of these are considered, as well as message-passing based distributed constraint satisfaction algorithms (Yokoo & Hirayama, 2000). In both categories, the SON design principles are followed, and the proposed algorithms are distributed in nature. The distributed local search graph coloring algorithms discussed here are fully distributed with no dedicated message-passing during the execution phase. The local interactions by the cells consist of the observation of the colors of neighboring cells. These interactions guide the actions taken by the individual cells. Thus, the local interactions between the cells result in a colored graph in a self-organized way, provided that the algorithm converges. The constraint satisfaction algorithms also involve local interactions, albeit with message-passing. Nevertheless, both local search and constraint satisfaction algorithms have self-organization properties, which make them suitable for application in self-organizing cellular networks.

Local Search Algorithms

The local search principle involves local changes by individual agents to move in the configuration space from one configuration to another, satisfying constraints or avoiding conflicts locally. The search space is explored by making perturbations to the existing configuration, known as local moves. For graph coloring, a local move is the change of color by one vertex. Two states that are connected by a local move are neighbors in the configuration space. The configuration space is searched by a sequence of local moves taken by vertices. A local move that improves the cost is known as a downhill move. For graph coloring, the local move that reduces the number of conflicts of a given vertex is the downhill move. A key feature of the configuration space of colorings on undirected graphs is the existence of plateaus, i.e., neighboring states with the same number of conflicts. A greedy local search in which each vertex picks the best move gets trapped in the local minima, which are often located at plateaus. A main feature of local search algorithms is that they can move on the plateaus and avoid entrapment in local minima. In this context, a plateau move is a local move that keeps the number of conflicts remains unchanged. They are important for escaping from local minima on plateaus, and reaching a conflict-free state, see Figure 3 for an example.

An effective strategy to further improve the performance of local search algorithms is to use simulated annealing (SA) dynamics, which involve occasional uphill moves, i.e., accepting local moves that increase the number of conflicts (Johnson, Aragon, McGeoch, & Schevon, 1991). The probability of uphill moves is controlled by a noise parameter, which is often called a temperature. There is a cooling schedule according to which the temperature is lowered. The performance of SA strongly depends on both the value chosen for the temperature and the cooling schedule. Best choices are influenced by the characteristics of problem instances, and the details of the search strategy. Consequently, finding the optimal set of parameters is a problem of interest and has been addressed in the literature (McAllester, Selman, & Kautz, 1997). As examples, two local search algorithms are discussed below, which can be used for SON problems in cellular networks. In both cases, asynchronous and periodic updates are considered, where each vertex in the conflict-graph runs the algorithm on its turn to update.

Figure 3. Enabling local (plateau) moves for coloring conflict-graphs

1. **Stochastic Local Search (SLS):** This algorithm is based on simple plateau moves and downhill moves. Different variants of this algorithm have been discussed for the graph coloring problem by Ahmed et al. (2010), and also Ahmed and Tirkkonen (2015). Consider a state where vertex x is using color c. Let $F(c)$ be the number of conflicts of cell x, i.e. number of neighbors using same color. The cell may or may not be in conflict with neighbors while using color c. At its turn to update, it randomly picks a color $c' \in C \setminus c$ and evaluates its cost in terms of conflicts given by cost function $F(c')$. If $\Delta = F(c') - F(c) \leq 0$, cell x selects color c'. Thus, all downhill moves and plateau moves are accepted. An important variant of this algorithm is SLS with focused plateau (SLS-FP) moves, in which a cell accepts plateau moves only if it is in a conflict. This leads to a reduction in the reconfigurations required to color the conflict-graph.
2. **Simulated Annealing (SA):** The standard SA algorithm has been discussed for channel selection problem in (Chen et. al, 2008). However, to make it distributed, temperature should either be kept constant or a temperature synchronization protocol is required. Consider a state where vertex x is using color c. At its turn to update, it randomly picks a color $c' \in C$ and evaluates its cost in terms of conflicts given by cost function $F(c')$. If $\Delta = F(c') - F(c) \leq 0$, vertex x selects color c', otherwise selects color c' with the probability $e^{-\Delta/T}$. Thus, the higher is the cost of taking the uphill move, the lower is the probability of it being accepted. The temperature parameter T is reduced according to some cooling schedule such as $T = T_0 / \log_2(2 + t)$.

Complete Distributed Algorithms

From the perspective of SON, complete algorithms for distributed CSPs are of interest. Local search algorithms are not complete, in that they do not guarantee convergence. More involved message-passing algorithms can be created to guarantee completeness, i.e. returning a solution if it exists. Such algorithms have been discussed for generic distributed CSPs by Yokoo and Hirayama (2000), and for distributed graph coloring by Peltomäki, Koljonen, Tirkkonen, and Alava (2012). A distributed CSP comprises of multiple variables and constraints distributed among nodes. Such problems are common in the areas of multi-agent systems and distributed networks. Applications of the concept include problems such as distributed resource allocation, distributed scheduling, and truth management systems (Yokoo, 2012). The underlying principles are generic and can be applied to solve a broad range of problems related to self-organized resource allocation in wireless networks. This makes them important from a practical stand point of designing SON algorithms for cellular networks. Graph coloring is a typical example of CSP. Therefore, complete distributed algorithms provide an attractive alternative for distributed graph coloring, and can be considered for several SON use-cases. Each vertex of the graph can be considered as a variable, whereas the colors represent the values to be assigned under the constraints dictated by the adjacency matrix of the graph. A number of complete distributed algorithms can be applied for self-organized graph coloring. These include backtracking algorithms and iterative improvement algorithms. In what follows, two well-known backtracking algorithms by Yokoo and Hirayama (2000) are briefly discussed.

1. **Asynchronous Back Tracking (ABT):** The ABT algorithm is essentially based on the backtracking principle, which entails the construction of a partial solution by the assignment of values to a subset of variables, such that all constraints are met. The partial solution is then expanded to a full

solution by adding more variables, one by one. When no value satisfies all the constraints of a given variable, the value of the last added variable to the partial solution is changed. This procedure is known as backtracking, and is a key principle of many constraint satisfaction algorithms. ABT is a distributed and asynchronous version of the backtracking algorithm, in which agents interact via message-passing. In ABT, agents communicate their current values using *ok* messages and new constraints using *nogood* messages. Each agent maintains a view of the values of other agents, which is known as *agentview*. There is a priority order of the agents which is determined by an alphabetical order of agent identifiers. An agent changes its value if it is not consistent with the assignments of agents with higher priorities. If there is no value that can be used to achieve consistency with higher priority agents, a *nogood* message is generated and sent to the higher priority agents. The higher priority agents then change their value.

2. **Asymmetric Weak Commitment Search (AWC):** A downside of ABT is that agents have a fixed priority, set by the alphabetical order. Thus, if the value selection of a high priority agent is bad, execution time increases because the lower priority agents have to perform an exhaustive search to reverse the bad decision. To tackle this issue, the AWC algorithm introduces the min-conflict heuristic to change the priorities leading to the avoidance of bad decisions. Thus, AWC dynamics enable reversal of the bad decision without resorting to the exhaustive search. To reflect the agent priorities, a parameter *priorityvalue* is introduced, which is communicated through the *ok* messages.

Example: Graph Coloring Algorithms for PCI Assignment

The PCI configuration problem is an important self-configuration SON use-case for LTE/LTE-A, and has been discussed extensively from both technical and standardization perspectives (Hämäläinen et al., 2011). Its potential is underscored by the fact that ultra-dense deployment scenarios are currently being considered for enabling high capacity in future networks. In such networks, self-organized assignment of cell IDs will be of paramount importance for ensuring seamless handovers and efficient operation (Szilagyi, Bandh, & Sanneck, 2013). It is well known that the PCI assignment problem can be effectively addressed using a graph coloring formulation (Bandh, Carle, & Sanneck, 2009; Diab & Mitschele-Thiel, 2013). Following this approach, different distributed graph coloring methods for self-organized PCI assignment are considered next. These methods are based on local search principles and constraint satisfaction algorithms. It is observed that distributed local search methods based on limited cooperation via local interactions can enable conflict-free and confusion-free assignments of PCIs. Based on the discussion in Ahmed et al. (2010), four simple distributed local search algorithms are discussed to color the conflict-graph. For brevity, further discussion on SA heuristics is omitted. The algorithms discussed here can be classified according to two characteristics. The first classification is related to the type of interference pricing or interference coupling between cells. When real-valued interference couplings are used, a real-valued price may be considered between neighbors using the same PCI. Such algorithms have been used for channel assignment in (Neel & Reed, 2006; Babadi & Tarokh, 2008). In contrast, when binary conflicts are considered, the decisions are made solely on the basis of the number of conflicting neighbors, and not on the strength of the conflicts.

In addition, algorithms considered here are classified according to the number of alternatives tried by a cell, when it is its turn to update the PCI. One alternative is to select the PCI randomly, as in Monte Carlo algorithms, for details see the work by Kardar (2007). A cell randomly selects one of the PCIs not used by itself at the moment. Then, it starts to use it, if the price (interference price, or number of conflicts), is less than or equal to the price with the earlier PCI. Multiple-try algorithms make use of the steepest descent principle (Zhang, Wang, Xing, & Wittenburg, 2005); the cell calculates the price for all PCIs, and randomly selects one of the PCIs with the lowest price. It is worth noting that the binary algorithms have focused plateau moves, which leads to an *absorbing local optimum*, that is, the cell does not change its PCI once it is in a conflict-free state (Zhang et al., 2005; Alava et al., 2008). This is important for avoiding the unnecessary reconfigurations of PCIs by the base stations. Moreover, when a cell is not in a local optimum, plateau moves are allowed. This means that if the price of the tested PCI (or with multiple-try, the lowest price of all PCIs) is the same as the price of the PCI that is being used, the cell changes the PCI. This enables the algorithm to break-out from a local optimum in search of a global optimum. Further details on the plateau moves can be found in the works by Zhang et al., (2005) and also Galinier and Hertz (2006). The four local search algorithms considered here are:

- Bin: binary pricing, random selection
- Real: real pricing, random selection
- Bin-multi: binary pricing, all candidates considered, random selection among the best candidates
- Real-multi: real pricing, all candidates considered, random selection among the best candidates

The real-pricing algorithms with random, and with best candidate selection were first discussed by Neel and Reed (2006), and Babadi and Tarokh (2008), respectively. The binary multiple-try algorithm is an asynchronous version of distributed stochastic search algorithm (DSA-D) algorithm, which was proposed by Zhang et al. (2005). The simplest algorithm, binary pricing with random candidate selection was first discussed by Ahmed et al. (2010). However, it is worth noting that similar principles are used for solving problems in statistical physics using the zero-temperature Markov chain Monte Carlo algorithm (Kardar, 2007). For comparison, two constraint satisfaction algorithms ABT and AWC are also considered. Both these algorithms involve message passing. The simulation results for a SCN comprising of 96 base stations deployed in an indoors WINNER scenario by Kyösti et al. (2008) are illustrated in the Figure 4. The algorithms are compared in terms of convergence probability for a large number of network instances. It is clear from the results that binary-pricing algorithms have a higher convergence probability compared to real-pricing ones. This is due to the reason that the binary algorithms can escape from the local-optima via the plateau moves, and thus have a higher chance of finding a global optimum. Overall, the AWC outperforms all other algorithms in terms of convergence properties, as it is able to color the conflict-graph of the network with nine colors only. On the other hand, the performance of ABT is constrained by the limit on the maximum number of iterations. In ABT algorithm, the agents have fixed order and the solution is found by exhaustive search through the backtracking principle, which makes it inefficient in terms of convergence speed. The AWC handles this problem through dynamic changing of priorities along with minimum conflict heuristics. Nevertheless, the ABT still has a higher convergence probability than real-pricing local search algorithms.

Example: Primary Component Carrier Selection

A closely related self-configuration problem, which can be mapped to a graph coloring problem, is PCC selection. In the PCC selection, interfering cells are configured with different component carriers to mitigate the mutual interference (3GPP, 2009a; 3GPP, 2009b). Graph coloring enables an efficient selection of component carriers, and thus yields an improvement in SINRs experienced by the users. For the

Figure 4. Comparison of graph coloring algorithms for PCI assignment

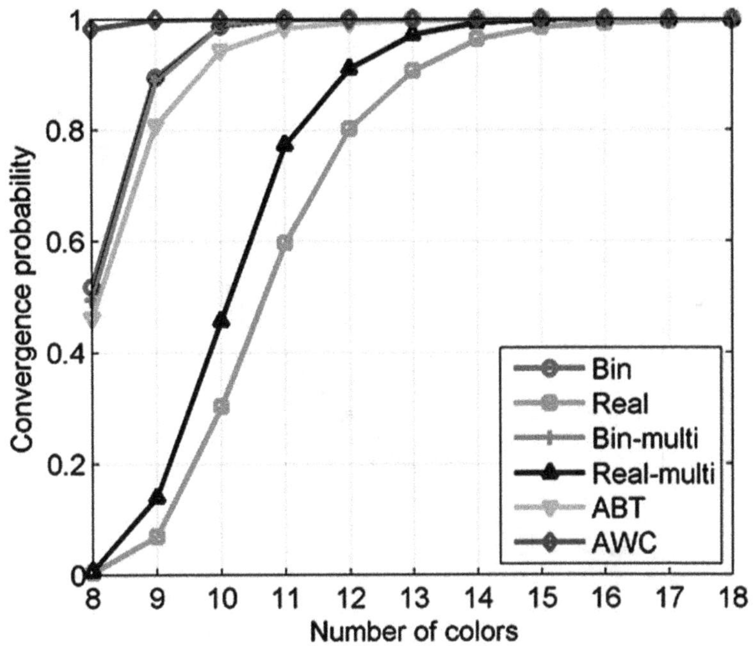

PCC selection problem, single-try and multiple-try variants of the binary-pricing and real-pricing SLS algorithms are considered for the same WINNER SCN simulation scenario used for PCI assignment. It is assumed that each cell randomly selects a PCC during the initialization phase and starts serving its users while acquiring handover measurements. This is followed by execution of the distributed graph coloring algorithm until convergence, where the maximum number of iterations is fixed. Finally, the statistics related to the carrier-to-interference (C/I) ratios experienced by the users in the system are collected. A comparison of the resulting CDFs is given in Figure 5, which illustrates that binary-pricing algorithms result in higher C/I ratios compared to their real pricing counterparts.

The gain is especially more pronounced for the users in the low C/I region. It is worth noting that the SLS algorithms discussed here do not involve any dedicated exchange of information, and the decisions by individual cells are based on passive observation of the values of interfering cells. On the other hand, the constraint satisfaction algorithms involve message-passing among the cells. However, the cooperative self-organizing framework employed by constraint satisfaction algorithms is more systematic, and it is applicable to a vast variety of problems. A main advantage of distributed constraint satisfaction algorithms is that completeness is guaranteed, i.e., the solution is found if it exists, or the algorithm terminates otherwise.

NETWORK UTILITY MAXIMIZATION AND PRICING ALGORITHMS

Utility maximization is a popular approach used for solving network optimization problems in a number of areas such as operations research, economics, and finance. In wireless networks, it has been applied mostly for generic distributed resource allocation problems in various types of networks such as *ad hoc*

Figure 5. Comparison of graph coloring algorithms for PCC selection

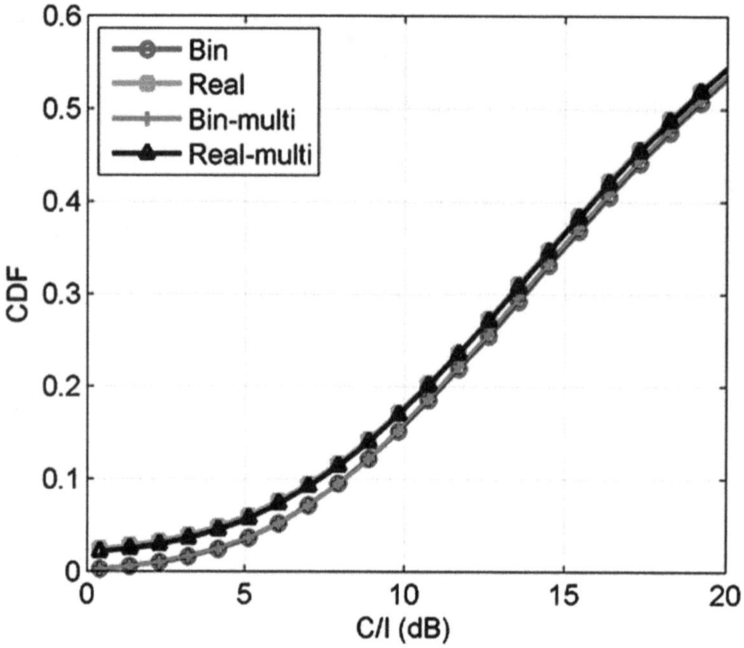

networks, as discussed by Chiang, Calderbank, and Doyle (2007), as well as Huang, Berry, and Honig (2006), WSNs considered in a work by Byers and Nasser (2000), and OFDMA based broadband networks see, for example, the work of Song and Li (2005). However, due to properties such as scalability and distributive nature, it is a model of choice for a systematic design of SON algorithms for current and future cellular systems, which comprise of a large number of small cells deployed randomly. Different approaches from optimization can be applied to construct algorithms for distributed NUM (Palomar & Chiang, 2006). In order to apply the NUM framework, each user in the network is assigned a utility function, which reflects its QoS or degree of satisfaction. Examples of utility functions with applications in wireless networks include functions of the expected data rate, like the proportional fair rate utility function. A complicating factor in the problem is the coupling that exists among the utility functions of users. For example, in the case of wireless networks, the SINRs of the users are coupled due to mutual interference. The aim is to maximize the sum utility of the network, i.e., the individual utilities summed over the users, under a set of equality and inequality constraints.

To enable self-organization, the algorithm must be distributed and involve limited communication overhead among the cells. The algorithms discussed here make use of a pricing exchange mechanism to solve the network optimization problem in a distributed manner. Thus, the local interaction among cells requires an active exchange of information, and is important to make intelligent and autonomous decisions. The systematic approach leads to (at least) locally optimal SON algorithms, and enables higher gains compared to other approaches in which cells make decisions merely based on passive observations of broadcast messages made by other cells (Ahmed & Tirkkonen, 2009). Typical SON use-cases that can be addressed using a NUM framework include ICIC, load balancing, capacity maximization, and energy savings. A number of papers have focused on NUM for ICIC problems in cellular networks (Stolyar & Viswanathan, 2009; Kasparick & Wunder, 2013; Rahman & Yanikomeroglu, 2010; Ahmed, Dowhuszko,

Tirkkonen, & Berry, 2013; Dowhuszko, Ahmed, & Tirkkonen, 2013). Stolyar and Viswanathan (2009) discuss a NUM based approach for self-organized fractional frequency reuse through distributed interference coordination, whereas Kasparick and Wunder (2013) focus on power control and scheduling for interference minimization. Moreover, Rahman and Yanikomeroglu (2010) have proposed heuristic fractional frequency reuse methods for interference avoidance and coordination. The work of Ahmed et al. (2013) deals with interference mitigation aimed at distributed network power minimization. Dowhuszko et al. (2013) discusses use of transmit beamforming for interference coordination. Load balancing and user-association is discussed by Son, Chong, and Veciana (2009). A NUM based joint optimization over multiple degrees of freedom for interference mitigation in multicell systems is considered in a number of works (Ahmed, Dowhuszko, & Tirkkonen, 2012; Ahmed, Dowhuszko, Tirkkonen, & Juntti, 2014).

Moreover, SON algorithms for efficient resource allocation can pave the way for an effective ICIC in SCNs (Hamza, Khalifa, Hamza & Elsayed, 2013). The existing ICIC techniques can be categorized as time domain, frequency domain, and power control techniques. For a detailed overview, see the work of Hossain, Rasti, Tabassum, and Abdelnasser (2014). Some algorithms developed in the following discussion are based on joint optimization over multiple degrees of freedom, where cells choose non-interfering modes of transmission via the allocation of resources in multiple dimensions (e.g. transmit powers over component carriers and multiuser scheduling). The performance of such algorithms is highly dependent upon the level of coordination between the cells. Assuming that a communication interface (such as X2) exists between adjacent cells, interference pricing messages may be exchanged, leading to high gains in network utility (Ahmed et al., 2012). For capacity optimization of the network, an expected rate function may be used as a utility function. Alternative utility functions of interest may include the logarithm of the expected rate (i.e. proportional fair), and the max-min rate function. The focus here is on a distributed interference pricing concept, and how it can be employed for designing self-organizing algorithms for different variants of the resource allocation problems relevant to SCNs. An effective way to mitigate inter-cell interference is to optimize the allocated resources across all available degrees of freedom (i.e., time, frequency, space, power). This reduces the mutual interference, and leads to an increase in the network utility. In order to solve the NUM problem in a distributed manner, a distributed version of the problem is formalized by decomposing the original problem into a number of sub-problems, one per cell. The distributed version of the problem is then solved by a pricing exchange mechanism, where each cell solves its sub-problem while taking into account the interference prices received from neighboring cells. Consequently, the individual cells are able to make more intelligent decisions regarding the allocation of resources, like transmit power, for mitigating the mutual interference. The exact nature of the decomposition method depends on the structure of the problem, nature of the utility function, and the constraints. This is elaborated in the following discussion, which reflects on different approaches that can be used for designing distributed NUM algorithms. The computational methods discussed here are general, and can be applied to different distributed resource allocation problems in wireless networks.

Decomposition Methods for Distributed NUM

The first step towards designing algorithms for distributed NUM involves the re-formulation of original NUM problem as a distributed problem. This can be achieved by decomposing the original problem into a number of sub-problems, one per network node, which are then solved iteratively in a synchronous or asynchronous manner. To further elaborate this point, a simple decomposition for network synchronization problem in cellular networks is discussed next.

Network Synchronization

Network synchronization is a fundamental issue in cellular systems, which has ramifications across a number of related RRM problems of prime importance for future wireless networks, e.g. flexible time division duplexing (TDD), coordinated multi-point transmission (CoMP), and other ICIC schemes (Amin, Ganesan, & Tirkkonen, 2012). Hot spots comprising of dense deployment of SCNs are predominantly indoors, where the satellite signals are weak due to high penetration losses that make satellite synchronization difficult. Thus, distributed run-time network synchronization with minimum network overhead is perceived as a key enabler for indoor 5G small cells (Mogensen et al., 2013).

Frame synchronization in wireless networks is a prototypical self-organization problem. In a centralized realization, there are problems related to conveying timing messages from a central node to leaf nodes. However, on the algorithmic level, centralized network synchronization is trivial. Complexity in self-organizing solutions is therefore, strictly due to the requirement of engineered self-organization. Self-organizing algorithms of network synchronization are usually based on neighborhood interactions, in which each cell synchronizes to its neighbors by listening to their transmissions.

In a network synchronization problem, the main aim of the base stations (cells) $i \in I$ is to agree on the value of the synchronization variable $\mathrm{x}_i = \hat{\mathrm{x}}$. For frame/event synchronization, the individual synchronization variables can be understood as points on a circle, represented as phases $\Phi = \{\varphi_i\}_{i\in I}$, where $\varphi_i \in \left[0, 2\pi\right]$. To apply the NUM approach to the network synchronization problem, the first step is to define a utility function for each base station (cell) as follows:

$$u_i(\Phi) = -\frac{1}{2}\sum_{j\in N_i} \| \varphi_i - \varphi_j \|_G^q, \tag{1}$$

where N_i is the set of base stations in the neighborhood of base station i, the exponent $q > 0$ determines how large or small agreements among the synchronization variables are taken into account, and the norm $\| \bullet \|_G$ is the geodesic norm given by:

$$\| \varphi_1 - \varphi_2 \|_G = \min(| \varphi_1 - \varphi_2 |, | 2\pi - \varphi_1 + \varphi_2 |). \tag{2}$$

The geodesic norm is the timing difference, or the measure of frame asynchrony between a pair of base stations. Then, the network utility is given by

$$U(\Phi) = \sum_{i\in I} u_i(\Phi). \tag{3}$$

Let us consider a greedy self-organizing approach for maximizing network utility, in which base stations update their phases at discrete time instants with the following best response update:

$$\varphi_i^* = \arg\max_{\varphi_i} u_i\left(\Phi\right). \tag{4}$$

Both synchronous and asynchronous updates are possible with this update rule. However, the synchronous rule may not converge due to oscillatory behavior induced by the best-response dynamic, in which each point on the circle representing a phase jumps to the other side as shown in the Figure 6.

In contrast to best-response, the following update rule is based on gradient descent principle and will result in a small step towards the solution

$$\varphi_i^* = \left[\varphi_i + \beta \nabla_{\varphi_i} U\left(\Phi\right)\right]_{\chi}, \tag{5}$$

where β is the step-size, $\nabla_{\varphi_i} U\left(\Phi\right)$ is the gradient of network utility, and $\left[\bullet\right]_{\chi}$ is the projection on the feasible set $\chi = \left[0, 2\pi\right]$. It can be seen that both best-response and gradient descent updates are simple to implement and can enable fully distributed self-organized synchronization in SCNs. However, depending on the initial state of the network, both of these methods can get trapped in a local optimum, where the network is not synchronized, and at the same time no local moves can improve synchronicity (Ahmed & Tirkkonen, 2014). This phenomenon is a fundamental consequence of the interaction of the topology of the configuration space (a collection of circles in this case), and the topology of the communication graph of the network.

Figure 6. Different update rules for NUM based network synchronization

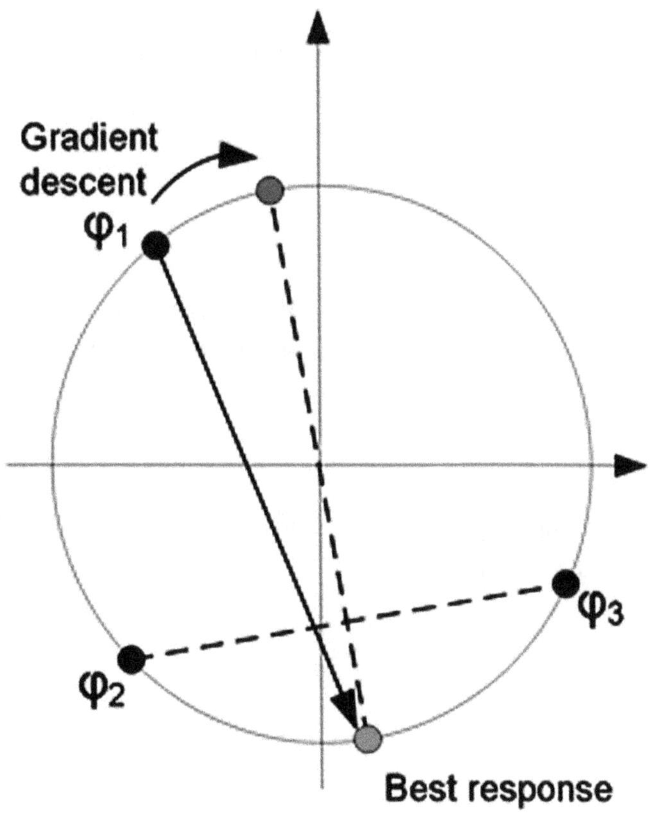

Decomposition based on KKT Conditions

An important method for solving constrained optimization problems is through the direct solution of so-called Karush-Kuhn-Tucker (KKT) conditions, which essentially are equations guaranteeing the vanishing of the gradient of the objective function. In this case, the decomposition involves formalizing the KKT conditions of the original NUM problem, and formulating a distributed problem with the same KKT conditions as the original NUM problem. Thus, the original problem is decomposed into a set of sub-problems which can be solved in an iterative manner.

Example: Interference coordination and capacity maximization

Interference mitigation is the fundamental issue in the design of efficient wireless systems, especially for future spectrum sharing SCNs envisaged for providing high data rates and ubiquitous connectivity to massive numbers of users. A NUM based approach can be effectively used in SON algorithms for interference coordination in SCNs. Consider a SCN comprising of a set of I cells and L users. Each cell $i \in I$ consists of a low-power base station $i \in I$ and subset of users $L_i \subseteq L$. All base stations use the same bandwidth B [Hz], which is divided into K equal sized carriers. The maximum power a base station can distribute over channels is denoted by $P_{\max}$ such that

$$\sum_{k=1}^{K} p_i^k \le P_{\max}. \tag{6}$$

The goal is to maximize the downlink system utility, which is the sum of utilities of individual cells. The utility of cell i is

$$U_i = \sum_{l \in L_i} u_l, \tag{7}$$

where u_l is the utility of user $l \in L_i$ served by base station i, which can be chosen depending on the performance metric to be maximized. Let us consider the proportional fair rate utility function

$$u_l\left(r_l\right) = \log\ \left(r_l\right), \tag{8}$$

where the user rate r_l is for simplicity taken as the sum of the Shannon rate of the user l on the K carriers. With interference treated as Gaussian noise,

$$r_l = \frac{B}{K}\sum_{k=1}^{K} w_l^k \log(1+\gamma_{i,l}^k) \quad l \in L_i \tag{9}$$

results, where $\gamma_{i,l}^k = \frac{p_i^k h_{i,l}^k}{I_{i,l}^k + n_0}$ is the SINR and $I_{i,l}^k = \sum_{j \in J_i} p_j^k h_{j,l}^k$ is the interference that user l faces on carrier k, while w_l^k is the fraction of resource that base station i allocates to user l on carrier k. This can be considered to be an average scheduling probability, or a scheduling weight. Each user in the network has its own local utility function, which represents its rate or QoS. The network level utility thus comprises of the sum of the utilities of individual users, where the aim is to maximize the network utility over transmit powers and scheduling weights of all the cells in the network. In order to enable this in a self-organized way, the coordination among cells should be minimal. In the method considered here, each cell engages in limited cooperation with the other cells that affect its total utility considerably. This cooperation entails exchange of pricing information, which is vital for achieving substantial gains in terms of network utility. To this end, the NUM problem is given by:

$$\begin{aligned} \underset{\mathsf{P},\mathsf{W}}{\text{maximize}} \quad & \sum_{i \in I} U_i\left(\mathsf{P}, \mathsf{W}_i\right) \\ \text{subject to} \quad & \sum_{k \in K} p_i^k \leq P_{\max},\ p_i^k \geq 0 \qquad \forall i \\ & \sum_{l \in L_i} w_l^k = 1,\ w_l^k \geq 1 \qquad \forall i,k \end{aligned} \tag{10}$$

where P and W comprise the powers and scheduling weights of all the cells in the network, and W_i denotes the scheduling weights of cell i. Let us define the term $\pi_{j,i}^k\left(\mathsf{P}, \mathsf{W}_j\right) = -\frac{\partial U_j\left(\mathsf{P}, \mathsf{W}_j\right)}{\partial p_i^k}$ as the power price on carrier k that cell j reports to cell i.

Then, it can be seen that the KKT conditions of (10) are the same as the KKT conditions of the following distributed problem, which comprises of $|I|$ sub-problems, one for each cell:

$$\begin{aligned} \underset{\mathsf{p}_i,\mathsf{W}_i}{\text{maximize}} \quad & \sum_{i \in I} U_i\left(\mathsf{p}_i, \mathsf{W}_i\right) - \sum_{k \in K} p_i^k \left\{\sum_{j \neq i} \pi_{j,i}^k\left(\mathsf{P}, \mathsf{W}_j\right)\right\} \\ \text{subject to} \quad & \sum_{k \in K} p_i^k \leq P_{\max},\ p_i^k \geq 0 \\ & \sum_{l \in L_i} w_{i,l}^k = 1,\ w_l^k \geq 1 \end{aligned} \tag{11}$$

Each cell solves its sub-problem and reports the updated prices to other cells. The updates are done in an asynchronous and periodic way, where the base station of each cell updates once in a given iteration at a unique time instant. First a base station solves its sub-problem and calculates the power and scheduling weights. This is followed by a pricing update step, in which new prices are reported to the other cells. In iteration n, each base station i performs the updates at a unique time instant $t_i\left[n\right]$. The results are shown in Figure 7, for a SCN comprising of four base stations deployed indoors in a building with WINNER office characteristics discussed by Kyösti et al. (2008). Base stations have closed subscriber groups in their respective coverage areas. In the coverage area of a base station, the probability of having a visiting user belonging to a different closed subscriber group is 20%. Four different cases related to power and scheduling weight optimization are compared for the proportional fair rate utility. The CDFs of the experienced user rates are plotted for different algorithms. In the baseline case, both power allocation (PA) and scheduling weights (SCH) are fixed, i.e. F-PA + F-SCH. This is followed by adaptive power allocation combined with fixed scheduling weights (A-PA+F-SCH). A-PA yields a higher mean data rate per user, and especially improves the situation of the users in low SINR region. The CDFs for other combinations namely fixed power and adaptive scheduling (F-PA + A-SCH), and adaptive power and adaptive scheduling (A-PA + A-SCH) are also shown. It is worth noting that (F-PA + A-SCH) is the baseline solution allowed by an LTE/LTE-A network in absence of ICIC. Joint allocation of power and scheduling weights results in a substantial improvements in the data rates of the low SINR users. This is due to the fact that the proportional fair rate utility results in network-wide fairness among the users in terms of data rates.

Primal and Dual Decomposition

Primal and dual decomposition constitute an important class of decomposition methods that allows distributed approaches for network optimization, and is applicable to vast range of problems. The KKT-based approach presented above cannot be straight forwardly used when there is a network-level constraint, binding the individual variable of the nodes. For such problems, a primal/dual approach is necessary. The performance of such methods mostly depends on the problem structure. For convex problems, both methods perform well and achieve an optimal solution. Selecting the appropriate decomposition method is an important step in the algorithm design procedure, and depends on the problem structure.

Figure 7. NUM based joint power allocation and scheduling for ICIC

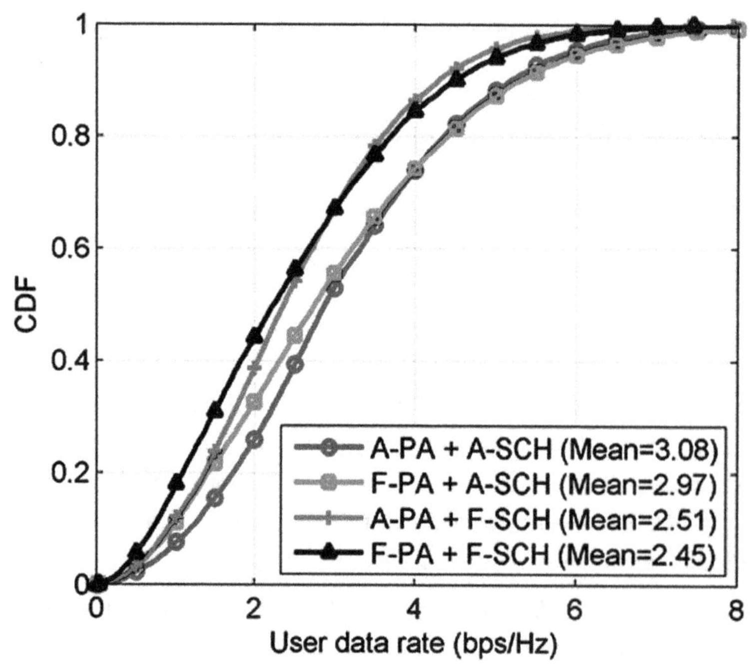

An example is discussed next, which involves an application of primal decomposition for ICIC problem in a cognitive radio network setting.

Example: Network Interference coordination for cognitive radio networks

Cognitive radio networking techniques are envisaged to play an important role in providing high capacity for 5G networks by enabling efficient use of spectrum. Cognitive radio based technologies have applications in many paradigms related to future 5G networks, such as heterogeneous networks, device-to-device communication, and coexistence with other networks (Lien, Chen, Liang & Lin, 2014). To unlock under-utilized spectrum to be used for mobile broadband communication, a number of spectrum sharing scenarios based on primary, licensed, and unlicensed spectrum sharing models are currently under consideration. Among the licensed modes, the Licensed Shared Access and Authorized Shared Access are highly promising for increasing the available spectrum under incumbent protection guarantees and regulatory control (Irnich, Kronander, Selén, & Li, 2013). Incumbent protection comprises mechanisms involving geographical constraints on the usage of spectrum through a geo-location database. Thus, efficient use of radio resources to maximize the usage of the unlocked spectrum within incumbent protection constraints is important for SCNs. In LTE/LTE-A based SCNs, this optimization should be enabled in a self-organizing manner, without compromising the potential gains. In this section, the focus is on self-organizing algorithms for maximization of network utility in SCNs, under network-wide constraints on transmit powers. It is observed that network level constraints in such NUM problems can be handled effectively by using primal and dual decomposition methods from optimization theory (Ahmed et al., 2014). For performance analysis, a modification of previously discussed SCN model is considered, where each base station has a single carrier (i.e., $|K|=1$) and a constraint exists on the sum of transmit powers of all the base stations in the network, i.e.,

$$\sum_{i \in I} p_i \leq P_{\text{net}}, \tag{12}$$

where p_i is the power of base station i and P_{net} is the maximum power that the network can use. Thus, in this case, the NUM problem can be expressed as

$$\begin{aligned} \underset{\mathrm{p,W}}{\text{maximize}} \quad & \sum_{i \in I} U_i\left(\mathrm{p}, \mathrm{w}_i\right) \\ \text{subject to} \quad & \sum_{i \in I} p_i \leq P_{\text{net}}, \; p_i \geq 0 \\ & \sum_{l \in L_i} w_l = 1, w_l \geq 0 \qquad \forall i, k. \end{aligned} \tag{13}$$

It is assumed that u_l is the proportional fair rate utility function as defined in equation (8). With the proportional fair rate utility function, the problem is convex, and therefore has a unique optimal solution. The next step is to apply the decomposition procedure to formulate a pricing based algorithm for finding the optimal solution in a distributed way. In this case, fixing the vector of transmission powers leads in decoupling of the optimization problem into $|I|$ sub-problems. Primal decomposition can thus be applied. Thus fixing p, yields the following lower level sub-problems, one for each cell:

$$\begin{aligned} &\underset{\mathsf{w}_i}{\text{maximize}} && U_i\left(\mathsf{w}_i\right) \\ &\text{subject to} && \sum_{l \in L_i} w_l = 1, w_l \geq 0. \end{aligned} \tag{14}$$

At a higher level, there is a master problem for updating the coupling variable p:

$$\begin{aligned} &\underset{\mathsf{p}}{\text{maximize}} && U_i\left(\mathsf{p}\right) \\ &\text{subject to} && \sum_{i \in I} p_i \leq P_{\text{net}}, p_i \geq 0 \end{aligned} \tag{15}$$

It can be seen that the power vector p is the complicating variable that couples the cells via both the utility functions and the constraints. There is network level constraint on the powers, which makes the problem more challenging than the conventional distributed power control problem. The next step is to use a pricing algorithm to solve the optimization over p by an iterative descent method. To this end, the network utility U_{sum} can be expressed as

Differentiating with respect to p_i gives

$$D_i = \pi_{i,i} + \sum_{j \neq i} \pi_{j,i} \tag{17}$$

where the terms $\pi_{i,i} = \dfrac{\partial U_i\left(\mathsf{p};\mathsf{w}_i\right)}{\partial p_i}$ and $\pi_{j,i} = \dfrac{\partial U_j\left(\mathsf{p};\mathsf{w}_j\right)}{\partial p_i}$ can be considered as a power benefit and a power price, respectively. Note that the power price indicates the negative effect that an increase of transmit power of base station in cell i has on utility of cell j. By exchanging power prices, the cells can know how much interference they are causing to neighbors, and can cooperatively maximize the total utility of the network over their respective powers. In absence of temporal changes in the channels, the optimal scheduling weight allocation in this case is the round robin allocation, in which all the users get an equal share of the resources. The optimization over p is carried out by using a distributed coordinate descent method, with asynchronous and periodic updates. In a given iteration n, each base station i performs the updates at a unique time instant $t_i\left[n\right]$.

It is also possible to apply the dual decomposition principle in this case, which will find the same solution as the primal. However, the dual algorithm will require an extremely small step-size leading to a large number of iterations. The performance of both algorithms in terms of user data rates is shown in Figure 8. Here, the base line case is the fixed power allocation, in which all the base stations use the same transmit power. Both primal and dual reach the same solution, which is significantly better than the base-line solution.

Moreover, the performance of primal algorithm with only power benefits message exchange is also shown. In this algorithm, no exchange of power price messages takes place among the base stations. The result is a moderate reduction in performance gains in terms of user data rates.

FUTURE RESEARCH DIRECTIONS

SON functionalities and concepts will enhance future cellular networks due to their potential in ensuring higher data rates, seamless connectivity, and reduced expenditures. Consequently, many new features will be added, especially for ICIC, energy savings, and other resource allocation problems. Both continuous and discrete optimization methods can be applied to tackle such problems in an effective way. In this regard, systematic approaches towards SON algorithm design such as graph coloring and NUM lead to significant gains and good convergence properties. These methods are generic and can be applied to a wide range of problems. Thus they are of key importance in the study of SON from a perspective of future networks. Moreover, in order to meet the design challenges and maximize the performance of future networks, different problems related to ICIC and resource allocation can be addressed jointly. This will make use of a high number of degrees of freedom thereby resulting in substantial performance improvement in future large-scale mobile networks of the future.

Figure 8. NUM based joint power allocation and scheduling for ICIC in a cognitive radio scenario

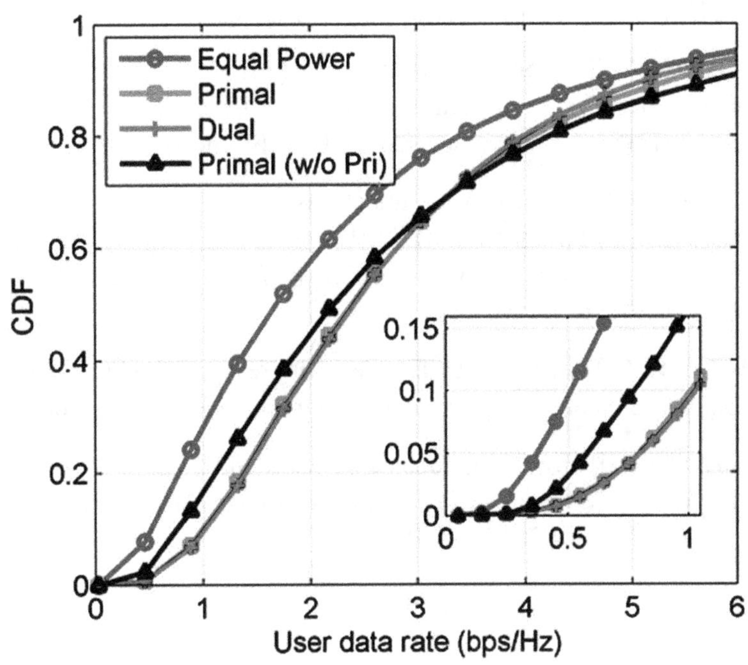

CONCLUSION

This chapter focused on a principled approach towards an algorithm design for enabling self-organization in future cellular networks. Different mathematical models from discrete and continuous optimization theory have been studied to develop distributed algorithms for SON problems for future cellular networks. Graph coloring algorithms based on local search and complete message-passing can be effectively used for SON problems with constraints on logical degrees of freedom, such as PCI assignment and PCC selection. Moreover, a number of SON problems can be modelled as NUM problems, and can be solved using pricing based decomposition methods. These include ICIC, capacity maximization, energy saving, and network synchronization problems. Exchange of dedicated information among cells results in substantially better performance and faster convergence, as compared to fully distributed solutions. When the constraints on the variables are local, direct distributed solution of KKT equations can be done, with limited message-passing. When there are global constraints, more involved decomposition methods have to be used. The discussed approaches are effective in designing SON algorithms for a multitude of SON problems relevant to future cellular networks.

ACKNOWLEDGMENT

This work was funded in part by the Academy of Finland (grant 284725).

REFERENCES

Ahmed, F., Dowhuszko, A. A., & Tirkkonen, O. (2012). Distributed algorithm for downlink resource allocation in multicarrier small cell networks. In *Proceedings of the International Conference on Communications*. Ottawa, Canada: IEEE. doi:10.1109/ICC.2012.6364716

Ahmed, F., Dowhuszko, A. A., Tirkkonen, O., & Berry, R. (2013). A distributed algorithm for network power minimization in multicarrier systems. In *Proceedings of the International Symposium on Personal Indoor and Mobile Radio Communications*. London, UK: IEEE. doi:10.1109/PIMRC.2013.6666456

Ahmed, F., & Tirkkonen, O. (2009). Local optimum based power allocation approach for spectrum sharing in unlicensed bands. In Spyropoulos, T., & Hummel, K. A. (Eds.), *Self-Organizing Systems: Proceedings of the 6th International Workshop (LNCS)* (*vol. 5918*, pp. 238-243). Berlin, Germany: Springer. doi:10.1007/978-3-642-10865-5_24

Ahmed, F., & Tirkkonen, O. (2014). Topological aspects of greedy self-organization. In *Proceedings of the International Conference on Self-Adaptive and Self-Organizing Systems*. London, UK: IEEE.

Ahmed, F., & Tirkkonen, O. (2015). Self Organized Physical Cell ID Assignment in Multi-operator Heterogeneous Networks. In *Proceedings of the Vehicular Technology Conference (VTC-Spring)*. Glasgow, UK: IEEE. doi:10.1109/VTCSpring.2015.7146077

Ahmed, F., Tirkkonen, O., Dowhuszko, A. A., & Juntti, M. (2014). Distributed power allocation in cognitive radio networks under network power constraint. In *Proceedings of the International Conference on Cognitive Radio Oriented Wireless Networks and Communications*. Oulu, Finland: IEEE. doi:10.4108/icst.crowncom.2014.255738

Ahmed, F., Tirkkonen, O., Peltomäki, M., Koljonen, J. M., Yu, C. H., & Alava, M. (2010). Distributed graph coloring for self-organization in LTE networks. *Journal of Electrical and Computer Engineering*, 5.

Alava, M., Ardelius, J., Aurell, E., Kaski, P., Krishnamurthy, S., Orponen, P., & Seitz, S. (2008). Circumspect descent prevails in solving random constraint satisfaction problems. *Proceedings of the National Academy of Sciences of the United States of America*, *105*(40), 15253–15257. doi:10.1073/pnas.0712263105 PMID:18832149

Aliu, O. G., Imran, A., Imran, M. A., & Evans, B. (2013). A survey of self organisation in future cellular networks. *IEEE Communications Surveys and Tutorials*, *15*(1), 336–361. doi:10.1109/SURV.2012.021312.00116

Amin, P., Ganesan, V. P. K., & Tirkkonen, O. (2012). Bridging Interference Barriers in Self-Organized Synchronization. In *Proceedings of the International Conference on Self-Adaptive and Self-Organizing Systems*. Lyon, France: IEEE.

Ashby, W. R. (1947). Principles of the Self-Organizing Dynamic System. *The Journal of General Psychology*, *37*(2), 125–128. doi:10.1080/00221309.1947.9918144 PMID:20270223

Babadi, B., & Tarokh, V. (2008). A distributed asynchronous algorithm for spectrum sharing in wireless ad hoc networks. In *Proceedings of the Annual Conference on Information Sciences and Systems*. NJ, USA: IEEE. doi:10.1109/CISS.2008.4558635

Balasundaram, B., & Butenko, S. (2006). Graph domination, coloring and cliques in telecommunications. In G. C. M. Resende & P. M. Pardalos (Eds.), *Handbook of Optimization in Telecommunications* (pp. 865–890). New York, NY: Springer. doi:10.1007/978-0-387-30165-5_30

Bandh, T., Carle, G., & Sanneck, H. (2009). Graph coloring based physical-cell-ID assignment for LTE networks. In *Proceedings of the International Conference on Wireless Communications and Mobile Computing*. Leipzig, Germany: ACM. doi:10.1145/1582379.1582406

Baran, P. (1964). On distributed communications networks. *IEEE Transactions on Communications Systems*, *12*(1), 1–9. doi:10.1109/TCOM.1964.1088883

Byers, J., & Nasser, G. (2000). Utility-based decision-making in wireless sensor networks. In *Proceedings of the Annual Workshop on Mobile and Ad Hoc Networking and Computing*. Boston, MA: IEEE. doi:10.1109/MOBHOC.2000.869228

Camazine, S., Deneubourg, J.-L., Franks, N. R., Sneyd, J., Theraulaz, G., & Bonabeau, E. (2003). *Self-Organization in Biological Systems*. Princeton, NJ: Princeton University Press.

Chen, J., Olafsson, S., Gu, X., & Yang, Y. (2008). A fast channel allocation scheme using simulated annealing in scalable WLANs. In *Proceedings of the International Conference on Broadband Communications, Networks, and Systems*. London, UK: IEEE. doi:10.1109/BROADNETS.2008.4769071

Cheng, M. X., Li, Y., & Du, D.-Z. (2006). *Combinatorial Optimization in Communication Networks*. New York, NY: Springer. doi:10.1007/0-387-29026-5

Chiang, M., Low, S. H., Calderbank, A. R., & Doyle, J. C. (2007). Layering as optimization decomposition: A mathematical theory of network architectures. *Proceedings of the IEEE*, *95*(1), 255–312. doi:10.1109/JPROC.2006.887322

Diab, A., & Mitschele-Thiel, A. (2013). Development of distributed and self-organized physical cell identity assignment schemes for LTE-advanced systems. In *Proceedings of the workshop on Performance monitoring and measurement of heterogeneous wireless and wired networks*. ACM doi:10.1145/2512840.2512852

Dowhuszko, A. A., Ahmed, F., & Tirkkonen, O. (2013). Decentralized transmit beamforming scheme for interference coordination in small cell networks. In *Proceedings of the International Black Sea Conference on Communications and Networking*, Batumi, Georgia. IEEE. doi:10.1109/BlackSeaCom.2013.6623394

Eisenblätter, A., Grötschel, M., & Koster, A. M. C. A. (2002). Frequency assignment and ramifications of coloring. *Discussiones Mathematicae Graph Theory*, *22*, 51–88. doi:10.7151/dmgt.1158

Galinier, P., & Hertz, A. (2006). A survey of local search methods for graph coloring. *Computers & Operations Research*, *33*(9), 2547–2562. doi:10.1016/j.cor.2005.07.028

Haken, H. (2006). *Information and Self-Organization: A Macroscopic Approach to Complex Systems (Springer Series in Synergetics)*. Berlin, Germany: Springer.

Hales, D., & Arteconi, S. (2006). SLACER: A Self-Organizing Protocol for Coordination in Peer-to-Peer Networks. *IEEE Intelligent Systems*, *21*(2), 29–35. doi:10.1109/MIS.2006.35

Hämäläinen, S., Sanneck, H., & Sartori, C. (2011). *LTE Self-Organising Networks (SON): Network Management Automation for Operational Efficiency*. West Sussex, UK: Wiley & Sons. doi:10.1002/9781119961789

Hamza, A. S., Khalifa, S. S., Hamza, H. S., & Elsayed, K. (2013). A survey on inter-cell interference coordination techniques in OFDMA-based cellular networks. *IEEE Communications Surveys and Tutorials*, *15*(4), 1642–1670. doi:10.1109/SURV.2013.013013.00028

Hossain, E., Rasti, M., Tabassum, H., & Abdelnasser, A. (2014). Evolution toward 5G multi-tier cellular wireless networks: An interference management perspective. *Wireless Communications, IEEE*, *21*(3), 118–127. doi:10.1109/MWC.2014.6845056

Huang, J., Berry, R. A., & Honig, M. L. (2006). Distributed interference compensation for wireless networks. *IEEE Journal on Selected Areas in Communications*, *24*(5), 1074–1084. doi:10.1109/JSAC.2006.872889

Hubaux, J.-P., Gross, T., Le Boudec, J.-Y., & Vetterli, M. (2001). Toward self-organized mobile ad hoc networks: The terminodes project. *IEEE Communications Magazine*, *39*(1), 118–124. doi:10.1109/35.894385

Irnich, T., Kronander, J., Selén, Y., & Li, G. (2013). Spectrum sharing scenarios and resulting technical requirements for 5G systems. In *Proceedings of the International Symposium on Personal, Indoor and Mobile Radio Communications (PIMRC Workshops)*. London, UK: IEEE. doi:10.1109/PIMRCW.2013.6707850

Johnson, D. S., Aragon, C. R., McGeoch, L. A., & Schevon, C. (1991). Optimization by simulated annealing: An experimental evaluation; part II, graph coloring and number partitioning. *Operations Research, 39*(3), 378–406. doi:10.1287/opre.39.3.378

Kardar, M. (2007). *Statistical physics of fields*. Cambridge, UK: Cambridge University Press. doi:10.1017/CBO9780511815881

Kasparick, M., & Wunder, G. (2014). Autonomous algorithms for centralized and distributed interference coordination: A virtual layer-based approach. *EURASIP Journal on Wireless Communications and Networking*, 1–20.

Kumar, V. (1992). Algorithms for constraint-satisfaction problems: A survey. *AI Magazine, 13*(1), 32.

Kyösti, P., Meinilä, J., Hentilä, L., Zhao, X., Jämsä, T., Schneider, C., Narandzić, M., ... Rautiainen, T. (2008). *WINNER II channel models*. Academic Press.

Lehser, F., Lei, L., Zecai, S., Zenshan, S., Djurica, M., Jorguseski, L., … Verschoor, A. (2008). *Next generation mobile networks recommendation on SON and O&M requirements*. NGMN Alliance.

Lien, S. Y., Chen, K. C., Liang, Y. C., & Lin, Y. (2014). Cognitive radio resource management for future cellular networks. *IEEE Wireless Communications, 21*(1), 70–79. doi:10.1109/MWC.2014.6757899

McAllester, D., Selman, B., & Kautz, H. (1997). Evidence for invariants in local search. In *Proceedings of the national conference on artificial intelligence and conference on Innovative applications of artificial intelligence of (AAAI/IAAI)*. Menlo Park, CA: AAAI press.

Mitton, N., Fleury, E., Lassous, I. G., & Tixeuil, S. (2005). Self-stabilization in self-organized multihop wireless networks, In *Proceedings of the International Conference on Distributed Computing Systems Workshops*. Columbus, OH: IEEE. doi:10.1109/ICDCSW.2005.122

Mogensen, P., Pajukoski, K., Tiirola, E., Lähetkangas, E., Vihriälä, J., Vesterinen, S., & Cattoni, A. F. et al. (2013). 5G small cell optimized radio design. In *Proceedings of the Global Communications Workshop on Emerging Technologies for LTE-Advanced and Beyond 4G*. Atlanta, GA: IEEE.

Neel, J. O., & Reed, J. H. (2006). Performance of distributed dynamic frequency selection schemes for interference reducing networks. In *Proceedings of the Military Communications Conference*. Washington, DC: IEEE. doi:10.1109/MILCOM.2006.302016

Olfati-Saber, R., Fax, J. A., & Murray, R. M. (2007). Consensus and Cooperation in Networked Multi-Agent Systems. *Proceedings of the IEEE, 95*(1), 215–233. doi:10.1109/JPROC.2006.887293

Palomar, D. P., & Chiang, M. (2006). A tutorial on decomposition methods for network utility maximization. *IEEE Journal on Selected Areas in Communications, 24*(8), 1439–1451. doi:10.1109/JSAC.2006.879350

Peltomäki, M., Koljonen, J., Tirkkonen, O., & Alava, M. (2012). Algorithms for self-organized resource allocation in wireless networks. *IEEE Transactions on Vehicular Technology, 61*(1), 346–359. doi:10.1109/TVT.2011.2151891

Prehofer, C., & Bettstetter, C. (2005). Self-organization in communication networks: Principles and design paradigms. *IEEE Communications Magazine, 43*(7), 78–85. doi:10.1109/MCOM.2005.1470824

Rahman, M., & Yanikomeroglu, H. (2010). Enhancing cell-edge performance: A downlink dynamic interference avoidance scheme with inter-cell coordination. *IEEE Transactions on Wireless Communications*, *9*(4), 1414–1425. doi:10.1109/TWC.2010.04.090256

3rd Generation Partnership Project. (2009a). *Primary component carrier selection, monitoring, and recovery*. Author.

3rd Generation Partnership Project. (2009b). *Feasibility study for further advancements of E-UTRA (LTE-Advanced)*. Author.

3rd Generation Partnership Project. (2010). *Evolved Universal Terrestrial Radio Access Network (E-UTRAN); Self-Configuring and Self-Optimizing Network (SON) Use Cases and Solutions*. Author.

Scully, N., Thiel, S., Litjens, R., Jorguseski, L., Nascimento, R., Linnell, O., … Schmelz, L. C. (2008). *Deliverable D2.1: Use Cases for Self-Organizing Networks*. SOCRATES.

Sohrabi, K., Gao, J., Ailawadhi, V., & Pottie, G. J. (2000). Protocols for self-organization of a wireless sensor network. *IEEE Personal Communications, 7*(5), 16-27.

Son, K., Chong, S., & Veciana, G. (2009). Dynamic association for load balancing and interference avoidance in multi-cell networks. *IEEE Transactions on Wireless Communications*, *8*(7), 3566–3576. doi:10.1109/TWC.2009.071140

Song, G., & Li, Y. (2005). Utility-based resource allocation and scheduling in OFDM-based wireless broadband networks. *IEEE Communications Magazine*, *43*(12), 127–134. doi:10.1109/MCOM.2005.1561930

Stolyar, A. L., & Viswanathan, H. (2009). Self-organizing dynamic fractional frequency reuse for best-effort traffic through distributed inter-cell coordination. In *Proceedings of the Conference on Computer Communications*. Rio de Janerio, Brazil: IEEE. doi:10.1109/INFCOM.2009.5062043

Strogatz, S. H. (2001). Exploring complex networks. *Nature*, *410*(6825), 268–276. doi:10.1038/35065725 PMID:11258382

Szilágyi, P., Bandh, T., & Sanneck, H. (2013). Physical Cell ID Allocation in Multi-layer, Multi-vendor LTE Networks. In A. Timm-Giel, J. Strassner, R. Agüero, S. Sargento, & K. Pentikousis (Eds.), *Mobile Networks and Management* (pp. 156–168). Berlin, Germany: Springer. doi:10.1007/978-3-642-37935-2_12

Yokoo, M. (2012). *Distributed constraint satisfaction: foundations of cooperation in multi-agent systems*. Berlin, Germany: Springer.

Yokoo, M., & Hirayama, K. (2000). Algorithms for distributed constraint satisfaction: A review. *Autonomous Agents and Multi-Agent Systems*, *3*(2), 185–207. doi:10.1023/A:1010078712316

Zhang, W., Wang, G., Xing, Z., & Wittenburg, L. (2005). Distributed stochastic search and distributed breakout: Properties, comparison and applications to constraint optimization problems in sensor networks. *Artificial Intelligence*, *161*(1), 55–87. doi:10.1016/j.artint.2004.10.004

Zhu, H., & Ray Liu, K. J. (2008). *Resource Allocation for Wireless Networks: Basics, Techniques, and Applications*. New York, NY: Cambridge University Press.

KEY TERMS AND DEFINITIONS

Continuous Optimization: A branch of optimization which deals with models involving variables that take values from a continuous range.

Discrete Optimization: A branch of optimization which deals with models involving variables that take values from a finite set.

Graph: A mathematical representation of a set of objects like network nodes, where some objects may be connected by links.

Network Utility Maximization: A concept involving maximization of the network level performance of a system, where the level of satisfaction of each node is represented by a utility function.

Self-Organization: A process in which local interactions among system components lead to an emergence of a global order.

Self-Organizing Networks: A network comprising of geographically separate nodes, where local interactions between nodes lead to network level goal such as connectivity and QoS maintenance.

Utility Function: A function which represents the degree of satisfaction of a system component.

Chapter 3
A Perspective on Self-Optimization in Next Generation Cellular Networks

Sumita Mishra
Amity University, India

Nidhi Mathur
IETE, India

ABSTRACT

During the past few decades wireless technology has seen a tremendous growth. The recent introduction of high-end mobile devices has further increased subscribers' demand for high bandwidth and quality of service. The number of nodes in future cellular systems will be too large to be configured and maintained manually. Further the mobility of users, the varying nature of the wireless channel and variation in user demand systems requires optimization of network parameters and delay in configuration may cause congestion and poor Quality of Service. Due to this increased complexity and the huge scale of wireless systems the network configuration, optimization and maintenance process performed by radio engineers has become inefficient and therefore, lot of research is being done to introduce self-optimizing capabilities within the network, which is expected to permit higher end user Quality of Service (QoS) and less operation cost and maintenance cost for telecom service providers. This chapter details the key aspects related to self optimization of next generation cellular networks.

1. INTRODUCTION

Wireless technology has seen a tremendous growth during the past few decades. Mobile & wireless applications have become an important part of our daily lives; as a consequence user demand for data consuming mobile applications like interactive TV, mobile video, advanced games and so on has increased. It is becoming increasingly difficult to accommodate the demand of high bandwidth and quality of service.

In today's mobile networks many network elements and associated parameters are manually configured, due to exponential increase in number of users this manual process is error-prone and time consuming

DOI: 10.4018/978-1-5225-0239-5.ch003

resulting in long delays which reduces Quality of Service. In addition training and maintaining skilled manpower is costly. Therefore, researchers are seeking to introduce self-organizing capabilities within the network, which is expected to permit higher end user Quality of Service (QoS) along with more spectrums to meet the rising mobile data demands from Subscribers. This chapter details key aspect related to self optimization of next generation cellular networks namely

- Automatic Neighbour Relations
- Load Balancing Optimization
- Handover optimization
- Random Access Channel Optimization
- Inter-Cell Interference Coordination
- Coverage and Capacity Optimization
- Energy Saving

It is expected that the development of novel self-optimization or self-healing methods will make the planning and management of future generation wireless networks much easier and minimize the needed human involvement. The next section gives a brief background of architecture of modern wireless networks subsequent sections outline the concept of self organization in next generation cellular networks.

2. BACKGROUND

4G is Fourth generation of mobile/wireless broadband services which provides mobile ultra-broadband Internet access. 4G is a collection of wireless standards, and can be realized using various technologies; one of them is Long Term Evolution(LTE).

LTE/LTE-A (Long Term Evolution Advanced) based wireless broadband technology is basically designed to support mobile Internet access. The main features of LTE are:

1. IP based system,
2. High peak data rate,
3. Flexibility of spectrum usage,
4. Low latency times,
5. Efficient radio usage,
6. Higher capacity per cell,
7. Cost-efficient deployment, etc.

4G supports IP (Internet Protocol) based packet switching communication system with OFDM (Orthogonal Frequency Division Multiplexing) multi-carrier transmission and other frequency domain schemes. This helps to obtain high data transfer rates, which are further increased by using Antenna arrays for MIMO (Multiple Input Multiple Output) communication to avoid multi-path propagation losses. The radio interface of LTE is based on Orthogonal Frequency Division Multiple Access (OFDMA) in the downlink and Single Carrier-Frequency Division Multiple Access (SC-FDMA) in the uplink to provide high data bandwidth along with high-degree resilience to reflections and interference. SC-FDMA has the advantage of smaller peak to average power ratio and more constant power to get high RF power

Table 1. Downlink and Uplink frames

Downlink Frame (OFDMA)						
Preamble	Header	Burst-1	Burst-2	Brust-3	-------------	
Uplink Frame (SC-FDMA)						
Preamble	Burst-1	Preamble	Brust-2	Preamble	Brust-3	----------

amplifier efficiency in the mobile handsets. Downlink peak rates are at least 100 Mbps while Uplink rates are at least 50 Mbps with RAN round-trip times of less than 10 ms.

Downlink and Uplink frames are shown in Table 1.

The OFDMA approach achieves high peak data rates in high spectrum bandwidth and also high flexibility in channelization. LTE therefore enables to boost spectral efficiency and operate in various radio channel sizes from 1.25 MHz to 20 MHz.

Further, MIMO (Multiple Input Multiple Output) technique is used for transmission and reception of signals. MIMO involves spatial multiplexing with pre-coding and transmit diversity; to address the problems of multiple signals arising from many reflections. MIMO technique increases the throughput via the additional signal paths. MIMO requires two or more different antennas with different data streams to distinguish the different paths, such as the schemes using 2 x 2, 4 x 2, or 4 x 4 antenna matrices.

Other important features of LTE-A include enhanced intercell interference coordination (eICIC), self-optimizing network capability (SON) and the deployment of heterogeneous network (HetNet). HetNet allows various Low Power Nodes (LPNs) to be distributed across a macro cell network as an underlay. The LPNs includes pico eNBs that are typically deployed by the operators or home eNBs which are randomly employed by consumers, relays, and distributed antenna systems (DAS). LTE-A also supports Device to Device communication (D2D). Thus, UE pair moving within close proximity to each other can establish a D2D link which can be operated either in the unlicensed spectrum band or with the assistance of eNB the link can operate in the same licensed band used by cellular UEs. Furthermore operator assisted D2D communications functions as controlled underlay to the existing LTE-A networks and has added features of transmit power control, peer discovery, physical resource blocks (PRBs) assignment and interference management.

For smooth and efficient functioning LTE architecture must have well-designed physical and transport channels for downlink and uplink. All packet transmissions involve both links and therefore dynamic resource utilization of channels is important.

Physical and transport channels for Downlink and Uplink are used for achieving different goals in transmission, and therefore need to be treated differently. These channels are:

Physical and Transport Channels for Downlink

Physical Channels

- Physical Broadcast Channel (PBCH): holds system information for UEs when requiring to access the network.
- Physical Control Format Indicator Channel (PCFICH): is used for managing the transmission format.

- Physical Downlink Control Channel (PDCCH): is used primarily to carry the scheduling information.
- Physical Hybrid ARQ Indicator Channel (PHICH): is used to report the Hybrid ARQ status.
- Physical Downlink Shared Channel (PDSCH): is used for unicast and paging.
- Physical Control Format Indicator Channel (PCFICH): supplies information to decode the PDSCH via UE.

Transport Channels

- Broadcast Channel (BCH): maps to Broadcast Control Channel (BCCH)
- Downlink Shared Channel (DL-SCH): is the main channel for Downlink data transfer, used by many logical channels.
- Paging Channel (PCH): conveys the Paging Control Channel (PCCH)
- Multicast Channel (MCH): transmit Multicast Control Channel (MCCH) information.

Physical and Transport Channels for Uplink

Physical Channels

- Physical Uplink Control Channel (PUCCH): is used for sending Hybrid ARQ acknowledgement
- Physical Uplink Shared Channel (PUSCH): is the UL counterpart of PDSCH
- Physical Random Access Channel (PRACH): is used for random access functions.

Transport Channels

- Uplink Shared Channel (UL-SCH): is similar to Downlink Shared Channel (DL-SCH).
- Random Access Channel (RACH): is used for random access requirements.

Apart from physical and transport channels LTE also use some logical channels to cover the data carried over the radio interface. These are:

Control Channels: Carry the control plane information:

Broadcast Control Channel (BCCH): provides system information to all mobile terminals connected to the eNodeB.

- **Paging Control Channel (PCCH):** *For paging information when searching a unit on a network.*
- **Common Control Channel (CCCH):** *For random access information, e.g. for actions including setting up a connection.*
- **Multicast Control Channel (MCCH):** *For Information needed for multicast reception.*
- **Dedicated Control Channel (DCCH):** *For carrying user-specific control information, e.g. for controlling actions including power control, handover, etc..*

Traffic Channels: Carry the user-plane data:

- **Dedicated Traffic Channel (DTCH):** *For the transmission of user data.*
- **Multicast Traffic Channel (MTCH):** *For the transmission of multicast data.*

LTE uses All IP based flat architecture. Flat architecture means reduced number of nodes which in turn reduces latency time and improves overall performance.

So, there are three components of this flat structure (Figure 1)

Logical connection between these is shown in Figure 2.

Evolved Packet Core (EPC) defines network functions based on flat architecture, thus reducing the number of nodes in the network. It uses IP (Internet Protocol) as the key protocol to transport all services; hence protocol conversion is not needed. EPC treats voice as one of many IP-based network applications namely VoIP. Further user plane and control plane are independent. These features ensure superior network performance and quality of services at reduced operation and maintenance cost.

EPC comprises of many network elements. These are described in next section (Figure 3).

User Equipment may refer to any equipment used for communication purpose by the end user;

Figure 1.

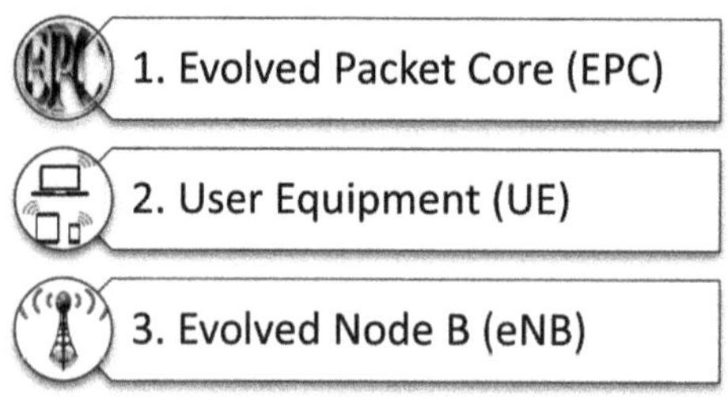

Figure 2.

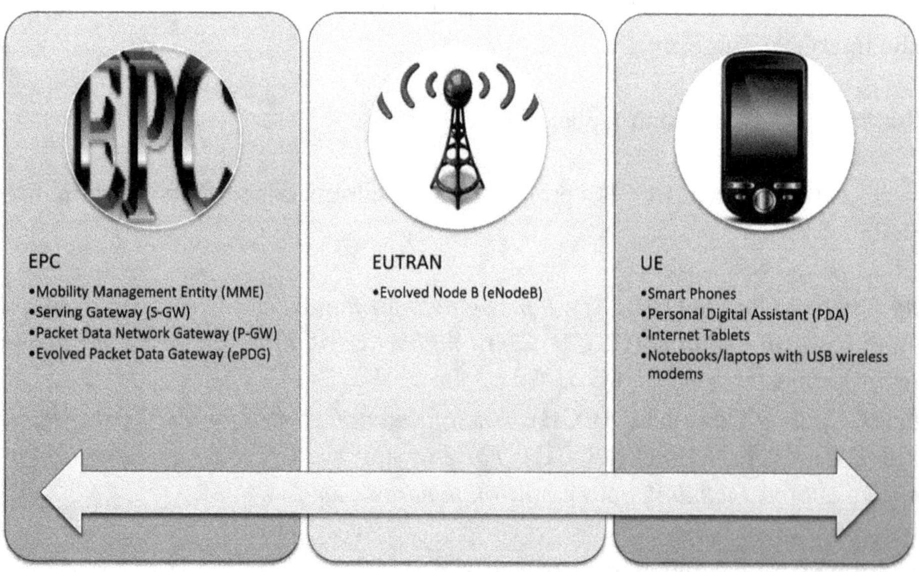

Figure 3.

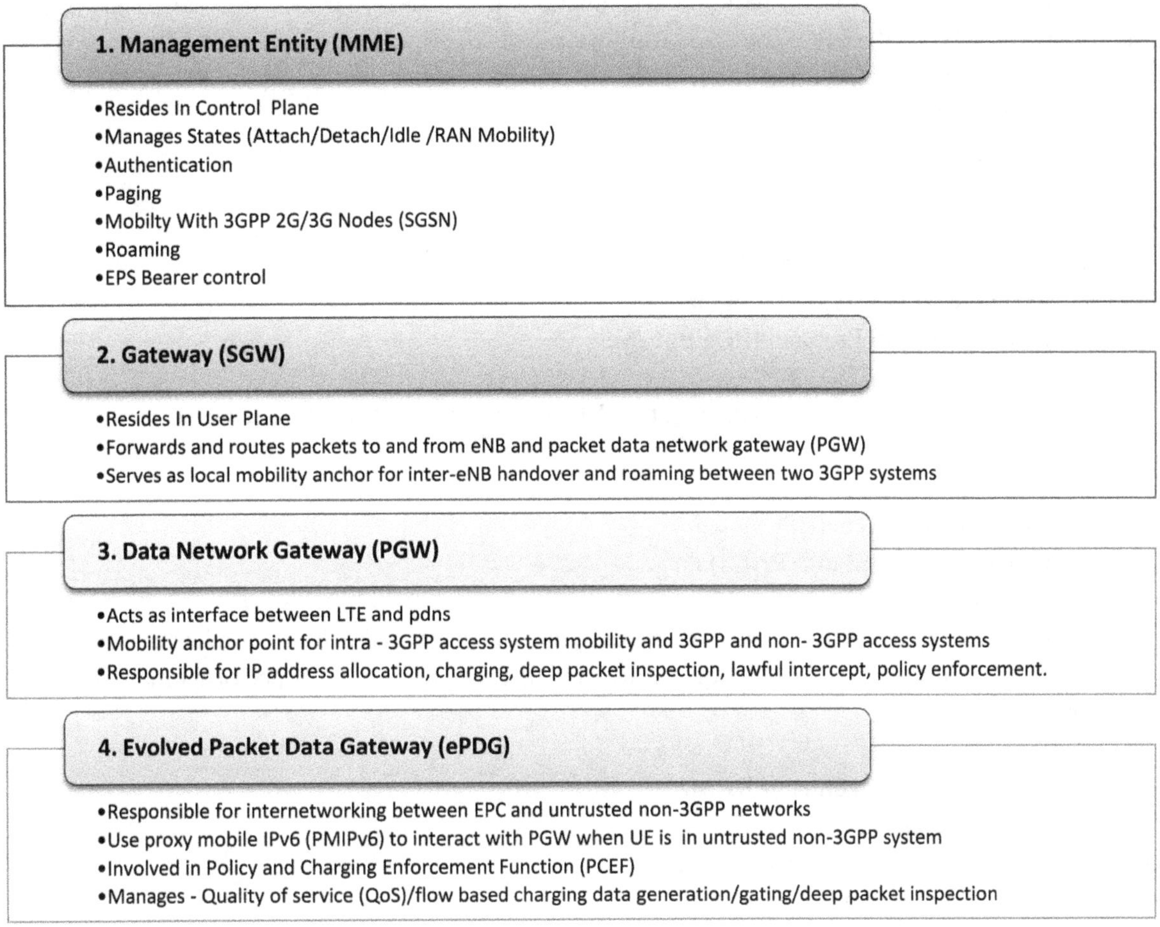

1. Smart Phones
2. Personal Digital Assistant (PDA)
3. Internet Tablets
4. Notebooks/Ultra books (Ultra Mobile PC)/laptops with USB wireless modems

These devices are usually carried by the user and can be connected to internet as and when required. UE contain a Universal Subscriber Identity Module (USIM), popularly known as SIM. USIM is used to:

1. Identify and authenticate the user
2. Derive security keys for protecting the radio transmission
3. Set up, maintain and remove network communication link
4. Help eNB for handovers and
5. Send Reports of other eNBs' locations to eNB of current cell.

UE can receive transmission from up to four antennas.

Evolved Node B (eNodeB/eNB) is equivalent to a radio base station in GSM/GPRS, but with certain additional features. It controls all radio related functions in the fixed part of the system and works as bridge between UE and the EPC. The eNodeBs are typically distributed throughout the coverage region.

Each eNB contains radio transmitter, receiver, control section, antenna system (typically a radio tower) and power supply.

eNB hosts the:

- PHYsical (PHY) Layer,
- Medium Access Control (MAC) Layer,
- Radio Link Control (RLC) Layer, and
- Packet Data Control Protocol (PDCP) Layer.

eNBs also contain resource management and logic control functions that have been traditionally separated into

1. Base station controllers (BSCs) and
2. Radio network controllers (RNCs).

This added capability allows eNBs to directly communicate with each other, eliminating the need for mobile switching systems (MSCs) or controllers (BSCs or RNCs). eNB performs many functions including:

1. Radio resource management - RRM,
2. Radio bearer control,
3. Radio admission control - access control,
4. Connection mobility management,
5. Resource scheduling between UEs and eNB radios,
6. Header compression,
7. Link encryption of the user data stream,
8. Packet routing of user data towards its destination (usually to the EPC or other eNBs),
9. Scheduling and transmitting paging messages (incoming calls and connection requests),
10. Broadcast information coordination (system information), and
11. Measurement reporting (to assist in handover decisions).

Each eNodeBs, is responsible for managing multiple cells. Integration of radio controller function into the eNodeB results in tight interaction between the different protocol layers of the radio access network (RAN). This reduces latency, improves efficiency and eliminates the need for a high-availability, processing-intensive controller.Furthermore, LTE does not support soft handover so there is no need for a centralized data-combining function in the network. Therefore as UE moves, all the information related to UE (UE context/buffered data etc) is transferred from one eNodeB to another.

The interface between two eNodeBs is known as *X2*. The interface between an eNB and EPC is known as *S1* interface or more specifically, *S1-MME* is the interface to the MME and *S1-U* is interface between eNB and SGW.

The protocols that run between the eNodeBs and the UE are known as the *AS protocols*. An eNodeB serves multiple UEs at a time but a UE connects to a single eNB at a time. Neighboring eNBs are also connected with each other for handovers. MMEs and S-GWs are pooled together. A set of these nodes is assigned to serve a particular set of eNodeBs. So;

- Each eNodeB is connected to multiple MMEs and S-GWs.
- Each UE is served only by one MME and S-GW at a time, and the eNodeB has to keep track of this association.
- This association changes only at the time of inter-eNodeB handover.

Protocol links of eNB are:

1. Air Interface Physical Layer
2. GPRS Tunnelling Protocol User Plane (GTP-U)
3. GTP-U Transport
4. Medium Access Control (MAC)
5. Non-Access-Stratum (NAS) Protocol
6. Packet Data Convergence Protocol (PDCP)
7. Radio Link Control (RLC)
8. Radio Resource Control (RRC)
9. S1 Application Protocol (S1AP)
10. S1 layer 1
11. S1 Signalling Transport
12. X2 Application Protocol (X2AP)
13. X2 layer 1
14. X2 Signalling Transport

eNodeB works as plug and play nodes. Once a new eNB is powered on; it is configured according to network parameters and starts providing services to the user. These services need to be optimized according to the usage. And often these eNBs need to change their parameters to provide users of the neighbouring cells in case the eNB in the neighbouring cell fails/come down due to some technical failure. So there are three main stages of work for an eNodeB (Figure 4).

Figure 4.

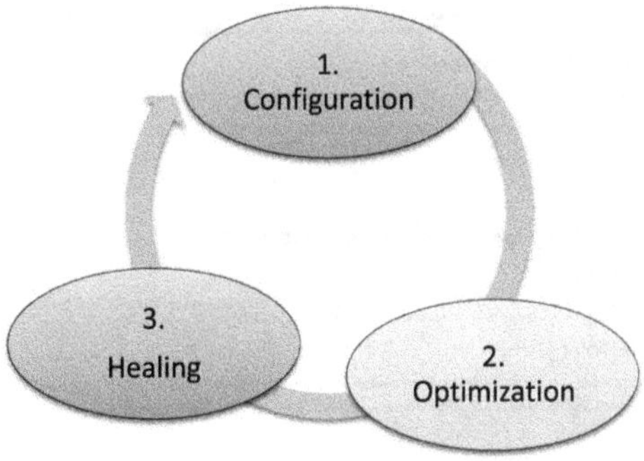

Earlier all these things were handled manually. Obviously it took time and expertise. More time means low Quality of service and as we all know that Expertise is costly. So, to cut the cost and save the time network providers have started automation for all these processes.

3. NETWORK AUTOMATION MECHANISMS

The process of network automation can be divided into three broad categories (Figure 5).

Self Organization ensures that whenever a new base station is added all the initial parameters are configured with the use of software without human intervention. Thus new node becomes operational in "plug-and-play" manner (Figure 6).

Figure 5.

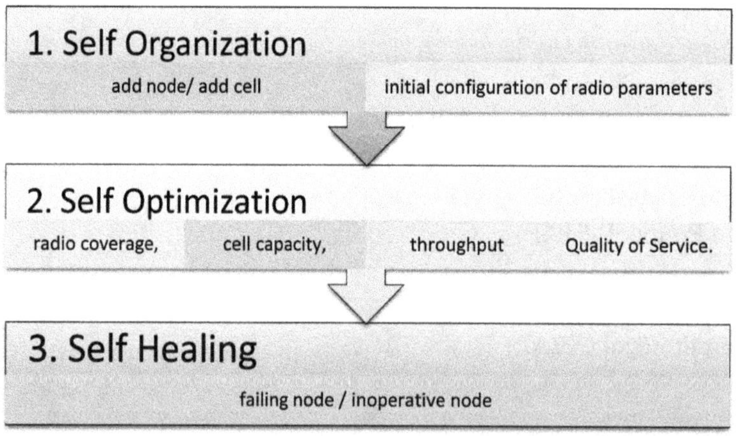

Figure 6.

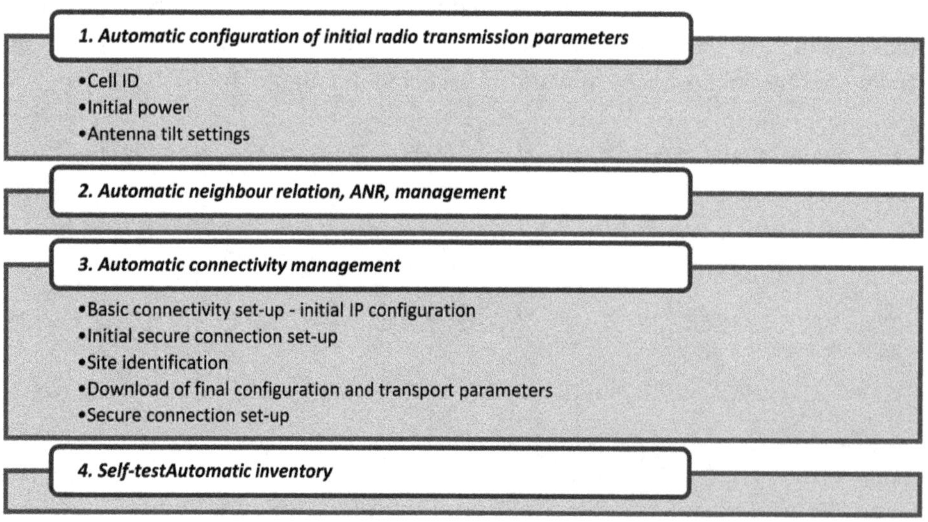

Whenever a new node/base station is introduced it is:

1. Immediately recognized and registered by the network.
2. Software establishes the connectivity of the node to the network.
3. Configuration parameters are downloaded autonomously.
4. Other nodes in the neighborhood automatically adjust all their parameters to avoid interference with the new node.

Performance optimization of a network actually requires the simultaneous optimization of a large number of different parameters and manual optimization is not feasible due to huge scale, heterogeneous operation and complexity of next generation radio networks. Consequently self optimization seems to be key enabling element towards cost effective operation and management of future mobile access networks. *Self Optimization* of network parameters then ensures the Quality of Service by adjustment of various network parameters which are discussed in detail in next segment.

Self Healing aims to reduce the effects of failure of a node arbitrarily by immediately identifying the defective node and adjusting the parameters of adjusting cells/nodes accordingly. The next section describes how these functions are implemented in automated mobile networks.

Three different architectures are used for Self Organization purposes:

1. Centralized architecture: In centralized approach optimization algorithms reside on the network management system (NMS) or a central SON server that manages all the eNBs. Salient features of this approach are:
    a. Central SON server has global information from all the eNBs,
    b. Centralized architecture can readily approach the global optimum.

But, parameter updates are rather slow, as central SON server collects information from all the eNBs which in turn receive the feedback from their UEs (user equipments). Thus optimization process cost much time because of the large operating interval. Therefore, purely centralized architecture is not preferred.

2. Decentralized architecture: In decentralized approach each eNB has a SON entity, i.e. each eNB runs the optimization process. Salient features of this approach are:
    a. Each eNB takes decision autonomously based on the feedback from its' UEs plus information from neighbour eNBs.
    b. allows for ease of deployment and
    c. Optimization process is faster.

This approach may prove fatal for the stability of the network and overall optimization performance is often compromised.

3. Hybrid architecture: is a combination of the centralized and the decentralized SON, which means:
    a. Part of a SON entity is located in the eNB, while the other is deployed in the NMS.
    b. It provides a tradeoff between the two architectures.

For example, the selection of eNBs participating in the optimization, the initial parameter setting, and the objective setting is done in the central NMS, while the core optimization algorithm runs within the involved eNB.

Each, architecture has its own advantages and disadvantages. Practical implementation depends on various parameters such as optimization objective, infrastructure deployment, pilot-strength measurements, traffic counters, and other specific conditions.

3.1 Automatic Neighbor Relations

In 2G and 3G systems Neighbour Relation Table (NRT) is created before the installation of a base station. All the handover regions in a coverage area are identified by radio engineers through drive/walk tests. Traditionally an operator maintains and updates this list for each cell. But with ever increasing number of cells it is a tedious process and there are always chances of error. Further entire procedure needs to be repeated if there are changes in interference pattern. This is costly, error –prone and time consuming procedure and new methods for automatically deriving NRTs are need of the hour. Automatic Neighbour Relation (ANR) function aims towards obtaining and managing Neighbor Relations (NRs) without the human intervention. Automatic Neighbour Relation minimizes the involvement of man power in this process to ensure less error and immediate updating of the NRT whenever a node joins or leaves the network and if implemented ANR function will automatically add and remove entries to/from NRT.

Each cell broadcasts a Physical Cell Identity (PCI), But PCI cannot be used for unique identification of cells. Cell also broadcast a globally unique cell identifier (CGI (Table 2)).

An existing Neighbor Relation from a source cell to a target cell means that eNB of the source cell:

- Knows the CGI and PCI of the target cell.
- Has an entry in the *Neighbour Relation Table (NRT)* for the source cell identifying the target cell.
- Has the attributes in this Neighbour Relation Table entry defined, either by Operations & Maintenance (O & M) or set to default values.

NRT contains:

- ECGI/CGI and PCI of the Identified neighbor cell.
- Three O&M attributes:
  - NoRemove: eNB shall not remove the Neighbor cell Relation from the NRT.
  - NoHO: Neighbor cell Relation shall not be used by the eNB for handover reasons.

Table 2. Neighbour Relation Table

Neighbour Relation		O&M Controlled Neighbour Relation		
NR	**ECGI/PCI**	**No Remove**	**No HO**	**No X2**
1	#1	Y		
2	#2			
3	#3	Y		Y

- NoX2: Neighbor Relation shall not use an X2 interface in order to initiate procedures towards the eNB parenting the target cell.

NRT is formed by using one of the below mentioned processes:

1. Intra-LTE ANR
2. Inter-LTE/Inter-RAT (Radio Access Technology) ANR

A. Intra-LTE ANR

Cells cannot themselves detect their neighbor cells themselves because they do not have any such facility; whereas user equipment (UE) can detect all the cells around it. So, eNB rely on the UE report to update its NR.

Suppose, eNB#1 sends Measurement command to UE by sending RRC Connection Reconfiguration message with measurement control information.

1. The UE sends a measurement report regarding eNB#2.
    a. This report contains PCI of eNB#2; but not its ECGI.
    b. eNB#1 checks if the reported PCI is already included in the NRT
    c. If reported PCI is not included in the NRT then eNB#1 proceeds to add the PCI to its NRT.
2. eNB#1 instructs the UE with another RRC Connection Reconfiguration Message, using the newly discovered PCI as parameter.Instruct UE to read the ECGI, the TAC and all available PLMN ID(s) of the related neighbor cell.
   To do so, the eNB#1 may need to schedule appropriate idle periods to allow the UE to read the
    a. ECGI from the SIB1 of the detected neighbor cell.
    b. The UE reads the requested information from SIB1 on PDSCH. UE needs to read MIB on PBCH,
    c. then DCI with SI-RNTI on PDCCH to read SIB1 on PDSCH.
3. When the UE has found out the new cell's ECGI, the UE reports the detected ECGI to the serving cell eNB. In addition the UE reports the tracking area code and all PLMN IDs that have been detected. If the detected cell is a CSG or hybrid cell, the UE also reports the CSG ID to the serving cell eNB.
4. The eNB decides to add this neighbor relation, and can use PCI and ECGI to:
    a. Look up a transport layer address to the new eNB.
    b. Update the Neighbor Relation List.
    c. If needed, setup a new X2 interface towards this eNB.

B. Inter-LTE/Inter-RAT (Radio Access Technology) ANR

eNB can instruct UE to detect cells on different frequencies/RATs also. UE detects and reports the *PCI (physical cell ID)* of such cells to eNB. Format of PCI depends upon the RAT of the Cell being measured (Table 3).

When eNB receives reports containing PCI, eNB instructs UE to;

Table 3. Measurement Info Reported By UE

System	Physical Cell Identity	Global Cell Identity
LTE Inter Frequency	Physical Layer Cell Identity	Global cell Id, TAC, PLMN ID
GSM	Band indicator, BSIC, RF carrier	Global cell Id, RAC
UMTS	RF carrier, Scrambling code	Global cell Id, LAC, RAC
CDMA 2000	PN offset	Global cell Id

1. Read CGI and RAC of detected cell if it is GSM EDGE Radio Access Network (*GERAN*)
2. Read CGI, Local Area Code (LAC) and Routing Area Code (RAC) if the detected cell is *Universal Terrestrial Radio Access Network (UTRAN)*

ANR extracts the information and updates its inter-frequency/RAT ANR accordingly.

3.2 Load Balancing Optimization

Mobile users and the resulting data usage are random, time varying, and often unbalanced, this make unequal load scenario for neighbour cells; consequently one cell may be overloaded while, other has much less users and its resources are not fully utilized; overloaded cell faces resource shortage, which affects access of new users and impacts the QoS of active users. Thus load imbalance seriously deteriorates the overall performance of the cellular network due to inefficient resource utilization; the problem is rectified by load balancing techniques (Figure 7).

Load of a cell is measured in terms of usage of different resources with respect to their available limits. These are:

1. Total transmit power
2. Total received power
3. Interference in a cell

Figure 7.

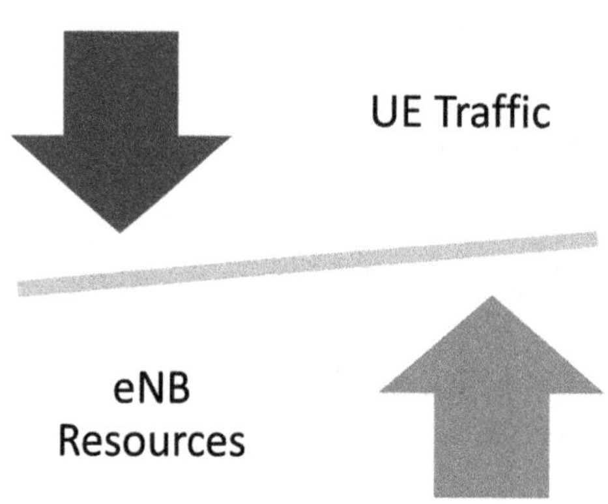

4. Cell throughput in downlink/uplink
5. Increase in blocking
6. Handover failure rate

Load balancing involves detecting network load imbalance by periodically exchanging information between neighbouring eNBs to compare the cells load and then, Load balancing optimization feature automatically change the network parameters to manage the unexpected loading conditions caused by sudden rise in data usage.

An efficient load balancing optimization algorithm ensures that

1. The system load is according to operator policy,
2. It should also "offload" users from one cell or carrier in order to achieve energy savings.

Because of coexistence of different technologies different load balancing optimization techniques are used:

A. Intra-LTE Load Balancing

UE is handed over to other eNB within the current LTE network. eNBs utilize X2 interface to exchange load information and load is transferred to neighbouring cells/ co-located cells on a different carrier frequency.

The load information consists of:

- Radio Resource Usage
- Uplink/Downlink Guaranteed Bit Rate (GBR)
- Physical Resource Block (PRB) usage
- Uplink/Downlink non-GBR PRB usage
- Uplink/Downlink total PRB usage
- Hardware (HW) load indicator
- Uplink/Downlink HW load: Low, Mid, High, Overload
- Transport Network Load (TNL) indicator
- Uplink/Downlink TNL load: Low, Mid, High,
- Overload Cell Capacity

B. Inter-RAT Load Balancing

UE handover occur between LTE and another network. S1 interface is used to exchange load information. A dedicated procedure for inter-RAT cell load request / reporting is provided with minimal impact using a generic SON container extension of the RAN Information Management (RIM) mechanism. Load information is provided in a procedure separated from existing active mode mobility procedures, which is used infrequently and with lower priority with respect to the UE dedicated signaling.

The load information consists of:

- Cell Capacity Class value
- Uplink/Downlink relative capacity indicator
- Uplink/Downlink available capacity for load balancing as percentage of total cell capacity

Load balancing techniques rely on the loads assessment by the eNBs by the feedback received from their UEs. An eNB caters many UEs, but all of them are not active all the time. This issue leads to different load balance techniques categorized as:

- Active Mode Load Balancing and
- Idle Mode Load Balancing

Active Mode Load Balancing

Active load balancing allows active mode UEs to be load balanced across cells to lower the overall congestion across cells. The advantage of active load balancing is that the system has a direct measurement mechanism and knowledge of each user's traffic requirements and radio conditions before deciding to load balance. Therefore, with the scheduler and interfaces to other base stations (X2 interface for intra-LTE and/or S1 interface for inter-RAT), it is possible to make accurate decisions for load-based HO. A "load-based HO" reason code is included during handover (HO) messaging to allow the target cell knowledge for admission control.

Idle Mode Load Balancing

Idle mode load balancing is more difficult to achieve since eNB cannot detect the idle mode users, eNB becomes aware of a user only when it become active or when the tracking area of the user changes and a Tracking Area Update (TAU) message is sent by the UE. The way around this immeasurable condition is for the system to adjust cell reselection parameters for the idle users based on the current active user condition, since, in LTE; idle mode inter-frequency load balancing is controlled by the cell reselection procedure. System parameters that control cell reselection and the operator's channel frequency preferences are transmitted to UEs in the System Information Blocks (SIBs). Thus as real-time traffic or QoS demands increase in a cell; it would be possible for the cell to adjust the cell reselection parameters in order to force users nearest the cell edge to select their strongest neighbour or to force a handover to a co-located carrier that has more available resources. Care must be taken to coordinate such parameter adjustments between cells (i.e. utilizing the X2 interface) in order to prevent coverage gaps, as well as to adjust active mode parameters to avoid immediate handover upon an idle to active transition.

Load balancing optimization comprises of a series of related functions. One of the important functions is the adaptation of handover configuration function, which enables requesting of a change of handover and/or reselection parameters at target cell, as a part of the load balance procedure. The source cell that initialized the load balancing estimates if the mobility configuration in the source and/or target cell needs to be changed. If the amendment is needed, the source cell initializes mobility negotiation procedure toward the target cell. This is applicable for both idle and active mobility cases. The source cell informs the target cell about the new mobility settings and provides cause for the change such as, load balancing related request. The proposed change is expressed as the difference (delta) between the current and the new values of the handover trigger. The handover trigger is the cell specific offset that corresponds to the threshold at which a cell initializes the handover preparation procedure. Cell reselection configuration may be amended to reflect changes in the handover setting. The target cell responds to the information from the source cell. The allowed delta range for handover trigger parameter may be carried in the failure response message. The source cell should consider the responses before executing the planned change of its mobility setting. All automatic changes on the HO and/or reselection parameters must be within the range allowed by OAM. Optimization software also detects the signs of movement of UE. Moving load between cells is achieved by adjusting the handover thresholds and hence; the process involves increasing/decreasing cell boundaries for load optimization. Position of cell boundary is changed by one of the two methods (Figure 8)

1. Coverage Adaptation based Load Optimization, and
2. Relay assisted Load Optimization.

1. **Coverage Adaptation based Load Optimization:** Coverage adaptation based load balancing is done by increasing/decreasing the coverage area of an eNB. This is possible either by remotely tilting the antenna by a certain degree to alter the cell boundary or by increasing the signal power to cover UEs at the cell boundary.
2. **Relay assisted Load Optimization:** Another method of load balancing is the introduction of relay nodes. Relay nodes are low power eNBs connected to the doner eNB via radio interface and shares its radio resources. Relay nodes enhance the coverage at the cell boundaries and thus increase the capacity of the cell.

Figure 8.

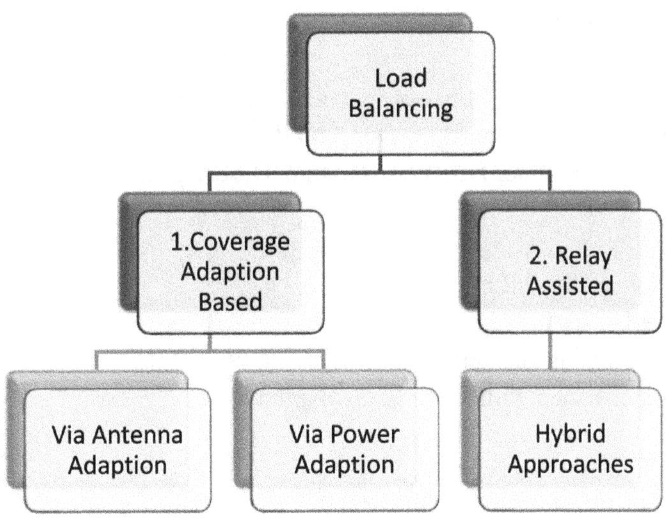

3.3. Handover Optimization

Once ANR and Load Balancing parameters are configured; handover optimization is required to avoid problems like ping-pong handover condition and radio link failure. Handover performance is a very critical quality measure for mobile communication systems. On one hand, mobile user wants seamless mobility support without service interruption or call dropping. On other hand, system load generated by handovers has to be kept within reasonable limits. In LTE the radio link performance at the cell border suffers from low SINR (Signal To Noise and Interference Ratio). Also, fast fading and shadowing of the radio signal leads to frequently change in the server conditions.

Choice of best server for the UE is a complicated process; a too early handover results in frequent handovers when the user is at the cell boundary causing extra burden generated by frequent handovers. This situation is described as ping-pong handover problem. But a bit too late handover can result in increase in radio link failure probability.

For this various performance indicators used are:

1. Handover failure rate per handover
2. Number of discarded PDUs per handover
3. Visit time distribution to target cell
4. Signal and interference measurement
5. MS location information

These parameters are analyzed periodically to identify the cells for which handover optimization is required. Optimization procedure alters

1. Handover thresholds and
2. Hysteresis values

Timers are used for synchronized activation of various parameters.

Handovers can be categorized in two types:

A. Intra-Frequency Handover

Intra frequency handover takes place between two cells that are using same carrier frequency. This occurs when signal strength of a new cell is greater than the current cell. Normally this type of handover is needed at cell boundary. As carrier frequency is same, UE can easily monitor the strength of two cells (Table 4).

- UE handover is done using X2 interface
- MME is unchanged
- Serving Gateway S-GW is unchanged
- HO procedure is performed without EPC involvement

Table 4.

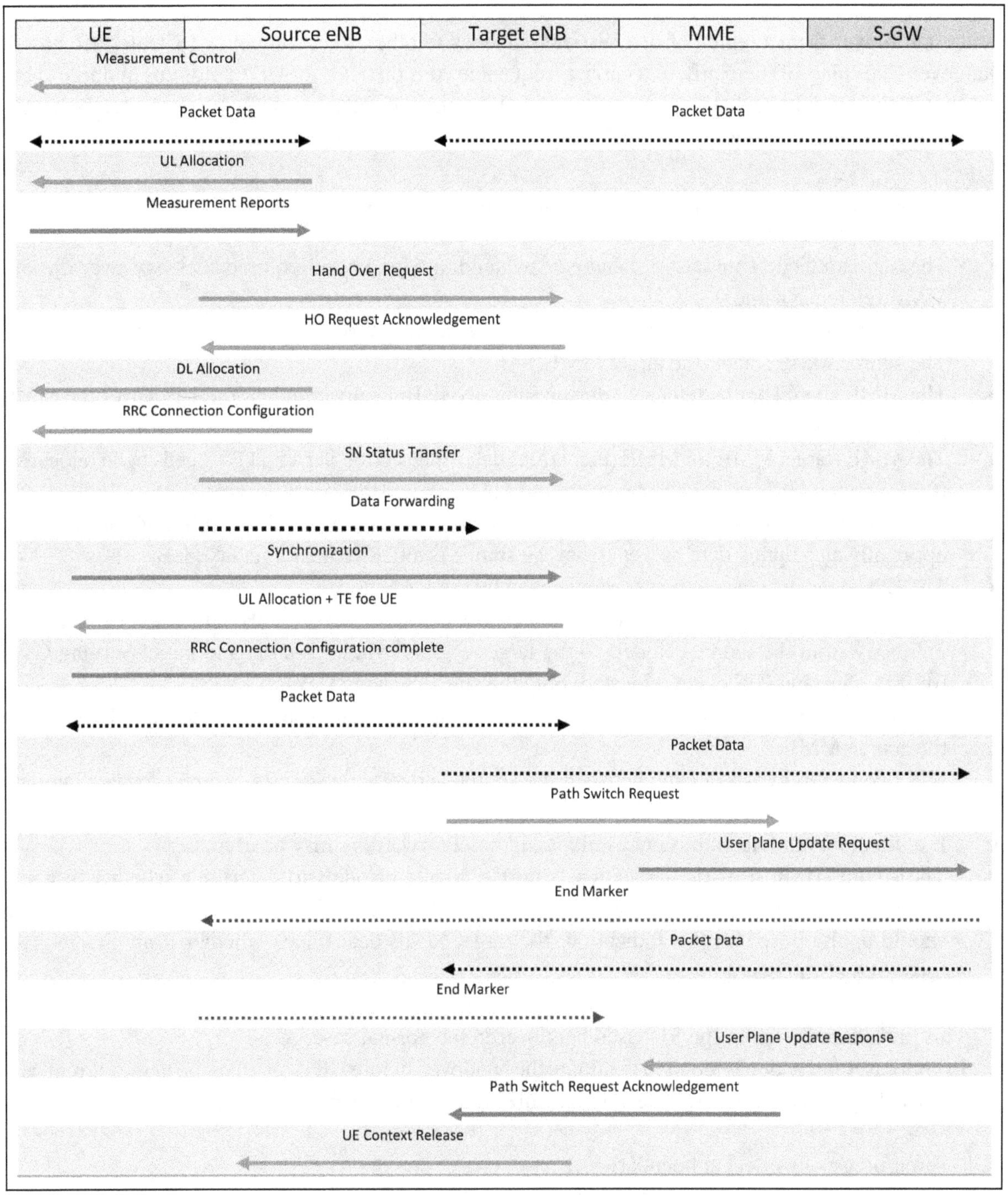
UE
Source eNB
Target eNB
MME
S-GW
Measurement Control
Packet Data
Packet Data
UL Allocation
Measurement Reports
Hand Over Request
HO Request Acknowledgement
DL Allocation
RRC Connection Configuration
SN Status Transfer
Data Forwarding
Synchronization
UL Allocation + TE foe UE
RRC Connection Configuration complete
Packet Data
Packet Data
Path Switch Request
User Plane Update Request
End Marker
Packet Data
End Marker
User Plane Update Response
Path Switch Request Acknowledgement
UE Context Release

B. Inter-Frequency Handover

When UE moves from a region of one carrier frequency to other, inter-frequency (S1 interface based) handover takes place. Due to different carrier frequencies it is difficult for UE to monitor multiple channels but in this case the probability of interference at the edge of cells is less. Features of inter frequency handover are listed below as:

- No connectivity between two eNBs between which UE moves.
- X2-based handover cannot be used.
- The source eNodeB initiates a handover by sending Handover Required message over the S1-MME reference point.
- This procedure may relocate the MME and/or the Serving GW.
- The source MME selects the target MME.
- The MME should not be relocated during inter-eNodeB handover unless the UE leaves the MME pool Area where the UE is served.
- The MME (target MME for MME relocation) determines if the Serving GW needs to be relocated.
- If the Serving GW needs to be relocated the MME selects the target Serving GW.
- The source eNodeB decides which of the EPS bearers are subject for forwarding of downlink and optionally also uplink data packets from the source eNodeB to the target eNodeB.
- The EPC does not change the decisions taken by the RAN node.
- Packet forwarding can take place either directly from the source eNodeB to the target eNodeB, or indirectly from the source eNodeB to the target eNodeB via the source and target Serving GWs (or if the Serving GW is not relocated, only the single Serving GW).
- The availability of a direct forwarding path is determined in the source eNodeB and indicated to the source MME.
- If X2 connectivity is available between the source and target eNodeBs, a direct forwarding path is available.
- If a direct forwarding path is not available, indirect forwarding may be used.
- The source MME uses the indication from the source eNodeB to determine whether to apply indirect forwarding. The source MME indicates to the target MME whether indirect forwarding should apply. Based on this indication, the target MME determines whether it applies indirect forwarding (Table 5).

This procedure describes the S1-based handover in the normal case,
In both cases fast action is required to initiate the handover, before UE is out of range of the current cell.
Thus self optimization solution need to optimize the cell boundaries and try to

1. Limit ping-pong effect at boundaries.
2. Accommodate changes due to addition of new cells.
3. Accommodate changes in propagation due to changes in environment (Construction activities/ weather changes etc…)

Table 5.

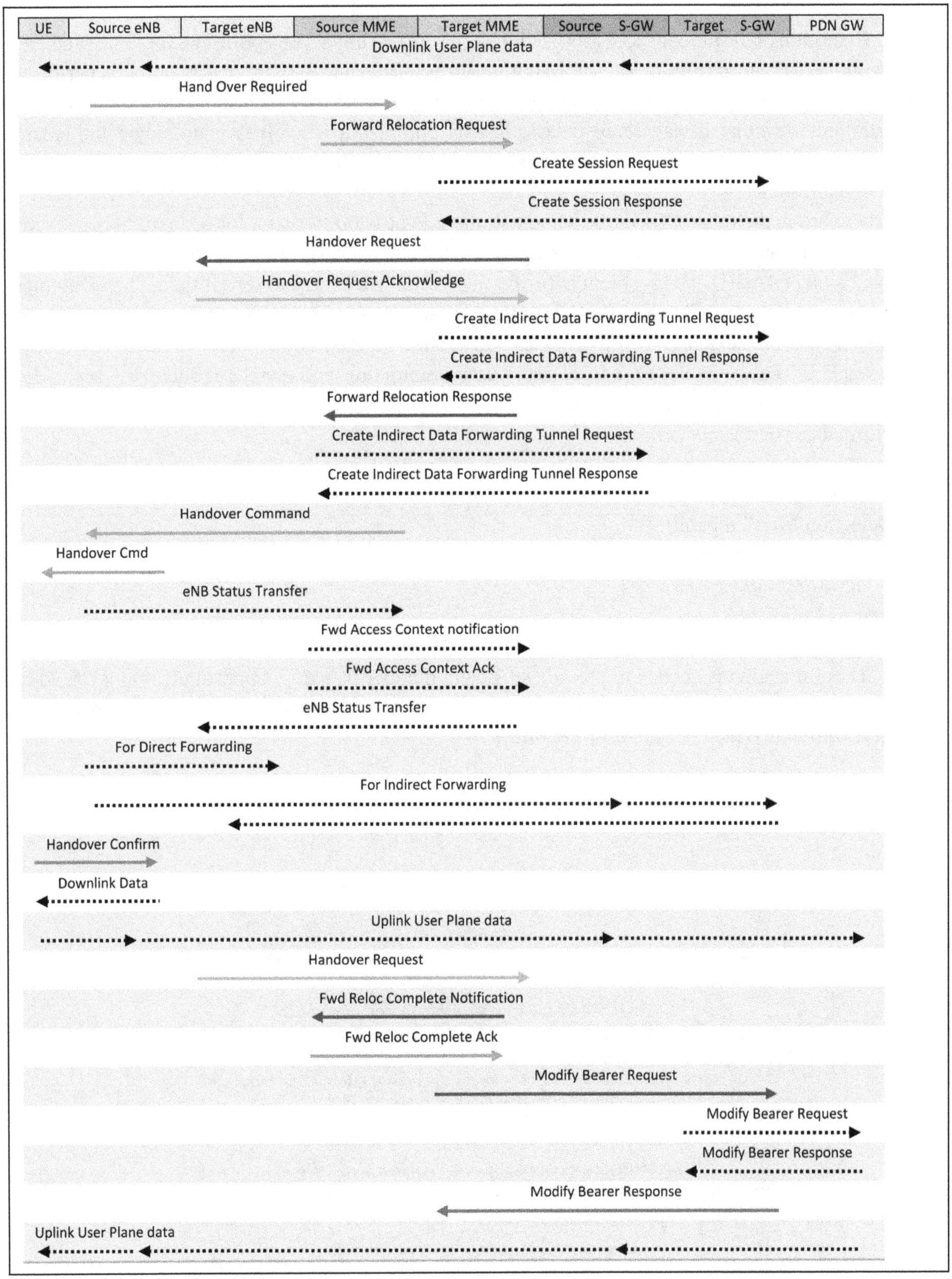
UE
Source eNB
Target eNB
Source MME
Target MME
Source S-GW
Target S-GW
PDN GW
Downlink User Plane data
Hand Over Required
Forward Relocation Request
Create Session Request
Create Session Response
Handover Request
Handover Request Acknowledge
Create Indirect Data Forwarding Tunnel Request
Create Indirect Data Forwarding Tunnel Response
Forward Relocation Response
Create Indirect Data Forwarding Tunnel Request
Create Indirect Data Forwarding Tunnel Response
Handover Command
Handover Cmd
eNB Status Transfer
Fwd Access Context notification
Fwd Access Context Ack
eNB Status Transfer
For Direct Forwarding
For Indirect Forwarding
Handover Confirm
Downlink Data
Uplink User Plane data
Handover Request
Fwd Reloc Complete Notification
Fwd Reloc Complete Ack
Modify Bearer Request
Modify Bearer Request
Modify Bearer Response
Modify Bearer Response
Uplink User Plane data

3.4. Random Access Channel Optimization

The configuration of the random access procedure has a critical impact on end-user experience and overall network performance (Figure 9). The Random Access Channel (RACH) optimization is performed at the evolved Node B (eNB) level. Incorrectly configured RACH may increase access setup time and access failures resulting in call setup delays. RACH optimization involves maintaining the balance between random accesses and services in the radio resource allocation while at the same time avoiding excessive interference.

As there are many other UEs in the same cell there is possibility that a lot of them become active at the same time and trying to get access to same eNB. This leads to severe overload on LTE Random Access Channel (RACH). The problem is further increased if UEs' try to repeat their access attempts, therefore Random Access Channel (RACH) optimization is important for overall network performance. RACH optimization balances random accesses and services in the radio resource allocation. Correctly configured RACH also enables eNB to decrease access setup time and thereby reduction in access failures and corresponding call setup delays.

Performance parameters are:

1. Access Probability
2. Detection Miss Probability
3. Detection Miss Ratio
4. Connection Ratio
5. Access Ratio

RACH optimization is needed to provide sufficient number of access opportunities to a UE within a cell. With increase in number of UE, number of opportunities for random access also increases. For self-optimization two methods are used (Figure 10):

Figure 9.

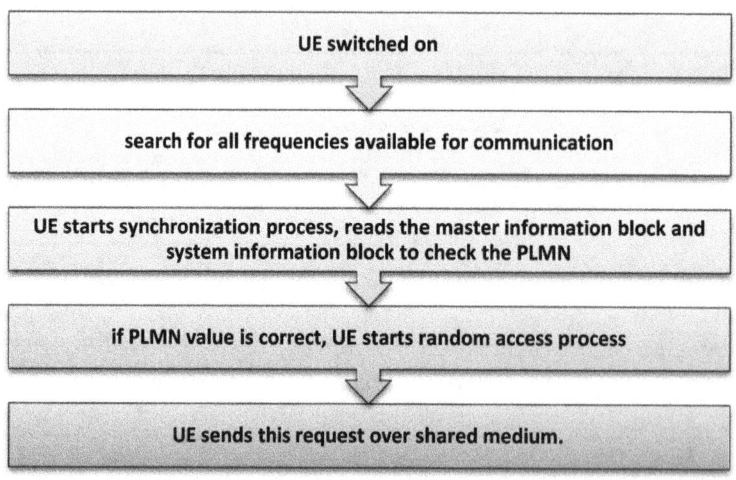

Figure 10.

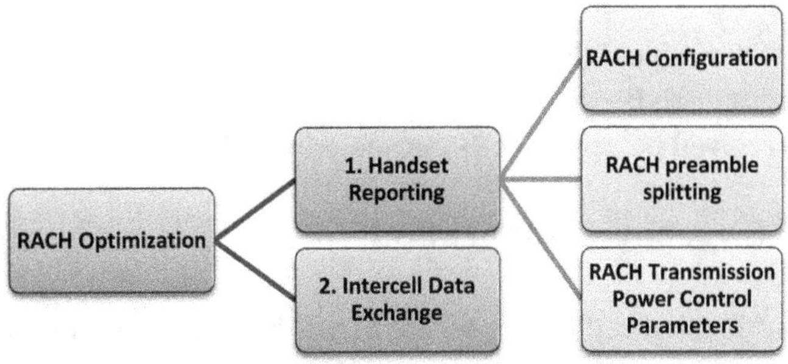

A. Handset Reporting

From recently obtained RACH reports from UEs, eNB determine the actual access delay. This report contains two parameters

1. Number of preambles sent, and
2. Contention detected

Based on this information Power control Parameter (PO) and Preamble format are modified to reach the set target delay. Optimization of PO increases the eNB's ability to read preamble but it also increases the chances of interference with neighbouring cells.

B. Inter Cell Data Exchange

As RACH configuration may affect other cells also; optimization software needs to exchange information between neighboring cells. This information is updated via eNB configuration update X2 messages. Allocation of same root sequence index is avoided by two neighbouring cells to avoid interference.

3.5 Inter-Cell Interference Coordination

Inter-Cell Interference Coordination (ICIC) involves the intelligent coordination of physical resources between various neighbouring cells to reduce interference from one cell to another. Each cell gives up use of some resources to improve performance for cell edge users which are impacted the most by inter-cell interference.

For implementation of ICIC;

1. Users are scheduled in units of Physical Resource Blocks (PRB)
2. Neighboring cells coordinate for portions of bandwidth and power across various frequency resource blocks for each cell.

Thus Inter-Cell interference is avoided in uplink and downlink.

Performance parameters are:

1. ICIC reporting thresholds/periods
2. Resource preferences in eNBs
3. Reference signal received power (RSRP) threshold for ICIC.

3.6. Coverage and Capacity Optimization

Cell coverage becomes critical at cell border where SINR drops down to levels below 0 dB because of neighbour cell interference. So the optimization target is to provide continuous coverage in which connections can be established with acceptable service quality.

Coverage and/or capacity problem is regional problem, and therefore, usually one or two neighbouring cells are involved in the optimization process. Hence, it is possible to run the algorithm in a decentralized manner to achieve the optimization. On the other hand, the capacity and coverage problem detection, the selection of eNBs, when to start/stop their optimization, and initial parameters of the algorithm need to be chosen in a central entity that can obtain the global information. Thus, hybrid architecture is used, i.e optimization, activation/deactivation and initial parameter setting is done by a central entity, and optimization algorithm is performed by every eNB in a decentralized manner.

Conventional method of coverage and capacity optimization involves testing the radio environment manually and adjusting the parameters such as base station transmit power and antenna tilt. Self optimization of coverage and capacity using Remote Electrical Tilt (RET) mechanisms will enable automatic configuration of parameters for optimized system performance.

Indicators for self-optimization tasks are:

- UE signaling reports.
- Pilot-strength measurements.
- Traffic counters.
- Spectral Efficiency (SE), referring to the information rate that can be transmitted per unit bandwidth.

Parameters used for optimization are (Figure 11):

Figure 11.

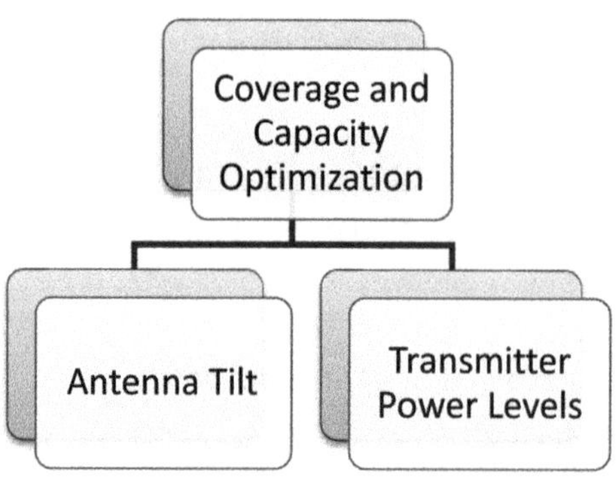

1. Antenna tilts
2. Transmitter power levels

CCO can be achieved by:

1. Adjustment of Antenna Parameters:
   For this a Remote Electrical Tilt (RET) antenna is required, adjustment is done by tilting the antenna remotely. Moving the antenna upwards increases the boundaries of a cell; whereas lowering decreases cell boundary. The tilt angle should be calculated carefully to avoid interference with neighbor cells in case of increase in cell boundary and coverage holes in case of decrease in boundary.
2. Adjustment of Power Level Parameters:
   It is possible to increase the transmitted power so that UE can receive the signal further away.

4. ENERGY SAVING

Cellular networks are designed for maximum throughput and excessive waste of energy occurs in low load conditions. eNBs situated in office/commercial area face heavy load conditions in office hours and low or almost zero traffic in night, whereas eNBs in residential area have more traffic after office hours. As we all know that, energy saving or efficient energy management is the need of the hour. The self organizing network (SON) supports selection of energy saving mechanism for consuming the lowest energy under low load conditions. In energy saving mode all the eNBs with very low traffic are switched to sleep mode; while the neighbouring eNBs cater the users if any of this node, by increasing their signal strength or range by changing their antenna parameters.

For optimization of power each eNB must have an energy saving mode with minimum power consumption allowing a restart of the eNB in less than 5 minutes triggered via the O&M or X2 interface.

Once loss of connection of the X2 or O&M interface is detected:

- eNB restarts without any further trigger.
- It gets information about the status of the neighbouring eNB.
- If additional capacity is needed, neighbour eNBs in energy saving mode are restarted via X2 interface immediately (less than 5 min)

Energy saving feature is also considered in other SON features such as load balancing, cell/service outage detection and mitigation. Further nodes in energy saving mode need to be considered for automatic Hand Over adjustment (via X2)

5. CONCLUSION

One of the critical areas in existing radio technologies that needs addressing is the development of self organizing capabilities in cellular networks. Traditional process employed for running the mobile network are manual and thus, costly, time consuming, use a lot of resources and manpower. SON can be utilized to reduce the number of manual test carried by network engineer for measuring service coverage and call stability. In future cellular networks, 2G, 3G, 4G will coexist, thus network optimization, maintenance and troubleshooting will be a key challenge for service providers. SON can simplify the HetNet configuration due to plug-and-play self-establishment of eNodeB network interfaces plus automated discovery of MME and neighbor eNodeB. Further network optimization of neighbour relations is a complicated task in modern radio networks due to large size with several hundred thousands of neighbour relations, for a single operator thus automation and self-optimization of neighbour relations is one of the most important features for upcoming mobile networks. Load balancing in SON LTE is often an autonomous function, and it is modeled as an optimization problem on matching between users and eNBs according to different metrics. One of LB function modules constantly monitors the load of the network. Unexpected loading conditions caused by sudden rise in data usage are managed by either adjusting network parameters or the surplus load is shifted from overloaded eNBs to the neighboring eNBs with free resources.

Self optimizing networks increase the efficiency and Quality of Service and at the same time reduce the operational cost. One of the critical factors that determine the overall cost for the operator is the energy expenses. Cuts on energy expenses could be realized if the capacity offered by the network matches the needed traffic demand at any point of time as close as possible. Thus the development of novel self-optimization or self-healing methods will make the planning and management of future generation wireless networks much easier and minimize the needed human involvement.

REFERENCES

3GPP. (n.d.). *Physical Layer Aspects for evolved Universal Terrestrial Radio Access (E-UTRAN), Technical Report TR 25.814*. Retrieved from http://www.3gpp.org

3GPP TS 32.500, "Self-organizing networks (SON); Concepts and requirements (Release 9)," V9.0.0, December 2009.

3GPP TR 36.902, "E-UTRA; Self-configuring and selfoptimizing network (SON) use cases and solutions (Release 9)," V9.3.1, March 2011.

3G Americas. (2011, July). *The benefit of SON in LTE: self-optimization and self-organizing networks.*

Atayero, & Luka. (n.d.). Adaptive Neuro- Fuzzy Inference System for Dynamic Load Balancing in 3GPP LTE. International Journal of Advanced Research in Artificial Intelligence, 1(1), 11–16.

Du, L. (2003, November). Towards Intelligent Geographic Load Balancing for Mobile Cellular Networks. *IEEE Trans. on Systems, Man, and Cybernetics, Part C: Applications and Reviews, 33*(4), 480–491. doi:10.1109/TSMCC.2003.818495

Lobinger, Stefanski, Jansen, & Balan. (2010). Load balancing in downlink LTE self-optimizing networks. In *Proc. of 2010 IEEE 71st Vehicular Technology Conference (VTC 2010-Spring)*. doi:10.1109/VETECS.2010.5493656

Sheng, Yang, Zhang, & Li. (2014). Zone-Based Load Balancing in LTE Self-Optimizing Networks: A Game-Theoretic Approach. *Vehicular Technology, IEEE Transactions on, 63*(6).

Siomina, P., Varbrand, P., & Yuan, D. (2006). Automated optimization of service coverage and base station antenna configuration in UMTS networks. *IEEE Wireless Communications*, *13*(6), 16–25. doi:10.1109/MWC.2006.275194

SOCRATES. (n.d.). *Self-optimisation and self-configuration in wireless networks*. European Research Project. Retrieved from http://www.fp7-socrates.eu

Viering, I., D¨ottling, M., & Lobinger, A. (2009). *A mathematical perspective of self-optimizing wireless networks*. IEEE International Conference on Communications 2009 (ICC), Dresden, Germany.

Warabino, T., Kaneko, S., Nanba, S., & Kishi, Y. (2012). Advanced load balancing in LTE/LTE-A cellular network. In *Personal Indoor and Mobile Radio Communications (PIMRC), 2012 IEEE 23rd International Symposium on*. doi:10.1109/PIMRC.2012.6362842

Zia, N., & Mitschele-Thiel, A. (2012). Self-organized neighborhood mobility load balancing for LTE networks. Wireless Days (WD), 2013 IFIP.

Chapter 4
Trust-Based Security Mechanisms for Self-Organized Networks (SONs)

S. Sivagurunathan
Gandhigram Rural Institute (Deemed University), India

K. Prathapchandran
Gandhigram Rural Institute (Deemed University), India

ABSTRACT

Self-Organized Networks (SONs) are an advancement of today's communication that arises to overcome the problems in traditional communications in terms of their planning, configuration, optimization, healing and management since the rate of utilizing communication technology is gradually increasing day by day hence an optimum mechanism is needed to cope up with all the changes in the communication era, the result is self-organized networks. The success of SONs depends on how it is effectively utilized without any compromise in its security. However its unique characteristics offer both opportunities and challenges. The aim of this chapter is to begins with the essential concept of SONs such as Mobile Ad Hoc Networks (MANET), Vehicular Ad Hoc Networks (VANET), Wireless Sensor Networks (WSN), Wireless Mesh Networks (WMN), Peer to Peer Networks (P2P), Cognitive Radio Networks, Bio-Inspired Networks and Internet of Things (IoT) and their limitations in different perspectives. As these networks have penetrated into the human life with an anytime anywhere capability, the security of the data being processed and communicated through these networks become vital. This chapter tries to bring out the issues and challenges in providing a trust based solution mechanisms for this type of networks.

1. INTRODUCTION

The world consisting of many different and connected parts and all are around us with complex process and system structures. Self-organization is an inherent feature that is being used by all living organisms and support to prevent the complex by the way it motivated by living world. The word self-organization was first proposed by the cyberneticist W.Ross Ashby in the year of 1940. Because of its inherent nature

DOI: 10.4018/978-1-5225-0239-5.ch004

and practice it is utilized by diverse fields such as physics, chemistry, biology, cybernetics, human society, psychology, education and networking system and is not exhaustive. The definition of self-organization is given by various scholars but everyone seemed to say clearly that evolution of a system is organized without the presence of a central process; the mechanism must equally constrain each other to perform as whole in order to attain a stable performance. Therefore self-organization is modeled mathematically, computationally or conceptually to help the interacting components and advances from a less organized state to a more organized state dynamically over some time, while exchanging energy, matter, and/or information with the environment (Prokopenko, 2009).The figure 1 shows the various interdisciplinary fields of self-organization.

In recent years mobile communication has enjoyed an unexpected rise in popularity. It is significantly increasing difficulty in terms of complexity of applications, heterogeneity of involved devices, radio access based technologies and dynamic nature etc., hence future mobile communication needs to have an adequate networking architecture that supports robustness, high degree of scalability, increasing performance, minimize the probability of failures and reliability in a cost effective manner without compromising quality of service and security. As a result, self-organized automation is introduced to develop Self-Organized Network (SON) in communication technology and as being a network themselves able to plan, optimize, operate, manage, monitor and self-heal by the way reduce administrative requirements for users and operators.

The SON is a cutting-edge advancement in communication technology and a promising solution for large scale networks to fight against increasing complexity, heterogeneity and dynamics in mobile communication systems. The motivation of self-organization for mobile communication technology comes from heterogeneity of devices, dynamics and scalable problem of networks, look forward to fast network communication services, autonomous reaction to problem, automation and distributed

Figure 1. Interdisciplinary concept of self-organization

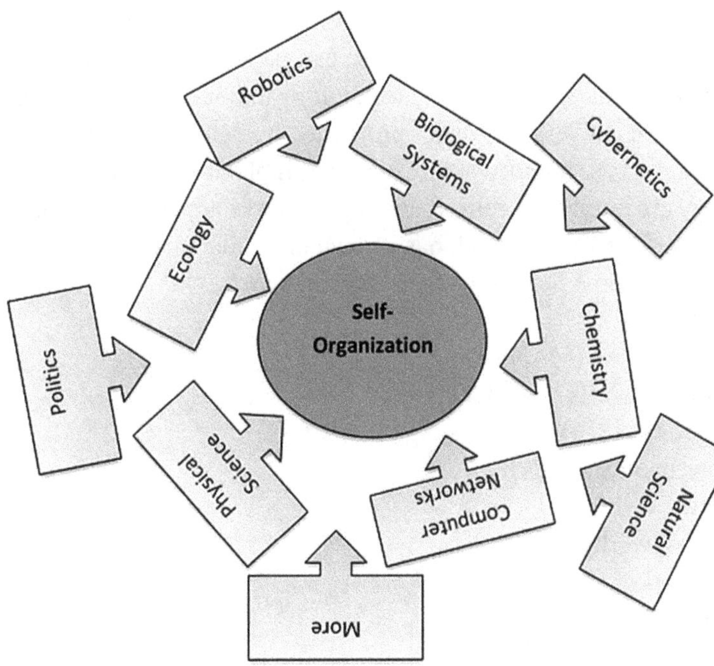

nature. Self-organization in mobile communication is defined by (liu,2009) as "*a network that can automatically extend, change their behavior, configure and optimize their topology, coverage, capacity, cell size, and channel allocation, based on changes in location, traffic pattern, interference, and the situation or environment*" as a consequence they offer plenty of technologies; mobile ad hoc networks, vehicular ad hoc networks, wireless sensor networks, peer to peer networks, cognitive radio, distributed information management, Internet of Things(IoT), Bio-inspired adaptive autonomy and any network with self-organizing capabilities and each of these technologies is offering plenty of applications that is used in various scenarios depending on the severity of problem domain for instances, disaster recovery, inter vehicle communication, firefighting, collaboration in business and education, military operation, managing hospitals, sensory networks etc.,

Contribution of this chapters is organized as follows; section 2 starts with basic introduction of various self-organized networks with their introduction and necessity, Section 3 talks about the limitations that are associated with SON because of its unique characteristics, Section 4 discusses the security related issues in SON, Section 5 discusses the traditional security schemes and their challenges, Section 6 discusses the common attacks that affect the SON, Section 7 talk about the impact of trust and its properties, section 8 converse the existing trust based mechanisms, section 9 discusses the future of SON with respect to trust based security and concludes the chapter.

2. VARIOUS SELF-ORGANIZED NETWORKS

According to the earlier cites, SON is an automation technology and has ability to establish networking whenever and wherever required so that it is also called as plug and play paradigm. Any SON based technologies consists of the three following main operations and also named as properties such as *self-configuration*, means planning and deployment carried out without human intervention hence reduction in capital expenditure, enable networking service in the absence of manual configuration and human administration; *self-optimization* means analysis of the performance and change operation of the network by the way maximum network performance can be achieved and *self-healing* means capable of recovering from faults if it happens, then any technology that has these three functions is called self-organized. In addition to the above, some major properties are self-maintenance, adapting, autonomous nature, decentralization, high scalability and resulting complexity. By making use of such properties that quoted above, there are number of self-organizing networks are evolved Figure 2 illustrate the various self-organized networks. Every network has its own or similar characteristic, architecture, working principle and applications. Some Self-Organized Networks are:

- Mobile Ad Hoc Networks (MANET)
- Vehicular Ad Hoc networks (VANET)
- Wireless Sensor Networks (WSN)
- Wireless Mesh Networks(WMN)
- Peer to Peer Networks(P2P)
- Cognitive Radio Networks
- Bio-Inspired Networks
- Internet of Things (IoT) etc.,

Figure 2. Self-Organized Networks

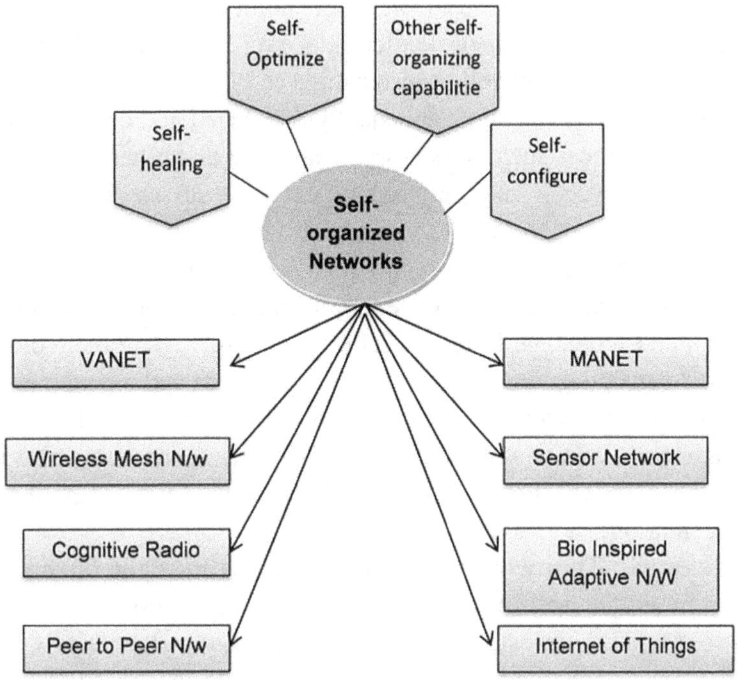

2.1. Mobile Ad Hoc Networks: Mobile Ad Hoc Network is also named as MANET, is a collection of nodes that are connecting together over a wireless medium without the support of any centralized devices. There is no dedicated router like in conventional network for packet forwarding instead each node acts as a router in its place. Nodes in the network are moving freely and randomly hence the topology of network is often changed. Due to node's movements, network partition occurs often therefore we cannot predict the reliability of nodes. At any time a node can join and leave the network and such type of network is used in military, disaster recovery etc.

2.2. Vehicular Ad Hoc Networks: A Vehicular Ad Hoc network is also named as VANET. Here instead of nodes the mobile cars are used to create a network obviously they are wireless. Hence every participating car will be acting as a router for exchange information between cars. Node movements happen in an organized manner instead of random manner like in MANET and nodes have interactions with the road site unit for better communication. This technology helps police and fire fighting vehicles to communicate with each other for safety purpose in an emergency situation, offering more enjoyable driving and travelling experience and offering traffic management, network gaming etc.,

2.3. Wireless Sensor Networks: A wireless sensor network is simply called as WSN where collection of spatially distributed sensor nodes connected with each other to monitor physical or environmental conditions and report to the user's base station. Sensor node comprises of several parts and each one performs unique task. The concept of sensor networks are motivated by military applications and now used in various industrial consumer applications and also in healthcare.

2.4. Wireless Mesh Networks: Wireless Mesh Network named as WMN where the communication network is formed by radio nodes that are organized in a mesh topology. Collection of mesh clients, routers and gateways form a WMN and working as a single network. The main advantage of WMN is providing redundancy means recover from single point of failure. This provides benefits in the fields of emergency situations, tunnels, battlefield surveillance and VoIP etc.

2.5. Peer to Peer Networks: This network is also named as P2P where the nodes are called peers. All the participating peers in the networks are equally privileged and equipped and each peer makes available their own resources such as memory, bandwidth and other resources to the entire peers in the network without the control of any centralized server. Each peer acts as both provider and buyer in the network in order to execute a particular task. It offers plenty of applications include file sharing network, intellectual property law and illegal sharing, multimedia and so on.

2.6. Cognitive Radio Network: It is cutting edge advancement in wireless technology where the radios are designed with an exceptional level of intellect and agility. Here the radios have the ability to monitor sense and detect the conditions of their environment and react upon. It consists of transmit-power; carry frequency and modulation strategy parameters and all these parameters will be changing depending on the incoming Radio Frequency (RF) stimuli. It offers more benefits when integrated with other devices. For instance, a mobile terminal with cognitive radio can sense the environmental condition.

2.7. Bio-Inspired Network: This network is typically derived from the biological concept and it is defined by class of approaches for efficient and scalable networking under indeterminate conditions. It consists of several intrinsic appealing characteristics and all these help to derive different algorithms and robust and resilient communication and information networking. There are so many biological principles and each one used in a specific application. For instance, artificial immune system, cellular signaling cascades and swarm intelligence etc.,

2.8. Internet of Things: It consists of variety of uniquely identifiable embedded computing devices that are interconnected with each other and exchange information without human intervention and participated in our daily lives. Moreover it covers variety of protocols, domains and applications such as infrastructure, energy and transportation monitoring, smart cities and agriculture etc.

3. LIMITATIONS IN SELF-ORGANIZED NETWORKS

The following section discusses the limitations of SONs and all these lead to various research avenues in SON (Robert et al., 2004; Mejri et al., 2014; Lopez et al., 2010; Wanga., 2012; Lidong. 1999).

- **Open Medium:** Typically the SONs have wireless architecture that can be accessible to any object. Mostly the architecture of SON is involved without administration and less human intervention.
- **Limited Resource Capabilities:** Compared with conventional network, SONs have limited resource capabilities in terms of memory, bandwidth, computation etc., because the devices involved in SONs are compact.
- **Limited Battery Power:** Wireless devices are having limited battery power only hence SONs also have due to its architecture.
- **Shared Wireless Medium:** Every device involved in SONs has able to share information without any centralized control there may be fixed architecture to control the network but most of the SONs have shared nature.

- **Lack of Human Administration:** SONs are also capable of not as much of human intervention because of their deployment nature. Initial deployment may require human intervention but later on Sons can work on their own.
- **Lack of Central Control:** Central administration is missing in SONs compared with conventional network. Here each involving device is going to act as its own administrator.
- **Dynamic Topology:** Devices involved in SONs have the ability to move so it always leads to dynamic topology. Device movement may occur purposely or may depend on the deployment environment.
- **Limited Physical Security:** conventional network has high security mechanisms because it can be easily controlled by any device. Whereas in SON, security is always in question because of its open and shared medium, lack of central control.
- **No Infrastructure:** SON has ad hoc architecture so that it can be easily deployed and destroyed whenever and wherever required because of its nature utilized by the applications.

4. SECURITY ISSUES IN SELF-ORGANIZED NETWORKS

Though SON is greatly in demand, there are still several challenges in implementing and needs research for improvements. Some challenges are configuration, discovery, routing, quality of service (QOS), scalability, reliability, conflict between parameters and goals, data measurement and processing, developing algorithms and security. Typically the basic security issues of SON comes from distributed and sprinkled network architecture, movement of nodes that lead to dynamic topology, shared wireless medium, frequent distribution of up-to-date locations and no proper pre-determined defense mechanisms despite the above discussed limitations. These properties make the SON become more vulnerable and network becomes affected by security threats in the form of attacks.

The security of SON has received tremendous attention because of the self-configuration and self-optimization and self-healing capabilities. Generally devices in the environment are operating on the basic assumption that all the devices are fully cooperating in performing communication related operations. Typically these type of environments are prone to the security threats of conventional networks, moreover they are also subject to some additional threats; these additional threats result from the self-organizing devices' fundamental characteristics, mainly the computation power, memory and battery capacity, and deployment environment and stability and predictability of nodes (Robert et al., 2004; Fei Wanga., 2012). In addition to, a fundamental vulnerability of this environment sometimes comes from open peer to peer architecture. Mobile devices in the environment are fully or partially in need of other devices so as to complete the communication successfully. It does not rely on any dedicated routers for forwarding the packets, in its place each device acts as a router and also there may be a base station for forwarding.

On the other hand, wireless channel is accessible to both legitimate network users and malicious attackers. The devices roaming in hostile environment with relatively poor physical protection have non negligible probability of being compromised. Therefore, we should not restrict ourselves to malicious attacks from outside the network, but also take into account the attacks launched from within the network by compromised devices. Trust relationship among devices also changes; because a new device may join or leave and that device may be compromised because of the frequent happenings of network partition and link failure. As a result, there is no clear line of defense in SONs from the security perspective.

4.1 Common Security Requirements

To secure a SON, the following common requirements are to be considered: Availability, Confidentiality, integrity, Authenticity and Non repudiation (Zhou et al., 1999; Stallings, 2012). *Availability* ensures the survivability of network services despite Denial of Services (DoS) attacks. *Confidentiality* ensures that certain information is never disclosed to unauthorized entities. *Integrity* guarantees that a message being transferred is never corrupted. *Authentication* enables a node to ensure the identity to the peer node with which it is communicating. *Non repudiation* ensures that the origin of a message cannot deny having sent the message and the receiver from denying the reception of the messages. Figure 3 shows the common security requirements for SON. Among the security requirements, authentication is important because communicating entities must be assured with the identity of each other and these identities should be authorized and recognized before the communication begins. Once an authentication is achieved, remaining security requirements can be achieved easily.

Authentication is a significant aspect in the security of SON that enables a device to ensure the identity to the receiver. As mentioned earlier once authentication is achieved, rest of the security requirements could easily be achieved. To ensure the authentication each device must trust other devices. Hence trust is a gateway and decision for mobile devices to authenticate each other. In order to complete the mission successfully each device should cooperate with each other. We can just say each device has trust on every other device in the same network. The SON acts as a decentralized distributed network; hence achieving cooperativeness among the devices is a complicated task. The reason is, each device in the network must trust every other device without any prior recommendations and interactions. But this blindness in communication will make it more vulnerable and affect the network's performance greatly. Hence cooperating devices must trust each other in order to achieve the desired security level. Moreover compared with traditional wired network, collecting trust evidences to evaluate a trustworthiness of a particular device in SON is crucial problem due to its dynamic nature.

Figure 3. Security Requirements

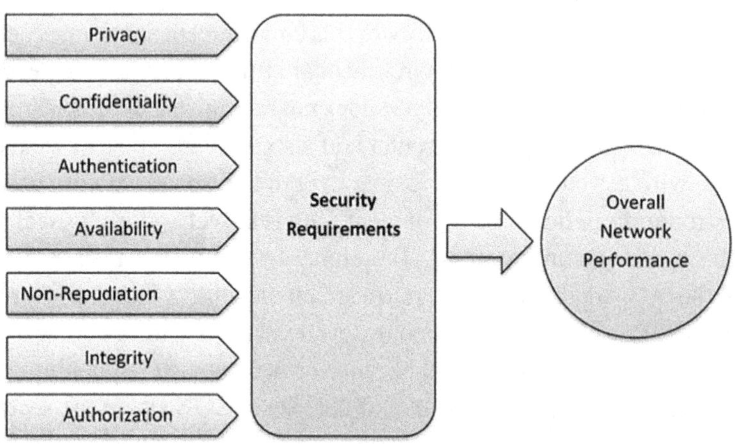

5. TRADITIONAL SECURITY SCHEMES AND ITS FLAW WITH RESPECT TO SON

In this section we discuss the basic security requirements and solutions that are widely used by the researchers. Traditional safety mechanisms such as secure multicasting methods, secure routing methods, privacy aware routing methods, intrusion detection methods, secure key distribution and for providing security requirements such as confidentiality, integrity, non-repudiation authentication, and availability are not efficient in SON where nodes have limited communication bandwidth and battery power. To achieve authentication, shared secret, Public Key Infrastructure [PKI], digital signature, digital certificate(Sivagurunthan et al., 2014) are used but these techniques are centralized, pre-determined depend on trusted third party thereby increasing computation power, memory and consumption of communication bandwidth and battery power. Moreover it requires a third party, network overhead dramatically increases and it leads to performance degradation in overall network's throughput, availability and robustness. The following Table 1 shows the solutions for the security requirements that we mentioned above.

6. ATTACKS IN SELF-ORGANIZED NETWORKS

In SON, due to the lack of central administration, open and shared wireless medium, it is vulnerable to various kind of attacks. Attack is nothing but actions that harm the network by obtaining information that is being transmitted, modification or deletion of transmitted information and disclosure of routing and other information. In general two types of attacks are possible in SON such as internal attack and external Attack. Internal Attacks are very difficult to predict and detect because it may be launched by any compromised or malicious node from inside of the network whereas external attack generates from outside of the network. It causes additional overhead on packets and tries to prevent the normal communication among the nodes and make problem. It is further classified into two categories such as passive attack and active attack. Passive attack is done to obtain information from the system that is being transferred and does not affect the system resources any way and active attack aims is to alter the system resources and affect its performance. It makes some modification of data streams and creation of false dreams by the way it affects the network. To meet the security requirements that we mentioned

Table 1. Traditional Solutions for the Security Requirements

SECURITY REQUIREMENTS	TECHNIQUES USED
Authorization	Password, Access Control List (ACL), Firewall
Authentication	Shared Secret, Public Key Infrastructure (PKI), Digital Signatures, Digital Certificate
Privacy	(Partially by) Encryption, Mechanisms to hide locations and routing protocols used
Data Confidentiality	Encryption
Availability	Intrusion Detection Systems/Intrusion Prevention Systems (IDS/IPS)
Data Integrity	Hash functions, Digital Certificate
Non-repudiation	Digital Signatures, System logs

above, we first analyze features of common attacks on self-organized networks. The following section describes specific attacks in SON (the list is representative, not exhaustive):

- **Flooding**: An attacker broadcasts lot of route request packets for a particular node ID who is not actually involved in the network in order to discover the route and sending lot of data packets by the way it consumes bandwidth (Chouhan et al., 2011).
- **Wormhole Attack:** This attack is accomplished by one or more node in the network. They disrupt the routing by short circuiting the normal flow of routing packets by creating own private network and forward through it. These types of attacks are easy to deploy but very hard to detect (Rutvij et al., 2010).
- **Black Hole Attack:** An attacker advertises itself as having a shortest path to the destination and get attention from other nodes in the network hence fake routes is created. It leads to packet drops and degrades network performance (Rai et al., 2010).
- **Replay Attack:** In this attack transmission of data is maliciously or fraudulently delayed or repeated which is originated by an adversary who intercepts the data and retransmits it. This attack is carried out by record and resending the old valid control message to all other nodes and force them to update, their routing table with stale route (Rai et al., 2010).
- **Byzantine Attack:** This attack is a more dangerous one, because it would seem the network to be operating normally in view of other nodes in the network but it is showing byzantine behavior. Here the compromised intermediate node or set of nodes work in collusion and carries out attacks such as creating routing loops, forwarding packets to invalid paths and dropping packets so it leads to lack of routing services (Rai et al.,2010).
- **Location Disclosure Attack:** This attack discloses the entire structure of the network or the physical location of the particular node or location of nodes that are adjacent to target nodes (Redwan et al., 2008).
- **Sinkhole Attack:** The attacking node offers extremely attractive links for the mobile nodes in the network. Hence lot of traffic bypasses this node resulting is congestion among (Saini et al., 2010).
- **Link Spoofing Attack:** Here a malicious node advertises itself that has a direct link to the far away node in its HELLO messages and intercepts the control services (Kannhavong et al., 2008).
- **Jamming:** An adversary interferes with the communication between nodes generally it's defined as DoS attack. This ultimate aim is to prevent the legitimate sender or receiver from the network (Sari et al., 2012).
- **Sybil Attack:** A single node impersonates several other nodes using various malicious means. In distributed environment there is no central control entity to ensure the guarantee and identity. Therefore it is possible for an unfamiliar entity to present more than one identity (Saha et al., 2010).
- **Denial of Service Attack:** An attacker can disrupt an entire network by disabling the network or by overloading it with false messages and also prevents or inhibits the normal use or management of communication facilities. This results in degrades the network performance (Aad et al., 2004).
- **Man-in-the-Middle Attack:** A malicious node simply occupies a place between a sender and a receiver and acts as a part of network and forwards false routing information to the destination as well as source (Femg et al., 2006).
- **Gray Hole Attack:** At the time of route discovery an adversary may behave like a genuine node. Thereafter it mutely drops some or all of the packets which are transmitted for forward. It is quite

harder to detect because packets may drop not only because of malicious nature but also overload, create congestion and may be of selfish nature (Banerjee et al., 2008).

- **On-Off Attack:** In this category of attack, a malicious node behaves well or badly alternatively depends on the environmental conditional change (Feng et al., 2013).
- **Bad Mouthing Attack:** This attack originate from recommendations, sometime there is possible for receiving dishonest recommendations hence framing good parties or boosting trust value of malicious peers (Feng et al., 2013).
- **Garnished Attack:** In this attack, a malicious node behaves well and badly alternatively like On-Off attack and aim is to remaining undetected while causing damage. In addition, these attacks suddenly conduct attacks as they collect higher trustworthiness (Li et al., 2013).
- **Confidentiality Attack:** Aim is to attempt to access the information that is stored in the network further it attacks on the confidentiality of communication and node information (Bankovic et al., 2011).
- **Identity Attack:** This attack occurs due to lack of source address verification and aim is to pretend to be other nodes. Clone, thief, mole and Sybil these are the examples of this attack (Bankovic et al., 2011).
- **Whitewashing Attack:** In this attack a malicious node purposefully joins and leave the network by the way enter with a new identity and aim is to create bad reputation record and accumulated its previous identity record (Meng et al., 2012).
- **Collusion Attack:** A malicious node can libel the normal node or providing false information during collaboration by the way negatively affecting the evaluations on that node (Meng et al., 2012).

7. IMPACT OF TRUST IN SON

To overcome the above problems that occur in the traditional security mechanisms trust comes into existence. The concept of trust management is introduced and identified as a separate component by (blaze et al., 1996). Trust is used in several fields the interdisciplinary concept of trust is shown in figure 4. Trust is a word which is originally derived from the social sciences. Trust is defined as "one entity (trustor) is willing to depend on another entity (trustee) (Mayer et al., 1995)" or "the trustor abandons control over the actions performed by the trustee (Bamberger et al., 2010)". Simply, trust is reliance on an object or an entity. According to the network and communication field trust is defined as "a set of relations among entities that participate in the protocol, which is based on the previous interaction of entities within the protocol" (Baras et al., 2005).

Cooperation among the nodes is fundamental for the operation of SON. In order to complete the mission successfully each node should cooperate with each other. We can just say each node has trust on other nodes for successful implementations. SON is a decentralized distributed network so that achieving cooperation among the nodes is a complicated task. The reason being, each node in the network must trust every other node without prior recommendations and interactions. But this blindness in communication will make it vulnerable in terms of security and affect the network performance greatly. Hence cooperating nodes must trust each other in order to achieve the desired security level. Compared with traditional wired network, collecting trust evidence to evaluate a trustworthiness of a particular node in SON is crucial problem due to its dynamic nature. In addition to that, traditional cryptographic methods are providing key based security where keys are pre-determined and also depend on any third party hence

Figure 4. Trust in Multidisciplinary

computational and network overhead dramatically increases and it leads to performance degradation in the network's throughput, availability and robustness (Gong et al., 2009). So in order to overcome such problems trust comes into existence.

7.1 Properties of Trust

Trust evaluation is a technical mechanism of expressive trust relationship for digital processing, in which the properties influencing trust will be evaluated (Yan et al., 2014). The following section describes the trust properties. The Figure 5 shows the trust properties.

Trust is *dynamic*, the trust evidences or information needed for trust establishment may change due to its topology changes hence result may be incorrect or inaccurate because a node can join and leave at any time (Eschenauer et al., 2002). Trust is *subjective* that nodes in the MANET may not have same level of trust with respect to other nodes because each node has different level of experience (Abdul-Rahman et al., 1997). The following figure shows the properties of trust.

Trust is *incomplete and transitive* that if *A* trusts *B*, *B* trusts *C* at the same time *A* does not required to trust *C*. In this case transitivity between *A* and *C* is based on two things such as *A* trusts in *C* or *A* trusts in the *C*'s recommendation of the third parties (Sun et al., 2006). Trust is *asymmetric* that each node has different resource constrain level so that some higher capacity node may not trust lower capacity and vice versa (Adams et al., 2005) and finally trust is *context-dependent* for instance if a given task requires high computational power, a node with high computational power is regarded as trusted at the same time a node that has low computational power is not malicious distrusted (Staab, 2004)

Figure 5. Trust Properties

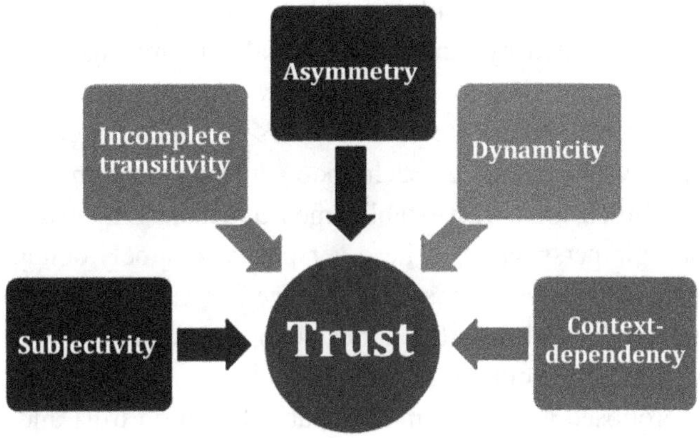

8. TRUST BASED SECURITY MECHANISMS IN SON

In this section we present a number of trust based mechanisms in the SON and highlight the relative merits and demerits of these mechanisms.

Decentralized architecture with limited resource constrained devices are involved in MANET. The basic vulnerability of MANET comes from open peer to peer, self-configuration architecture. Basically nodes are communicating with relatively poor physical protection. In addition to, multi-hop nature also make the MANET's security critical. Traditional security mechanisms are not well suited for MANET because of its distinct nature. The following section describes is described the trust based security mechanisms in MANET.

Lacharite et al. (2008) proposed a security trust monitoring layer to detect wormhole attack by examining test bed on OLSR routing protocol. The core concept of this model is Security Trust Monitoring (STM) where traffic analyzer and inter communication of distributed STMs. Each model performs its unique operations. Traffic analyzer is used to calculate local trust values by observing neighbors behavior. The calculated local trusts are shared among the neighbors using trust sharing mechanism, which is carried out in the inter-communication of distributed STMs module and that sharing process is accomplished by intercommunication process. Trust election and validation module is used to validate the trust values and after the validation of trust those messages are called global trust. Both the local and global trust values are stored and handled by the trust table module. Final module is called trust to trusted state, this will occur when any node under examination that moves malicious to suspicious or misbehaving to trusted.

Gong et al. (2009) proposed a trust vector model to identify the malicious nodes and also addresses the trust dynamic issue. The trust vector model is calculated based on evaluation of experience, knowledge and recommendations. To normalize the trust vector, as single value trust policy vector is used which also has same parameters like trust vector. Typically the sum of policy vector is equal to one. The experience is calculated by number of packets any node had actually forwarded divided by number of all packets any node responsible for forwarding. The knowledge is calculated by multiplication of packet loss probability from node A to node B and packet loss probability from node B to node A. Finally recommendation is calculated by node as evaluation to node B by collecting recommendations about node B from other nodes. Then trust dynamics is addressed by finding the time gap between the trust value of node A to node B at time t1 and decayed value of node A to node B at time t2.

Li et al. (2012) proposed security architecture for MANET through collaboration and trust mechanisms. In which the misbehavior nodes are identified by the Gossip based Outlier Detection algorithm and trustworthy nodes are evaluated by multi-dimensional trust management. This outlier algorithm consists of four steps such as local view formation, local view exchange, local view update and global view formation. Moreover local view update phase follows two update methods namely simple averaging method and trust based-weight method. Each node generates its own local view by observing the neighbors behavior. Multi-dimensional trust establishment and management are used to evaluate the node trustworthiness from multiple perspectives. These perspectives namely collaboration trust, behavioral trust and reference trust. Collaborative trust is used to judge how the node is participating in the network activities. Behavioral trust is derived from the nodes abnormal behaviors and finally reference trust is derived from other nodes opinion about the target node.

Bhalaji et al. (2012) proposed a security model that is based on trust and association status of all neighboring nodes to prevent the Grey Hole attack on dynamic source routing. Here the optimal is selected based on the nature of association between the nodes not by the first arrival of route reply packets. In this model, nodes are categorized into unknown, known and companion nodes and these are called association values which is based on their trust and appropriate threshold trust value. Unknown nodes are untrusted nodes with minimal trust values. Second companion nodes are fully trusted nodes with maximum trust values. Finally, unknown nodes are intermediate between known and companion nodes. In order to store the association values, association table is used. By using this table we can identify the malicious nodes. Moreover the values of association tables may change due to node forwarding and receiving activities. Trust value is calculated by adding the ration between the number of packets actually forwarded and number of packets to be forwarded, ratio of number of packets received from a node but originated from others to total number of packets received from it and acknowledgement bit.

Xia et al. (2012) proposed a dynamic trust prediction model to evaluate the trustworthiness by nodes historical behavior as well as nodes future behavior through extended fuzzy logic rules prediction. This model integrated into the source routing mechanism and form Trust based Source Routing protocol (TSR) to choose a shortest route with security. The trust prediction model takes nodes historical trust, nodes current trust and route trust into account for evaluating the final trust. In TSR, route discovery is accomplished with the help of FLOW-REQ message and route maintenance is carried out when route update, route handoff, node mobility and route error occurs. All these activities are based on the trust prediction mechanism.

Adnane et al. (2013) possess a trust based security for OLSR routing protocol constructed on trust specification language. Also offered trust based reasoning to allow each node to evaluate the behavior of other nodes. Typically, this model consists of three phases, neighborhood selection, Multi-Point Relays (MPR) selection and routing table calculation. MPR are selected nodes that forward broadcast messages in the network. Trust establishment is carried out in the neighborhood discovery phase itself using the trust specification language along with the HELLO message. The second phase is MPR selection that nodes with minimum number of neighbors enabling it to reach all the two hop neighbors those nodes are called MPR nodes. This phase also follows the trust specification language to establish trust with the help of TC message. In the routing table calculation, anode calculates the shortest path using Dijkstra shortest path algorithm also with the help of trust specification language. Trust based reasoning is used to validate the neighborhood discovery by whether the neighborhood nodes encountered the trust specification language and also is used to validate the MRP selection by checking the TC message and data packets generation. Wang et al. (2014) proposed a light weight trust based QOS routing to improve the network performance and prevent the attack from malicious nodes. This model compromises of three components first, trust computation based on direct trust that is successful packet forwarding and receiving and secondly called trust judgment that is indirect trust means recommendation from neighbors. Finally trust update done at regular interval of time. Moreover this trust method is integrated with routing protocol.

Typically VANET architecture comprises of the following three infrastructures for efficient communication such as infrastructure to vehicle, vehicle to infrastructure and vehicle to vehicle communications. VANET is offering plenty of applications and these applications are in need of exchange messages between vehicles in order to attain a smooth and safe driving. But the content of messages may be changed due to topological or malicious vehicles moreover the content of the message can have an impact on drivers behaviors. So we can conclude that securities in VANET are in question. Though traditional security mechanism are available to ensure security in VANET but it is only suitable for infrastructure based communication in VANET such as Infrastructure to vehicle or vehicle to infrastructure but in case of vehicle to vehicle communication traditional mechanisms are not well suited because of limitations in resources and the vehicle cannot authenticate directly via infrastructure (Chuang et al., 2013). In highly dynamic traffic, authentication cannot cope up efficiently.

Chuang et al. (2013) proposed a trust extended authentication mechanism called TEAM in order to satisfy the security requirements and to withstand against various attacks. Mainly they focus trust based security for vehicle to vehicle communication and utilize the transitive property trust relationship to improve the authentication procedure. They classify trust with three roles in VANET such as a law executer, a mistrustful vehicle and a trustful vehicle and the authentication is achieved based on the transitivity among the three roles. Chen et al. (2010) proposed a trust opinion aggregation scheme to evaluate the quality of information shared among nodes in the environment which is the extension of their previous work called identity-based signature algorithm. There are two types of identities that sign the messages one is sender who is the message originator another one is evaluator who gives trust opinions then trust opinion aggregation is executed again it follows two cases. The first case is a third party combines the original signed message with n signed trust opinion from n distinct nodes. In the second case, a third party combines two aggregate on the same message into a large aggregate. Finally verification of signature is done with three cases such as a signed message with or without trust opinion and merged aggregate from two aggregate.

Huang et al. (2010) proposed situation aware trust architecture which compromises and addresses the following issues such as security policy management is used to address the trust situations and number of policy on roads. Proactive trust management is used to manage the dynamic and inter vehicle trust and finally, social network trust is based on cryptography and it is used to promote application development and adoption for VANET. As we focus only on trust, we consider only the second phase so here trust evaluation is invoked in proactive manner and each vehicle can be able to evaluate to trust of other vehicles by distributing its moving trajectory to others. Mar'mol et al. (2012) proposed a trust and reputation infrastructure based method called TRIP. Their ultimate aim is to quickly and accurately distinguish malicious or selfish nodes spreading false or bogus messages throughout the network. Trust is calculated when each vehicle receives any messages or warning from other vehicles. Three factors are considered when evaluating trust such as direct reputation score is given by vehicle i to vehicle j at a given interval of time, recommendation is given by vehicle k to vehicle j and reliability of recommendation is given by road side unit for vehicle j. Based on the overall trust, three levels of trust is established for vehicles high, medium and low. In performance metrics, they are considered scalability and accuracy factors.

Zhang et al. (2010) proposed a scalable and efficient trust based framework for message propagations and evaluation when nodes share information regarding road condition, safety and providing opinion about others. This framework consists of several components such as message evaluation and message propagation. The message evaluation consists of two modules; analysis and action. Analysis module is used to generate trust opinions and action module is used to make local decision. Message propagation module is consisting of cluster cooperation and relay control. Cluster cooperation serves as the foundation for message propagation and trust opinion aggregation and relay control model works as a filter that controls the relay of messages. Peer to peer module is used to manage the trustworthiness of nodes based on role based and experience based trust. In performance they consider scalability and effectiveness.

Due to the inherent nature such as low cost rapid deployment, self-organized and spatially autonomous, sensor nodes play a vital role in assessing environmental condition. At the same time limited communication range and multi-hop transmission of sensors leads unreliable communication in such network. As pointed out earlier, traditional mechanisms always consume bandwidth and other resource again leads to vulnerability.

Feng et al. (2013) proposed a trust management scheme based on D-S Evidence theory where trust is calculated based on both direct and recommendation trust and D-S Evidence theory is used to make right decision when many uncertain information occur. In addition, direct trust is calculated both directly monitoring the behavior of objects and historical records of that object. Finally both direct and recommendation values passed over the flexible synthesis method to attain an overall trust by the way reduction in energy consumption when calculating recommendation trust. This method withstands against On-Off and Bad mouthing attack. Zhang et al. (2010) proposed a novel hierarchical trust management scheme to minimize the communication and storage overheads. Here trust evaluation is based on trust, indirect trust and energy associated with particular node. A trust varying function is used to consider the dynamic aspect of trust that gives the most recent value for trust calculation. The performance of the proposed work compared with group based trust management scheme in terms of computation, memory and communication overhead.

Reddy (2012) proposed an agent-based trust approach in detecting malicious nodes. In this model, node's trust is evaluated based on rating that means the ratio of packet forwarded to packet received. In addition to, the rating is also evaluated by voting mechanism means each node votes the successive node depends on the packet forwarded by and received by that node. The updating of trust is achieved by Sporas or Molina's formula. No specific attack is considered in this model. Bao et al. (2012) proposed a Hierarchical trust management protocol to deal with selfish and malicious nodes. Instead of single attribute, multidimensional trust attribute is used which is derived from communication and social networks. In addition to, stochastic Petri net technique is used to analyze the performance of the network. Overall trust is calculated either by direct or indirect trusts by using four trust components such as intimacy, honesty, energy and unselfishness. Finally this model is applied to trust based geographical routing as an application.

Duan et al. (2014) proposed a trust aware secure routing framework called TSRF to fight against various attacks. The routing framework combines both trust metrics and QOS metrics to attain desired security level. Trust evaluation involved with both direct and indirect trust and in addition to secure trust based path selection is also involved in this model. This model works well against tampering attack, On-Off attack and bad-mouth attacks. Anita et al. (2014) proposed a two way acknowledgement based trust framework to minimize the communication overhead and memory. Addition to it, minimizing the number of recommendations gathered from neighbors for trust established by the way communication overhead can reduce and avoid the promiscuous mode of operations by putting sensor node in sleep mode. Direct trust is calculated based on the observation of number of successful and failed transactions. The performance of this model is analyzed by various performance metrics such as control overhead, packet loss and network lifetime.

Li et al. (2013) proposed a lightweight and dependable trust for clustered wireless sensor networks. The aim is to reduce communication overhead and eliminate the bad-mouthing attack, reducing networking consumption and overcoming the limitations of traditional weighting methods. Here trust is calculated in two levels one is inter-cluster another one is intra-cluster. Both the levels are involved in direct trust calculation which is based on the past interactions and feedback values. In performance consideration, this model takes into account, the communication overhead, average storage overhead and successful packet delivery radio.

WMN is used in day today life typically in the form of personal digital assistance, mobile phones and laptops etc. These devices are generally having limited computing power, battery, bandwidth and mobility so there is a possibility of attack; additionally lack of central control makes the network highly vulnerable. As we come to know from the literature, trust is established in WMN based on pre-configuration of nodes with secret key and also in some occasions central trust authority is used. Apart from this, the following section describes some trust based mechanisms in WMN:

Xiao et al. (2014) proposed a an access authentication protocol for trusted handoff based on multi-technologies such as hierarchical network model, Elliptic curve cryptography, trust evaluation and gray relevance analysis. In this model, for handoff the mobile platform's configuration must be measured before joining the network and legitimate user only takes part in the network. To ensure this, trusted handoff protocol is presented. They have also presented trust based authentication protocol to ensure security when handoff from one network zone to another occurs. Then trust model is used to evaluate the starting and runtime system states where the trust is established by integrity measurements such as hardware, operating system, shared libraries, configuration documentations etc., and these trusts are validated by trusted computing group only in the starting state. After the trust evaluation, there are three types of trust is classified such as extremely trusted, critically trusted and untrusted. In performance, this model takes success rate, hand off delay into accounts.

Bankovic et al. (2011) proposed a security framework to isolate ill-behaved nodes by repudiation systems rating mechanism and one more technique unsupervised learning algorithm is called self-organizing map is used to detect deviation from normal behaviors and moreover this model detects and confines new, undetected attacks in the network. Initially the trust value of newly joined node is set to zero. Overtime it is recalculated based on previous trust value and neighbors measurements of node. In performance, they are considered as redundancy problem and routing behavior attacks. Saxena et al. (2011) proposed a hierarchical architecture for detecting selfish behavior and enforce co-operation among the mesh routers by using reputation management typically reputation is derived from trust. In addition, it monitors the behaviors of mesh routers by periodically sending reports to mesh gateway it consists of sink agents. Trust is calculated based on actual forward count of the mesh router and expected forward count of the mesh router. In performance, they evaluated the packet delivery ratio, false positive rate and hop length of the reporting packets.

Peer to peer networking offers numerous services such as file sharing, content retrieval and communication with social network and more. With the rapid development of this network portable devices have inspired extending the characteristics and futures of P2P. But in terms of security, there is always a space because of devices mostly portable hence limited processing capabilities so it leads to vulnerabilities.

Anand et al. (2013) proposed a unified trust management strategy for content sharing that means sharing of digital data, computer programs, multimedia files and electronic documents and so on. The contribution of this model follows peer availability, controlled scalability, content availability, heterogeneous environment and polluted and poisoned content prevention. In this, trust evaluation is associated with content availability where trust is calculated based on the contribution of the peer in the network by the way non cooperating peers are eliminated from the network. After the evaluation of trust, each peer is classified into high, medium and new comer or anonymous modes. In performance, they have compared the proposed model with BitTorrent network. Qureshi et al. (2012) proposed a reputation and trust management scheme named as M-Trust to reduce the computation complexity and to combat the collusion attacks to minimize the impact of false rating from malicious peers. In addition, distributed algorithm is used for reliable trust rating aggregation and management. The reputation of each peer is measured by personal opinion from other peers' interaction with that peer in certain period also considers recommendation from other peers. Accuracy, reliability, convergence speed, detection rate under various constrains mobility are some of the performance metrics they have considered.

Meng et al. (2012) proposed a trust model based on feedback arbitration to withstand against malicious peers. This model follows a punishment mechanism to improve the trust decision and resist the repeated attack of the malicious peers. The trust evaluation is carried out by service, feedback and arbitration credibility and after that trust decision is made. When performance, evaluation of credibility, success rate, validation of trust decision making, punishment mechanism and resisting attack such as whitewashing, collusion attacks are considered. Tian et al. (2014) trust based incentive mechanism called super-trust to motivate cooperation among the peers in p2p networks by the way fake, misleading and unfair referrals can be eliminated from the network and the nodes are grouped based on the interest of participating peers. This model represents three types of trust relationships such as trust relationship between two regular peers, trust relationship between two super peers and trust relationship between regular and super peers where super peer is a node and each group has only one super peer. In the first, trust is evaluated based on direct trust calculated by number of satisfied transactions and number of un-satisfied transactions, referral from others and the credibility of other peers. In second relationship, aggregation of all direct trust is used. Final one is the combination of two that we have discussed earlier. In performance they have considered so many factors such as overhead of topology, searching performance, collusion and denigrating.

Kang et al. (2014) proposed a trust based model to prevent pollution attack. They introduced dynamic confidence factor to dynamically adjust the weight of the direct and indirect trust values and they calculated direct trust based on the unique features of pollution attack. Here direct trust is calculated based on previous direct transactions such as number of received clean chunks and number of received polluted chunks. Indirect trust is calculated based on third parties peers' recommendations such as credibility of the third peer and recommendation value of the trustee. This model withstands against bad-mouthing, persistent, on-off and collaborative attacks. In performance, they have compared this model with existing trust model.

As pointed out earlier IoT is a novel paradigm in wireless technology and defined by variety of devices or objects around us and communicate, cooperate with each other to attain a common goal. The objects represent mobile phones, laptops, sensors etc., every wireless technology has a weakness in security perspective hence IoT is not an exception. The vulnerability of IoT comes from open wireless medium, easily affected by physical attack, because most of the time components of IoT are unattended, just like other self-organizing networks it also has limited resource constrains in terms of bandwidth, energy and computational resources. The following section describes the trust based mechanisms available in this domain.

Saied et al. (2013) proposed Trust Management System called TMS to manage cooperation in a heterogeneous IoT architecture, nodes with different resource capabilities is calculated by one another based on operation of a set of collaborative services such as quality of recommendation, the weighting factor, current context and resources capabilities. After the decision is made based on all received reports (positive) and past interactions, and eliminates the attacks. This model shows its performance by preventing bad-mouth, ballot stuffing attacks and selective behavior attacks.

Cognitive radio network is with a sensor, a radio, a knowledge database, a learning engine and a reasoning engine and all these elements integrated with infrastructure mode or an ad hoc mode depending on the working nature. When it goes with ad hoc nature vulnerability of attacks are high due to its open policies, programming interface of cognitive radio and its configuration. In the following we discuss some trust based mechanisms that resist vulnerabilities.

Parvin et al. (2013) proposed a conjoint trust assessment for secure communication to ensure the secure spectrum sharing and withstand against attacks such as selfish, malicious and faultless nodes. Here trust establishment is done in two ways such as trust value of the spectrum requesting secondary user in secondary user network and trust value of the spectrum requesting secondary user in primary user network then the overall trust is calculated by combining these trusts. Based on this decision is made whether the secondary user access the spectrum or not. By the way malicious users are separated from the network. Trust value is calculated by secondary user's base station by accumulated trust value which is given by every node in the secondary user's network. Likewise primary user base station will also calculate the trust value. Finally conjoint trust is calculated by combining these two values. In performance, they are compared with the trustworthiness of primary and secondary networks with different nodes.

Pei et al. (2013) proposed a sensing and etiquette reputation-based trust management to enhance the robustness and fairness by the malicious behaviors can be reduced effectively. Robustness can be achieved by both trust management mechanism and trust based inner user evaluation mechanisms. Fairness can be achieved by trust based resource allocation mechanism. Three important factors are considered while evaluating reputation such as context, sensing evidence, scope and its importance and time window for collecting sensing reputation evidence. In performance they leave considered, sensing and etiquette values. Sazia et al. (2012) proposed a trust based mechanism to authenticate the secondary users and measures the throughput in CRN. Here trust is calculated from both previous interactions and recommendations and risk is also calculated based on data fusion method. Depending on the trustworthiness of both primary and secondary user trust, throughput is calculated based on this trust is classified as medium, second medium and maximum trusts.

- **Inferences:** Trusts based mechanisms are utilized by various self-organized technologies because of these inherent features. To make decision regarding other devices before any communication comments or over some period of time, trust is used as a key factor. Hence the evaluation of trust in SONs is based on the following techniques such are feedback from others, services done by devices over a period of time, reputation of the devices in the network, direct and indirect observations, experience gathered in the network, identity of devices, social based trust, policy based mechanisms, opinion from their neighborhood and weighting factors such as energy, bandwidth, lifetime etc., and all these mechanisms are solely implemented or integrated with the existing routing mechanisms to withstand against various attacks and also improve the overall performance of such networks.

9. FUTURE DIRECTIONS AND CONCLUSION

Self-Organized Networks are likely enablers of networking anywhere anytime concept without much support from any centralized control and almost less human intervention, which is the current trend of information sharing mechanisms. While these networks are rapidly deployable at instance, because this usage has increased. At the same time security of SONs is also significant. Past investigations show that prevention alone and it consists of traditional security mechanisms are not enough because of the unique characteristics and properties of SONs. Moreover most of the techniques used at present were originally developed for wired network. An alternative method to overcome problems in conventional security, trust based security mechanisms are engaged by the researchers.

In this chapter we have given a detailed introduction about SON and its enabled networking types. As explained in this chapter, security is fundamental one for any SON but achieving security in the presence of attackers is a highly trivial task. Especially, when the number of attackers cross certain limit, the overall performance of the networks in question. Hereby, we have discussed all the available attacks that affect the network both internally and externally. From this discuss, most of the security schemes that are derived for wired, so it is not support SONs in all the times. So we discussed the limitations of traditional security mechanisms and some proposed mechanisms and the impact of trust and its properties. Finally we presented some trust based mechanisms for SONs that proposed by various authors in the last decade.

In future self-organized network will offer services to a huge number of applications and services to its clients in terms of voice and video communication, rapid data access and access information via cloud or virtualized environment. But these facilities and services will have to look forward various challenges that are imposed by the future SONs that are dynamic, seamless, reconfigurable, with variable load conditions, and different quality of services. But withstanding such challenges is highly complicated task and future of SONs must have a high degree of network operation with robustness and reliability.

REFERENCES

Aad, I., Hubaux, Y., & Knightlyz, E. (2004). Denial of Service Resilience in Ad Hoc Networks. ACM.

Abdul-Rahman, A., & Hailes, S. (1997). Using Recommendations for Managing Trust in Distributed Systems. *Proc. IEEE Malaysia Int'l Conf.on Communication.*

Adams, W. J., Hadjichristofi, G. C., & Davis, N. J. (2005). Calculating a Node's Reputation in a Mobile Ad Hoc Network. *Proc. 24th IEEE Int'l Performance Computing and Communications Conference.*

Adnane, A., Bidan, C. & Junior. (2013). Trust based security for the OLSR Routing Protocol. Elsevier, 36, 1159-1171.

Anand, P. M., & Bhaskar, V. (2013). A unified trust management strategy for content sharing in Peer-to-Peer networks. *Applied Mathematical Modelling*, *37*(4), 1992–2007. doi:10.1016/j.apm.2012.04.050

Anita, X., Martin, J., Maickam, L., & Bhagyaveni, M, A. (2014). Two Way Acknowledgment based trust framework for wireless sensor networks. *International Journal of Distributed Sensor Network*. doi: 1155/2013/952905

Bamberger, W. (2010). Interpersonal Trust – Attempt of a Definition. *Scientific Report.*

Banerjee, S. (2008). Detection/Removal of Cooperative Black and Gray Hole Attack in Mobile Ad-Hoc Networks. *Proceedings of the World Congress on Engineering and Computer Science.*

Bankovic, Z., Fraga, D., Moya, J., Vallejo, J., Malago, P., Araujo, A., & Nieto-Taladriz, O. et al. (2011). Improving security in WMNs with reputation systems and self-organizing maps. *Journal of Network and Computer Applications*, *34*(2), 455–463. doi:10.1016/j.jnca.2010.03.023

Bao, F., Chen, I., Chang, M., & Cho, J. (2012). Hierarchical Trust Management for Wireless Sensor Networks and its applications to Trust Based Routing and intrusion detection. *IEEE eTransactions on Network and Service Management*, *9*(2), 169–183. doi:10.1109/TCOMM.2012.031912.110179

Baras, J. S., & Jiang, T. (2005). Managing Trust in Self-Organized Mobile Ad Hoc Networks. In *Proc. 12th Annual Network and Distributed System Security Symposium Workshop.*

Bhalaji, N., & Shanmugam, A. (2012). Dynamic Trust Based Method to Mitigate Greyhole Attack in Mobile Ad Hoc Networks. *Procedia Engineering, Elsevier*, *30*, 881–888. doi:10.1016/j.proeng.2012.01.941

Blaze, M., Feigenbaum, J., & Lacy, J. (1996). Decentralized trust Management. In *Proceedings of IEEE Symposium on security and Privacy.*

Chen, C., Zhang, J., Cohen, R., & Ho, P. (2010). Secure and Efficient Trust Opinion Aggregation for Vehicular Ad-hoc Networks. *IEEE Digital Library.*

Chouhan, N., & Yadav, S. (2011). Flooding Attacks Prevention in MANET. *International Journal of Computer Technology and Electronics Engineering*, *1*, 68–72.

Chuang, M., & Lee, J. (2013). TEAM: Trust-Extended Authentication Mechanism vehicular Ad-Hoc Networks. IEEE.

Duan, J., Yang, D., Zhu, H., Zhang, S., & Zhao, J. (2014). TSRF: A Trust aware secure routing framework in wireless sensor networks. *International Journal of Distributed Sensor Network.*

Eschenauer, L., Gligor, V. D., & Baras, J. (2002). On Trust Establishment in Mobile Ad Hoc Networks. *10th Int'l Security Protocols Workshop.*

Femg,H., & Liu,C. (2006). Design of a Joint Defense System for Mobile Ad Hoc Networks. *IEEE Digital Library*, 742-746.

Feng, R., Che, S., Wang, X., & Yu, N. (2013). *Trust management scheme based on D-S Evidence Theory for Wireless Sensor Networks. International Journal of Distributed Sensor Network.* doi:10.1155/2013/948641

Gómez Mármol, F., & Martínez Pérez, G. (2012). TRIP, a trust based reputation infrastructure–based proposal for vehicular ad hoc networks. *Journal of Network and Computer Applications*, *35*(3), 934–941. doi:10.1016/j.jnca.2011.03.028

Gong, W., You, Z., Chen, D., Zhao, X., Gu, M., & Lam, K. (2009). Trust based malicious nodes Detection in MANET. *IEEE Digital Library.* 10.1109/EBISS.2009.5138139

Huang, D., Hong, X., & Gerla, M. (2012). Cognitive radio network security: A survey. *Journal of Network and Computer Applications*, *35*(6), 1691–1708. doi:10.1016/j.jnca.2012.06.006

Kang, X., & Wu, Y. (2014). A trust-based pollution attack prevention scheme in peer-to-peer streaming networks. *Computer Networks*, *72*, 62–73. doi:10.1016/j.comnet.2014.07.012

Kannhavong, B., Nakayama, H., Nemoto, Y., & Kato, N. (2008). *SA-OLSR: Security Aware Optimized Link State Routing for Mobile Ad Hoc Networks*. IEEE. Doi.10.1109/ICC.2008.283

Lacharite, Y., QuanNguyen, D., Wang, M., & Lamont, L. (2008). A Trust-based Security Architecture for Tactical MANET. *IEEE Digital Explorer*. 10.1109/MILCOM.2008.4753215

Li, W., Parker, J., & Joshi, A. (2012). Security through Collaboration and Trust in MANETs. *Mobile Networks and Applications*, *17*(3), 342–352. doi:10.1007/s11036-010-0243-9

Li, X., Zhou, F., & Du, J. (2013). LDTS: A Lightweight and Dependable Trust System for Clustered Wireeless Sensor Networks. *IEEE Transactions on Information Forensics and Security*, 8.

Liu. (2009). *Security and Trust Management in Self-organizing Wireless Adhoc Networks*. Retrieved from http://citeseerx.ist.psu.edu

Lopez, J., Roman, R., Agudo, I., & Fernandez-Gago, C. (2010). Trust management systems for wireless sensor networks: Best practices. *Computer Communications*, *33*(9), 1086–1093. doi:10.1016/j.comcom.2010.02.006

Mayer, R. C., Davis, J. H., & Schoorman, F. D. (1995). An integrative model of organizational trust. *Academy of Management Review*, *20*, 709–734.

Mejri, M., Ben-Othman, J., & Hamdi, M. (2014). Survey on VANET security challenges and possible cryptographic solutions. *Vehicular Communications*, *1*(2), 53–66. doi:10.1016/j.vehcom.2014.05.001

Meng, X., Ding, Y., & Gong, Y. (2012). Trust: A trust model based on feedback-arbitration in structured P2P network. *Computer Communications*, *35*(16), 2044–2053. doi:10.1016/j.comcom.2012.06.007

Miorandi, D., Sicarib, S., & De Pellegrini, F. (2012). Internet of things: Vision, applications and research challenges. *Ad Hoc Networks*, *10*(7), 1497–1516. doi:10.1016/j.adhoc.2012.02.016

Parvin, S., Hussain, F., & Hussain, O. (2013). Conjoint trust assessment for secure communication in cognitive radio networks. *Mathematical and Computer Modelling*, *58*(5-6), 1340–1350. doi:10.1016/j.mcm.2013.01.001

Parvin, S., Hussain, F., Hussaina, O. K., & Faruquea, A. A. (2012). Trust-based Throughput in Cognitive Radio Networks. *Procedia Computer Science, Elsevier*, *10*, 713–720. doi:10.1016/j.procs.2012.06.091

Pei, Q., Yuan, B., Li, L., & Li, H. (2013). A sensing and etiquette reputation-based trust management for centralized cognitive radio networks. *Neurocomputing*, *101*, 129–138. doi:10.1016/j.neucom.2012.08.005

Prokopenko. (2009). Guided self-organization. *HFSP Journal*, *3*, 287-289.

Qureshi, B., Min, G., & Kouvatsos, D. (2012). A distributed reputation and trust management scheme for mobile peer-to-peer networks. *Computer Communications*, *35*(5), 608–618. doi:10.1016/j.comcom.2011.07.008

Rai, A., Tewari, R., & Upadhyay, S. (2010). Different Types of Attacks on Integrated MANET-Internet Communication. *International Journal of Computer Science and Security*, *4*, 265–274.

Reddy, Y. B. (2012). Trust Based Approach in Wireless Sensor Networks using an Agent to Each Cluster. *International Journal of Security. Privacy and Trust Management*, *1*, 1.

Redwan, H., & Kim, K. (2008). Survey of Security Requirements, Attacks and Network Integration in Wireless Mesh Networks. *IEEE Digital Explorer*. .10.1109/FCST.2008.15,3-9

Robert, W. (2004). Self-organizing innovation networks: Implications for globalization. *Technovation*, *24*(3), 187–197. doi:10.1016/S0166-4972(03)00092-0

Rutvij, H., Ashish, Patel, Parmar, & Shah. (2010). MANET Routing Protocols and Wormhole Attack against AODV. *International Journal of Computer Science and Network Security*, *10*, 12–18.

Saha, H., Bhattacharyya, D., & Banerjee, P. K. (2010). Semi-Centralized Multi-Authenticated RSSI Based Solution to Sybil Attack. *International Journal of Computer Science & Emerging Technologies*, *1*, 338–341.

Saied, Y., Olivereau, A., Zeghlache, D., & Laurent, M. (2013). Trust management system design for the Internet of Things: A context-aware and multiservice approach. *Computers & Security*, *39*, 351–365. doi:10.1016/j.cose.2013.09.001

Saini, A., & Kumar, H. (2010). Comparison between Various Black Hole Detection Techniques in MANET. *Proceedings of National Conference on Computational Instrumentation*.

Sari, A., & Necat, B. (2012). Securing Mobile Ad-Hoc Networks against Jamming Attacks through Unified Security Mechanism. *International Journal of Ad Hoc, Sensor & Ubiquitous Computing*, 3.

Saxena, N., Denko, M., & Banerji, D. (2011). A hierarchical architecture for detecting selfish behavior in community wireless mesh networks. *Computer Communications*, *34*(4), 548–555. doi:10.1016/j.comcom.2010.04.040

Sivagurunthan, S., & Prathapchandran, K. (2014). *Trust based Security schemes in Mobile Ad Hoc Networks – A Review*. IEEE Digital Explorer; doi:10.1109/ICICA.2014.67

Staab, S. (2004). The Pudding of Trust. *IEEE Intelligent Systems*, *19*, 74–88.

Stallings, W. (2012). *Network Security Essentials*. Pearson.

Sun, Y. L., Yu, W., Han, Z., & Liu, K. Jr. (2006). Information Theoretic Framework of Trust Modeling and Evaluation for Ad Hoc Networks. *IEEE Journal on Selected Areas in Communications*, *24*(2), 305–317. doi:10.1109/JSAC.2005.861389

Tian, C., Yang, B., Zhongand, J., & Liu, X. (2014). Trust-based incentive mechanism to motivate cooperation in hybrid P2P networks. *Computer Networks*, *73*, 244–255.

Wang, B., Chen, X., & Chang, W. (2014). A Light Weight trust based QOS routing algorithm for Ad Hoc Networks. *Journal of Pervasive and Mobile Computing*, *13*, 64–180. doi:10.1016/j.pmcj.2013.06.004

Wanga, F., Wanga, F., Huang, B., & Yang, T. (2012). SONR: A reliable reputation system of self-organized network. *Journal of Network and Computer Applications*, *35*(3), 914–926. doi:10.1016/j.jnca.2011.03.008

Xia, H., Jia, Z., Li, X., Ju, L., & Edwin, H. M. (2012). Trust Prediction and Trust based Source Routing in Mobile Ad Hoc Networks. *Ad Hoc Networks*, *11*(7), 2096–2114. doi:10.1016/j.adhoc.2012.02.009

Xiao, P., He, J., & Fu, Y. (2014). An access authentication protocol for trusted handoff in wireless mesh networks. *Computer Standards & Interfaces*, *36*(3), 480–488. doi:10.1016/j.csi.2013.08.016

Yan, Z., Zhang, P., & Vasilakos, A. V. (2014). A survey on trust management for Internet of Things. *Journal of Network and Computer Applications*, *42*, 120–134. doi:10.1016/j.jnca.2014.01.014

Zhang, J., Chen, C., & Cohen, R. (2010). A Scalable and Effective Trust-Based Framework for Vechicluar Ad Hoc Networks. *Journal of Wireless Mobile Networks Ubiquitous Computing and Dependable Applications*, *1*, 3–15.

Zhang, J., Shankaran, R., Orgun, M. A., Varadharajan, V., & Sattar, A. (2010). A trust management architecture for Hierarchical Wireless Sensor Network. *IEEE Digital Library*.

Zhou, L., & Haas, Z. J. (1999). Securing Ad Hoc Networks. *IEEE Digital Library*.

KEY TERMS AND DEFINITIONS

Active Attack: To alter the system resources and affect its performance. It makes some modification of data streams and creation of false dreams by the way it affects the network.

Attack: An action that harms the network by obtaining information that is being transmitted. Modification or deletion of transmitted information and disclosure of routing and other information.

Authentication: A node to ensure the identity to the peer node with which it is communicating.

Authorization: Enables users to access the resources based on the users' identity and policy.

Availability: The survivability of network services despite Denial of Services (DoS) attacks.

Confidentiality: Certain information is not disclosed to unauthorized entities.

Integrity: A message being transferred is not corrupted.

Non Repudiation: The origin of a message cannot deny having sent the message and the receiver from denying the reception of the messages.

Passive Attack: To obtain information from the system that is being transferred and does not affect the system resources any way.

Privacy: It ensures the appropriate use of information and means information can be exchanged without anyone seeing them and will not be shared without permission from the owner of information.

Self-Organized Networks: A network that can automatically extend, change its behavior, configure and optimize its topology, coverage, capacity, cell size, and channel allocation, based on changes in location, traffic pattern, interference, and the situation or environment.

Trust (According to Network): A set of relations among entities that participate in the protocol, which is based on the previous interaction of entities within the protocol.

Trust: One entity (trustor) is willing to depend on another entity (trustee) or the trustor abandons control over the actions performed by the trustee.

Chapter 5
SON in Location-Based Services

Ramprasad Subramanian
University of Technology Sydney, Australia

Kumbesan Sandrasegaran
University of Technology Sydney, Australia

Farhana Afroz
University of Technology Sydney, Australia

Xiaoying Kong
University of Technology Sydney, Australia

ABSTRACT

This chapter forays into the literature of emerging field of Self-Organization Network (SON) and its application in the area of location-based services. SON is a widely applied technology in the ad hoc networks, wireless sensor networks and automatic computer networks. This chapter deal exclusively about the projects, standards and literature of SON in the context of the cellular networks and how it can aid to achieve higher position accuracy information of the subscriber. Additionally, in this chapter a clear understanding of SON taxonomy and the guidelines for the design of SON has been presented and a comparison of SON with the existing solutions about its strength and weakness has been made in this chapter. The key research areas for further development is also highlighted in this chapter. This literature will provide comprehensive guidance for anyone interested to delve into the research of SON in cellular wireless communication and location based services.

1. INTRODUCTION

In the past few decades, the growth of wireless communications evolved from 2G to 3G to LTE and further developing to 5G. The mobile communication has taken the field of telecommunication to the next paradigm and the entire world has become mobile. Owing to this, fresh and real time, first-hand information has become a need of the hour coupled along with the location information. The continuous need of these first hand and real-time position information has created enormous business opportunity and it has resulted in the development of new features, higher coverage, flexibility and higher interoperability. These developments have become so visible, and we have reached a situation where life without this has become unthinkable. To cope up with the competition the mobile network operators are expanding the network very aggressively and thereby extending the telecommunication networks to variable terrain. The expansion of telecom networks has provided a quantum leap to the area of position information service. Location information plays a very important role in the modern communication technologies

DOI: 10.4018/978-1-5225-0239-5.ch005

and has become crucial in several important applications such as rescues operations, emergency and navigation, handover, delivery of high-speed data to the subscribers etc.

Positioning can be defined as a clear and unambiguous location of an individual or object with reference to a coordinate system or with reference to a known reference point. Although the information regarding the position is a very important but its association with certain time makes it more useful to understand the higher derivatives like position of an individual or an object at a particular instance of time, acceleration, speed etc. With the aid of time and position information, tracking of an individual can be performed. Thus, tracking can be defined as a position information of an individual or object with a function of time. Thus, tracking can be mathematically defined as

$$T = \int_{0}^{\infty} p(x)$$

T: Is defined as the tracking window.
p(x): Is defined as the position information obtained at regular interval of time.

The integral *0 to* can be defined as the continuous tracking window with start and end time.

The integral 0 to ∞ can be defined as the integration of position of an individual or object between the start time and end time of the tracking window. Hence from the above equation we can understand the tracking as the series of positioning dots of a individual or object obtained in a given frame of time window.

Equipping the position information alongside the communication technology is a very hot topic. Although the necessity of the position information has existed for many years and centuries. Actually, the art of obtaining the position information has existed for many centuries. Some millennia ago people used compass, wind direction, sea currents and celestial constellations like stars and sun to identify and estimate position and orientation. Technology of knowing position of sun and identifying the orientation and direction towards sun can be still used even today in the absence of all modern day tools to make a rough estimate of one's position. The research activities in the past decade has evolved in the course of integrating modern day communication and positioning technology in a single system that resulted in the development of standards by 3GPP. This standardization has resulted in providing a combined technology of communication and positioning that can benefit the whole network and individual users and the service providers. The location information is a basic requirement for many new technologies like cooperative communication system, delivery of new protocols and applications, etc.

The location information system can be classified into positioning system and positioning solutions. The former works by integration of hardware and software that analyses the properties of the communication links to estimate the position of the user. Positioning systems are typically integrated into the communication technologies like GPS integration into the mobile phones and the base stations (BS). The positioning solution is a software-based approach where the users' position information is obtained in an indirect way in an opportunistic fashion from the mechanism that is already existing in the communication technology, thus exploiting the communication technologies to obtain the position information. The first method listed above has some disadvantages in terms of costs. Since it requires additional hardware the cost of the solution increases, hence the software based approach is proffered. The positioning solution makes use of the power control and synchronization mechanism to measure the wireless channel that can be used to obtain the user's position. Examples of measurements that is used are received signal strength and propagation delay of signals.

2. SON IN LOCATION BASED SERVICES

The concept of SON is originally developed from inspiration obtained from nature where certain biological creatures demonstrates a certain traits of unique organizational behavior in order to achieve certain goals and adopting intelligently to the changing dynamics of immediate environment. Certain examples from nature like shoaling fish, swarming insects and herding sheep, synchronous flashing of fire flies, human body's complex system and flocking of birds can be analyzed from nature to suit our study.

SON is not new, and it has existed before. The idea of SON is already implemented in biology, computer science and cybernetics. Even though the concept of SON existed for a long time but there is no single explanation of SON exist till date. A visionary statement on SON was provided by Hayken terming it as a new discipline for the next generation of cellular networks. The SON is defined as an adaptive functionality where the network changes can be deducted automatically and based on that decisions are made intelligently to changes. The notion of SON was viewed as a cooperative network where the nodes cooperate with each other to achieve the common addaptiveness to the changes. These definitions of SON makes it very important candidate to be used in location-based services to improve the accuracy of the position information of the subscriber or an object under surveillance. In our chapter we focus our attention on improving the accuracy of position information using SON obtained through positioning solution. The usage of SON in positioning solution gives operator the required freedom to move or relocate the site as and when required without worrying about the network changes. When an operator moves the sites such as cell on wheels sites (COWs) due to some commercial reasons to provides some temporary coverage to a location the SON updates the neighbor relations and other necessary information so that these updated information would be present in T1 & T2 dumps provided by the operator which will enable the LBS service provider to provide better position accuracy of the object under surveillance (Figure 1).

2.1 Characteristics of Self Organizing Solutions

1. **Scalability:** Strict scalability can be defined as the solution complexity should not increase along with the scale and size of the system. Therefore the system should remain operational if certain nodes enter or leave the system. The same approach can be followed in the engineering system and especially in the wireless environment. The operators tend to modify the network based on the requirements and demands. Hence, the adaptive algorithms should be mindful of this fact and the changes should be taken cared properly without any breakdown of network. The algorithms should be austere in terms of time, space and other hardware resources required to be implemented. The lower the complexity of the solution the better its scalability. The algorithms should not require a global coordination to take care of the modifications in the network. Instead, the solution should be local. This approach reduces the overhead and improves the flexibility. As an example, the adaptive algorithm used in the wireless environment like LTE, any changes in the antenna tilt of the nodes with the objective of providing better coverage or load balancing should not depend on centrally implemented schemes to take care of the changes. Instead, the local cooperation among the nodes

Figure 1. SON architecture

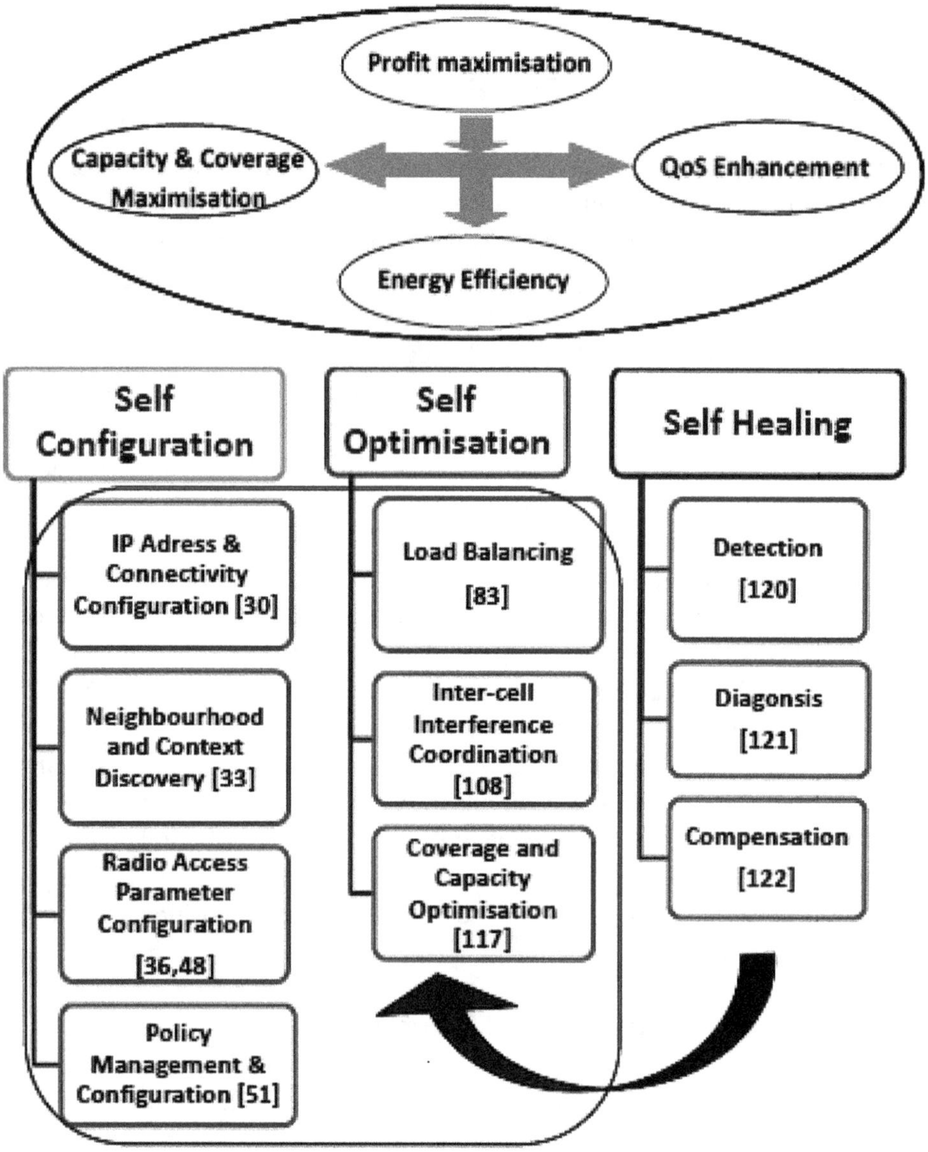

can create better self organization. Thus a perfectly scalable system should be less complex as the nodes increases. With increase in computational power even centralized complex system can be afforded (Dressler, 2008).

2. **Stability:** The stability can be defined as an algorithm or adaptation mechanism that can traverse consistently a finite number of states within a acceptable finite time. The stability also offers robustness to the system. Stability also ensures that system does not oscillate between states infinitely. Perfect stability ensures that the system is elastic and self-healing. The readers can refer (Gazi & Passino, 2003) for the details on stability.

3. **Agility:** A system be stable and scalable but still not perfectly adaptive and it might be sluggish. Agility is one of the important factor affecting self organizing system. The agility of the system can be defined on the basis of how much adaptive a change can be for the changes. But realizing a perfectly agile system is not possible. The agility of a system depends on feedback, processing and decision making. Depending on the objectivity, the threshold for the agility for desired system can be set by operator. It is worth mentioning that distributive control is a key to enhance the agility of the system, since it reduces feedback delays in system. Further reading on the agility can be found in (Samaan & Karmouch, 2009). Figure 2 defines the SON time scale of the operation to achieve agility.

3. POSITIONING ARCHITECTURE AND SIGNALING IN LTE

In LTE positioning architecture LCS client, LCS target and LCS server are three important network elements that plays a very important role. LCS server is a physical or logical entity managing positioning for an LCS target device by obtaining measurements and other location information, providing assistance data to assist UEs in measurements and computing or verifying the final position estimate. LCS client can be a software or hardware entity and it interacts with server to obtain the position information of a target or send a position query to the server. LCS client may reside inside the LCS targets. The position query for a LCS target is sent through LCS client. The client makes as location request to the LCS server and the server intern honors the request and sends the position of the LCS target.

The gateway mobile location center (GMLC) is the first node an external LCS client accesses in a public land mobile network (PLMN), through Le interface. After performing registration authorization, it sends positioning requests to MME and receives final location estimates from corresponding entity via SLg interface. A location retrieval function (LRF), which may or may not be co-located with the

Figure 2. Time scale for SON elements (Samaan & Karmouch, 2009)

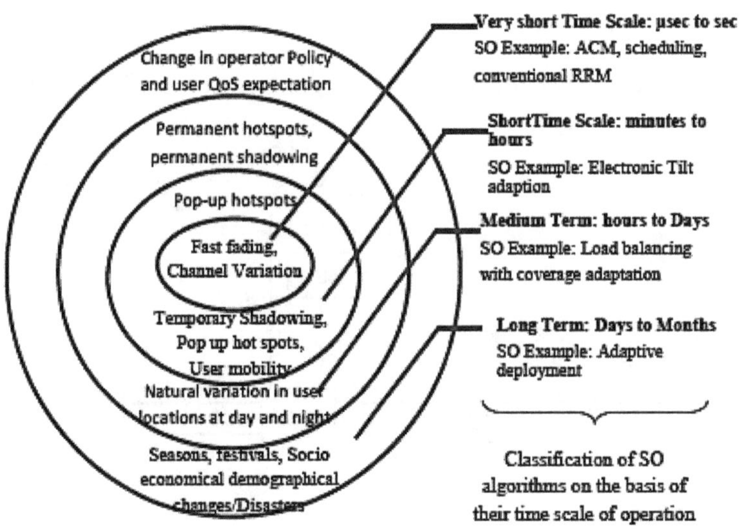

GMLC, is responsible for retrieving or validating location information, providing routing and/or correlation information of a UE that has initiated an IP multimedia subsystem (IMS) emergency session. Two new protocols have been standardized specifically to support positioning in LTE: the long - term evolution positioning protocol (LPP) (Third Generation Partnership Project [3GPP], 2010) and the long - term evolution positioning protocol annex (LPPa) (3GPP, 2010). The LPP (3GPP, 2010) is a point - to - point protocol between an LCS server and LCS target device, used in order to identify position the target device. The following transactions have been specified that happens in LPP protocol: capability transfer procedure (request/provide messages), assistance data transfer procedure (request/provide messages), and a location information transfer procedure (request/provide messages). Multiple LPP procedures of any of the aforementioned types can be used in series and/or in parallel. LPP is used both by control - plane.

LPPa is specified only for control - plane positioning procedures. However, with user plane and control - plane interworking, LPPa can also assist user - plane solution by querying eNodeBs for information and eNodeB measurements not related to a UE connection.

3.1 Control Plane

To support LCS, at least two functional nodes must be present in LTE control - plane architecture (refer to Figure 3) the evolved serving mobile location center (E - SMLC), which controls the coordination and scheduling of resources required to locate mobile device, and GMLC, which controls the delivery of position data, user authorization, charging, and more. The LPP messages are transmitted transparently to the MME, using RRC as a transport over the LTE - Uu interface between the UE and the E - SMLC, using S1 - AP over the S1 - MME interface, and using LCS - AP over the SLs interface between the eNodeB and the E - SMLC. LPPa conducts the LPPa location information transfer procedures for positioning - related information. LPPa is also transparent to the MME, which routes the LPPa packets transparently over the S1 - MME and SLs interfaces without knowledge of the involved LPPa transaction. The LTE positioning architecture for the control plane is shown in Figure 3. To describe the operation of the architecture, consider the case where the MME receives a positioning request for some LCSs associated with a particular LCS target (e.g., a UE). The MME then sends an LCS request in an LCS - AP location request message (3GPP, 2010) to an E - SMLC. The E - SMLC processes the LCSs request to perform

Figure 3. LTE control plane architecture for location based services (3GPP, 2010)

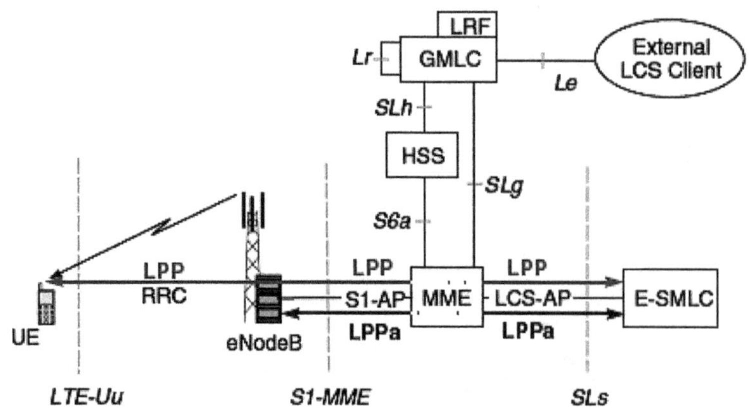

a positioning of the target UE. The E - SMLC then returns the result of the LCS back to the MME. The MME then forwards the result at least to the requesting node.

3.2 User Plane

In general, secure user plane location (SUPL) (Open Mobile Alliance, 2009) supports and complements control (refer to Figure 4) - plane protocols to enable LBS support with the least possible impact on the control plane and the deployed network. The LTE user - plane positioning architecture is shown in Figure 4. SUPL uses established data - bearing channels (i.e., the LTE user plane) and positioning protocols (i.e., LPP) for exchanging the positioning - related data between an LCS target and an LCS server. In the general user - plane protocol stack, SUPL occupies the application layer with LPP transported as another layer above SUPL. After establishing a TCP/IP connection and initiating the SUPL and then LPP sessions, the flow of LPP messages can be the same as in the control - plane version of LPP, just with the SUPL enabled terminal (SET) as the LCS target and the SUPL location platform (SLP) as the LCS server. The SLP implements the SUPL location center (SLC) and the SUPL positioning center (SPC) functions with the latter either being integrated in the E - SMLC or attached to it with a proprietary interface. The SLC system coordinates the operations of SUPL in the network and implements the following SUPL functions as it interacts with the SET over the user - plane bearer: privacy function, initiation function, security function, roaming support, charging function, service management, and position calculation. The SPC supports the following SUPL functions: security function, assistance delivery function, SUPL reference retrieval function (e.g., retrieving data from a GPS reference network), and SUPL position calculation function.

Figure 4. LTE user plane architecture for location based services (3GPP, 2010)

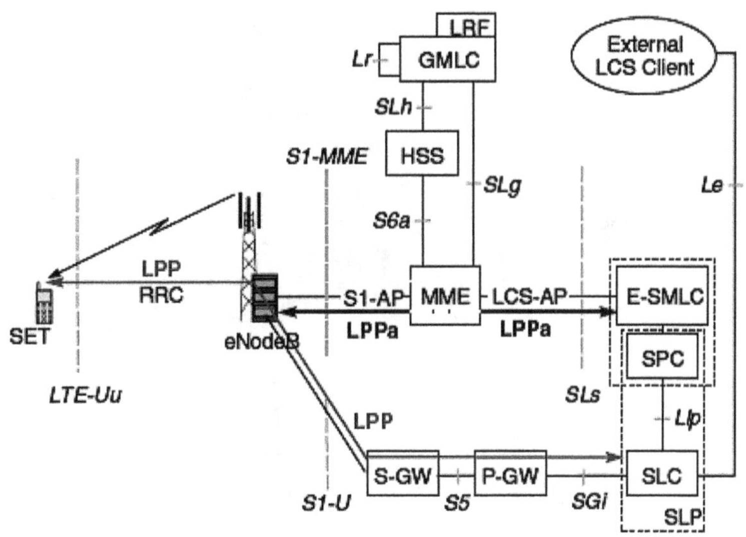

3.3 Positioning Methods in LTE

In LTE based localization, the coordinates of the users location is determined by the measurements obtained from the network. The obtained time difference of arrival (OTDOA) is a technology introduced in release 9 of 3GPP specification (refer to Figure 5).

In this technique the user equipment (UE) measures the difference in time between the several eNodeB (eNB) and uses trilateration to find its location. But because of certain condition the UE may not be able to communicate with sufficient eNB and resulting in ambiguity in position information. To overcome this issue cooperative approach is taken so that UE not only communicate with the eNB but also with other UEs to obtain its location (Shladover & Tan, 2006).

In Uplink time difference of Arrival (U-TDOA) (refer to Figure 6) which is developed by True Position (Hofmann-Wellenhof, Lichtenegger & Collins, 2001), which relies on multilateration. It is a network based solution and no additional requirements like chips or software to be installed into mobile device. This method obtains the location by calculating the time difference at which signals reaches the LMUs installed in the BTs. Accuracy depends on the network layout and density of the LMUs. This technology works well in urban but suffers in the rural because of the density of the networks (Parker & Valaee, 2007).

Approximate point in triangulation (APIT) algorithm is frequently applied in sensor networks. In this algorithm anchors are selected among the sensing nodes. These anchors broadcasts its location information and the node who information needs to be computed receives the beacon from N audible anchors. Then the node tests whether the device is inside or outside the triangle and this is called as APIT testing. The location or point where the most triangles overlap is the location of the node and this is called as APIT referencing (SAI Global, 2003). In scan approach, a particular area is selected and from that area the maximum possible areas where mobile devices can be located. Once this area is determined, the

Figure 5. OTDOA technique (Shladover & Tan, 2006)

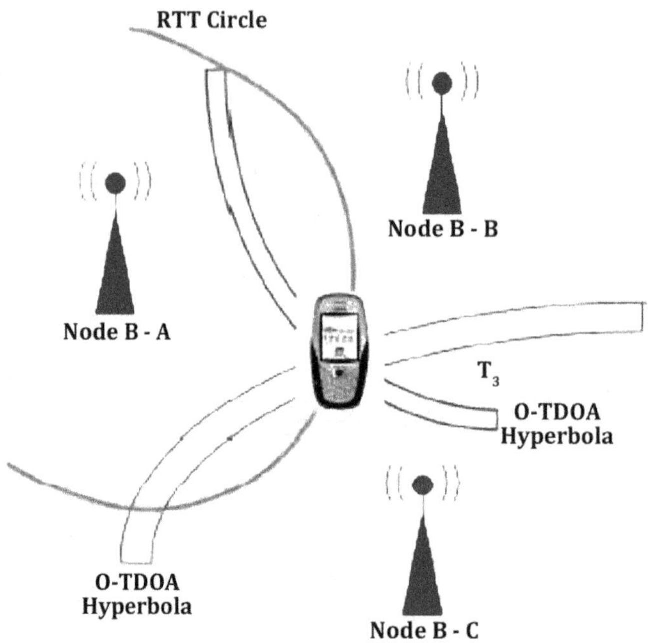

Figure 6. U-TDOA Technique (Hofmann-Wellenhof, Lichtenegger & Collins, 2001)

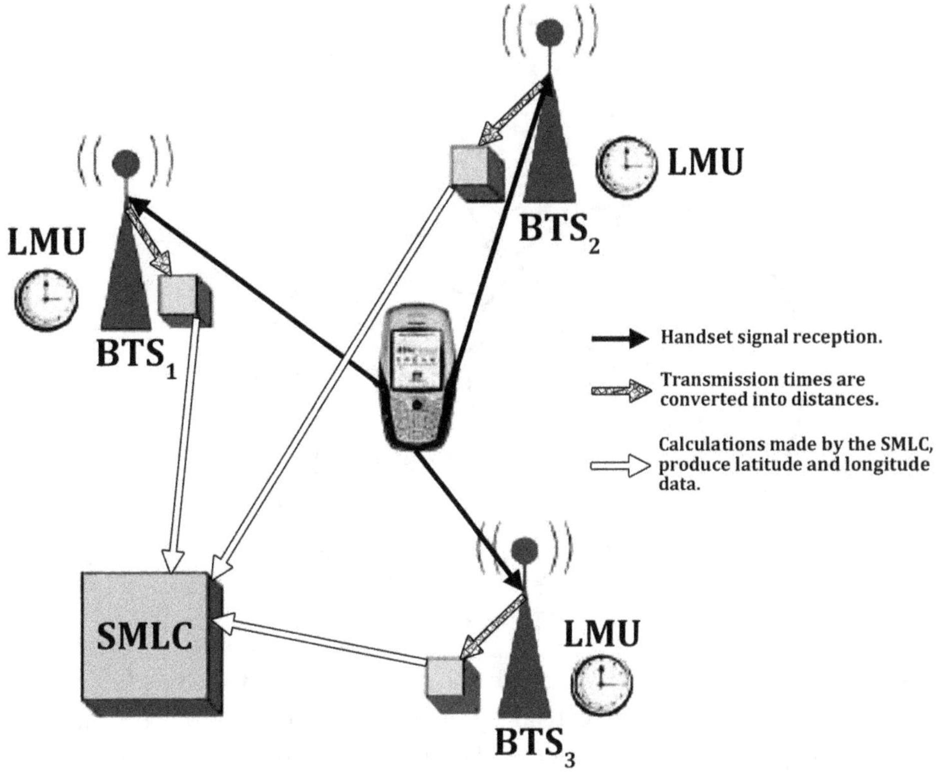

centre of gravity of the area under investigation is calculated and that area is considered to be location of the mobile. Another algorithm called as reference point algorithm (Amundson & Koutsoukos, 2009). The radio maps stores the RSS values obtained during the offline phase. This method uses the reference tags or RFID tags. Reference tags are active tags used for calibrating the environmental parameters. Tags that were associated with tracking objects is called tracking tags. In (Vaghefi, 2014) triangulation techniques based on the regression has been proposed. With pre-determined regression equations and RFID signals the distance between the tracking tag and RFID readers 11, 12 and 13 is obtained and with this the location of the tag is estimated.

In path loss models the distance between the node and the three base station is calculated. This models generally uses the triangulation technique (refer to Figure 7) to calculate the position of the vehicle node. The RSS (Received signal strength) based approach works by measuring the received signal level at one. This technique also employees the triangulation technique to determine the location of the node in two dimension (Vaghefi, 2014). This scheme works excellently in urban and indoor. The RF signals bounces off when it hits a solid objects and causes multipath before it reaches the destinations. The receiver in the destination receives the same signal multiple times because of this effect. This phenomena is of multipath is utilized to calculate the position of the nodes. The location service provider must collect the finger printing and must be stored it in the data base and by comparing it with this database the position of the nodes can be estimated.

Angle of arrival (AOA) based positioning technique (refer to Figure 8) measures the angles of the signal arrived at the nodes. To compute the position of the nodes it is enough to measure the angles of

Figure 7. Locating tag with the triangulation technique (Vaghefi, 2014)

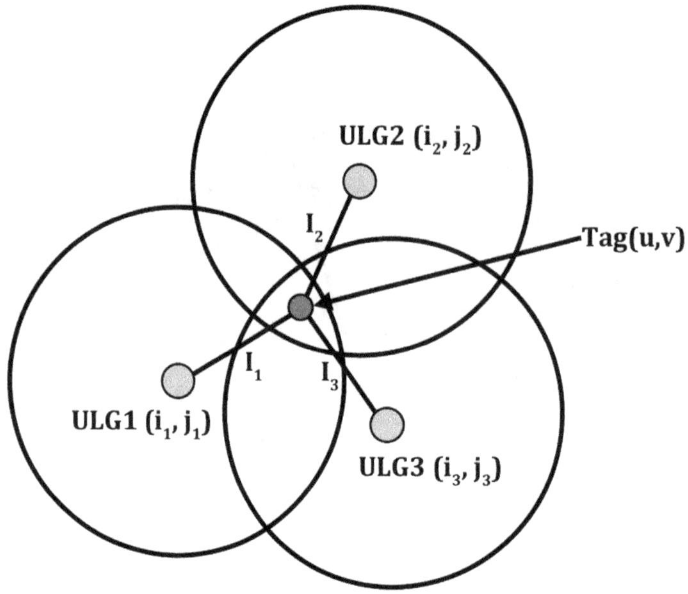

the straight lines that connects the node to the reference nodes. In this method multiple antennas is used at the base station to determine the incident angle of the arriving signal. If a node transmits a signal in the LOS condition the antenna array can determine the direction of the signal coming from. The second base station working in same technology can also locate the node and the compare the obtained position with the first base station (TruePosition, 2004; QUALCOMM 2001; Caffery, 2000).

In the algorithm, ToA (Time of arrival method) the signal travel times between the node to the base station is used to calculate the location of the vehicle node (He, Huang, Blum, Stankovic & Abdelzaher, 2003) (refer to Figure 9). There are two ways of measuring the location of the nodes in ToA. The first one is by measuring the phase of the received signal and the second method is by measuring the arrival time of a wideband narrow pulse. The first method has a limitation in the urban condition because of heavy multipath effect and for these environments the second solution is best suited. In the second solution when the bandwidth is increased the distance measurement error is reduced. This makes the distance

Figure 8. Estimating mobile position using AoA (He, Huang, Blum, Stankovic & Abdelzaher, 2003)

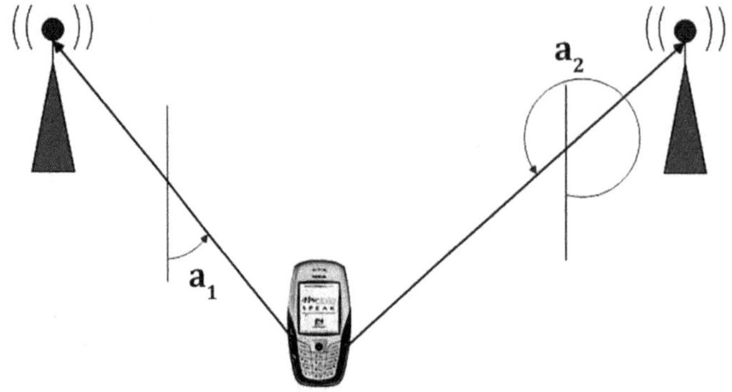

error caused by the erroneous estimates of ToA will be insignificant in the UWB. This method requires to be synchronised with the cellular network so that when the nodes have the common clock and when it is receiving the signal the ToA can be estimated for the incoming signal. In time difference of arrival algorithm the two signal travelling between the vehicle node and the reference nodes determines the location of the vehicle node in the hyperbola with two foci. But to estimate more accurate position of the vehicle node third base station is required. In this algorithm the reference nodes must be synchronized in time but if there is no synchronization then the RTT between the two nodes can be measured to the estimate the distance (Ladd, Bekris, Rudys, Marceau, Kavraki & Wallach, 2002; Yun & Kim, 2007; Chen, Ting, Maa & Chen, 2004; Li & Liu, 2005].

The idea of CP was proposed for location determination in wireless ad hoc and sensor networks. In non-CP condition the individual nodes estimates the position of its own. But the main idea behind the CP is to share the information between the neighbour nodes and thereby improving the position accuracy of the nodes. In cooperative communication, the terminals tries to help each other and there collectively achieves the better accuracy (Caffery, 2000; Lee & Scholtz, 2004; Adams, Gregorwich, Capots & Liccardo, 2001). Potentially in LTE the cooperation can be achieved by using relay nodes (RNs) which can assist in communication between the sources and the corresponding destinations, and/or allowing the nodes to help each other in the communication network to communicate with the corresponding destinations.

4. TECHNICAL CHALLENGES IN LOCATION BASED SERVICES

Accuracy of information is one of the major challenges faced in the location-based services (LBS). The infrastructural growth in the cities is so intense that the cities landscape is filled with numerous skyscrapers and high rise buildings. This has resulted in the creation of the word called as urban canyons. These developments pose a very serious threat and challenges to the RF signal penetration and transmission.

Figure 9. Time difference of arrival of the signal (Caffery, 2000)

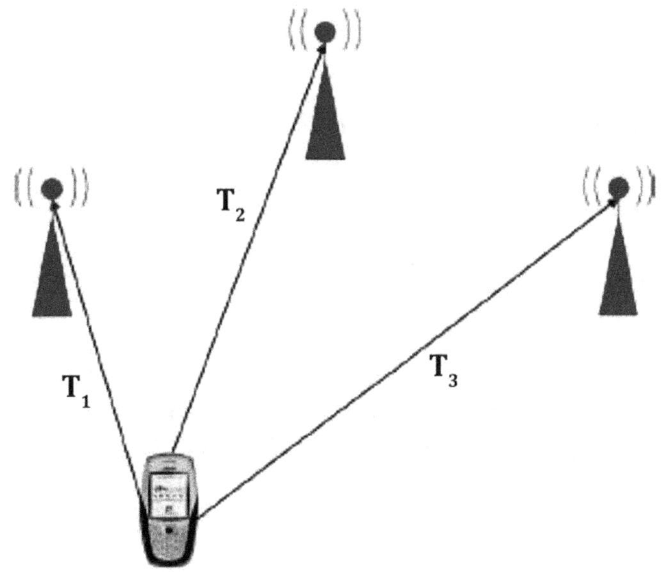

Apart from this due to the various activities like fund raising, sports activities and other types of entertainment activities results in constant movement of mobile sites that is called as cell on wheels (COWs) by the operator to the required location to temporarily cope up with the sudden influx of people to these venues to attend these activities. So as a result of this rapid movements, the operators find it difficult to maintain this information in the T1 dump and sometimes the information provided by them in T1 dump may not be accurate. These dumps are very much essential to the LBS service providers to generate the RF spectral database. The SMLC/SAS (serving mobile location center and standalone SMLC) requests the position information from the mobile and the received signal strength and multipath information obtained from the mobile can be computed and compared with the with the RF spectral database based on the T1, T2 dumps and offline drive tests conducted by the mobile network operator to obtain the latitude and longitude of the mobile. But the challenge here is the maintenance of the proper data in the T1 and T2 dumps. Due to the aforementioned reasons, the constant changes cellular network makes it difficult for the operators to maintain proper database of the network. So, in order to improve the accuracy the database should be maintained properly without much difficulty to the operator. This is possible with the usage of SON algorithms which can keep the database changes up to date.

The statistics in Figure 10 refers to the automatic neighbor relation algorithm activities in the network for the automatic neighbor relationship (ANR) management. In this, the ANR SO algorithm keeps track of the every changes in neighbor relationships and update the network operation and maintenance (OAM) system. The algorithm creates and deletes the neighbors relationships automatically without human intervention. This algorithm improved the T1 and T2 dumps provided by the operators and thus the position accuracy. The plotting of the vehicles has been illustrated in the Figure 11 below. The ANR SO algorithm improves the agility of the system and the changes are recorded quickly and the impact are minimized. The Figure 11 shows the plotting of the location of the vehicles position calculated after implementing the ANR algorithm which manages the neighbors relations automatically without any human intervention.

Figure 10. Automatic neighbor relation (ANR) stats

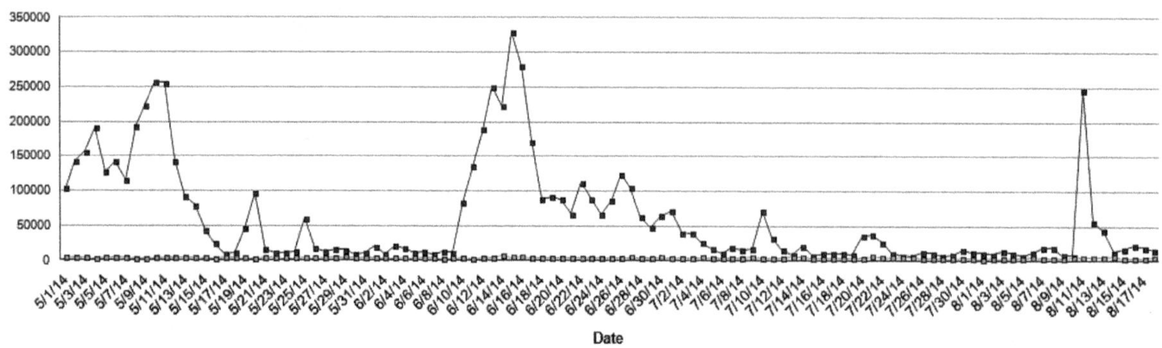

Figure 11. SON based location services simulation

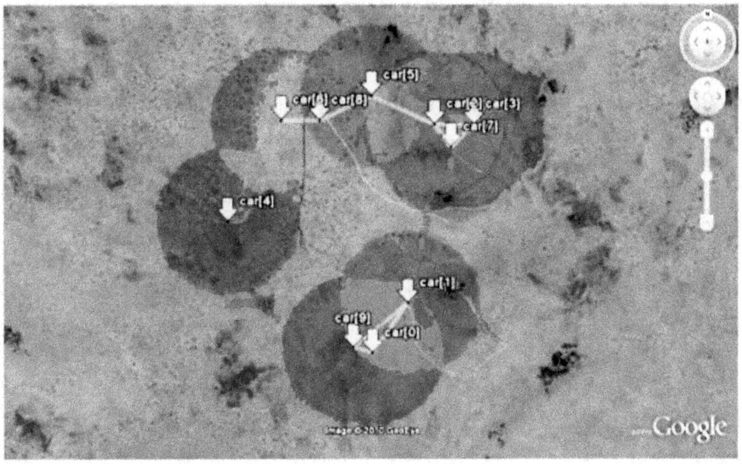

5. CONCLUSION

In this chapter, the SON and its application in the location based services is reviewed. The benefit of using SON in LBS and how SON can help to increase the position accuracy of the object under surveillance is also reviewed in this chapter. Apart from SON, the various positioning methodologies is also reviewed in this chapter. This chapter will provide as a guide to the researchers who is interested to pursue the research in the area of LBS and SON.

REFERENCES

Adams, J. C., Gregorwich, W., Capots, L., & Liccardo, D. (2001). Ultrawideband for navigation and communications. *Proceedings IEEE Aerospace Conference.*

Amundson, I., & Koutsoukos, X. D. (2009). A survey on localization for mobile wireless sensor networks. In *Proc. 2nd Int. Conf. Mobile Entity Localization Tracking GPS-Less Environ., MELT*. doi:10.1007/978-3-642-04385-7_16

Caffery, J. J. (2000). *Wireless location in CDMA Cellular Radio Systems*. Boston: Kluwer Academic.

Chen, J.-C., Ting, P., Maa, C.-S., & Chen, J.-T. (2004). Wireless geolocation with TOA/AOA measurements using factor graph and sum-product algorithm. *IEEE 60th Vehicular Technology Conference.*

Dressler, F. (2008). A study of self-organization mechanisms in ad hoc and sensor networks. *Elsevier Computer Communications*, *31*(13), 3018–3029. doi:10.1016/j.comcom.2008.02.001

Gazi, V., & Passino, K. M. (2003). Stability analysis of swarms. *IEEE Transactions on Automatic Control*, *48*(4), 692–697. doi:10.1109/TAC.2003.809765

SAI Global. (2003). *Standard Specification for Telecommunications and Information Exchange Between Roadside and Vehicle Systems—5.9 GHz Band Dedicated Short Range Communications (DSRC) Medium Access Control (MAC) and Physical Layer (PHY) Specifications.* ASTM E2213 -03(2010).

3GPP. (2010). *Evolved Universal Terrestrial Radio Access (EUTRA); LTE Positioning Protocol (LPP).* 3GPP TS 36.355, Release 9 (V9.0.0).

3GPP. (2010). *Evolved Universal Terrestrial Radio Access (EUTRA); LTE Positioning Protocol A (LPPa).* 3GPP TS 36.455, Release 9 (V9.0.0).

3GPP. (2010). *LCS Application Protocol (LCS - AP) between the MME and E - SMLC; SLs interface.* 3GPP TS 29.171, Release 9 (V2.0.0).

He, T., Huang, C., Blum, B., Stankovic, J., & Abdelzaher, T. (2003). Range-Free Localization Schemes for Large Scale Sensor Networks. *International Conference on Mobile Computing and Networking: Proceedings of the 9th annual international conference on Mobile computing and networking.* doi:10.1145/938985.938995

Hofmann-Wellenhof, B., Lichtenegger, H., & Collins, J. (2001). *Global Positioning System Theory and Practice* (5th ed.). Wien, Austria: Springer-Verlag. doi:10.1007/978-3-7091-6199-9

Ladd, A. M., Bekris, K., Rudys, A., Marceau, G., Kavraki, L. E., & Wallach, D. S. (2002). *Robotics-based location sensing using wireless Ethernet.* ACM MOBICOM.

Lee, J.-Y., & Scholtz, R. A. (2004). Ranging in a dense multipath environment using an UWB radio link. *IEEE Transactions on Selected Areas in Communications*, *20*(9), 1677–1683.

Li, W., & Liu, P. (2005). 3D AOA/TDOA emitter location by integrated passive radar/GPS/INS systems. *Proceedings of 2005 IEEE International Workshop on VLSI Design and Video Technology.*

Open Mobile Alliance. (2009a). *Secure user plane location architecture.* OMA - AD - SUPL - V2_0, Candidate Version 2.0.

Open Mobile Alliance. (2009b). *Secure user plane location protocol.* OMA - TS - ULP - V2_0, Candidate Version 2.0.

Parker, R., & Valaee, S. (2007). Vehicular node localization using receivedsignal-strength indicator. *IEEE Transactions on Vehicular Technology*, *56*(6), 3371–3380. doi:10.1109/TVT.2007.907687

QUALCOMM. (2001). *Synchronous vs. Asynchronous Deployment of WCDMA – Effect on Communication System Performance.* Author.

Samaan, N., & Karmouch, A. (2009). Towards autonomic network management: An analysis of current and future research directions. *IEEE Communications Surveys and Tutorials*, *11*(3), 22–36. doi:10.1109/SURV.2009.090303

Shladover, S. E., & Tan, S. K. (2006). Analysis of vehicle positioning accuracy requirements for communication-based cooperative collision warning. *J. Intell. Transp. Syst.: Technol., Plan. Oper.*, *10*(3), 131–140.

TruePosition. (2004). *An Examination of U-TDOA and Other Wireless Location Technologies: Their Evolution and Their Impact on Today's Wireless Market*. Author.

Vaghefi, R. M. (2014). Improving positioning in LTE through collaboration. *11th Workshop on Positioning, Navigation and Communication (WPNC).*

Yun, K., & Kim, D. (2007). Robust location tracking using a dual layer particle filter. *Pervasive and Mobile Computing*, *3*(3), 209–232. doi:10.1016/j.pmcj.2006.10.001

Section 2

Wireless Access and Network Architecture

Chapter 6
Cognitive Radio Networks:
Issues and Solutions

Deepti Singhal
International Institute of Information Technology Hyderabad, India

Kunal Sankhe
International Institute of Information Technology Hyderabad, India

Chandan Pradhan
International Institute of Information Technology Hyderabad, India

Rama Murthy Garimella
International Institute of Information Technology Hyderabad, India

ABSTRACT

The ever increasing demand for communication bandwidth has led to spectrum scarcity. We need to manage spectrum as a scare resource. Better utilization of spectrum is the key requirement for the radio communication systems. The inefficient usage of the existing spectrum can be improved through opportunistic access to the licensed bands without interfering with the primary users. This introduces the concept of dynamic spectrum access. Cognitive radio is must for providing quality of service in dynamic spectrum access. To deal with the cognition at physical layer, upper layers are also modified to support cognition and adaptation which make a network cognitive network. This book chapter describes the technical challenges and approaches for cognitive radio networks. Spectrum access models from regulatory point of view are presented. It also explains the challenges and solutions from literature for physical, MAC and network layer of cognitive networks. In the end of the chapter, real hope for designing protocols for such networks, i.e. cross layer protocol design is discussed.

INTRODUCTION

Traditional wireless networks are running with fixed spectrum assignment policies regulated by government agencies. Spectrum is assigned to service providers on a long term basis for large geographical regions. The spectrum is allowed to be used by licensed users. Federal Communications Commission (FCC) measurements have indicated that 15-85% of the time, many licensed frequency bands remain unused while some other bands are highly over-crowded. These overcrowded bands face the issue of spectrum scarcity. In order to better utilize the licensed spectrum, FCC has launched a secondary mar-

DOI: 10.4018/978-1-5225-0239-5.ch006

kets initiative, whose goal is to remove regulatory barriers and facilitate the development of secondary markets in spectrum usage rights among the wireless radio services. The inefficient usage of the existing spectrum can be improved through opportunistic access to the licensed bands without interfering with the primary users. This introduces the concept of Dynamic Spectrum Allocation (DSA), which implicitly requires the use of cognitive radios to improve spectral efficiency.

Cognitive radio is an intelligent radio that is aware of its surrounding environment and dynamically adapts to the transmission or reception parameters to achieve efficient communication without interfering with primary users. Thus, a Cognitive Radio Network (CRN) (Akyildiz, 2006) consists of primary and secondary users. The primary users are the licensed users and hence have exclusive rights to access the radio spectrum, whereas the cognitive users are the unlicensed users that can opportunistically access the free spectrum bands, without causing harmful interference to primary users. This introduces cognition and adaptation to physical layer of the network. These networks work on multi-channel environment with dynamic availability of channels based on primary user activity. These channels can be of different frequencies and/or different bandwidth. The operation of these networks is very different from simple multi-channel environments as different frequencies have different characteristics, like transmit power levels, transmission distance and multipath effects.

The first step to accommodate cognitive radio and dynamic spectrum sharing is to open the TV whitespaces for other users. Best practices for sharing these bands are currently being studied by International Telecommunication Union (ITU) Radio-communication Study Groups 1, 5 and 6. ITU-R Study Group 1, working on Spectrum Management, is expected to provide a Report on best practices in Spectrum management for cognitive radio systems by mid-2014.

CRNs generate new challenges to a variety of stakeholders. For RF equipment vendors, the challenge is to build equipment which can be operated on any spectrum band and capable of doing on the fly changes in the radio transmission parameters. For wireless service providers, the challenge is to use their own resource efficiently and opportunistically use the other bands while maximizing their profit. As other providers can use their band opportunistically, security is also a major concern from user's perspective. Ensuring the licensing policies which enable the secondary usage of the spectrum is a challenge for regulatory bodies. Licensing policies should provide benefit to licensed users as well as cognitive users.

To deal with the cognition in radio (i.e. physical) layer, upper layers should also be modified to adapt to the reconfiguration done at physical layer. Thus, adaptation and cognition should also be included at upper layers to achieve the benefits of cognitive radio technology. This makes a network as cognitive network (Thomas, 2006). According to (Thomas, 2005) cognitive network is defined as:

A cognitive network is a network with a cognitive process that can perceive current network conditions, and then plan, decide and act on those conditions. The network can learn from these adaptations and use them to make future decisions, all while taking into account end-to-end goals.

As cognition should be incorporated in upper layers to effectively utilize the benefits of cognition at physical layer cognitive networks and CRNs are used interchangeably in literature. In this book chapter also we use these term to refer to network that includes cognition at upper layers also.

Initially the research community focused on the issues assuming that the Cognitive User (CU) is unlicensed user (ISM band user). ISM band was initially reserved internationally for industrial, scientific and medical purposes other than telecommunications. Now ISM band is shared with license-free communication applications such as wireless sensor networks, wireless LANs, cordless phones etc.

These applications have made ISM band overcrowded and hence opportunistic access to other bands may solve the spectrum scarcity problem in these bands. Now research community has also started working on Cognitive Cellular Networks (CCN). In CCN, the secondary user is also a licensed user of some other band and opportunistically accessing PU's band. Simplest model from the regulatory point of view is where cellular users are allowed to opportunistically use TV white spaces. IEEE 802.22 is the standard based on CR technology for WRAN that uses UHF/VHF TV bands. Paper (Junfeng, 2013) discusses about using these TV bands for LTE network. Cognitive networks can be broadly classified as infrastructure-less cognitive networks and infrastructure-based cognitive networks as shown in Figure 1. In the infrastructure-based cognitive networks, the cognitive base-station collects the spectrum information acquired locally by the cognitive users and makes the decision. According to this decision, each cognitive user reconfigure its communication parameters. Infrastructure-less cognitive networks or Cognitive Radio Ad Hoc Networks (CRAHNs) (Akyildiz, 2009) require distributed protocols. In CRAHNs, cognitive users need to have the cognitive capabilities and they are responsible for determining the actions based on the local observation. Since cognitive users cannot predict the influence of their actions on the entire network, cooperative schemes are essential. In cooperative schemes, the observed information can be exchanged among cognitive users to broaden the knowledge on the network.

This chapter describes the technical challenges of cognitive networks and existing strategies for service providers and regulatory bodies. Rest of the book chapter is organized as follows: Second section presents the regulatory models for spectrum access. Third section discusses the cognitive cycle for dynamic spectrum access. Section 4 presents the physical layer functionalities including spectrum sensing, geo-location based method for primary user activity and spectrum handoff. MAC layer functionalities and approaches are discussed in section 5. Section 6 gives the details about network layer routing in cognitive radio networks including some of the existing schemes from literature. Section 7 briefly talks about transport layer in cognitive networks. Finally section 8 presents the real hope for designing protocols in cognitive radio networks, i.e. cross layer solutions.

Figure 1. Classification of Cognitive Networks

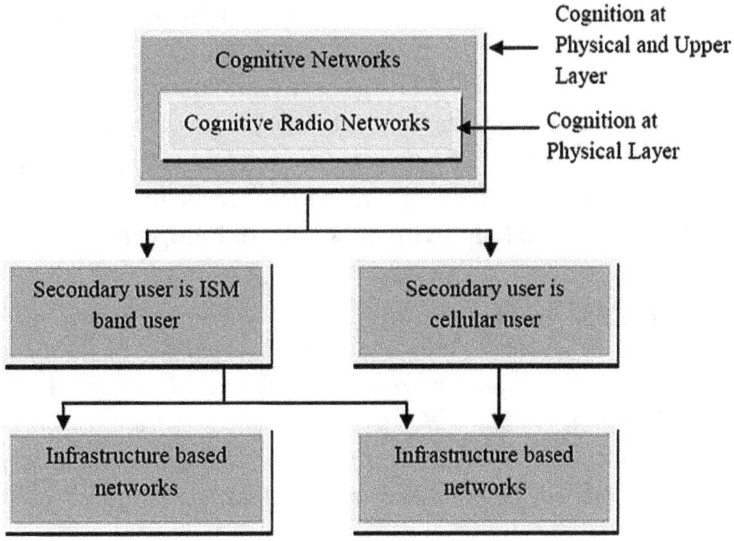

REGULATORY MODELS FOR SPECTRUM ACCESS

When we talk about dynamic spectrum access, the first question comes is "What are the accessibility models for spectrum?"

Simplest model from the regulatory point of view is where cellular users are allowed to opportunistically use TV white spaces. IEEE 802.22 is the standard based on CR technology for WRAN that uses UHF/VHF TV bands. General accessibility models which allow the opportunistic access of any band are discussed below.

The first model is Collective Use of Spectrum (CUS). CUS does not provide exclusive licenses to spectrum users and hence does not provide protection in interference. CUS application must share the spectrum with other users. However, some technical restriction can be applied by the national regulatory bodies. Traditionally cellular networks are assigned exclusive licenses to mobile network operators (MNO) and hence guarantee QoS to licensed users. This is not possible in CUS and hence MNOs are not interested in this kind of shared access of spectrum. Radio Spectrum Policy Group (RSPG) of European commission defines CUS as:

Collective Use of Spectrum allows an unlimited number of independent users and/or devices to access spectrum in the same range of designated CUS frequencies at the same time and in a particular geographic area under a well-defined set of conditions.

Authorized Shared Access (ASA) and Licensed Shared Access (LSA) are the models that allow shared use of licensed spectrum. Idea is to share the spectrum when and where primary licensee is not using it. According to ASA, primary user has exclusive rights for the spectrum band and could share its underutilized spectrum with other users with predetermined rules and conditions. These conditions may be static based on location and time or may be dynamic based on frequency, time, location and traffic of primary users. ASA provides QoS to the primary users as well as increases the spectrum utilization with non interference bases. Nokia and Qualcomm proposed the concept of ASA. They claim that this model is well suited for small cells due to their lower transmit power levels. Nokia and Qualcomm are working and providing solutions for LTE ASA, they claims that ASA can be applied to existing LTE-Advanced infrastructure and there will be no special impact on the device side beyond implementing the ASA frequency band. In September 2013, Nokia Solutions and Networks demonstrated world's first authorized Shared access field trial with TD-LTE spectrum (Matinmikko, 2013). USA based company, Spectrum Bridge Inc, is also giving solutions for this kind of shares access.

ASA is generalized for sharing spectrum between different types of systems in LSA. LSA basically focuses on the regulatory aspects of ASA. European and USA regulators are working to develop architecture/rule for LSA. Radio Spectrum Policy Group (RSPG) of European commission defines LSA as:

A regulatory approach aiming to facilitate the introduction of radio-communication systems operated by a limited number of licensees under an individual licensing regime in a frequency band already assigned or expected to be assigned to one or more incumbent users. Under the Licensed Shared Access (LSA) approach, the additional users are authorized to use the spectrum (or part of the spectrum) in accordance with sharing rules included in their rights of use of spectrum, thereby allowing all the authorized users, including incumbents, to provide a certain Quality of Service (QoS).

Figure 2. LSA Architecture

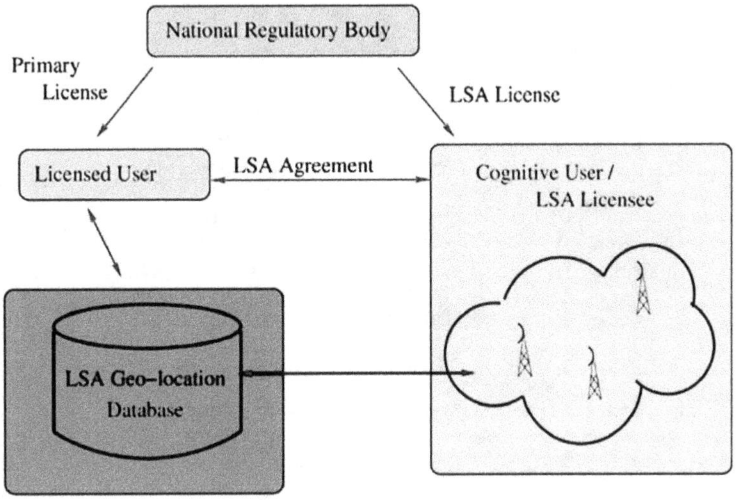

LSA regulatory architecture is shown in Figure 2. National Regulatory Body (NRB) provides primary licenses and LSA licenses to primary user and cognitive user/LSA licensee respectively. LSA agreement between primary user and LSA licensee outlines the terms and conditions of shared use of spectrum including details about geographical area, technical conditions, when and how to vacate the spectrum. LSA geo-location database provides information about the spectrum availability to LSA licensee. Using secure access protocols and policies, the database ensures protection and enforcement of national regulatory policies and objectives, while minimizing restrictions to shared use by LSA licensees. This database may be operated by NRBs, the primary users or a trusted third party.

COGNITIVE CYCLE FOR DSA

Cognitive cycle for dynamic spectrum access is shown in Figure 3. The cognitive cycle builds the spectrum opportunity map identified by the sensing block and schedules the resources dynamically among the cognitive users. Furthermore, the cognitive cycle allows cognitive users to vacate the selected channel when a primary user becomes active on that channel.

Following are the main features of spectrum management in cognitive cycle:

- **Spectrum Sensing:** The cognitive users are allowed to use the unused portion of spectrum. Therefore, a cognitive user should monitor the available spectrum bands to detect spectrum opportunities. Spectrum sensing can be done by the cognitive users or a trusted third party. In case of third party sensing, third party shares the geo-location database of spectrum with the cognitive users. Some efforts are already taken by companies to provide geo-location database of spectrum. FCC has approved Spectrum Bridge Inc.'s for providing spectrum database services for television white spaces in USA since January 26, 2012. Google is also working with industry and regulators for TV whitespaces database. Google spectrum database had been certified by the FCC and is available to wireless devices that are approved by the FCC for TV white space bands. Google's

Figure 3. Cognitive Cycle

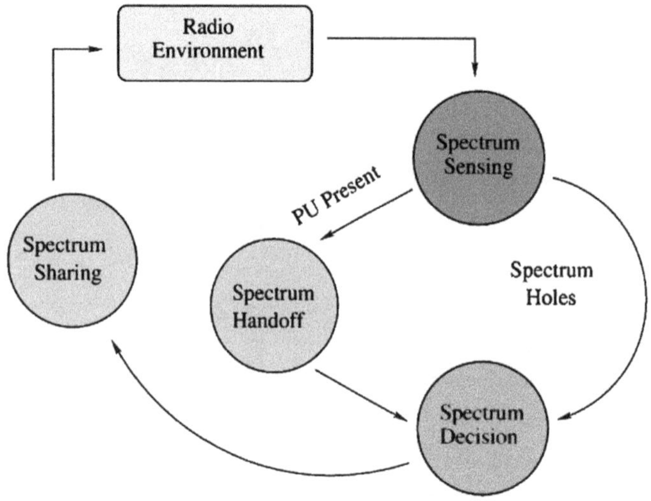

database can be accessed by researchers, device manufacturers and anyone interested through Google API for identification.

- **Spectrum Decision:** Once the available spectrum bands are identified, cognitive users select the most appropriate band according to their QoS requirements.
- **Spectrum Handoff:** If primary user presence is detected in the specific portion of the spectrum in use, cognitive user should vacate that spectrum and continue their communication in some other vacant band. Spectrum handoff process needs to communicate with spectrum sensing process to find the vacant band.
- **Spectrum Sharing:** As multiple cognitive users may be looking for the spectrum at same time, transmission between cognitive users should be coordinated to avoid collisions. This spectrum sharing capability is taken care by spectrum sharing process. Spectrum sharing can be classified as centralized spectrum sharing and distributed spectrum sharing. Spectrum sharing can also be classified as cooperative spectrum sharing and non-cooperative spectrum sharing. In cooperative sharing, communication effect to other cognitive users is considered while selecting the channel. While in non-cooperative sharing, channel is selected based on local policies. For networks where centralized infrastructure is present, spectrum sharing is done through the base-station and hence comes under cooperative sharing. For infrastructure-less networks, where the controlling infrastructure is not present, spectrum sharing is done based on local policies. More about spectrum sharing is discussed in MAC layer functionalities.

PHYSICAL LAYER FUNCTIONALITIES

Primary user detection, spectrum handoff and designing a transmitter and receiver for cognitive radio are the main issues for cognitive radio physical layer. Energy consumption in transmitting a signal depends on the transmit power level of the signal, thus power control is also required in CRNs. Selecting physical layer parameters based on the channel selected at MAC layer is also a functionality of physical layer. The well known sensing techniques from the literature with their pros and cons are discussed in this section. An emerging alternative to spectrum sensing, i.e., spectrum database with its key feature is also presented. Various strategies for spectrum handoff are also discussed.

Spectrum Sensing

Opportunistic access to the unused frequency band across licensed frequency bands is the key concept of DSA. It aims to maximize spectral efficiency and capacity within congested wireless environment. Availability of ideal radio spectrum varies with time, frequency and location resulting in spectrum opportunities called spectrum holes. Cognitive radio (CR) users can utilize these spectrum holes opportunistically, when licensed primary user (PU) is not operating on that frequency band. Spectrum sensing is an important technique for implementation of DSA. It helps CR users to identify unutilized spectrum band reliably and also determine quickly if PU becomes active, so CR users can vacate the band immediately to minimize the interference caused to PU. Generally, spectrum sensing techniques can be classified into three groups, namely, non-cooperative sensing, cooperative sensing, interference-based sensing as shown in Figure 4. These are described below.

Non-Cooperative Sensing

Generally, CR users do not have real-time interaction with primary transmitter and receiver. Therefore, CR users need to detect the signals from primary transmitter before using the spectrum bands. Non-cooperative sensing is based on a well-known technique called signal detection, which can be described as a method to detect the presence of signal in a noisy environment. The model for signal detection at time t can be described as (Yucek, 2009):

$$\begin{aligned} H_0 &: x(t) = w(t) : \textit{signal is absent} \\ H_1 &: x(t) = h * s(t) + n(t) : \textit{signal is present} \end{aligned} \tag{1}$$

where $x(t)$ is the received signal of a CR user, $s(t)$ is the transmitted signal of a licensed PU, $n(t)$ is additive white Gaussian noise (AWGN) and h is the channel gain. Here, H_0 and H_1 are two hypothesis of the absence and presence of signal transmission from a licensed PU respectively.

Figure 4. Classification of spectrum sensing

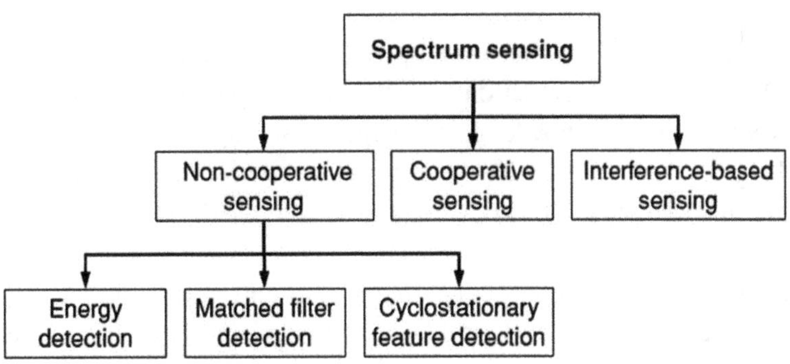

The performance of a spectrum sensing technique is generally measured in terms of the probability of correct detection (P_d), the probability of false alarm (P_f), and the probability of miss detection (P_m). Mathematically,

$P_d = Prob\{decision = H_1 \mid H_1\}$, $P_f = Prob\{decision = H_1 \mid H_0\}$ and
$P_m = Prob\{decision = H_0 \mid H_1\}$.

The three well-known techniques of non-cooperative sensing with their advantage and disadvantage are discussed below.

- Matched Filter Detection: A matched filter (Barry, 2004) is the optimal linear filter which maximizes the signal to noise ratio for a given input signal in the presence of stochastic additive white noise. The matched filter correlates a known signal or template with an unknown signal to detect the presence of the template in the unknown signal. This is equivalent to convolving the unknown signal with the filter whose impulse response is the mirror and time shifted version of a reference signal. The operation of matched filter detection is expressed as below:

$$y[n] = \sum_{k=-\infty}^{\infty} h[n-k]x[k] \tag{2}$$

where 'x' is unknown signal, 'h' is impulse response of the matched filter that is matched to the reference signal to maximize the SNR.

Most of the wireless technologies include transmission of pilot sequences, preambles or spreading codes to allow channel estimation and synchronization. CR user can exploit this information to detect presence of signal transmission from primary user. Matched filter detector achieves optimal performance, if pilot signal is already known to CR user. However, lack of synchronization between primary user and CR user can significantly deteriorate the detection accuracy.

Major advantages and disadvantage of matched filter are given below:

Advantage:

- It needs less time to achieve high processing gain and probability of false alarm and missed detection due to coherent detection.

Disadvantages:

- It requires prior information of primary user signal which is very difficult to be available at the CRs.
- It requires dedicated sensing receiver for all primary user signal types.
    - Energy Detector: Energy detection is non-coherent spectrum sensing technique, which does not require any a priori knowledge of primary user signal (Urkowitz, 1967). Energy of the received signal is compared with the threshold to detect the presence of signal transmission of primary user (Digham, 2007).

To measure the signal energy in a particular frequency band in time domain, energy detector simply needs a bandpass filter, an analog to digital (A/D) converter, squared law device and an integrator.

First, input signal is passed through bandpass filter (BPF) to band limit the signal. Then filtered signal is squared and integrated over an observational time T. Finally, output of integrator is compared with the threshold to decide presence of primary user's transmission in the corresponding frequency band.

Let $x(n)$ be the sample of input signal. The decision rule can be stated as follow:

$$decide\begin{cases} H_0 & if\ E_{time} < \gamma \\ H_1 & if\ E_{time} > \gamma \end{cases} \tag{3}$$

where $E_{time} = E[|\ x(n)\ |^2]$ is the estimated energy of received signal, γ is the threshold derived from the statistics of the noise. In practice, energy detector uses approximated $\hat{E}_{time} = \sum_{n=1}^{N} |\ x(n)\ |^2$ The block diagram of energy detection in time domain is shown in figure 5.

The advantages and disadvantages of energy detection are listed below:

Advantages:

- Energy detector is simplest spectrum sensing technique and it does not require any knowledge of the signal of primary user.

Disadvantages:

- It is susceptible to the uncertainty of noise power.
- It can only detect the presence of the signal but cannot differentiate the type of signal (e.g. signals from CR users sharing the same channel with the primary user). Therefore, the detection error would be high in presence of signal sources other than the licensed user.
    - Cyclostationary Feature Detector: Cyclostationary feature detection relies upon periodic redundancy introduced into a signal by sampling and modulation (Gardner, 1991). It uses non-random statistics to detect and possibly even classify a signal of interest.

A random process is called cyclostationary when its mean and autocorrelation show periodic behavior in time. When a transmitted signal is modulated using sinusoid carriers, cyclic prefixes (as in OFDM) and code or hoping sequences (as in CDMA), modulated signal shows cyclostationary behavior, while noise still remains random and does not exhibit any periodicity. Hence, cyclostationary feature detector can effectively detect signal even in extremely low SNR environments.

In practice, a signal property called Spectral Correlation Function is measured to detect the signal. A cyclic detector calculates the cyclic autocorrelation function (CAF) given by (Enserink, 1994):

Figure 5. Block diagram of energy detection in time domain

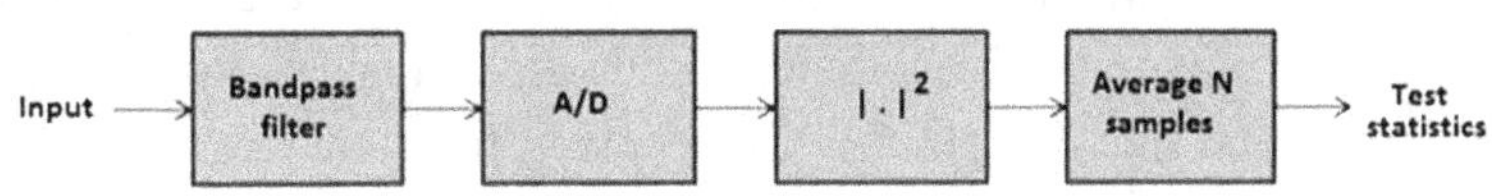

$$R_X^\alpha(\tau) = \lim_{T\to\infty} \int_T x(t+\frac{\tau}{2})x^*(t-\frac{\tau}{2})e^{-j2\pi\alpha t}dt \quad (4)$$

The Fourier transform of the CAF, which is equal to the spectral correlation function (SCF), is given by:

$$S_X^\alpha(f) = \int_{-\infty}^{\infty} R_X^\alpha(\tau)\, e^{-j2\pi f\tau} d\tau \quad (5)$$

To derive a normalized version of SCF, spectral coherence function (SOF) is given by:

$$C_X^\alpha(f) = \frac{S_X^\alpha(f)}{[S_X^0(f+\frac{\alpha}{2}) \times S_X^0(f-\frac{\alpha}{2})]^{1/2}} \quad (6)$$

The detection of the presence and absence of signal is performed based on scanning the cyclic frequencies of spectral coherence function. At a given cyclic frequency, if spectral coherence function is above the threshold level, the signal is present otherwise signal is absent.

The advantages and disadvantages of cyclostationary feature detection are given below:

Advantages:

- Cyclostationary detection can provide a more accurate sensing result and it is robust to variations in noise power.

Disadvantages:

- The detection is complex and requires long observation periods to obtain the sensing result. Also, the performance of the detector degrades in the presence of timing and frequency jitters (which smear out the spectral lines), and RF non-linearities (which induce spurious peaks).

In order to improve the overall performance of spectrum sensing, multiple sensing techniques can be integrated together. E.g. energy detector can be used for quick scanning over entire frequency band and frequency bands which have high energy densities can be eliminated. Then, cyclostationary feature detection can be used for accurate detection of signal transmission of primary user over remaining frequency band.

Cooperative Sensing

Although, non-cooperative sensing techniques discussed earlier perform well in an AWGN channel, these techniques provide poor results in multipath fading environment. Also, these techniques may not detect the signal from the primary transmitter due to hidden node problem as shown in figure 6. Additionally, CR transmitter can have a good line of sight to a CR receiver, but it may not able to detect the primary transmitter due to shadowing effect as shown in figure 7. Hence, non-cooperative sensing alone does

Figure 6. Transmitter detection problem: hidden node problem

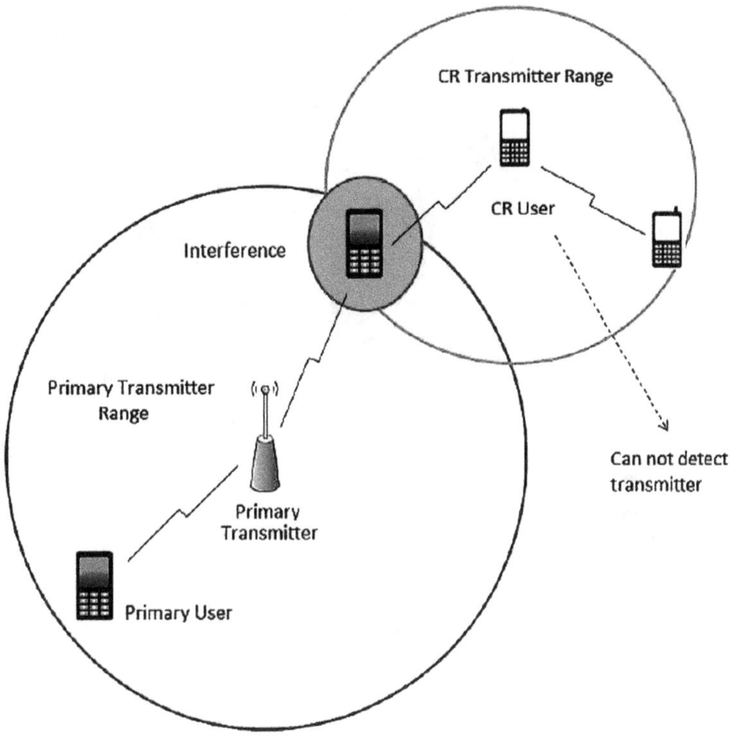

not avoid causing interference to the primary user. To overcome these issues, cooperative sensing can be used, where multiple CR users share spectrum sensing information among each other to detect the presence of the primary user. The cooperative sensing significantly improves the accuracy of detection by mitigating multipath fading and shadowing effect and it also solves the hidden node problem. However, this incurs a greater communication and computation overhead compared with non-cooperative sensing.

The cooperative spectrum sensing can be centralized or distributed (Ganesan, 2005). In centralized cooperative spectrum sensing (CCSS), each CR user performs spectrum sensing independently through sensing channel and makes hard or soft decision on whether the primary user is present or not. All CR users forward this information to central entity through reporting channel. Central entity fuses the CR decisions using some fusion logic and makes a decision about the presence or absence of primary user. Then, the central entity broadcasts this information to all CR users on broadcast channel. As, CCSS requires to be gathered all the information at the central location, CCSS is less scalable, less fault tolerant and has redundant reporting cost in large cognitive network. To overcome this problem, distributed or decentralized cooperative spectrum sensing (DCSS) can be used. In DCSS, CR users share information through local communication, in order to make their own decisions. DCSS is more reliable and scalable as it does not require any backbone infrastructure. Moreover, control overhead in DCSS is comparatively less than CCSS. Few state-of-the–art cooperative techniques are discussed in (Akyildiz, 2011).

Advantages of cooperative sensing: The effect of multipath fading, shadowing and hidden node problem can be significantly reduced using cooperative sensing. It reduces false alarm probability and increases detection accuracy.

Figure 7. Transmitter detection problem: shadowing effect

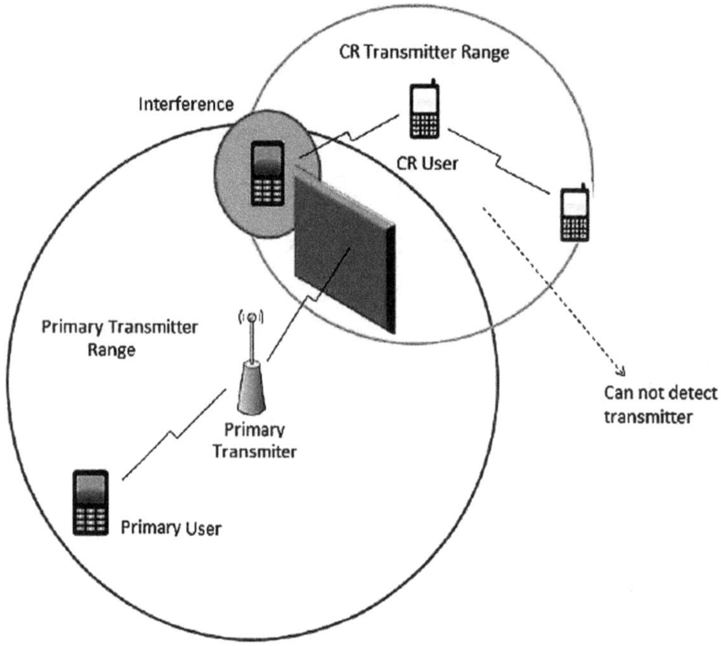

Disadvantages of cooperative sensing: The CR user need to perform sensing periodically as sensed information become obsolete fast due to mobility and channel impairments etc. This considerably increases data overhead. It also requires geographical spread of cooperating nodes to get benefits of cooperation.

Interference-Based Sensing

In this case, the sensing algorithm will measure the noise/interference level (from all sources of signals) at the receiver of the licensed user. This information is used by an unlicensed user to control the spectrum access (e.g. by computing expected interference level) without violating the interference temperature limit. Alternatively, an unlicensed transmitter may observe the feedback signal from a licensed receiver to gain knowledge on the interference level.

Geo-Location Database

An important issue in spectrum sensing is the reliability of the partial measurements performed by a single user. Mobile radio systems suffer from distortions like multipath fading, shadowing, causing severe degradation to the signal, which may lead opportunistic secondary users not to detect the primary activity in a certain moment or location. The sensors may suffer from the hidden node problem in which the primary signal strength at the non-licensed user position is weak but opportunistic transmission will still interfere with the licensed operation. Cooperative spectrum sensing can prove to be the useful solution to alleviate the above mentioned problem. In cooperative sensing, the decision on channel status is based on measurements from multiple cognitive users (or sensors) rather than a single user. The reliability of the decision is proportional to the number of cognitive users participating in channel measurement.

Cooperative sensing shortens the sensing time of the spectrum while improving overall sensitivity. In cognitive wireless sensor networks (CWSNs), cooperative sensing is facilitated by selecting a cluster head which combines fuses the sensing data from multiple nodes to give a reliable decision.

In cellular networks, along with the channel status, the geo-localization of these measurements can be used for development of a geo-location database (Geo-DB) (Osa, 2012). The Geo-DB stores the spectrum holes in a certain frequency band at a given time and location. One of the example for spectrum holes in frequency band is that of TV white space (or spectrum holes), denoted by TVWS. The white space is the result of releasing of frequency band resulting from the analog-to-digital TV switchover, referred as digital dividend (DD) in literature. The Digital TV (DTV) band is a typical example of inefficient spectrum use since, depending on the geographical location, only certain channels are occupied. The information about these TVWS with respect to a given geographical location and time can be stored in a Geo-DB and used to opportunistically access them by the cellular network. The latest cellular network standard being deployed is LTE Advanced (LTE-A) where LTE stands for Long Term Evolution. Discussion on LTE-A is beyond the scope of the chapter.

In order to maintain and update the database, cognitive users sense the channel and report measurements using IEEE 802.21 Multimedia Independent Handover (MIH) standard (Taniuchi, 2009). The network retrieves and compiles the cooperative sensing measurements and their geographical location in order to decide upon the channel vacuity in a certain location and thereby updating the Geo-DB. The geo-location information is used to propose a power control mechanism for opportunistic access to the primary band, aiming to minimizing the interference to licensed and other secondary users. Figure 8 shows the basic architecture to implement and maintain Geo-DB. A new node in the LTE-A referred to as the cognitive resource manager (CRM) is introduced, which coordinates the opportunistic specification access (OSA) to the unlicensed spectrum based on the notifications reporting the channel status perceived by the cognitive user equipments (UEs). LTE-A specification considers UE localization through the LTE positioning protocol (LPP) and LPP annex (LPPa) (3GPP TS 36305). Several different positioning methods are mentioned in the standard, namely: observed time difference of arrival (OTDoA), assisted-Global navigation Satellite system (A-GNSS) and enhanced-cell ID (E-CID). All of these positioning methods are based on measurements collected by the UE or the eNodeB (.i.e., the Base station in LTE-A). The mobility management entity (MME) is the entity that receives the request for the localization of a UE from another entity such as another UE, eNodeB (connected through S1 link) or other nodes. Then, the MME sends a location service request to the enhanced serving mobile location centre (E-SMLC), which will execute the positioning procedure through LPP and LPPa protocols. The SLs interface between E-SMLC and MME serves as a tunnel for E-SMLC to transparently carry LPP and LPPa. The CRM and E-SMLC are interconnected using the MME. CRM requests the location service to the MME, which will activate the E-SMLC service. The resulting location calculated by the E-SMLC is sent to the MME that finally forwards it to the CRM. With the obtained information, the CRM can map sensing reports and location to build the Geo-DB.

Owing to user's positioning capability, the Geo-DB will contain valuable information about which frequency bands can be used by a given eNodeB at a specific moment of time and the maximum coverage range in order not to interfere with the primary system. The CRM collects the sensing information from the UEs and the positioning information from the location service provided by the MME and updates the database after the cooperative decision has been taken. Once this process is finished, the CRM possesses the location of every opportunistic UE and which spectrum bands are suitable for OSA and which not. The Geo-DB will contain information about the occupation of different spectrum bands in the licensed

Figure 8. Model for implementing and maintaining Geo-DB (Osa, 2012)

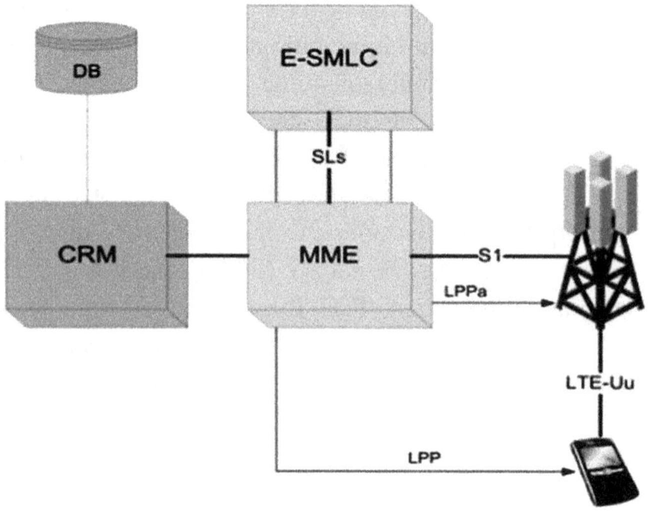

spectrum band on a per-cell basis, including also the maximum coverage distance from the eNodeB. This way, an eNodeB, identified in the Geo-DB, is able to opportunistically use those spectrum bands with reduced transmission power in case the maximum ranged is detailed in the corresponding register. Otherwise, the corresponding resource can be used without restrictions regarding the transmission power. The Geo-DB also contains the final decision concerning the different resources. The information contained in the Geo-DB must be periodically updated in order to take account the potential utilization patterns in the licensed spectrum especially if those changes are due to the primary system activity. A cooperative decision taking into account the information provided by all the users inside a certain range would allow increasing the sensing accuracy. An up-to-date database will reduce the collision probability that could be caused by the lack of synchronization between real state of primary spectrum and the available information stored in the database.

The cooperative decision making mechanism is implemented in the CRM. Without loss in generality, we will carry out analysis on per spectrum band basis. The LTE-A specific per-resource block (RBs) based analysis can be studied in (Osa, 2012) by the interested readers. The input data considered by the decision mechanism includes all the channel state reports from the UEs served by the same eNodeB and the geographical position of the UEs obtained through the aforementioned location services. Every channel state notification creates a new entry in the CRM containing the UE identifier, the estimated geographical location of the UE, the channel or resource monitored, the licensed activity state sensed in that resource and the time in seconds when the channel state report was received. Once all this data is collected from different UEs, the CRM can make decisions about the vacuity of the monitored resources at different locations. The collected sensing reports are classified by their distance to the eNodeB in different ranges. For each range and spectrum band, an independent cooperative decision will be made. Multiple samples of the sensing data from different resources and UEs are combined in order to update the Geo-DB. For the same UE and sensed resource, only the most recently collected data is used in the cooperative decision calculation.

There are two hard-decision rules for cooperative decision making are introduced in (Osa, 2012). The conservative rule declares the resource as free from licensed activity if all the UEs report such state; otherwise the channel is considered to be occupied. On the other hand, the aggressive strategy declares the resource as idle provided a single UE senses the channel as free. However, due to the abovementioned single sensor measurement uncertainty, the different measurements reported over time must be considered in the final decision. That is to say, old measurements must not have the same importance in the final decision as the newest reports due to the fast-changing radio-channel state conditions. Soft-based cooperative decision stands on this idea and performs better than the abovementioned hard-decision rules.

In soft-based cooperative decision, each reported measure has an associated weight. The CRM will combine the information of the spectrum band state with the weights and will make a decision by comparing the result with a defined threshold. Each spectrum band state notification is weighted according to the elapsed time between the moment the notification was received by the CRM (i.e., the ith notification is received at time t_i) and the instant when the channel state decision is made- t_{now} -as seen in Equation (7). T_{MAX} is the time elapsed between two consecutive measures of a specific primary spectrum band. In addition to this linear weight equation, the quadratic and the square root version of the formula is analyzed in (Osa, 2012).

$$w_i = \frac{T_{MAX} - (t_{now} - t_i)}{T_{MAX}} \tag{7}$$

The decision regarding the resource availability for opportunistic access will be taken according to the value of the spectrum decision metric on a given spectrum band, defined as:

$$U = \sum_{i=0}^{N-1} \frac{(-1)^{d_i} N_{d_i} w_i}{N^2} \tag{8}$$

where d_i is the state of the monitored band reported in the ith notification, equal to 0, if spectrum band is sensed as idle or equal to 1, if spectrum band is sensed occupied. N is the number of notification events considered in the decision making mechanism, including free spectrum band notifications, N_0, and occupied spectrum band notifications, N_1. N_{d_i} is the number of measurements that agree with the state of the spectrum band notified in the ith measurement report: N_0 if the spectrum band was sensed occupied, N_1 otherwise. As reflected by (8), the spectrum decision metric U will depend on the number of measurements taken into consideration, the weight of each measurement and the most reported single-sensor measurement. If most of the measurements agree on the spectrum band vacuity, a positive value of U is obtained. On the contrary, if the majority of measurements reports that the spectrum band was occupied, a negative value of U is obtained. Note that, in order to normalize the equation, it is needed to divide it by the second power of N, in such a way that $-1 \leq U \leq 1$. Depending on the value of U, the considered spectrum band is stated as a candidate for OSA (H_0) if that result is greater than a certain threshold, denoted as γ. Otherwise, the resource is not available for opportunistic usage (H_1). The decision threshold γ must be tuned in order to provide the largest OSA probability without exceeding the interfering limit.

$$Decision = \begin{Bmatrix} H_0 & if\, U < \gamma \\ H_1 & if\, U > \gamma \end{Bmatrix} \tag{9}$$

Spectrum Handoff Techniques in Wireless Network

Spectrum handoff is referred to the process where cognitive user switches the spectrum band physically and reconfigures the communication parameters (e.g. operating frequency, modulation type) according to the new spectrum. Handoff techniques largely depend on the wireless network where it is being deployed. We here would analysis handoff techniques in two major commercially used wireless networks: 1) Infrastructure less network and 2) Cellular network which is an infrastructure based network.

Cognitive Radio based Infrastructure Less Network

Spectrum handoff in infrastructure-less networks are done with local knowledge of spectrum holes. Spectrum handoff here can be implemented based on two strategies: reactive spectrum handoff and proactive spectrum handoff. In reactive handoff, cognitive user performs spectrum switching after detecting the presence of primary user. Hence, immediate switching is needed in this kind of strategy. In proactive spectrum handoff, cognitive user predicts the primary user activity and decides the new spectrum. Once the spectrum decision is done, cognitive user switches the spectrum before the link failure happen. Thus, spectrum handoff could be soft handoff or hard handoff as in traditional networks. As discussed in cognitive cycle spectrum decision block decides the spectrum to be used.

Cognitive Radio based Cellular Networks

The basic working principle of cognitive radio deployed in cellular network is same as that of the WSNs. The spectrum handoff process in cellular networks differs from the WSNs in the fact that the handoff process is controlled by the eNodeB or the base station (BS). The system architecture for spectrum handoff in cellular networks, discussed here, is adopted from (Lee, 2012) and can be applied to any standard in cellular network (including LTE-A). The architecture defines a spectrum pool structure to facilitate spectrum mobility. The spectrum pool is a set of contiguous licensed spectrum bands. A cell can have more than one spectrum pool operating at different frequencies (figure 9). To facilitate seamless communication, the architecture defines two types of cell coverage as shown in figure 9(a). 1) Basic area (BA): a basic coverage not overlapped with that of neighbor cells. 2) Extended area (EA): a large coverage overlapped with basic areas of neighbor cells. A cell has multiple basic spectrums accessible only within BA and at least one extended spectrum band supporting both BA and EA. The EA of current cell is adjacent to the cells having the same spectrum pool (extended neighbors). There are four kinds of spectrum handoff defined in (Lee, 2012) for cellular networks, they are Intracell/intrapool handoff, Intracell/interpool handoff, Intercell/intrapool handoff and Intercell/interpool handoff. This is shown in figure 9(b).

Figure 9. a) Spectrum pool-based CR network architecture b) Different handoff types in CR network (Lee, 2012)

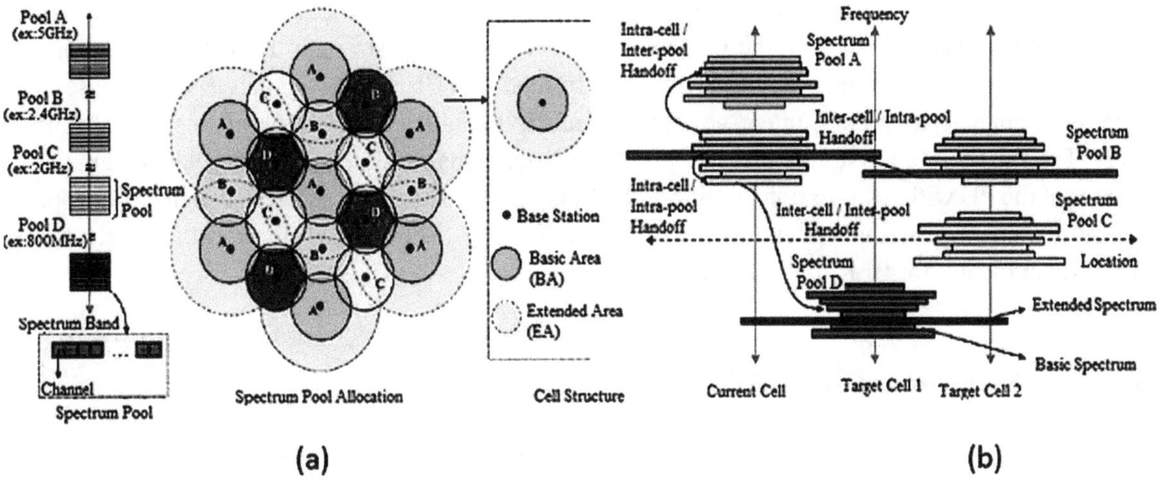

The problem of co-channel interference (CCI) arises, when the CR user (or secondary UE) uses the licensed spectrum when PU (or primary UE) is active in the band. This initiates spectrum handoff procedure to eliminate CCI. Once it is decided which users would undergo spectrum handoff, the next step is to choose the optimal handoff scheme. The eNodeB (or BS) controls all the spectrum handoff activities (Lee, 2012). The spectrum handoff in Cellular network follows the following procedure (Pradhan, 2014):

1. When a CR user or secondary UE is made to switch the spectrum band, the first option is intracell/intrapool handoff. In this case, the UE looks for the available spectrum band in the same spectrum pool in the current cell.
2. In case, there is no available unused spectrum band in the spectrum pool, then the UE undergoes intercell/intrapool handoff by switching to the extended neighbor through the EA of the cell. The extended neighbor has the same spectrum pool as that of the current cell and hence this handoff technique doesn't require RF reconfiguration (Lee, 2012).
3. In the situation, where the spectrum of EA of the cell is occupied, intercell/intapool handoff is not possible. The ideal handoff in that case will be intracell/interpool handoff, where the UE shifts to another spectrum pool operating at different frequency range within the same cell. This requires reconfiguration of the RF-front end.
4. In worst case scenario, where all the spectrum pool is unavailable in the current cell, the only option left is intercell/interpool handoff. This has tremendous overhead as it not only involves switching of eNodeB but also RF-front end reconfiguration.

Both the intercell/intrapool and intercell/interpool handoff techniques are proactive methods, .i.e., when secondary UE undergo handoff events, they perform handoff procedures while maintaining communication. After secondary UE make all decisions on the handoff, they cut off communication channels and switch to a new spectrum band or a new eNodeB. Since each cell has time-varying wireless resource because of the dynamic nature of underlying spectrum in CR networks, it cannot have a permanent extended spectrum. The extended spectrum of the current cell cannot be used in its extended neighbors, to avoid co-channel interference, leading to decrease in their capacity. As a result, CR networks necessitate an intercell resource allocation scheme to select and maintain the extended spectrum (Lee, 2012).

When secondary UE is selected for intercell handoff, the selected UE needs to find the proper target cell. Unlike the classical handoff, the UE cannot observe the signal strength from other neighbor cells while maintaining the connection to current cells. Instead, the UE selects the new eNodeB based on a stochastic connectivity model. Interested readers can find detailed explanation of this model in (Lee, 2012). The figure 10 analyses the bit error rate (BER) performance of a BPSK modulated signal at the receiver of the PU with CR user transmitting at the same licensed band. It can been seen that the proposed spectrum handoff procedure discussed above has 5.4dB gain over the traditional cellular architecture with respect to CCI for BER of 10 (Pradhan, 2014).

MAC LAYER FUNCTIONALITIES

Here we discuss the MAC layer issues and existing solutions for CRNs. Functionality of the cognitive cycle discussed above is the main challenge of MAC layer in cognitive radio networks. MAC protocol for cognitive networks should support multi-channel operation. In cognitive networks, availability of channels is dynamic and also the channels are heterogeneous, i.e. they could be of different bandwidth and different frequencies with different propagation characteristics. In multichannel networks, MAC layer is responsible for channel assignment and medium access control. Due to the dynamic nature of cognitive networks some other challenges also arise in the design of MAC layer. Classifications of cognitive MAC protocols based on various factors are listed in the figure 11.

Based on spectrum access method, MAC protocols can be classified into contention based protocols, TDMA based protocols, hybrid protocols and OFDMA based protocols. To avoid interference to primary user in TDMA based spectrum access method, all secondary users should be synchronized and should follow the same timeslot division as of the primary user. Thus, TDMA based approach cannot be used for more than one kind of primary user. In cooperative spectrum allocation based MAC protocols, the

Figure 10. BER curve for PU with both PU and CR user at spatial space (Pradhan, 2014)

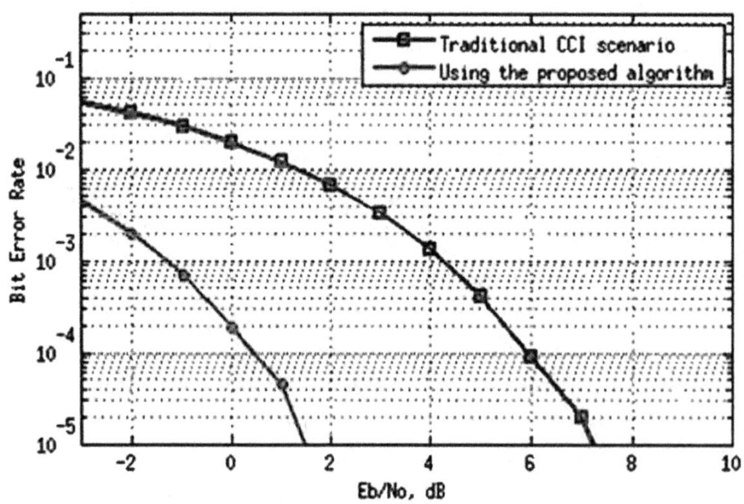

Figure 11. Cognitive MAC Protocol classification

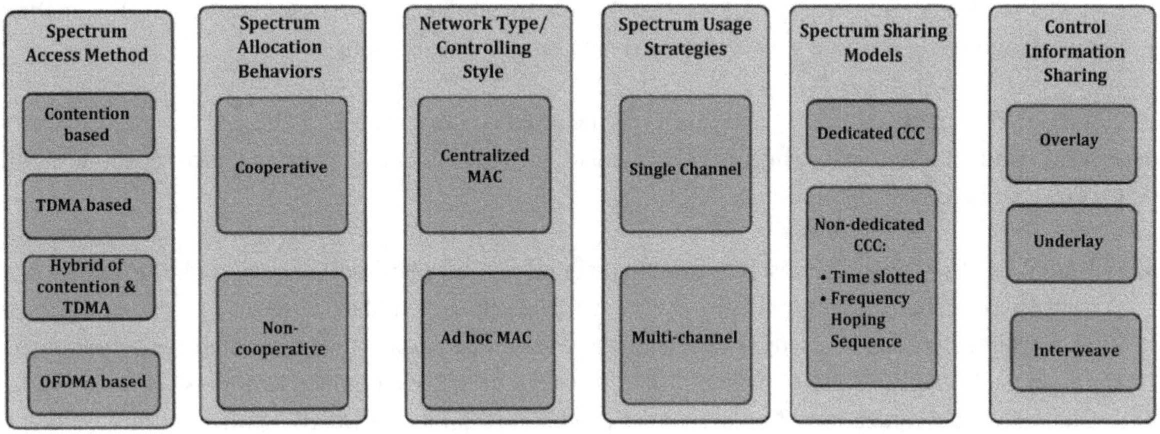

cognitive user also considers requirements from other cognitive users and fairly uses the channels. While in Non-cooperative spectrum allocation based MAC protocols, the cognitive user only considers its own requirement and voraciously access the channel whenever available. The MAC protocols can be classified into underlay spectrum sharing model based MAC protocols, overlay spectrum sharing model based MAC protocols and Interweave spectrum sharing model based MAC protocols according to spectrum sharing models. In the underlay spectrum sharing model based MAC protocols, a cognitive user can operate in the primary user's band if the interference caused to the primary user is below some threshold. Cognitive user is restrictive to short distance communication because of interference constraint of such protocols. In the overlay spectrum sharing model based MAC protocols, the cognitive user has access to the primary user codebook. Primary user codebook information is used at the cognitive receiver to cancel the interference caused by the primary user. Overlay and underlay spectrum sharing model based MAC protocols allow cognitive user and primary user to transmit simultaneously. Interweave spectrum sharing model based MAC protocols use primary spectrum band in spectrum opportunities. Below are some of the main issues of MAC layer in cognitive networks.

- Channel Assignment: In multi-channel environments, channel assignment is critical in achieving high channel utilization. Ideally, it is best to have the traffic equally divided among channels. Cognitive networks work with heterogeneous spectrum bands. Thus, the change in transmission channel also changes the transmission parameters and the rate of data transmissions in cognitive networks. Also, availability of channels is not certain due to primary user activity. This makes the selection of transmission channel from the spectrum holes identified by the spectrum sensing more challenging.
- Control information sharing: Control signaling is necessary for every communication protocol. It is much more important in cognitive environment to support transmission coordination and spectrum related information exchange between the cognitive users. It helps in neighbor discovery, spectrum sensing coordination, exchange of local measurements and other control signaling needed between the cognitive users. It can be done in following manner:
    - Common Control Channel (CCC): Portion of spectrum band is allocated for control signaling. CCC can be allocated as in-band CCC or out-of-band CCC depending on whether it shares the data channel or uses a dedicated spectrum, respectively.

    - Time Slot Based: Portion of time slot in all the channels are used for control signaling.
    - Frequency Hopping Sequence: Sequence of frequencies is used for control signaling. Cognitive users hop between these bands for control signaling.

For networks operated in ISM band assuming the global common control channel can be justified by using ISM band for control signaling. In other kind of networks, identification of control channel should be part of cognitive MAC protocol.

- Neighbor Discovery: The distance at which signal can be heard will change with the variation in transmission channel and transmit power level used, thus neighbor discovery is also a big issue. Two approaches can be used for neighbor discovery, one to use control channel for neighbor discovery and other is to broadcast neighbor discovery messages to all the channels.
- Channel Synchronization: Single channel protocols cannot be directly applied to multi-channel environment because the sender and the receiver have to be operating on the same channel. Here the focus is on the situation where each node is equipped with a single interface, so a node can only listen to one channel at a time. When two nodes are operating on the same channel, we say that their channels are synchronized. Thus, before the actual data transmission, both the transmitter and receiver should synchronize their respective radios to the same channel. The process of node moving from one channel to another is called channel switching. In cognitive networks, channel switching is more frequent as cognitive users must vacate the channel whenever primary user comes back. This requirement make channel synchronization as one of the important feature of cognitive MAC protocol.
- Medium access control: When a node transmits a packet onto the channel, the signal reaches all nodes in the transmission range of the sender. If a node receives multiple packets on the same channel at the same time, it cannot properly decode the packet because of interference. Thus, medium access control among the nodes is needed to avoid collision similar to simple wireless MAC strategies.
- Multi-channel Hidden Node Problem: In a multi-channel environment with single half-duplex transceiver, a node can only hear RTS and CTS transmitted on the tuned channel. In such case, a node is not able to hear other channel RTS and CTS, which leads to multi-channel hidden node problem.
- Channels Scheduling: When a node has packets queued for multiple destination nodes with different channels, channels scheduling is needed to schedule the packets on the radio interface based on the operating channel. Some basic approaches for channels scheduling are:
    - Fixed channel scheduling: The simplest method is to schedule channels with fixed period in a round robin manner. This scheme is not suitable where load is different among channels.
    - First-in first-out (FIFO): A node switches the channel based on the first packet in the queue. If we schedule packets using FIFO algorithm, in worst case, channel switching could be needed for every single packet which will lead to higher transmission delays.
    - Threshold based approach: To minimize the channel switching delay, this method tolerates a packet delay up to a certain bound. A node can stay on a particular channel, until a packet to be sent on a different channel has waited more than the threshold. Then it switches the channel.

Now we discuss some of the cognitive MAC protocols from literature:

- Dynamic Open Spectrum Sharing (DOSS) (Liangping, 2005): DOSS uses a busy tone broadcasting based approach for collision avoidance. Whenever a node transmits or receives data on a channel, it also emits a busy tone signal in the corresponding busy tone band. Spectrum sensing support is provided. Packet flow sequence is REQ "REQ -> REQ-ACK -> DATA -> ACK". REQ & REQ-ACK messages also negotiate about the channel to be used. Whenever a node wants to transmit or receive data on a channel, it senses the corresponding busy tone channel, this ensures that the transmission of the other cognitive users is not mistaken with the primary user activity. The drawback of DOSS is that it requires multiple radio transceivers, and separate frequency bands for the busy tone channel and control channel.
- C-MAC (Cordeiro, 2007): In C-MAC, all the channels have super-frames containing a beacon period and a data transfer period with the assumption that beacon period is non-overlapping for all the channels. Data transfer period is accessed in TDMA manner by all the users. C-MAC also uses rendezvous channel for control signaling. In this protocol, spectrum switching is not instantaneous which may cause primary user interference. Also, the selection method of non-overlapping beacon periods is not presented in paper.
- HC-MAC (Juncheng, 2008): In HC-MAC, each time slot is divided in three parts: contention, sensing and transmission. In contention period, C-CTS, C-RTS are used to reserve a slot for the sensing. In the sensing period, cognitive sender and receiver both sense channels and communicate to get a common free channel, followed by data transfer. In this protocol C-RTS message cannot be received by a node already busy in some communication, and can lead to multi-channel hidden node problem.
- MMAC-CR (Timmers, 2010): MMAC-CR is an improvement of MMAC, hence it includes multi-channel operation. Time frame is divided into an ATIM window and a data transfer window. The ATIM window is used to negotiate the channel between transmitter and receiver, it also solves the issue of channel synchronization. It uses the IEEE 802.11 Timing Synchronization Function (TSF) for time synchronization. Packet flow sequence is "Beacon -> Scan Result Packet (SRP) -> ATIM -> ATIM-ACK-> ATIM-RES" on control channel and "RTS -> CTS -> DATA -> ACK" on data channels. In MMAC-CR, the transmitter inserts a list of its spectral opportunities and queue status in the ATIM frame. The receiver compares this list with its own list and selects the common opportunity for data communication. The selected channel is then inserted in the ATIM-ACK frame. To inform the neighbors of the transmitter that were not able to hear the ATIM-ACK frame, the transmitter sends an ATIM-RES frame, which includes the selected data channel. Nodes that did not send or receive ATIM frame; enter a sleep state for the beacon interval to save energy. This protocol does not provide a silent period for spectrum sensing. The silent period for spectrum sensing insures that the cognitive user's data is not mistaken as primary user's data. Also, this protocol suffers from spectrum wastage due to the insertion of the complete list of spectral opportunities and queue status in the ATIM frame.
- Cognitive Radio-EnAbled Multi-Channel MAC (CREAM-MAC) (Zhang, 2011): The CREAM-MAC protocol maintains a Channel Usage List (CUL) and a Free Channel List (FCL). For medium access, it uses "RTS -> CTS -> RES -> Data" sequence flow. RTS, CTS and RES are done on the control channel. RTS carries the FCL and based on FCL, the receiver selects a channel to be used which is then reserved by a RES message. It uses single transceiver that alternates between

monitoring the CCC and the data spectrum bands. It does not provide support for spectrum sensing and channel synchronization.

- Cognitive-Radio-based Carrier Sense Medium Access with Collision Avoidance (CR-CSMA/CA) (Liqiang, 2011): In CR-CSMA/CA, Prepare-To-Send (PTS) frame is sent before the RTC-CTS mechanism of CSMA/CA for spectrum sensing support. Once other cognitive uses receive PTS frame, they keep quiet for the following duration so that cognitive user signal is not mistaken as a primary user signal. This protocol also does not give any information about channel synchronization.
- Channel Aggregation Diversity (CAD) (Ren, 2014): In CAD MAC, cognitive users select a group of channels from the available channels based on sensing results, and select an upper bounded power level for the selected channels. Then, channel aggregation technology is used to simultaneously use the selected channels for data transmission. CAD MAC assumes that all cognitive nodes are synchronized and have the same time slot division as of primary users. This assumption is not practical for more than one kind of primary users.

Summary of these protocols with respect to the issues discussed above is presented in table 1. Busy tone broadcasting based protocols require multiple radio transceivers and separate frequency bands for busy tone channel and control channel. To avoid interference to primary user in TDMA based protocols (like C-MAC), all secondary users should be synchronized and follow the same time-slot division as of the primary user. Thus, for more than one primary user network, TDMA based approach cannot be used. Most of the contention based protocols do not provide a solution to multi-channel hidden node problem.

As described above, most of the proposed cognitive MAC protocols assume multiple radio support in devices which is not always possible. Most of them also lack in channel synchronization and channel scheduling. A cognitive MAC protocol which solves all the challenges of cognitive networks is still needed. Selection of channel at MAC layer will not always give an efficient end-to-end solution. Hence, it is suggested to select channel at upper layers in cognitive networks.

Table 1. Classification of MAC Strategies

Protocol Name	Spectrum Access Method	Medium Access	Interface Requirement	Multichannel Hidden Node	Channel Synchronization	Spectrum Sensing Support	Channel Scheduling
DOSS	Busy Tone Broadcasting	REQ-REQACK-DATA-ACK	Multiple	Yes	Yes	Yes	No
C-MAC	TDMA Based	TDMA Based	Single	Yes	Yes	No	No
HC-MAC	Contention Based	RTS-CTS	Single	No	Yes	Yes	No
MMAC-CR	Contention Based	ATIM Window with RTS-CST	Single	Yes	Yes	Partial	No
CREAM-MAC	Contention Based	RTS-CTS	Single	No	Yes	No	No
CR-CSMA/CA	Contention Based	RTS-CTS	Single	No	Yes	Yes	No
CAD MAC	Channel Aggregation Based	RTS-CTS-RES	Single	No	Yes	Yes	No

ROUTING LAYER FUNCTIONALITIES

Routing in cognitive networks targets the creation and maintenance of wireless multi-hop routes among secondary users by deciding both the spectrum to be used and the relay nodes of the route. Figure 12 lists the challenges for designing effective routing metrics for cognitive networks.

Paper (Khalife, 2009) argues that no general routing protocol can be proposed for cognitive networks. According to the paper, cognitive networks can be classified into three categories based on primary user activity models; each requiring specific routing solutions. In the case of a static primary user activity model, traditional routing algorithms can be used as primary user activity and inactivity is for longer periods like in case of TV broadcasts. In the case of a dynamic primary user activity model, specific routing protocols for cognitive networks should be developed. While in case of highly dynamic primary user activity model, the availability of such frequency bands for the whole communication duration becomes an unrealistic assumption. Therefore, complete opportunistic routing protocol, where every packet can be forwarded opportunistically over available channels with no route setup, constitutes a potential solution. This thesis considers dynamic primary user activity model for routing.

In literature, various routing algorithms have been proposed. Paper (Youssef, 2014) provides a survey of routing metrics used in cognitive networks. The routing metrics are classified based on the support for single path and multiple paths. These metrics can further be classified as follows:

- Delay
- Hop Count
- Power Consumption
- Location based
- Spectrum Availability
- Route Stability
- Probabilistic
- Route Bandwidth Capacity
- Route Closeness
- Dead Zone Penetration
- Metrics that Capture SU Interference

Figure 12. Routing metric design challenges.

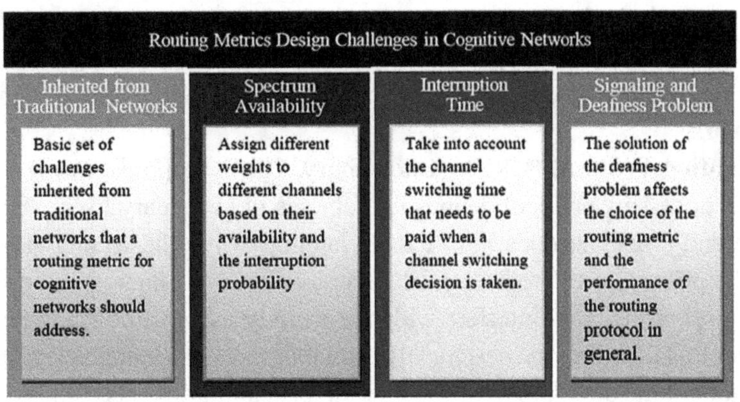

Delay, hop count, power consumption, location based and probabilistic routing metrics are inherited from traditional networks. Spectrum availability, route stability, route bandwidth capacity, route closeness, dead zone penetration and metrics that capture secondary user interference are specific to cognitive networks. The routes closeness metric favours routes that are far away from each other. Selecting non-close routes makes them less vulnerable to primary user's activities as it would be difficult for an active mobile primary user to interrupt all the routes at the same time. The Dead Zone Penetration metric (DZP) is a route maintenance metric and can be used in conjunction with any of the existing routing. This metric uses multipath routing to penetrate active primary user's zones by using the relay to cooperatively send data to the next hop. These different metrics perform differently in different network conditions and/or requirements.

In this section we discuss some of the existing solutions for routing.

- **Cognitive Ah-hoc On-demand Distance Vector (CAODV):** (Cacciapuoti, 2010) (Cacciapuoti, 2012) routing modifies the existing AODV protocol such that it avoids primary user activity areas and applies joint path and channel selection. This protocol broadcasts the route request to neighbors on each channel and for each available channel it stores multiple paths for a destination.
- **Dual Diversity Cognitive Ad-hoc Routing Protocol (D^2CARP):** (Rahman, 2012) is a modification of CAODV. CAODV individually exploited the path and spectrum-diversities while D^2CARP jointly exploits both. It also shows that joint exploitation of path and spectrum diversities gives better results. Both CAODV and D^2CARP apply the shortest path solution without considering channel parameters, and hence do not give optimal results.
- **Ant-based Spectrum Aware Routing (ASAR):** (Bowen Li, 2009) adapts to path and spectrum availability. In this protocol, forward ants are broadcast on the control channel. The forward ants follow the spectrum feasible routes while storing the node addresses in itself. The backward ants update the statistic history and route quality in routing table. A deterministic policy for route selection is used in place of probabilistic forwarding to minimize channel handoff time. Each ant contains a list of all the nodes of the route which results in high overhead on control channel leading to control channel saturation. Also, the algorithm does not take mobility of secondary users in consideration.
- **SpEctrum Aware Routing protocol for Cognitive ad-Hoc networks (SEARCH):** (Chowdhury, 2009) uses the shortest path based greedy approach for route selection. The objective of the protocol is to minimize hopcount and switching delay which in turn minimizes end-to-end delay. Every intermediate hop adds its ID, current location, time stamp and flag status in the route request packet. This results in high routing overhead. Also, it requires location awareness which is not a good option as GPS service availability to all nodes in the network might not be feasible. Power consumption of GPS will also reduce the battery life of the nodes, reducing the effective lifetime of entire network.
- **Cognitive Routing Protocol (CRP):** (Chowdhury, 2011) proposes a metric for joint spectrum and node selection. This protocol defines two classes of secondary users. The first class tries to minimize the end-to-end secondary user route latency, while the second class gives greater importance to the primary user protection. The second class is of interest in opportunistic networks as secondary users should not interfere with the primary users while using licensed spectrum. In CRP, each user first identifies the best locally available spectrum band. Selection of spectrum band is defined as an optimization problem considering probability of spectrum availability, bit arrival

variance, multipath fading and switching latency. After spectrum band selection, next hops are selected based on minimum forwarding delay. The drawback of the protocol is that the spectrum characteristics such as bandwidth are not considered while selecting the spectrum band. Results in the paper also show that the performance of CRP is average for flows with second class.

- **Optimal Primary-awarE Route quAlity (OPERA):** (Marcello, 2012) proposes the metric which calculates the link delay and end-to-end delay for any pair of source and destination nodes. Link delays are calculated considering the probability of link availability which in turn is measured by primary user characteristics, expected throughput, and characteristics of sensing process. This metric gives the theoretical bounds to end-to-end delays. OPERA combined with Dijkstra (Dijkstra, 1959) and Bellman-Ford (BellmanFord,) algorithms for searching for the shortest path in a weighted graph gives optimal results. The drawbacks of OPERA are (i) it lacks a route recovery process, (ii) for each sensing cycle, OPERA should calculate the weighted metric which requires search for the shortest path using Dijkstra or Bellman-Ford algorithm. This results in high routing overhead, (iii) OPERA can interfere with primary users as it considers probability of primary user activity instead of actual sensing output.
- **Cognitive Cross-layer Multipath Probabilistic Routing (CCMPR):** (Singhal, 2014) proposes a cross layer routing for cognitive networks. CCMPR uses spectrum holes, identified by the MAC layer to transmit data between hops. It gives joint solution for channel selection, routing and transmit power control. Transmit power is selected for each hop in the route based on channel history. CCMPR uses multi-objective optimization function as a cost function for routing. Paper shows that the proposed routing metric is convex and gives pareto optimal solution. This work can be extended to include other parameters from physical layer.

Most of the presented protocols were inherited from traditional routing protocols, i.e. the routing metrics are old metrics like hop-count or delay, and algorithms send data only in spectrum opportunities. The goals for routing in cognitive networks are:

- Use spectrum opportunities effectively to benefit secondary users without causing harmful interference to primary users.
- Save energy of nodes to increase the network lifetime.
- Minimize end-to-end delay.
- Increase throughput.
- Improve route stability.

TRANSPORT LAYER IN COGNITIVE NETWORKS

The performance of TCP depends on the packet loss probability and the round trip time (RTT). Link errors, packet loss probability and RTT depend on the access technology, interference level, channel bandwidth and frequency of the channel in wireless communication. Thus, the performance of transport protocol is dependent on the physical layer parameters. Therefore, the transport protocols that are designed for existing networks cannot be used in cognitive networks. The spectrum sensing results, activity of the primary users, large-scale bandwidth variation based on spectrum availability, and the channel switching process should to be considered in the transport protocol design. Paper (Felice, 2009) proposes a window-based transport protocol for cognitive networks, TP-CRAHN, which distinguishes each of these events by a combination of explicit feedback. This protocol closely interact with physical layer channel information, link layer spectrum sensing and buffer management, and a predictive mobility framework that is developed at the network layer to achieve desired TCP rate.

CROSS LAYER SOLUTION: HOPE FOR DESIGNING PROTOCOLS IN COGNITIVE NETWORKS

In the traditional OSI model of communication, strict boundaries between layers are enforced and meta-data for layers are kept unavailable to other layers. While in cross layer design strict boundaries are removed and exchange of meta-data is permitted across the layers. The layered OSI model for communication was built for wired communication and later adapted to the wireless communication. Issues with wireless links and opportunistic behavior of cognitive networks are the main motivations for cross layer design in cognitive networks. In cognitive wireless environment, several problems cannot be handled by layered communication. Below we discuss specific needs for cross layer design with respectto network layer and transport layer.

Need for Cross Layer Design with Respect to Network Layer

In cognitive networks, physical layer parameters should be changed based on the available spectrum. This change in the channel parameters may affect the QoS of the path selected. For example, an optimized MAC protocol may provide the best channel/power/rate assignment for a particular link but such an assignment can be quite inefficient for end-to-end path of the flow. This necessitates that the decisions at network layers should consider and reflect the parameters of lower layer. Hence, lower layer optimization without considering network layer protocols can lead to sub-optimal solution. This necessitates the use of cross layer protocols with hybrid routing metrics.

Need for Cross Layer Design with Respect to Transport Layer

Link unavailability due to the primary user activity could be mistaken as network congestion by transport layer protocol. In case of TCP as a transport layer protocol, this may cause to increase the size of congestion window. As discussed earlier, the performance of TCP depends on bandwidth and frequency of spectrum. In cognitive networks, bandwidth changes with the spectrum switching at MAC layer. The size of contention window should also be updated with the change in transmission spectrum bandwidth. Thus, appropriate changes are needed in the design of transport layer for cognitive networks which takes feedback from lower layers.

To summarize, cognitive networks use spectrum opportunities for communication where dynamic adaptability to the transmission or reception parameters is required. Paper (Srivastava, 2005) also shows that the cross layer protocols are more effective than layered protocols where dynamic adaptation of parameters is required. Cognitive networks use spectrum opportunities for communication where dynamic adaptability to the transmission or reception parameters is required. For cross layer design of the network, objectives from different layers should be combined to one optimization problem to ensure QoS at each layer. For example, if one wants to minimize bit error rate, minimize the transmit power, increase throughput of the link, minimize the end-to-end delay. All these objectives lead to tradeoffs for each other. Hence, these objective functions should be combined to one optimization function using multi-objective optimization techniques and that combined function should be minimized with the specified constraints. For cross layer design of the network, objectives from different layers should be combined to one optimizations problem. This will ensure the QoS at each layer.

REFERENCES

Akyildiz, I. F., Lee, W.-Y., & Chowdhury, K. R. (2009, July). CRAHNs: Cognitive radio ad hoc networks. *Ad Hoc Networks*, *7*(5), 810–836. doi:10.1016/j.adhoc.2009.01.001

Akyildiz, I. F., Lo, B. F., & Balakrishnan, R. (2011). Cooperative spectrum sensing in cognitive radio networks: A survey. *Physical Communication*, *4*(1), 40–62. doi:10.1016/j.phycom.2010.12.003

Akyildiz Ian, F. (2006, September). Won-Yeol Lee, Mehmet C. Vuran, & Shantidev Mohanty. NeXt generation / dynamic spectrum access / cognitive radio wireless networks: A survey. *Computer Networks*, *50*(13), 2127–2159. doi:10.1016/j.comnet.2006.05.001

Barry, J. R., Lee, E. A., & Messerschmitt, D. G. (2004). *Digital communication*. Springer. doi:10.1007/978-1-4615-0227-2

Bellman, R. L. (n.d.). On a Routing Problem. *Quarterly of Applied Mathematics*.

Cacciapuoti, A., Calcagno, C., Caleffi, M., & Paura, L. (2010). CAODV: Routing in mobile ad-hoc cognitive radio networks. In *Wireless Days*.

Cacciapuoti, A. S., Caleffi, M., & Paura, L. (2012). Marcello Cale_, & Luigi Paura. Reactive routing for mobile cognitive radio ad hoc networks. *Ad Hoc Networks*, *10*(5), 803–815. doi:10.1016/j.adhoc.2011.04.004

Chowdhury, K., & Di Felice, M. (2009). SEARCH: A routing protocol for mobile cognitive radio ad-Hoc networks. In *Sarnoff Symposium, 2009. SARNOFF '09*. IEEE. doi:10.1109/SARNOF.2009.4850323

Chowdhury, K. R., & Akyildiz, I. F. (2011). Crp: A routing protocol for cognitive radio ad hoc networks. Selected Areas in Communications. *IEEE Journal on*, *29*(4), 794–804.

Chowdhury, K. R., & Felice, M. D. (2009). Search: A routing protocol for mobile cognitive radio ad-hoc networks. *Computer Communications*, *32*(18), 1983–1997. doi:10.1016/j.comcom.2009.06.011

Cordeiro & Challapali. (2007). C-mac: A cognitive mac protocol for multi-channel wireless networks. In *New Frontiers in Dynamic Spectrum Access Networks, 2007. DySPAN 2007. 2nd IEEE International Symposium on*.

Digham, F. F., Alouini, M.-S., & Simon, M. K. (2007). On the energy detection of unknown signals over fading channels. *IEEE Transactions on Communications*, *55*(1), 21–24. doi:10.1109/TCOMM.2006.887483

Dijkstra, E. (1959). A note on two problems in connexion with graphs. *Numerische Mathematik, 1*(1), 269-271.

Enserink, S., & Cochran, D. (1994). A cyclostationary feature detector. In *Signals, Systems and Computers, 1994. 1994 Conference Record of the Twenty-Eighth Asilomar Conference on*. IEEE.

Felice, M. D., Chowdhury, K. R., & Akyildiz, I. F. (April 2009). Tp-crahn: a transport protocol for cognitive radio ad-hoc networks. In *INFOCOM 2009* (pp. 2482–2490). IEEE.

Ganesan, G., & Li, Y. (2005). Cooperative spectrum sensing in cognitive radio networks. *New Frontiers in Dynamic Spectrum Access Networks, 2005. DySPAN 2005. 2005 First IEEE International Symposium on*. IEEE. doi:10.1109/DYSPAN.2005.1542628

Gardner, W. A. (1991). Exploitation of spectral redundancy in cyclostationary signals. *Signal Processing Magazine, IEEE*, *8*(2), 14–36. doi:10.1109/79.81007

3GPP TS 36305: Evolved Universal Terrestrial Radio Access Network (E-UTRAN)

Jia, J., Zhang, Q., & Shen, X. (2008, January). Hc-mac: A hardware-constrained cognitive mac for efficient spectrum management. *Selected Areas in Communications. IEEE Journal on*, *26*(1), 106–117.

Khalife, H., Malouch, N., & Fdida, S. (2009). Multihop cognitive radio networks: to route or not to route. *Network, IEEE, 23*(4).

Lee, W.-Y., & Akyildiz, I. F. (2012). Spectrum-aware mobility management in cognitive radio cellular networks. *Mobile Computing, IEEE Transactions on*, *11*(4), 529–542. doi:10.1109/TMC.2011.69

Li, B., Li, D., Wu, & Li. (2009). Asar: Ant-based spectrum aware routing for cognitive radio networks. In *Wireless Communications Signal Processing, 2009. WCSP 2009. International Conference on.*

Ma, L., Han, X., & Shen, C.-C. (2005). Dynamic open spectrum sharing mac protocol for wireless ad hoc networks. In *New Frontiers in Dynamic Spectrum Access Networks, 2005. DySPAN 2005. 2005 First IEEE International Symposium on.*

Marcello Calefi, I. F. (2012). Akyildiz, & L. Paura. Opera: Optimal routing metric for cognitive radio ad hoc networks. *Wireless Communications. IEEE Transactions on*, *11*(8), 2884–2894.

Matinmikko, M., Palola, M., Saarnisaari, H., Heikkila, M., Prokkola, J., Kippola, T., … Yrjola, S. (2013). Cognitive Radio Trial Environment: First Live Authorized Shared Access-Based Spectrum-Sharing Demonstration. *Vehicular Technology Magazine, IEEE, 8*(3).

Osa, V., Herranz, C., Monserrat, J. F., & Gelabert, X. (2012). Implementing opportunistic spectrum access in LTE-advanced. *EURASIP Journal on Wireless Communications and Networking*, *2012*(1), 1–17. doi:10.1186/1687-1499-2012-99

Pinyi, R., Wang, Y., & Du, Q. (2014, February). Cad-mac: A channel-aggregation diversity based mac protocol for spectrum and energy efficient cognitive ad hoc networks. Selected Areas in Communications. *IEEE Journal on*, *32*(2), 237–250.

Pradhan, C., Sankhe, K., Kumar, S., & Murthy, G. R. (2014). Cognitive base station design for efficient spectrum utilization in cellular network. In *Wireless and Optical Communications Networks (WOCN), 2014 Eleventh International Conference on*. IEEE. doi:10.1109/WOCN.2014.6923095

Rahman, M., Caleffi, M., & Paura, L. (2012). Joint path and spectrum diversity in cognitive radio ad-hoc networks. *EURASIP Journal on Wireless Communications and Networking*, *2012*(1), 1–9. doi:10.1186/1687-1499-2012-235

Singhal, D., & Garimella, R. M. (2014). Cognitive cross-layer multipath probabilistic routing for cognitive networks. *Springer's Wireless Networks*. DOI: 10.1007/s11276-014-0847-2

Srivastava, V., & Motani. (2005). Cross-layer design: a survey and the road ahead. *Communications Magazine, IEEE*, *43*(12).

IEEE Std 802.21-2008. (2009). *IEEE Standard for Local and metropolitan area networks - Media Independent Handover Services*. IEEE.

Thomas, R. W., DaSilva, L. A., & Mackenzie, A. B. (2005). Cognitive networks. In *Proc. of IEEE DySPAN 2005*.

Thomas, R.W., Friend, D.H., DaSilva, L.A., & MacKenzie, A.B. (2006). Cognitive networks: adaptation and learning to achieve end-to-end performance objectives. *Communications Magazine, IEEE*, *44*(12).

Timmers, M., Pollin, S., Dejonghe, A., Van der Perre, L., & Catthoor, F. (2010). A Distributed Multichannel MAC Protocol for Multihop Cognitive Radio Networks. *Vehicular Technology, IEEE Transactions on*, *59*(1).

Urkowitz, H. (1967). Energy detection of unknown deterministic signals. *Proceedings of the IEEE*, *55*(4), 523–531. doi:10.1109/PROC.1967.5573

Xi & Su. (2011, February). Cream-mac: Cognitive radio-enabled multi-channel mac protocol over dynamic spectrum access networks. *Selected Topics in Signal Processing. IEEE Journal of*, *5*(1), 110–123.

Xiao, Hu, Qian, Gong, & Wang. (2013). Expanding LTE network spectrum with cognitive radios: From concept to implementation. *Wireless Communications, IEEE*, *20*(2). doi:10.1109/MWC.2013.6507389

Youssef, M., Ibrahim, M., Abdelatif, M., Chen, & Vasilakos, A.V. (2014). Routing Metrics of Cognitive Radio Networks: A Survey. *Communications Surveys & Tutorials, IEEE*, *16*(1).

Yucek, T., & Arslan, H. (2009). A survey of spectrum sensing algorithms for cognitive radio applications. *IEEE Communications Surveys and Tutorials*, *11*(1), 116–130. doi:10.1109/SURV.2009.090109

Zhao, L., Zhang, W., Song, W., & Zhang, H. (2011). Cognitive radio csma/ca protocol for femto-wlans. In *Communication Software and Networks (ICCSN), 2011 IEEE 3rd International Conference on*.

Chapter 7
WiMAX and Wi-Fi Services through Mobile Networks:
Issues on Network Planning, Optimization, and Sustainability

Sajal Saha
Narula Institute of Technology, India

Asish K. Mukhopadhyay
S. R. Group of Institutions, India

Angana Chakraborty
Indian Institute of Engineering Science and Technology (IIEST), India

Anup Kumar Bhattacharjee
National Institute of Technology, India

ABSTRACT

WIFI and WiMAX became popular worldwide due to simplicity of installation and cost reduction compared with traditional DSL cable. The key challenge of WiFi is the restricted mobility of the user whereas mobile WiMAX (IEEE 802.16m) provides ubiquitous environment to the user. The recent challenges of WiMAX is managing radio resources (both licensed and unlicensed) efficiently through different QoS service classes, implementing proper modulation and coding schemes at PHY, proper antenna selection in MIMO to achieve better QOS. User can avail the benefits of speed and mobility through an integrated architecture comprising WiFi and WiMAX. This chapter provides an in depth performance evaluation of Mobile WiMAX and WiFi under various Modulation and Coding Schemes (MCS), cyclic prefixes and different path-loss models under integrated environment. Moreover, we present an optimized adaptive modulation scheme that senses the SNR and adaptively switches to required MCS to achieve desired level of QoS and link stability.

1. INTRODUCTION

The demand for connectivity and bandwidth is increasing day by day. WLAN or WiFi(1999) and Long Term Evolution (LTE), Worldwide Interoperability for Microwave Access (WiMAX) which is also known as Wireless MAN,emerge as promising technology in the market to meet the requirement of the user in a ubiquitous environment.

DOI: 10.4018/978-1-5225-0239-5.ch007

LTE proposed in 2004 where as WiMAX and LTE proposed in 2001. Among them WIFI and WiMAX became popular worldwide among them due to simplicity of installation and cost reduction compared with traditional DSL cable.

There are three types of WiMAX systems are (Díaz, J. M., Del Río, L. V., Cárdenas, O. Á., &Baldivia, 2014) available:

A. Fixed WiMAX (IEEE 802.16- 2004)
B. Mobile WiMAX (IEEE 802.16e/m-2005)
C. Mesh WiMAX (IEEE 802.16j- Mesh)

Fixed WiMAX(16-2005, 2005) system having coverage area 5-7 km and speed up to 48 Mbps (fixed downlink) and 7 Mbps (fixed uplink). It provides fixed connection between a base station (BS) and multiple subscriber station (SS) via point to multipoint link (PMP).

Mobile WiMAX(16-2005, 2005) is broadband wireless access solution(Jeffrey G. Andrews, ArunabhaGhosh, 2007), (J. Sachs, L. Muñoz, R. Aguero, J. Choque, G. Koudouridis,R. Karimi, L. Jorguseski, J. Gebert, 2004) that supports the connectivity of mobile node (MN) under the coverage of BS and provides mobile telephone service via VoIP. Mobile WiMAX provides speed upto 9.4 Mbps for downlink and 3.3 Mbps for uplink across the coverage area of 3 km (J. Sachs, L. Muñoz, R. Aguero, J. Choque, G. Koudouridis,R. Karimi, L. Jorguseski, J. Gebert, 2004).

In Mesh WiMAX network WiMAX BS is connected within its coverage area with additional BS's with smaller coverage area to relay the connectivity. Sometimes received signal strength reduced due to Non Line of Site(NLOS) effect of obstacles and the additional BS's which are called multihop relay stations(Narasimha & Sivalingam, 2009) having minimal functionality used to relay signals between BS and subscriber station(SS) to enhance the quality of the link to the end user. Figure 1 give an outline how fixed, mobile and mesh WiMAX works.

Figure 1. Fixed and Mobile WiMAX & Mesh WiMAX

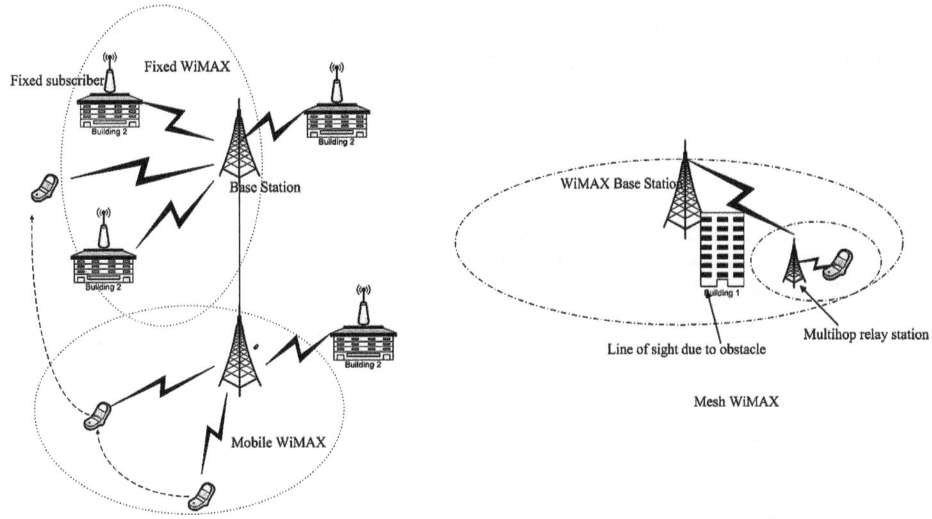

WiFi or WLAN offered very high data rate (from 1Mbps to 55 mbps) within its limited range. It is based upon IEEE 802.11 standard. WLAN gets popularized in 2004 for providing internet to private and corporate network. Initially WLAN used for data connectivity. WLAN recent standard family (IEEE 802.11n) is providing VoIP over WIFi hotspots as low cost voice services.

In this chapter, we have analyzed the performance of mobile WiMAX (IEEE 802.16m) and WiFi(IEEE 802.11n)and WiMAX - WiFi integrated network model.

WLAN and WiMAX standards

WLAN or WiFi evolution and standardization controlled by IEEE 802.11 working group and WiFi alliance. WiMAX evolution and standardization controlled by IEEE 802.16 working group and WiMAX forum.

1.1. IEEE 802.11 Standards Family

First generation WLAN (introduced in 1985) started operating in 900 Mhz ISM band. Later IEEE standard working committee approved amendment of OFDM and spread spectrum technology separately as a part 802.11a and 802.11b respectively for secure and higher data transmission through wireless medium.

IEEE 802.11a and IEEE 802.11b faced interoperability issues with other wireless technologies evolved at the same time. Moreover, IEEE 802.11a and IEEE 802.11b became a legacy system as they failed to support real time multimedia applications. In 2003, IEEE 802.11g standard was evolved to support both DSSS and OFDM at the PHY layer and resolved interoperability issues. Earlier, IEEE 802.11 standard working group draft a new standard and sent it to WiFi alliance. WiFi alliance would develop the system for testing and certification. The process usually takes longer period of time with respect to technology change and vendor requirement. Consequently, IEEE 802.11 standard working group and WiFi alliance decided to work parallelly coordinating with each other to release new standard as quickly as possible. In 2009 IEEE standard WiFi alliance introduced MIMO technology in IEEE 802.11n. Additional transmit and receive antenna in MIMO enhanced QoS through increased throughput, lower latency and constant jitter. Evolution of IEEE 802.11 standards family is shown in Table 1.

1.1.1 IEEE 802.11n

WiFi or WLAN(Standerd, 2007) introduced in the ‘90s offered very high data rate (from 1Mbps to 55 mbps) within its limited range. To combine the benefits of high coverage of the cellular systems and the

Table 1. IEEE 802.11 standard family

Standard	Amendment	Release year	Bandwidth
IEEE 802.11	Direct Sequence Spread Spectrum (DSSS), Frequency-Hopping Spread Spectrum(FHSS)	1997	2.4 Ghz
IEEE 802.11a	Orthogonal Frequency Division Multiplexing(OFDM)	1999	5Ghz
IEEE 802.11b	DSSS	1999	2.4 Ghz
IEEE 802.11g	OFDM, DSSS	2003	2.4 Ghz
IEEE 802.11n	OFDM	2009	2.4 Ghz and 5Ghz
IEEE 802.11ac	OFDM	2013	5Ghz
IEEE 802.11ad	Single Carrier(SC),OFDM	2012	60Ghz

high data rate of WiFi, it is necessary to integrate the WiFi with Broadband Wireless Access (BWA) system like WiMAX where the concept of overlay network is introduced.

1.1.2. WiFi or WLAN(IEEE 802.11n)Network Architecture:

WiFi operates on two modes: ad-hoc and infrastructure. In infrastructure mode an Access Point (AP) coordinates the transmission among the nodes within its radio coverage, called Basic Service Set (BSS). Several BSS constitute an Extended Service Set (ESS) .ESS is connected to the wired network through distribution system. In ad-hoc mode, nodes can communicate with each other without any AP. Infrastructure mode is more relevant for integration with cellular or other network. A Mobile Node (MN) can only associate with one AP at a time. All the MNs associated with an AP communicate with each other either through the AP or directly coordinated by the AP. Roaming across APs is supported in layer-2 through Inter-AP Protocol (IAPP). The APs generate beacons periodically that contain the network-id (or Extended Service Set Identifier, ESSID) and cell-id (which is the AP's MAC address) in addition to other information. On power-up, a MN associates with the AP by sending associate request frame to the AP. When the MN moves to a new cell where it receives a beacon with the same network-id but a new cell-id, it associates with the new AP by sending re-associate request frame that includes the MAC address of the old AP. The new AP can communicate with the old AP through IAPP to download the context information. Table 3 shows the difference between WiFi and WiMAX.

1.2. IEEE 802.16 Standards Family

Unlike WiFi alliance, WiMAX forum formed in 2003 by several stakeholders. Industry and standardization forums and agencies in telecommunication such as Internet Engineering Task Force (IETF), 3rd Generation Partnership Project (3GPP2), DSL forum, Open Mobile Alliance (OMA) and Intel joined hands to form WiMAX forum (more than 470 members till date) to conceptualize WiMAX towards deployment. Figure 3 shows the timeline of past and current projects of the IEEE 802.16 working group(Carl Eklund, Roger B. Marks, Subbu Ponnuswamy, Kenneth L. Stanwood, 2007).

WiMAX forum is a nonprofit organization that certifies and promotes the compatibility and interoperability of IEEE 802.16 compatible product and IEEE 802.16 working group provides (Etemad & Corporation, 2008),(L. Nuaymi, 2007), (Recommendation, 2003), (Yarali & Rahman, 2008),(J. Jee, S. Madanapalli, 2008),(IEEE, Pareit, Lannoo, Moerman, & Demeester, 2012), (Nakamura, 2008),(Alvarion, n.d.) various specifications and functionalities to IEEE standard 802.16 such as point to multipoint access, fixed access, mobile access. WiMAX forum (Singh, 2012) chooses some of the important functionalities and parameters among those based on market demand for mobile WiMAX. IEEE working group reviews sound technical contribution for standardization that may conflict with industry requirement monitored by WiMAX forum. Based on industry requirement, WiMAX forum created fixed WiMAX and Mobile WiMAX profiles in order to sustain backward compatibility and interoperability between different technologies. Evolution of IEEE 802.16 standards family is shown in table 2.

1.2.1. Mobile WiMAX(IEEE 802.16m) Network Architecture

SassanAhmedi briefly discussed the overview of Mobile WiMAX (IEEE 802.16m) in (Ahmadi, 2008) as an amendment of IEEE 802 standard family. Each Mobile WiMAX network architecture as shown in

Table 2. IEEE 802.16 standard family

Standard	Amendment	Release Year	Bandwidth	Release	Status
IEEE 802.16	Fixed Broadband Wireless Access	1999	LOS 10-66 Ghz		Suspended
IEEE 802.16a	System Profile	2000	10-66 Ghz		Suspended
IEEE 802.16c	Physical layer and MAC definition	2002	NLOS 2-11 Ghz NLOS 2-11 Ghz		Suspended
IEEE 802.16d	Maintenance and system profile	2003	LOS 10-66 Ghz NLOS 2-11 Ghz	1.0	Merged
IEEE 802.16e	Mobile Broadband wireless access system	2003	LOS 10-66 Ghz NLOS 2-11 Ghz	1.5	Superseded
IEEE 802.16f	Management Information Base(MIB)	2004			Superseded
IEEE 802.16g	Management plane procedures and services	2005			Superseded
IEEE 802.16i	Mobile MIB	2005			Merged
IEEE 802.16j	Multihop relay(Mesh WiMAX)	2006			Suspended
IEEE 802.16k	Bridging of 802.16	2006			Current
IEEE 802.16h	Improved Coexistence Mechanisms for License-Exempt Operation	2010			Suspended
IEEE 802.16m	Advance air interface with data rates of 100mbits/sec(Mobile WiMAX)	2011	Less than 6 Ghz	2	Current
IEEE 802.16p	Air Interface for Broadband Wireless Access Systems	2012			Current
IEEE 802.16n	Air Interface for Broadband Wireless Access Systems	2013			Current

Figure 2, consists of two prime entities: ASN (Access Service Network) and CSN (Connectivity Service Network). ASN provides the radio interfaces that connect MNs with the network. ASN handles the radio interface and contains the Base Station (BS). ASN gateway which has a one-to many relationship with BS. ASN gateway handles the mobility between BS.

Figure 2. WiMAX network architecture & security architecture as specified in IEEE 802.16 standard (802.16-2004),(16-2005, 2005), (Finneran, 2004)

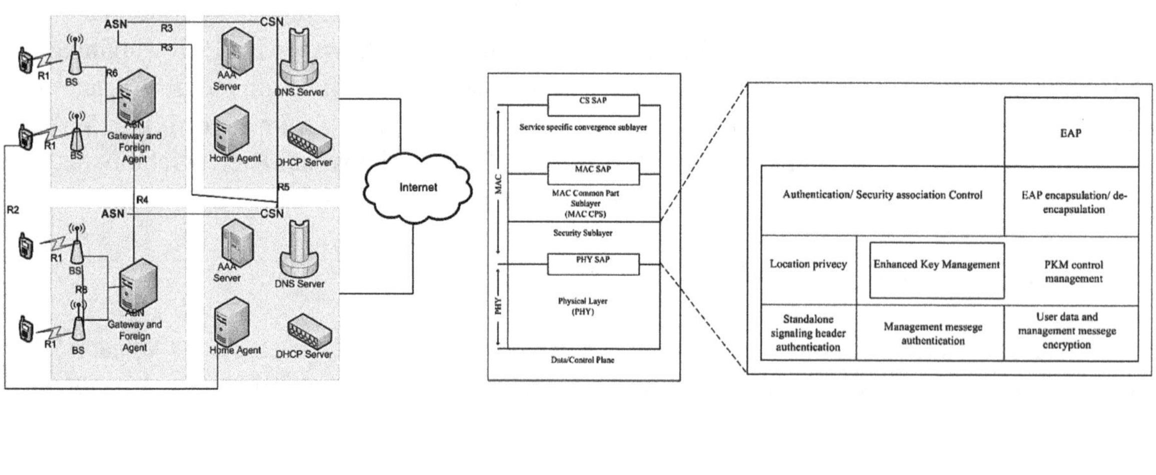

Table 3. 802.16m interfaces

R1 interface	MN to BS connection, radio link
R2 interface	MN to HA connection, supports roaming
R3 interface	ASN to CSN connection, supports authentication, billing and MIP messages
R4 interface	ASN to ASN connection, defines mobility procedures when a MN crosses from one ASN to another ASN.
R5 interface	CSN to CSN connection, supports roaming.
R6 interface	MN to ASN connection, supports mobilty messages.
R7 interface	Internal to ASN gateway, not shown in figure2.
R8 interface	BS to BS connection, supports handoff.

ASN gateway provides QoS, authentication, accounting and authorization (AAA) and mobility management to the MN via BS through R6 interface. ASN gateway provides IP connectivity to MN via CSN gateway through R3 interface. ASN gateway also connects to other ASN gateways for mobility management through R4 interface. The The CSN contains gateways for internet access, routers, servers or proxies for AAA, IP-allocation, user databases, and internetworking devices. HA and AAA together provide authentication as shown in figure 2. AAA also provides billing records. CSN handles admission and policy control, mobility between ASN and specific WiMAX services such as location based services.

Interfaces are used to communicate between different entities.Specifications of the different interfaces are shown in Table 3.

Figure 3. IEEE 802.16 project timeline

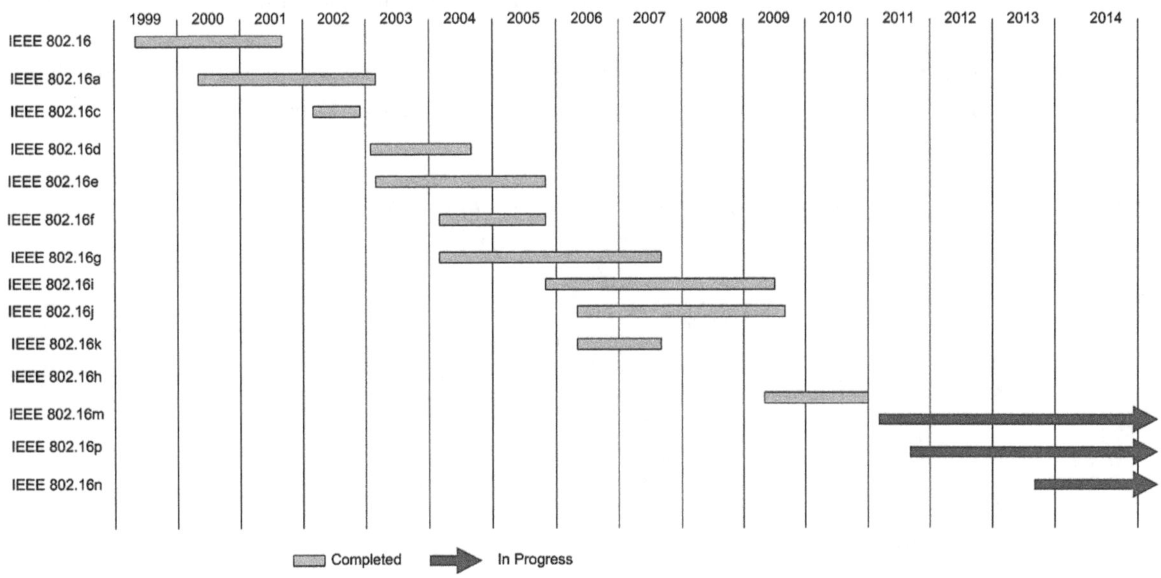

1.3 WiFi and WiMAX Security Standard

1.3.1. WiFi Security Standard

WiFi provides security in layer 1 and layer 2 as shown in figure 4. Layer 1 uses two well known spread spectrum technologies, a) Frequency hopping spread spectrum (FHSS) b) Direct sequence spread spectrum (DSSS). Layer 2 provides security through

- Authentication.
- Encryption/Decryption
- Eavesdropping

Authentication of MN is done through Wireless Equivalent Privacy(WEP) with AP. WEP uses 128 bit key(Standard, 2001) in IEEE 802.11a/b. WEP has some flaws as hackers can easily break the encryption recognizing its pattern. IEEE 802.11n replaces WEP with WiFi Protected Access (WPA2). Nowadays WiFi AP manufacturer provides per user, per session dynamic WEP/WPA2 key tied with WiFi login.

1.3.2. WiMAX Security Standard:

WiMAX provides security in Layer2 (Broadband, n.d.), (Alcatel-Lucent, 2004).WiMAX MAC layer consists of three sub layersas shown in Figure 2.

- Privacy security sub layer performs authentication, encryption, and key exchange from MN to CSN gateway via BS.
- MAC common part sub layer (MAC CPS) performs bandwidth allocation, connection establishment and connection maintenance.
- Convergence sub layer (CS) transform data, schedule service classes to maintain QoS.

Private Key (PKM) control management as shown in figure 3 provides user authentication by exchanging certificates between MN and AAA server attached to CSN gateway. User authentication of mobile

Figure 4. IEEE 802.11n network architecture & security architecture

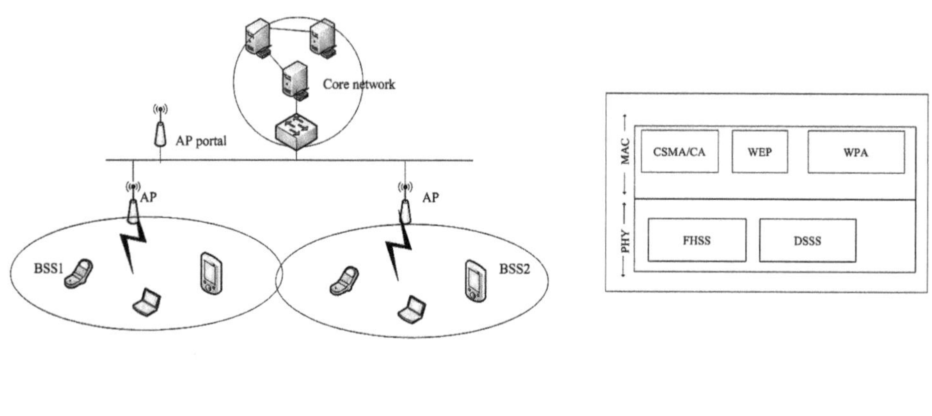

WiMAX is done through private key management version 2(PKMV2) with extensible authentication protocol (EAP) with key caching mechanism.EAP provide robustness to the WiMAX security system by asking several credentials like user id, password, and digital certificates. This mechanism is used to optimize the security service and reduce user authentication time during handover (table 4).

User data encryption is done using Data Encryption Standerd(DES) and Advance Encryption Standerd(AES). Both are much more robust then WEP used by WLAN.

1.4 WiFi and WiMAX: A Brief Comparative Study

Unlike WLAN, WiMAX provides robust QoS protection through grant request mechanism in MAC sub layer which allows the subscriber to provide required bandwidth for real time multimedia services like audio, video streaming etc. service provider can charge a premium (G.S.V. Radha K. Rao, 2007) for guaranteed QoS. WiFi network uses CSMA/CA protocol in MAC sub layer where if the number of subscriber is less supposes 10, then occasional collision may occur that results back off and retransmis-

Table 4. Comparison between WiFi and WiMAX

	WiFi			WiMAX
Current Standard	IEEE 802.11n	IEEE 802.11ac	IEEE 802.11ad	IEEE 802.16m
Primary application	Wireless LAN	Wireless LAN	Wireless LAN	Broadband wireless access
Frequency(GHz)	2.4	60	5	2.66-11
Spectral efficiency	2.7 bps/Hz			5 bps/Hz
Evolved in	2009	2013	2012	2011
Max Speed(Mbps)	54	500	7Gbps	100
Coverage area	100m	33m	>10m	50km
Advantage	Speed, low cost	Wider Channel, Denser Amplitude Modulation	Higher Bandwidth	Speed, range
Disadvantage	Range	Range	Range	Comparatively higher equipment cost
Access technology	OFDM(64 channels)	OFDM(256 channels)	OFDM(256 channels)	OFDM(256 channels)
Security	WEP	WEP	WEP	PKMV2 with EAP
Encryption	WPA,WPA2	WPA2 Personal, enterprise protection	128 bit AES	3DES,AES
Mobility	No	No	No	Yes
Primary application	Wireless LAN	Wireless LAN	Wireless LAN	Broadband wireless access
Bandwidth	Unlicensed	Unlicensed	Unlicensed	Licensed and unlicensed both
MAC protocol	CSMA/CA	CSMA/CA	CSMA/CA	Grant request mechanism
Access technology	MIMO/OFDM(IEEE 802.11n)	MIMO/OFDM(IEEE 802.11ac)	SC/OFDM(IEEE 802.11ad)	MIMO/OFDM(IEEE 802.16e)/SOFDM(IEEE 802.16m)
QoS	Limited QoS support	Limited QoS support	Limited QoS support	Pre-connection QoS
Interoperability	less	less	less	robust

**Michael F Finneran,"WiFivsWiMAX- a comparison of technology, markets and business plans" sion of the packets. This additional overhead may not effect significantly on overall throughput as the number of subscriber is less. If the number of subscriber increases and goes beyond 50 then frequent collision may occur resulting back off and retransmission of data and drop in throughput. User may experience significant delay during multimedia streaming. WiMAX avoids such issues using grant request mechanism that allocates a small portion of each transmitted frame as a contention slot. When a subscriber asks for a service to BS through contention slot, BS evaluates the contention slot through subscriber service level agreement (SLA). BS allocates bandwidth based on the service classes (UGS, BE, rTPS, nrTPS) associated with SLA of subscriber. Bandwidth allocation changes dynamically based on number of subscribers and available resources which is controlled by the service class scheduling algorithm discussed in section 3.1. Another advantage of WiMAX is to maintain link stability during mobility using adaptive modulation and coding scheme (MCS) as discussed in 2.1.2.

2. RECENT CHALLENGES IN 802.16M

802 families are based on OSI reference model. Upper four layers (above network layer)of OSI model are remaining same for IEEE 802.11g (WLAN) and IEEE 802.16m (WiMAX). Significant contributions and changes are made in the lower two layers to optimize the performance of WiMAX and WiFi. WiMAX-QoS and performance also depends on the network layer issues like routing and mobility management (Akyildiz, Xie, & Mohanty, 2004), (George, Kumar, Cavalcanti, & Agrawal, 2008), (Sun, 2001). In this chapter we focus on the issues marked by the dotted lines as shown in the figure 5.

Figure 5. Focus area

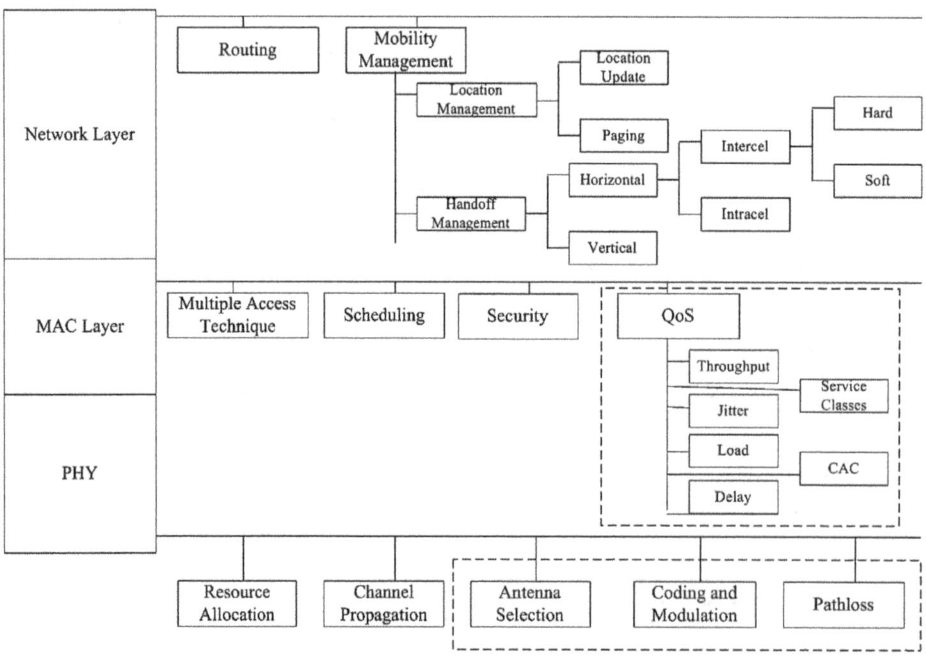

2.1 PHY Challenges

WIMAX physical layer performance depends on proper channel allocation of licensed and unlicensed bands, Antenna selection, and proper selection of coding and modulation scheme and minimized effect of path loss.

2.1.1. Antenna Selection in MIMO

Multiple Input Multiple Output (MIMO) antennas are now used to operate at higher data rates with potential diversity. The array of antennas at each end of the communications circuit has the combined effect to minimize errors and optimize data speed.

MIMO helps to improve wireless access in WiMAX and WiFi system by

- Connecting 802.11n WiFi hotspots with each other and to the Internet.
- Connecting ASN gateways of WIMAX.
- Smooth interoperability between WiFi,WiMAX and other networks.

Compared to a conventional single antenna system, the channel capacity of a multiple antenna system with N_T transmit and N_R receive antennas can be increased by the factor of min (N_T, N_R), without using additional transmit power or spectral bandwidth. Due to the increasing demand of faster data transmission speed in telecommunication systems, the multiple antenna systems are studied (Hochwald & Sweldens, 2000), (Foschini, 2002) and implemented with success for the ever increasing broadband wireless access networks.

MIMO uses three technologies named Beam forming, Spatial Multiplexing and Spatial or Antenna Diversity.

Beam forming deals with the numbers of transmitter and receiver antennas.

Multiple replicas of the radio signal from different directions in space give raise to spatial diversity which can increase the transmission reliability of the fading radio link. A basic idea of the diversity techniques is to convert Rayleigh fading wireless channel into more stable Additive White Gaussian Noise (AWGN)-like channel without any catastrophic signal fading.

In the spatial multiplexing techniques, on the other hand, the multiple independent data streams are simultaneously transmitted by the multiple transmit antennas, thereby achieving a higher transmission speed. MIMO is able to transmit multiple data streams simultaneously over same bandwidth, separated spatially; hence the term is Spatial Multiplexing.

When the spatial-multiplexing techniques are used, the maximum achievable transmission speed can be the same as the capacity of the MIMO channel; however, when the diversity techniques are used, the achievable transmission speed can be much lower than the capacity of the MIMO channel.

Channel capacity (Etemad & Corporation, 2008) of MIMO channel can be derived as

$$C_{MIMO} = W \sum \log_2(1+ S/N \,.\, \lambda_i /N_T) \tag{1}$$

Where N_T = No. of transmit antenna. W is the channel bandwidth. At high SNR, SNR becomes infinity. But in practice MIMO channel (Foschini, 2002) changes randomly which is an ergodic process and channel state information (CSI) is not available at transmitter side. Figure 6 shows the ergodic capacity of the random MIMO channel under different antenna selection. It is observed that increasing number of transmit and receive antennas the channel capacity of MIMO increases thus reliability increases as well as the complexity of the system. So there is a tradeoff between channel capacity and complexity of the system and the breakeven point lies between 2x2 to 4X4 combination of transmit and receive antennas where we can achieve 10bps/ Hz spectral efficiency as per the above graph shown in figure 6. Spectral efficiency up to 15bps/Hz is considered to be efficient spectral efficiency (Liang & Columbia, n.d.) according to Bell lab standard.

2.1.2. Modulation and Coding Schemes (MCS)

MN achieves high throughput with 64 QAM used as MCS in both the end (MN and BS) when signal strength is high with line of sight (LOS) condition. In Mobile WiMAX scenario changes dynamically with respect to distance, signal strength, channel availability etc. Physical layer needs to change the MCS scheme dynamically as per the dynamic requirement of the MN. We introduce an Adaptive Modulation and Coding(AMC) model that senses the SNR and adaptively switches to required MCS to achieve desired level of QoS and link stability.

2.1.2.1. AMC Model

The aim of this model is to create a universal machine that can be adapted easily based on MN mobility. Antenna gain of the WiMAX Base Station (BS) is fixed at 15dbi. When signal strength is high, highest modulation scheme is used (64 QAM) to increase the system capacity. As MN moves out to the periphery of the coverage area causes signal strength fading, AMC model moves to lower modulation scheme to maintain QoS and link stability. Change of MCS depends on the threshold SNR received by MN that continuously sense SNR and change MCS accordingly.

Figure 6. Ergodic capacity of random MMO channel vs SNR under different antenna selection

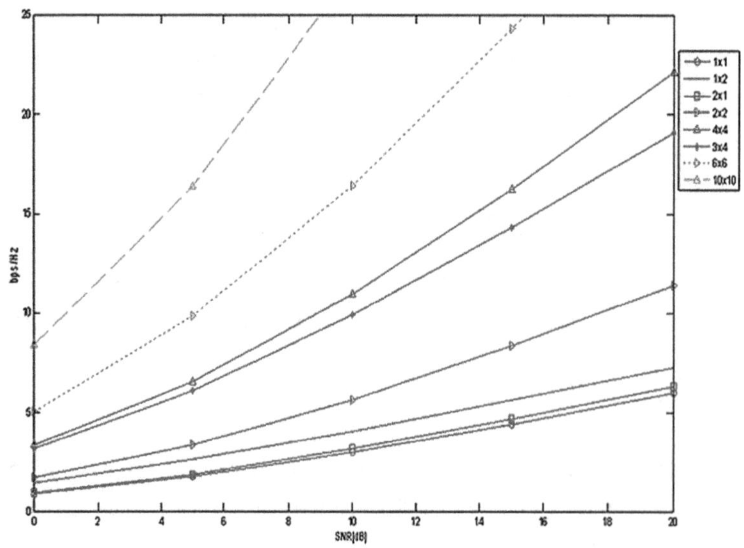

Figure 7. Channel state diagram

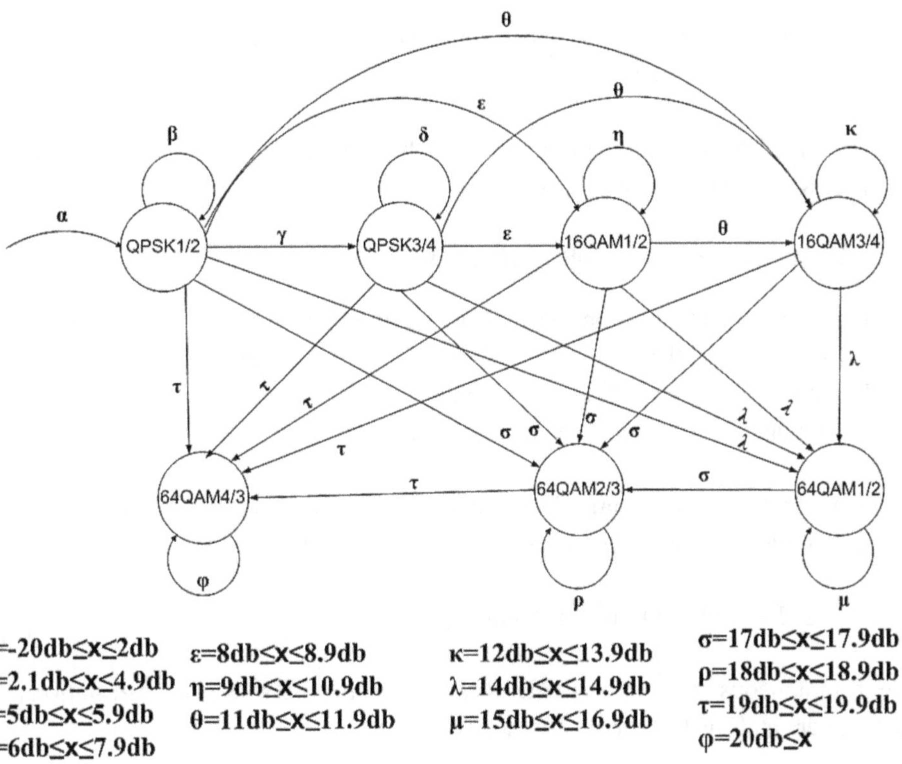

Working function of AMC model is shown by the state diagram as shown in figure 7. Selection of threshold SNR value is a challenging problem. Two techniques to select the threshold SNR is given in (Marabissi, Tarchi, Fantacci, & Balleri, 2008).

2.1.3. Path-Loss Models

In Communication system path loss means attenuation of electromagnetic waves due to physical obstacle during transition from sender to receiver. In WiMAX receiver exploits reflected, scattered, diffracted signals. Proper network planning and use of path loss model (Milanovic, 2007), (Access & Technologies, 2008), (Kumari, Yadav, Yadav, & Sharma, 2011),(Hata, 1980) may reduce the signal attenuation. Some well-known path loss models are free space model, suburban fixed model, Outdoor to indoor pedestrian pathloss model, Vehicular environment. This models are based on detailed geometric information on terrain model, dimension of the obstacle etc.

1. **Free Space Model:** It hardly used in real time environment (V.S. Abhayawardhana, I.J. Wassell, D. Crosby, M.P. Sellars, n.d.), (ItiSahaMisra, 2009) as it does not consider the multipath fading effect.

Mathematical formulation: $P_{rx}(r)=P_{tx}G_{tx}G_{rx}\lambda^2/((4\pi)^2r^2L)$.

P_{rx} and P_{tx} are received power and transmitted power in watts, respectively; G_{rx} and G_{tx} are the gain of the receiving and transmitting antennas, respectively; L is the system-loss factor.

2. **Suburban Fixed Model:** Considering cell with large coverage area and higher antenna height considering MN having low mobility (Erceg et al., 1999).

Mathematical formulation: $PL=H+10\log_{10}(d/d_0)+X_f+X_h+s$.

PL is the instantaneous attenuation in dB, H is the intercept and is given by free space path-loss at the desired frequency over a distance of d_0=100 m.γis a Gaussian random variable over the population of macro cells within each terrain category. X_f andX_h are the correlation factors of the model for the operating frequency and for the MS antenna height, respectively.

3. **Outdoor to Indoor Pedestrian Path Loss Model:** Small cell size with low antenna height. Pedestrians are located in streets, inside buildings (M.1225, 1997).

Mathematical formulation: $PL=40\log_{10} R+30\log_{10} f+49$.

PL is the instantaneous attenuation in dB, R is the distance between the base station and the mobile station in kilometers and f is the carrier frequency.

4. **Vehicular Environment:** Larger cell size and higher antenna height considering MN having mobility (60kmph to 180kmph)(V.S. Abhayawardhana, I.J. Wassell, D. Crosby, M.P. Sellars, n.d.), (Aphiraksatyakul, Seet, & Lau, 2008).

Mathematical formulation: $PL=40(1-4\times10^{-3}\times\Delta h_b)\log_{10}R-18\log_{10}\Delta h_b-21\log_{10}f+80\text{dB}$.

R is the distance between the base station and the mobile station, f is the carrier frequency and Δh_b is the base station antenna height in meters.

3. SIMULATION RESULTS AND ANALYSIS

We used OPNET simulator (Zakrzewska, Berger, & Ruepp, n.d.) and MATLAB for simulation. Optimized Network Engineering Tool (OPNET) provides a development environment supporting the communication networks. Both behavior and performance of the modeled system can be analyzed through discrete event simulator. OPNET provides graphical editor to enter network model details. It consists four such types of editor names network editor, node editor, process editor and parameterized editor organized in a hierarchical way. We design the optimized network model scenario in OPNET, run the model under different condition, collected data and plot the data in MATLAB.

Some important parameters considered during simulation is shown in Table 5.

We set up fifteen mobile nodes from Mobile_1_1 to Mobile_15_1under base stations BS_1 to BS_15 respectively as shown in Figure 8. Although WiMAX standards specify that WiMAX supports large

Table 5. Parameter taken in Simulation

Min. MN velocity	60 km/hr	
Max. MN velocity	100 km/hr	
Frequency Bandwidth used	2.3-2.4 Ghz	
Channel Bandwidth	8.75/10 Mhz	
Antenna type	Omni directional	
MIMO	Matrix 4/4	
Transmission power	Max. 20watt	
RCV Buffer size	64kbytes	
Antenna Gain	15dBi	
Base Station Parameters:		
	Maximum Number of Mobile Node (MN)	15
	Default Service class Name	Gold(rtPS)
	Mac address	Auto assigned
	Maximum Power transmission	0.5W
	PHY profile	wireless OFDMA 20 MHz
	PHY profile Type	OFDM
Mobile Node	Handover Parameters:	
	MN handover Retransmission timer	30ms
	Maximum handover request retransmit	6ms
	Maximum sustained traffic rate	4MBPs
	Minimum reserved traffic rate	0.5MBPS
No. of simulation runs for each scenerio		10
Simulation time		3600 ms

coverage area but in real time it supports approximately 3 km. We consider different modulation and coding schemes(MCS) like QPSK1/2, QPSK3/4, 16QAM1/2, 16QAM3/4, 64QAM1/2, 64QAM2/3, 64QAM3/4,Adaptive(Bhunia, Misra, Sanyal, & Kundu, 2011), (Access & Technologies, 2008), (Islam, Rashid, & Tarique, 2011). Different simulation results have been shown in the following figures by considering 15 nodes at 100kmph.Simulation time is 3600 sec. After simulation, we collected the data values and plot the data values in MATLAB.

Figure 9 variation of voice packet delay at different MCS. Peak VoIP packet delay of Adaptive Modulation and Coding (AMC) is minimum (0.5 msec) among other MCS. Peak delay is 3msec (64QAM3/4) that is much below the threshold level of 150msec as recommended by ITU-T (M.1225, 1997). Figure 10 variation of data packet delay at different MCS. Figure 11 variations of loads (packets/second) at different MCS. Figure 12 change of throughput at different MCS. Empirical analysis also shows that Adaptive Modulation and Coding (AMC) and QPSK1/2 achieve 90% throughput (marked by 2 lines) as maximum sustainable traffic rate of the communication link is 4mbps. 64QAM2/3 and 64QAM ¾ fails to achieve minimum benchmark of 60% throughput.

Figure 13 and Figure 14 represents change of uplink SNR and downlink SNR respectively at different MCS. Figure 15, Figure 16, Figure 17, Figure 18depicts variation of path-losses at different speed(60-

Figure 8. Mobile WiMAX network model considering higher traffic (15 MNs)

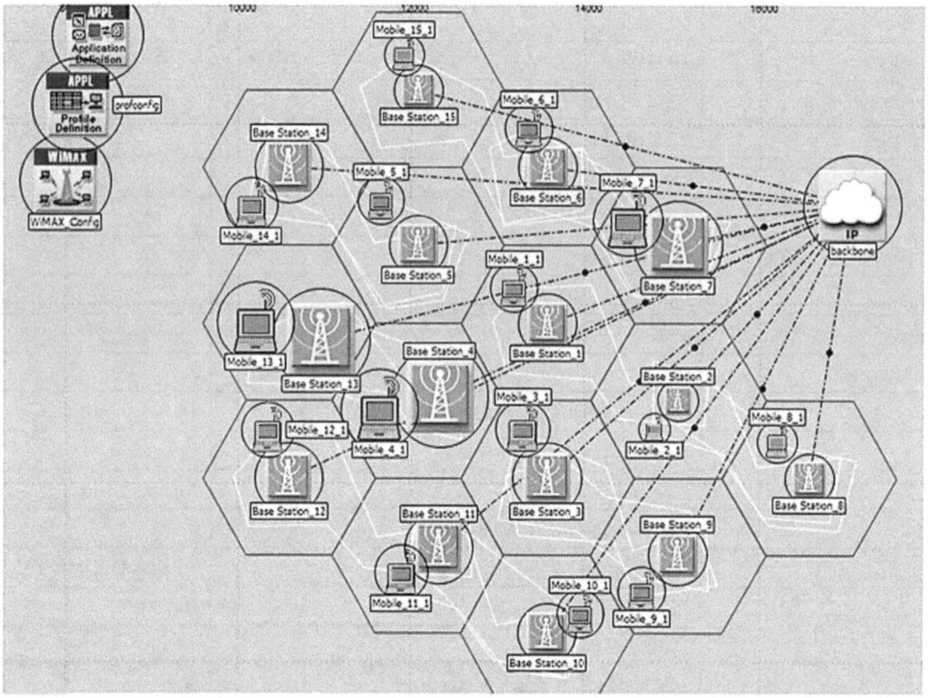

90kmph). Figure 15 represents variation of path-losses in free space(ideal situation). Figure 16 represents variation of path-losses at vehicular condition of the MN at 90km/h. Figure 17 represents variation of path-losses of MN in suburban region. Figure 18 represents variation of path-losses when the mobile handheld device is carried by a pedestrian. It has been observed from Figure 15 -18 that path- loss for free space propagation model is minimum even with high mobility (90kmph). Path- loss for pedestrian outdoor to indoor is maximum, this may be due to the radio wave component reflected and diffracted

Figure 9.

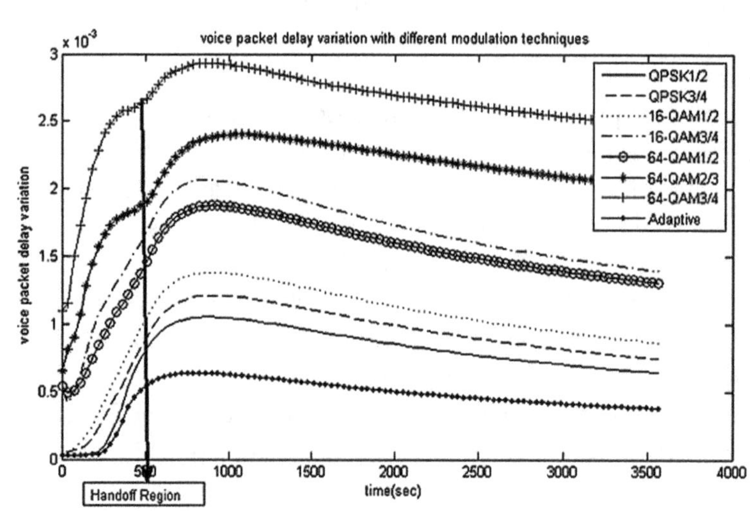

Figure 10.

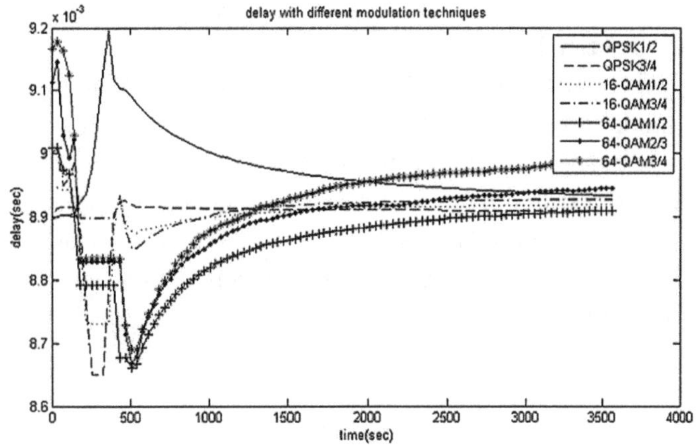

Figure 11.

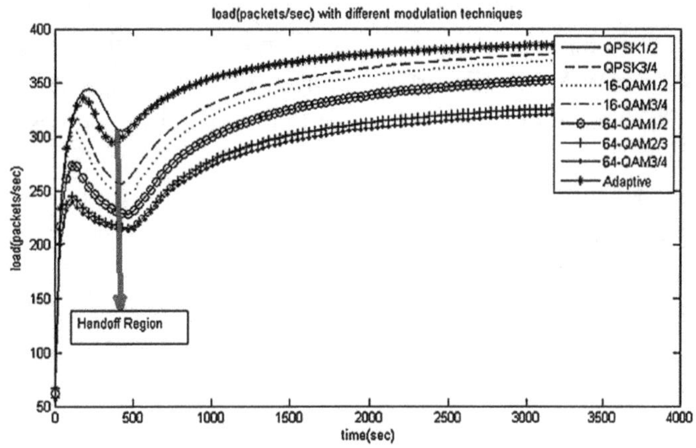

Figure 12.

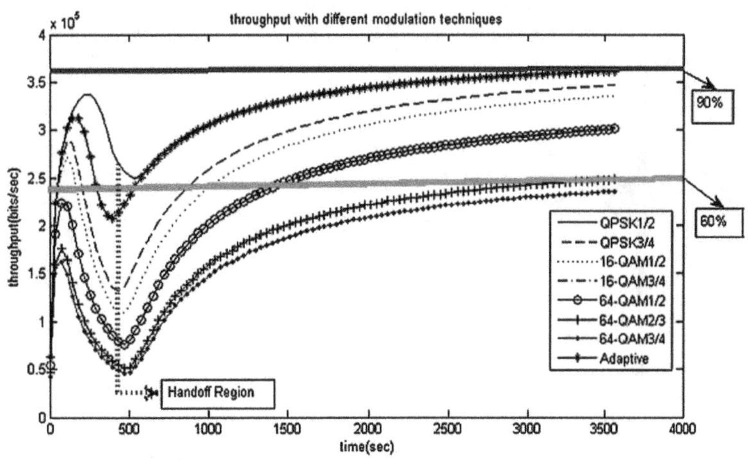

on building reaching the receiving antenna. Figure 19 represents delay variation at various path-loss models. Delay for the free space model is minimum and it is maximum for pedestrian outdoor to indoor model, this may be due to the radio wave component reflected and diffracted on building reaching the receiving antenna. Figure 20 and Figure 21 represents load and throughput variation respectively under various pathloss models. Load (packets/sec) is maximum for the free space propagation model whereas it is minimum for pedestrian outdoor to indoor model. Throughput is maximum for the free space propagation model and it is minimum for pedestrian outdoor to indoor model. Although all path loss models achieve the minimum throughput benchmark (60%) in the proposed WiMAX system except handoff duration as shown in the Figure12. As far as 60% throughput is concerned, the rate is adequate to support applications like VoIP and video transmission. Figure 22 represents downlink BER(Bit Error Rate) at various cyclic prefixes.

3.1. Performance Analysis of QoS Service Classes

WiMAX forum offers four different QoS service classes for optimum (Saha, 2013) use of radio resources and efficient control of traffic. Through service classes, WiMAX provide robust QoS offerings to customer based on their demand and available radio resources. For example, if a customer wants VoIP service then WiMAX can identify the corresponding service classes and map it with the radio resources through a scheduling algorithm. Proper WiMAX network planning requires an efficient scheduling algorithm that will serve (Whitepaper, 2010) customer demand using minimum radio resources. The service classes are:

1. Unsolicited Grant Service (UGS)
2. Real-time Polling Service (rtPS)
3. Non-real -time Polling Service (nrtPS)
4. Best effort (BE)

Figure 13.

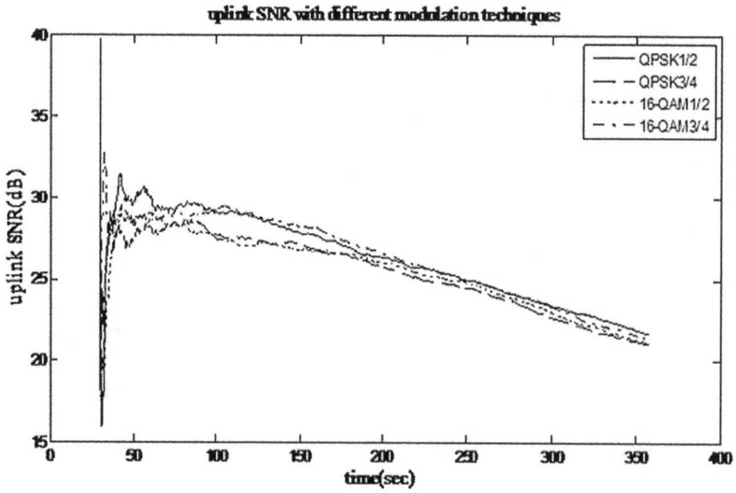

Figure 14.

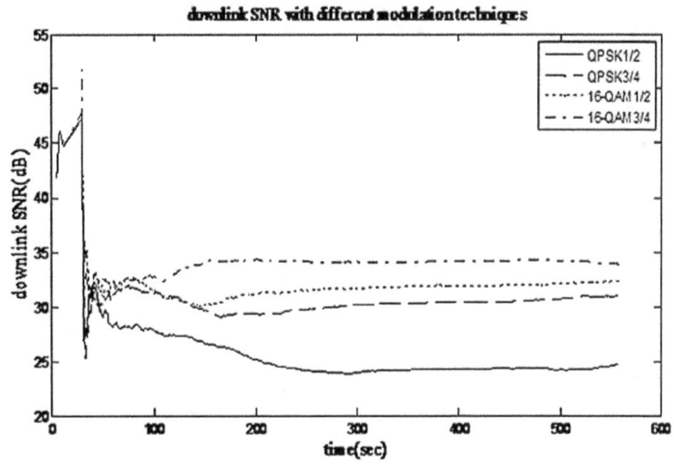

Figure 15.

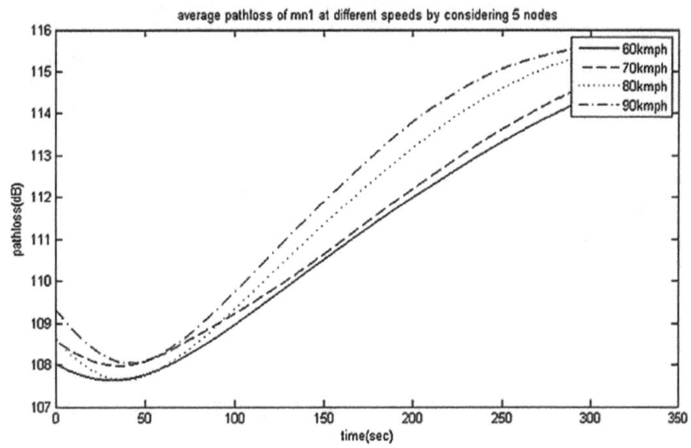

Figure 16.

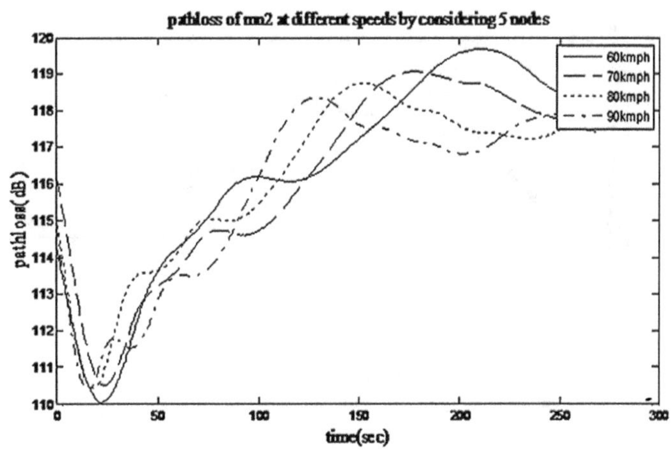

Figure 17.

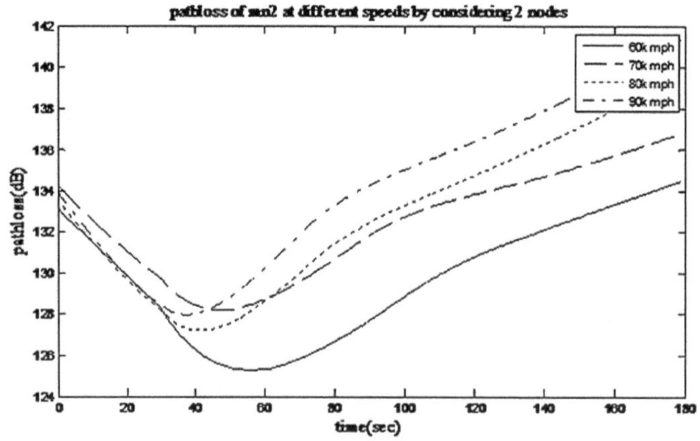

Figure 18.

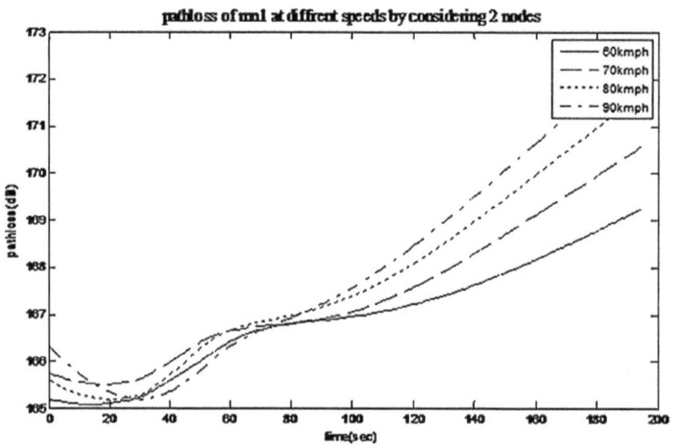

Figure 19.

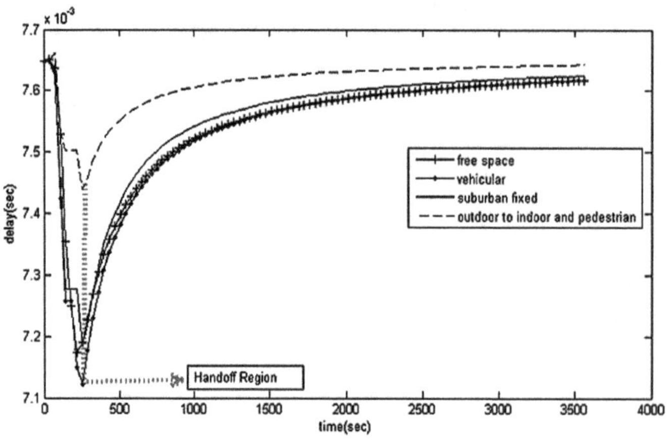

Figure 20.

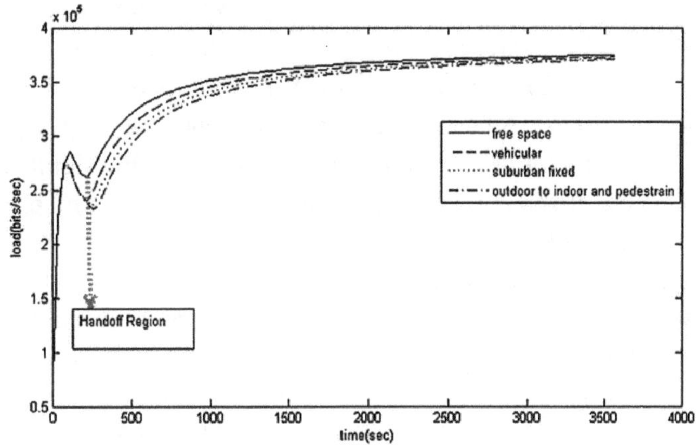

Figure 21.

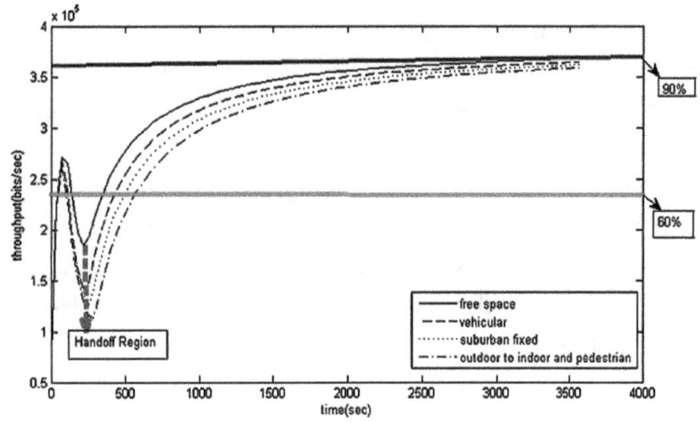

Figure 22.

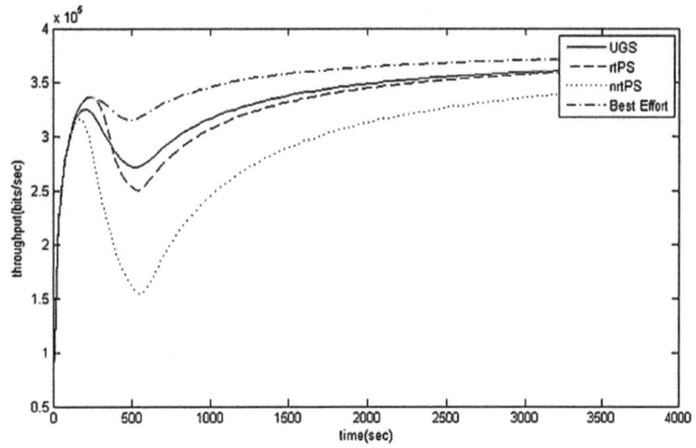

QoS issues are handled at MAC sub layer. The IEEE 802.16m slandered MAC sub layer provides QoS differentiation for the different types of applications through five different scheduling service types also called QoS service classes (Galshetwar, Jayakumar, & Mittal, 2012), (So-in, Jain, & Al-tamimi, 2010), (So-in, Member, Jain, & Tamimi, 2009).Khalil (Khalil & Ksentini, 2007), Jain(Jain & Verma, n.d.) are compared different scheduling services like Proportional Fair Scheduling algorithm, Cross-layer Scheduling Algorithm, and TCP aware uplink Scheduling algorithm. Liu(Liu, Member, Wang, & Giannakis, 2006) tries to optimize the resources in Cross layer scheduling algorithm by providing fixed slots k to the UGS out of the n slots (n>k) of radio resources. (n – k) slots are allocated to the rtPS, nrtPS and are dynamically based on an algorithm defined in (Liu et al., 2006), (MarinaAguado, Eduardo Jacob, Marion Berbineau, 2010), (Rashwan, Elbadawy, & Ali, 2009), (Chowdhury, Misra, & Sanyal, 2012), (Huang & Subramanian, 2009).UGS designed to support constant bit rate which is not used in real life situation. rtPS designed to support real time services with variable bit rates and delay i.e. VoIP on a periodic basis. nrtPS designed to support non real time applications with variable data rates which require guaranteed data and delay i.e. VoIP with silence suppression. nrtPS non real time and delay tolerant services that require variable size data grant burst types on a regular basis such as FTP.BE designed to support data streams that do not require any guarantee in QoS such as HTTP.

A comparative analysis of throughput has been made with other four available scheduling algorithms like UGS, rtPS, nrtPS, BE as shown in Figure 23.

When we are dealing with throughput, Best Effort algorithm seems to be the best for the architecture taken into consideration. Throughput of nrtPS is minimum. A comparative uplink packet drop analysis of different scheduling algorithms(UGS,rtPS,nrtPS,Best Effort) has been made in Figure 23. In this case, uplink packet drop for UGS is minimum. Downlink packet drop of different scheduling algorithms(UGS,rtPS,nrtPS,Best Effort) has been compared in Figure 24.

From Figure 24,we can say that downlink packet drop of UGS is maximum and Best Effort is minimum.

We can conclude from Figure 22and Figure 24 that we can use Best Effort scheduling algorithm except for some specific conditions.

Figure 23.

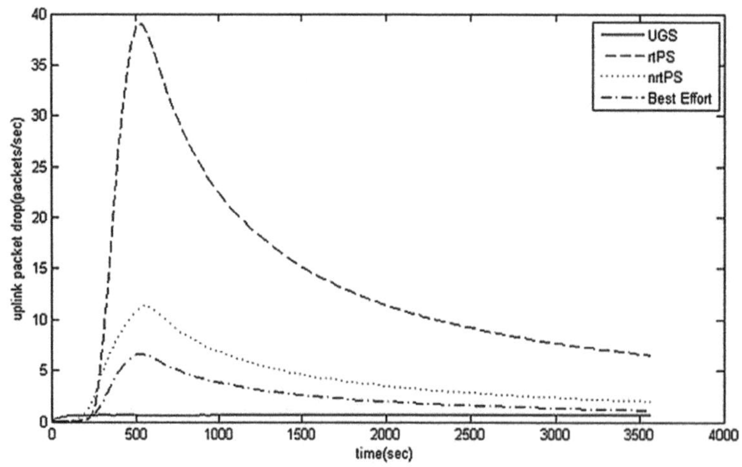

Figure 24.

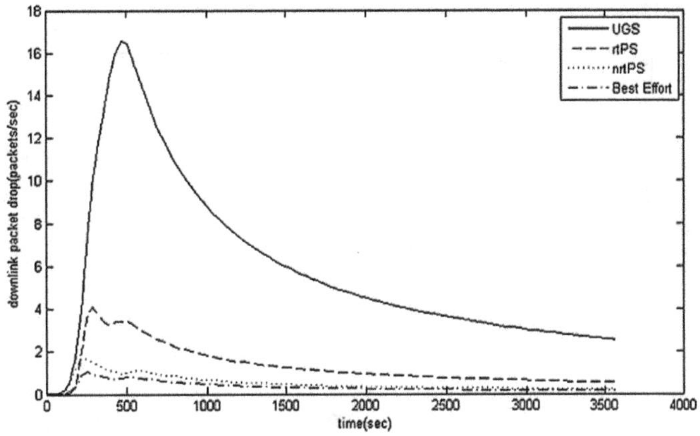

3.2. Performance Analysis of 802.11n

We configured WLAN IEEE802.11n standard in OPNET. Here two mobile nodes Mobile_1_1 and Mobile_2_1 are located under Access Points AP_1 and AP_2 respectively as shown in Figure 28. Common parameters are considered as specified in Table 5. Channel bandwidth considered as 22 MHz each. Access point transmitted power is 5 mli watt.

We also considered two scenarios a) Static condition b) moving condition of the MN.

Scenario1: MN's are in static condition

Figure 25 shows delay under static condition. Figure 26 shows load (bits/sec) under static condition. Figure 27 shows throughput under static condition. The data transfer rate is constant at approx 150 kbps.

Figure 25.

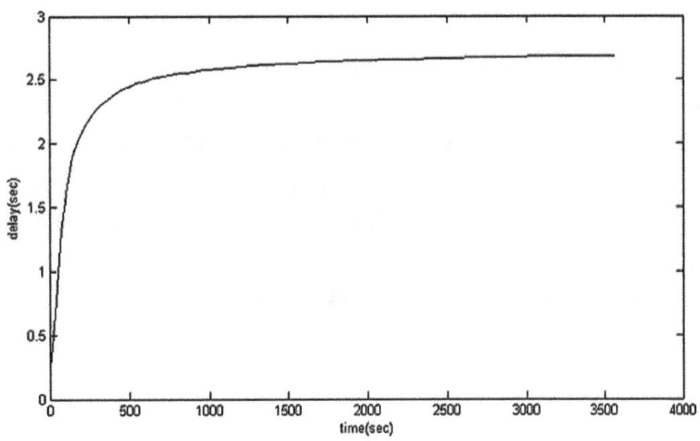

Figure 26.

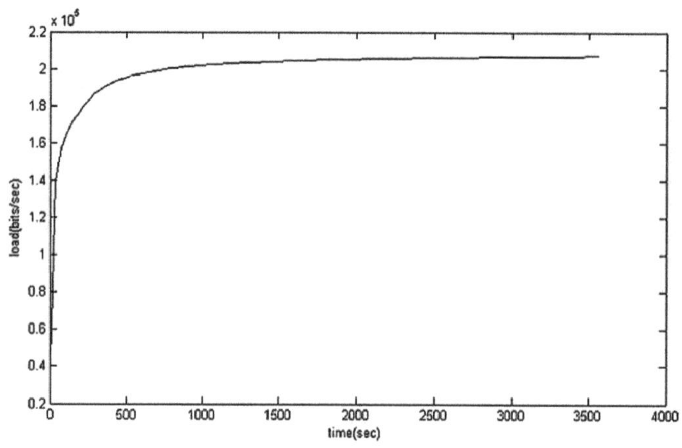

Figure 27.

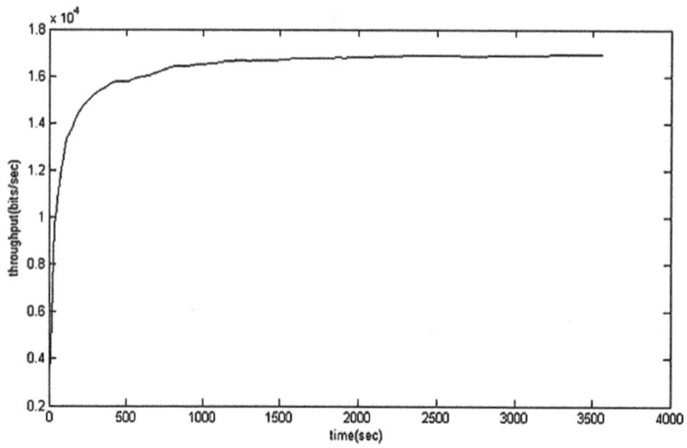

Scenario 2: MN's are Moving

We defined the movement path of two MN's as shown in the figure. MN's Mobile_1_1 is going to the cell under access point AP_2 from AP_1 and Mobile_2_1 is going to the cell under AP_1 from AP_2. We checked the results by considering 100 KMPH speed.Figure 29 shows delay under node moving condition at 100 KMPH speed. Under node moving condition delay is about 2.3 sec which is less than delay under static condition (2.5 sec).Figure 30 shows throughput at 100 KMPH speed. The data transfer rate is approx 280 kbps. So, throughput increases at a high mobility because under static scenario, throughput is 150 kbps. But, throughput dropped drastically to 64.28% because there is a handoff at about 0.3 msec.

Figure 28. Node movement path

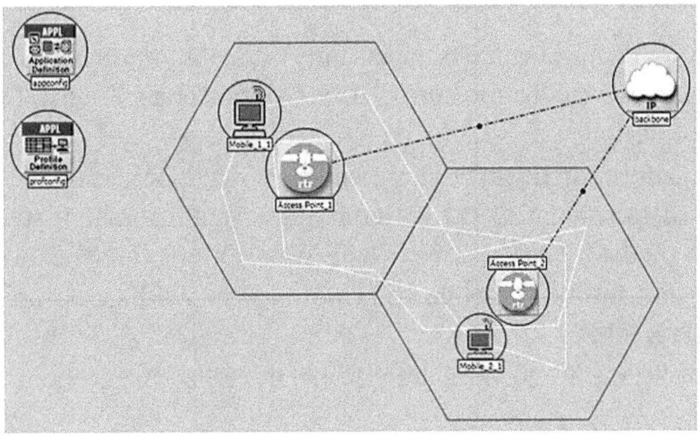

Figure 29.

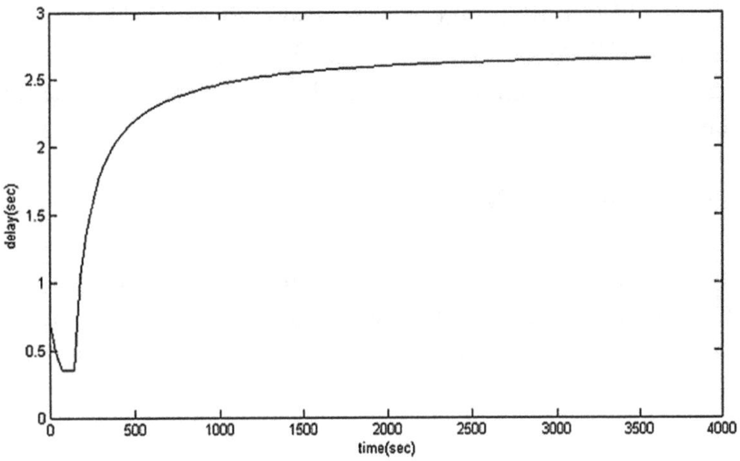

Figure 30.

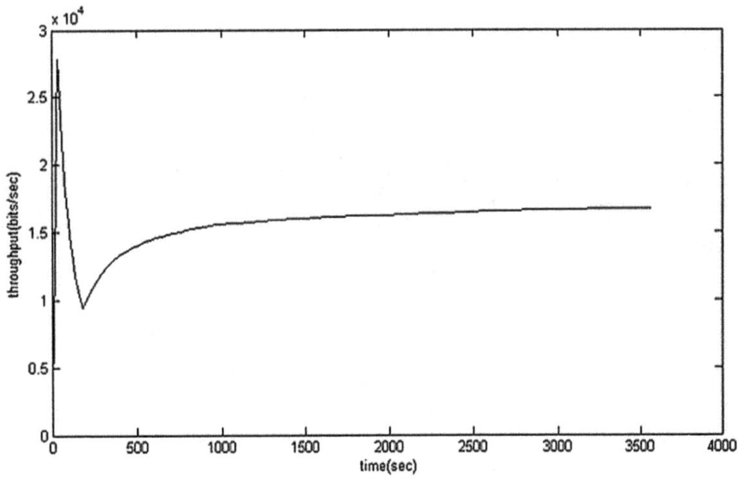

4. WIMAX_WIFI INTEGRATED ARCHITECHTURE

Integration of WiMAX_WiFi has been described in (Cavalcanti, Agrawal, Cordeiro, Xie, & Kumar, 2005), (Karthika, Sumithram, & Shanmugam, 2013), (Allan Borges Pontes, Diego Dos Passos Silva, José Jailton, Jr., OtavioRodrigues, Jr., 2008).We created WiMAX_WiFi integrated model(Figure 31) in OPNET simulation platform. In this model, we choose Cisco MDS 9216 switch to connect WiMAX base station and WiFi access point. MD 9216 is a multilayer intelligent high-speed switch, is connected to backbone network. A trajectory has been set where WiMAX mobile node moves from WiMAX base station to WiFi access point and WiFi mobile node moves from WiFi access point to WiMAX base station at a high speed (100 KMPH).

Simulation results of WiMAX and WiFi are given in the following sections.

4.1 WiMAX Simulation Results

Figure 32 shows delay of WiMAX at 100 KMPH, delay is about 0.000165 sec. Figure 33 shows load (bits/sec) of WiMAX. Load of WiMAX is at about 90,000 bits/sec.Figure 33 shows throughput of WiMAX at 100 KMPH. Throughput of WiMAX is at about 90,000 bits/sec in the integrated model. There is no handover in the integrated model. From this it is clear that,this integrated model prevents the problem of call drop and call block.

4.2 WLAN or WiFi Simulation Results

Figure 34 shows delay of WiFi at 100 KMPH. In the integrated model, delay is about 3.1sec.Figure 33 shows load (bits/sec) of WiFi. Load of WiFi is at about 90,000 bits/sec in the integrated model.

Figure 35 shows throughput of WiFi at 100 KMPH. Throughput of WiMAX is at about 100 bits/sec in the integrated model.

Figure 31. WiMAX_WiFi Integrated Model

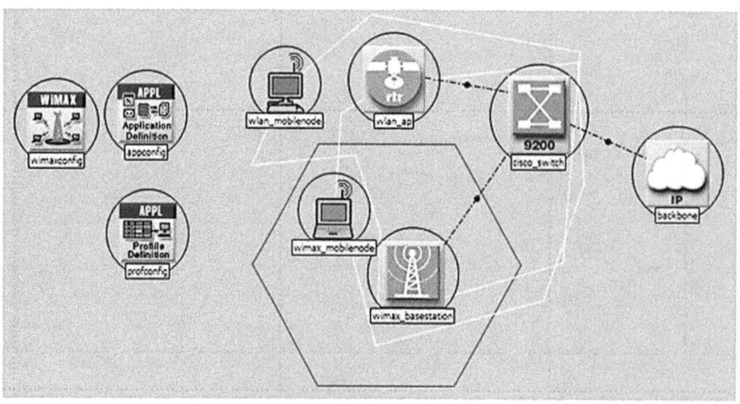

Figure 32.

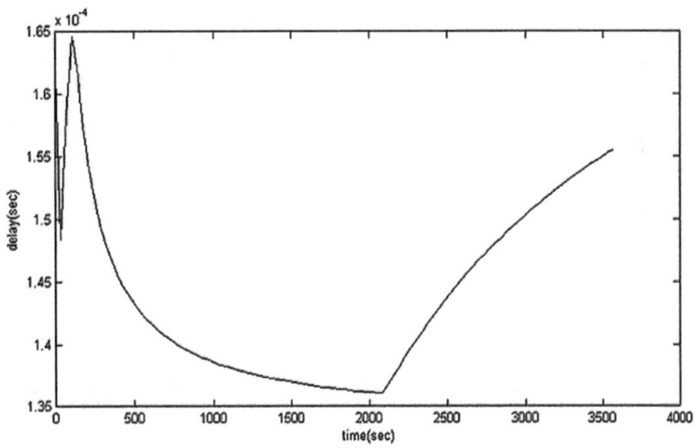

Figure 33.

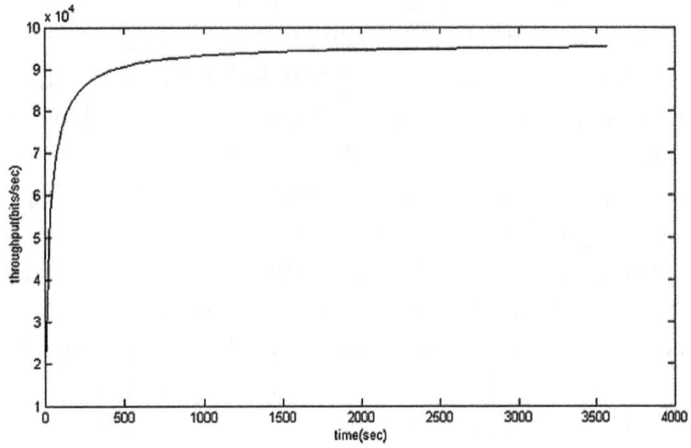

Figure 34.

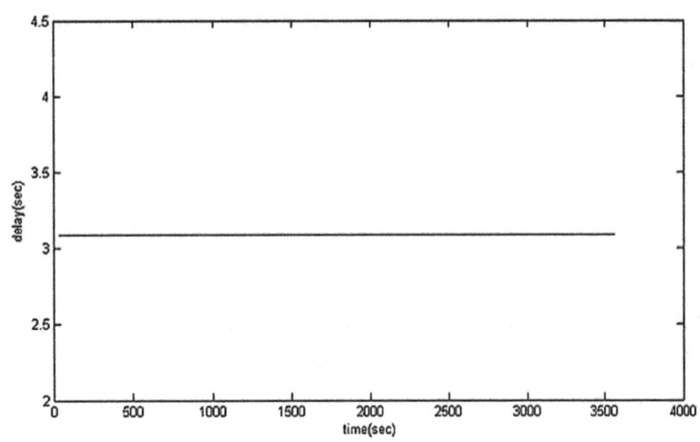

Figure 35.

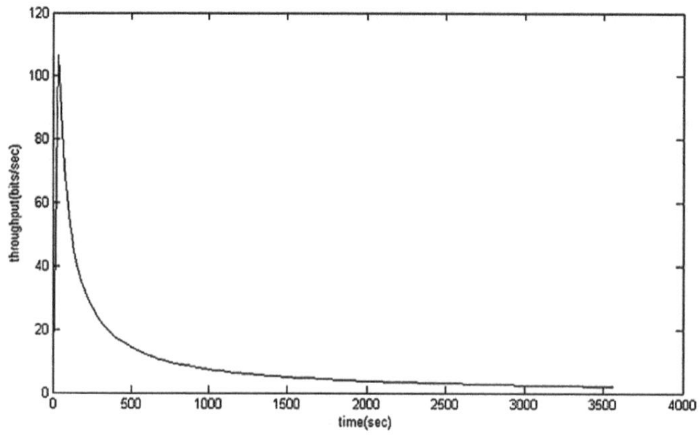

5. CONCLUSION

WiMAX is conveniently deployable due to ease of installation and low cost. It is very important to guarantee QoS to the end users to avail and sustain the diversifying applications of WiMAX. An optimized architecture conforming to the recently standardized IEEE 802.16m framework is needed for integrating both mobility management and QoS to provide broadband service in a ubiquitous environment to end users. In (Sajal Saha et. el., 2013), author proposed a mobility management model THMIP at network layer to achieve better throughput with minimal handoff latency. In this chapter, our analysis on the performance of the WiMAX considered following parameters of the physical layer.

a) Different MCS with constant speed (100kmph) b) Different speed with proposed AMC c) Different path loss models with proposed AMC at a constant speed (90kmph).

Performance matrices considered are delay, load, SNR, throughput. From the empirical studies, it is evident that proposed AMC allows the WiMAX network to yield higher throughput covering longer distance. We have also analyzed throughput, uplink packet drop and downlink packet drop of different service classes of mobile WiMAX. It has been observed that we can use BE service class except from some specific conditions. We tried to figure out the fluctuation in throughput and packet drop at 60KMPH, 90KMPH, 100KMPH speed to define the complete agility of the system. After simulation, it is evident that 90KMPH is the maximum achievable speed for a MN under vehicular condition. Besides WiMAX, we have also verified performance of WiFi under static and node moving condition.Performance matrices considered are delay, load, and throughput. It has been observed that WiMAX gives far better performance than WiFi. We have configured a WiMAX_WiFi Integrated model and analyzed performance of it. In this model, there is a less chance of handover,i.e. call drop and call block.

REFERENCES

16-2005, I. S. 802. (2005). *IEEE standard for local and metropolitan area networks part 16: air interface for Fixed broadband Wireless access system- physical and medium access control layer for combined mobile and Fixed Operation in licensed bands.*

Abhayawardhana, V. S., Wassell, I. J., Crosby, D., & Sellars, M. P. (2005). Comparison of Empirical Propagation Path Loss Models for Fixed Wireless Access Systems. In *Vehicular Technology Conference, 2005. VTC 2005-Spring. 2005 IEEE*. doi:10.1109/VETECS.2005.1543252

Access, R., & Technologies, O. (2008). *Introduction to WiMAX Modeling for Network R & D and Planning*. Author.

Aguado, Jacob, & Berbineau. (2010). QoS Challenges in Handover Process. In *WiMAX Security and Quality ofService: An End-to-End Perspective*. Academic Press.

Ahmadi, S. (2008). An Overview of Next Generation Mobile WiMAX Technology. *Communications Magazine, IEEE, 47*, 84–98.

Akyildiz, I. F., Xie, J., & Mohanty, S. (2004). A survey of mobility management in next-generation all-IP-based wireless systems. *IEEE Wireless Communications, 11*(4), 16–28. doi:10.1109/MWC.2004.1325888

Alcatel-Lucent. (2004). *WiMAX, making ubiquitous high-speed data services a reality*. Retrieved from http://www1.alcatel-lucent.com/publications/abstract.jhtml?repositoryItem=tcm:172-44851635

Alvarion. (n.d.). *Wireless Connectivity Breaking the line-of-sight boundaries*. Author.

Andrews, & ArunabhaGhosh, R. (2007). Fundamentals of WiMAX: Understanding Broadband Wireless Networking. Prentice Hall.

Aphiraksatyakul, D., Seet, B. C., & Lau, C. T. (2008). Evaluation of terrain effects on Mobile WiMax in a Vehicular Environment. *Proceedings - 2008 8th International Conference on Intelligent Transport System Telecommunications, ITST 2008*. doi:10.1109/ITST.2008.4740290

Bhunia, S., Misra, I. S., Sanyal, S. K., & Kundu, A. (2011). *Performance study of mobile WiMAX network with changing scenarios under different modulation and coding*. http://doi.org/<ALIGNMENT.qj></ALIGNMENT>10.1002/dac

Broadband, M. (2007). *Welcome to Your Internet Future*. Author.

Cavalcanti, D., Agrawal, D. P., Cordeiro, C., Xie, B., & Kumar, A. (2005). Issues in integrating cellular networks, WLANs, and MANETs: A futuristic heterogeneous wireless network. *IEEE Wireless Communications, 12*(3), 30–41. doi:10.1109/MWC.2005.1452852

Chowdhury, P., Misra, I. S., & Sanyal, S. K. (2012). *Cross Layer QoS Support Architecture with Integrated CAC and Scheduling Algorithms for WiMAX BWA Networks*. Academic Press.

Díaz, J. M., Del Río, L. V., Cárdenas, O. Á., & Baldivia, M. G. (2014). Analysis of Propagation Models, Delay, and Throughput for WiMAX in Urban Environments. In *Broadband Wireless Access Networks for 4G: Theory, Application, and Experimentation* (pp. 1–13). Information Science Reference. http://doi.org/doi:10.4018/978-1-4666-4888-3

Eklund, C., Marks, R. B., Ponnuswamy, S., & Stanwood, K. L. (2007). WirelessMAN: Inside the IEEE 802.16 Standard for Wireless Metropolitan Area Networks. IEEE.

Erceg, V., Member, S., Greenstein, L. J., Tjandra, S. Y., Parkoff, S. R., Gupta, A., & Bianchi, R. et al. (1999). *An Empirically Based Path Loss Model for Wireless Channels in Suburban Environments*. Academic Press.

Etemad, K., & Corporation, I. (2008). *Overview of Mobile WiMAX*. Author.

Foschini, G. J. (2002). Layered space-time architecture for wireless communication in a fading environment when using multi-element antennas. *Bell Labs Technical Journal*, *1*(2), 41–59. doi:10.1002/bltj.2015

Galshetwar, G. M., Jayakumar, A., & Mittal, Y. (2012). Comparative Study Of Different Scheduling Algorithms For Wimax MAC Scheduler Design. *International Journal of Engineering*, *2*(2), 1031–1037.

George, A., Kumar, A., Cavalcanti, D., & Agrawal, D. P. (2008). Protocols for mobility management in heterogeneous multi-hop wireless networks. *Pervasive and Mobile Computing*, *4*(1), 92–116. doi:10.1016/j.pmcj.2007.03.002

Hata, M. (n.d.). Empirical formula for propagation loss in land mobile radio services. *IEEE Trans. Veh. Tech., 29*, 317–325.

Hochwald, B. M., & Sweldens, W. (2000). Differential unitary space-time modulation. *IEEE Transactions on Communications*, *48*(12), 2041–2052. doi:10.1109/26.891215

Huang, J., Subramanian, V., Agrawal, R., & Berry, R. (2009). Joint scheduling and resource allocation in uplink OFDM systems for broadband wireless access networks. *Selected Areas in*, *27*(2), 226–234. doi:10.1109/JSAC.2009.090213

I. S. 802. 1.-2009 (Revision of I. S80216-2004). (n.d.). *IEEE Standard for Local and metropolitan area networks Part 16: Air Interface for Broadband Wireless Access Systems*. http://doi.org/<ALIGNMENT.qj></ALIGNMENT>10.1109/IEEESTD.2009.5062485

Islam, S., Rashid, M., & Tarique, M. (2011). *Performance Analysis of WiMax/WiFi System under Different Codecs*. Academic Press.

ITU-T Recommendation. (2003). One-way transmission time. *ITU-T Recommendation*, G.144.

Jain, A., & Verma, A. K. (n.d.). *Comparative Study of Scheduling Algorithms for WiMAX*. Academic Press.

Jee, & Madanapalli. (2008). *IP over IEEE 802.16 Problem Statement and Goals*. ETF RFC 5154.

Karthika, A. L., Sumithram, G., & Shanmugam, A. (2013). *Performance of voice in integrated WiMAX-WLAN and UMTS-WLAN*. Academic Press.

Khalil, A., & Ksentini, A. (2007). *Classification of the Uplink Scheduling Algorithms*. Academic Press.

Kumari, M., Yadav, T., Yadav, P., & Sharma, D. (2011). *Comparative Study of Path Loss Models in Different Environments*. Academic Press.

Liang, Y. W., & Columbia, B. (2010). *Ergodic and Outage Capacity of Narrowband MIMO Gaussian Channels*. Academic Press.

Liu, Q., Member, S., Wang, X., & Giannakis, G. B. (2006). *A Cross-Layer Scheduling Algorithm With QoS Support in Wireless Networks*. Academic Press.

M.1225, R. I.-R. (1997). *Guidelines for evaluation of radio transmission technologies for imt-2000.*

Marabissi, D., Tarchi, D., Fantacci, R., & Balleri, F. (2008). Efficient Adaptive Modulation and Coding Techniques for WiMAX Systems. *2008 IEEE International Conference on Communications*. doi:10.1109/ICC.2008.636

Milanovic, J. (2007). *Comparison of Propagation Models Accuracy*. Academic Press.

Misra. (2009). *Wireless Communications and Networks 3G and beyond*. Tata McGraw Hill Education Private Limited.

Nakamura, V. M. (2008). *Standardization Activities for Mobile WiMAX*. Academic Press.

Narasimha, S., & Sivalingam, K. M. (2009). *Improved opportunistic scheduling algorithms for WiMAX Mobile Multihop Relay networks*. IEEE. doi:10.1109/HIPC.2009.5433228

Nuaymi, L. (2007). *WiMAX: Technology for Broadband Wireless Access*. John Wiley and Sons. doi:10.1002/9780470319055

Pareit, D., Lannoo, B., Moerman, I., & Demeester, P. (2012). *The History of WiMAX : A Complete Survey of the Evolution in Certification*. Academic Press.

Rashwan, A. H., Elbadawy, H. M., & Ali, H. H. (2009). *Comparative Assessments for Different WiMAX Scheduling Algorithms*. Academic Press.

Sachs, J., Muñoz, L., Aguero, R., Choque, J., Koudouridis, G., Karimi, R., . . . Gebert, J. F. M. (2004). Future wireless communication based on multi-radio access. In *Proceedings Wireless World Research Forum Meeting*.

Saha, S. (2013). *Performance Analysis of Service Classes*. IEEE.

Saha. (2013). QoS and Mobility Management Issues on Next Generation Mobile WiMAX Networks. In *Broadband Wireless Access Networks for 4G: Theory, Application, and Experimentation* (pp. 298—323). Information Science Reference, IGI Global. http://doi.org/doi:10.4018/978-1-4666-4888-3.ch16

Singh, R. K. (2012). *WiMAX Networks: Services and their Quality Matrics*. Academic Press.

So-in, C., Member, S., Jain, R., & Tamimi, A. (2009). Scheduling in IEEE 802. 16e Mobile WiMAX Networks. *Key Issues and a Survey*, *27*(2), 156–171.

Sun, J. (2001). Mobility management techniques for the next-generation wireless networks. *Proceedings of the Society for Photo-Instrumentation Engineers*, 155–166. doi:10.1117/12.445223

TWT Whitepaper. (2010). *WiMAX QoS Classes*. TWT.

Yarali, A., & Rahman, S. (2008). WiMAX broadband wireless access technology: Services, architecture and deployment models. *2008 Canadian Conference on Electrical and Computer Engineering*. doi:10.1109/CCECE.2008.4564499

Zakrzewska, A., Berger, M. S., & Ruepp, S. (2008). *Modeling Multistandard Wireless Networks in OPNET*. Academic Press.

Chapter 8
Joint Source Channel Coding and Diversity Techniques for 3G/4G/LTE-A:
A Review of Current Trends and Technologies

Surajit Deka
Gauhati University, India

Kandarpa Kumar Sarma
Gauhati University, India

ABSTRACT

There has been a tremendous growth in the joint source channel coding techniques of recent wireless communication system. The main objective of this technique is to simplify the architecture which eliminates the network errors and instability. This chapter mainly focuses on Joint Source Channel Coding (JSCC) in Multiple Input Multiple Output (MIMO) diversity to minimize the effects of noise, distortion and related phenomena for better reception of the signal with increased efficiency and spectrum utilization with link reliability. JSCC techniques in presence of 3G/4G/LTE-A for high data rate communication with the carrier aggregation is an important element for communication over wireless channel. In this chapter, the authors shall discuss the current state of research, the challenges related to the deployment of JSCC, the likely solutions and the usefulness as part of upcoming communication networks. A major part of the work shall focus on their suitability in LTE-A frameworks so essential for sustaining the current growth in wireless communication.

INTRODUCTION

The expansion of wireless networks and increasing use of high data rate mobile communication has constantly driven new innovations. One primary factor is the increasing demand of channel capacity. Adding to this demand is an even present requirement of link reliability and continuously improving quality of

DOI: 10.4018/978-1-5225-0239-5.ch008

service (QoS). Qos is an issue which suffers for wireless communication channel due to the stochastic nature of it. Stochastic behaviour of wireless communication channel is linked to fading which becomes severe with mobility of the transceiver sets. Diversity schemes like multi-input multi output (MIMO) have been successful in mitigating the ill-effects of fading and have also enhanced capacity of channels. Yet there are certain issues like application of coding which can be used effectively to improve QoS and link reliability. Of late joint source and channel coding (JSCC) has been accepted as important addition to the wireless/mobile communication framework which have shown considerable amount of promise to improve QoS and mitigate the ill-effects of stochastic behaviour of the wireless channel. One of the benefits of the use of JSCC is the optimization achieved while being part of transceiver systems used in LTE/LTE-A frameworks. JSCC have received attention due to the promise demonstrated by them in improving the channel capacity utilization in current and upcoming topologies (Guillemot & Christ, 2003).

Basically for the data compression process, source coding is necessary that eliminates possible redundancy from the corresponding input source signals. On the other hand, perfect redundancy insertion is related to channel coding. Perfect redundancy insertion is a process that creates a special kind of protection from the ill-effects of noise related to the channel. Shannon's theorem of joint source-channel coding consists of two parts (Shannon, 2001): first part depicts that if a specific source with a source coding rate R is less than the channel capacity C, i.e., $R < C$ then a reliable means of transmission of the source signal through the channel is possible; second part or the converse part states that if $R > C$ then reliable communication of the source signal through the channel is impossible. These two parts of Shannons' theorem depicts the separate consideration scenario for both the source and channel coding for overall system performance without any loss of information (AZAMI, Duhamel & Rioul, 1997).

This separability for both the source and channel coding holds if the communication corresponds to point to point communication, i.e. single channel. However, with multi-path fading broadcasting communication this phenomenon is not realistic. In a broadcasting source-channel coding system, the clear distinction between the source coding problem and the channel coding problem already disappears, and hence the usual concept of achievable rates so popular in the field of multi-user information theory becomes irrelevant.

Moreover, a tandem source-channel coding may, in practice necessitate very long blocks of source symbols and very complex coders. Suppose that for a simple transmission scheme, a *binary schematic channel* (BSC) is associated with the transition probability, $\in = 0.01$. For this purpose, transmission of an input signal from a binary symmetric source (BSS) through this channel of average distortion, $D \leq 0.1$is required. With the consideration of Shannon's theorem for joint source-channel coding, design of a specific source code for the binary symmetric source (BSS) with average distortion ≈ 0.1 is essential first, and then design an analogous channel code for the binary symmetric channel (BSC) with small error probability (AZAMI, Duhamel & Rioul, 1997).

Inclusion of both the source and channel coding modules together in the same processing block in order to reduce the overall system complexity, compared to the tandem scheme is the main objective of the joint source channel coding techniques.

The two basic advantages of using MIMO in wireless communication is that it can enhanced capacity of channel and mitigate the affects of fading. Diversity can be combined with MIMO spatial multiplexing to improve performance. Due to the simultaneous communication capability of MIMO, channel capacity is enhanced. For these advantages of MIMO technology, recently it draws the attention of scientists and researchers and become a very promising and attractive technique for future mobile wireless communication systems especially with the benefits of combined JSSC.

Global network architecture which is based on joint source channel coding is a more challenging wireless and mobile communication technique such as 4G. To effectively achieve the optimal allocation of user and system resources, consideration of the co-operative optimization of communication system components is necessary. This approach, generally known as a promising technique of joint source channel coding and decoding (JSCC/D) that allows for strategies in which techniques the source coding, channel coding, modulation with possible network parameters are combinely determined to yield the effective end-to-end system performance in LTE and LTE-A techniques.

This chapter mainly focuses on joint source channel coding in multiple input multiple output (MIMO) diversity to minimize the affects of noise, distortion and related phenomena for better reception of the signal with increased efficiency and spectrum utilization with link reliability. Joint source/channel coding techniques in presence of 3G/4G/LTE-A for high data rate communication with the carrier aggregation is an important element for reliable communication over wireless channel. The application of these joint source and channel coding techniques is limited by several relevant practices and emerging options. In this chapter, we shall discuss the current state of research, the challenges related to the deployment of joint source/channel coding, the likely solutions and the usefulness as part of upcoming communication networks. A major part of the work shall focus on their suitability in LTE-A frameworks so essential for sustaining the current growth in wireless communication.

In this chapter, we discuss some of the basic theoretical aspects of joint source channel coding techniques in section 1. JSCC techniques are mainly used to improve QoS and link reliability in recent wireless communication system. Then in following section some of the recent techniques which is used for joint source channel coding incorporation with MIMO are discussed. From the literature review a simple and generic system model for the implementation of the technique is presented in the fourth section. Finally the conclusion is given in the last section.

BASIC THEORITICAL ASPECTS

In order to get a proper understanding of the joint source channel coding problems, it is necessary to study the basic theoretical considerations of JSCC and other such related topics. Here we discuss certain basic theoretical concepts related to the work.

JSCC Application and its Role in Communication

JSSC is mainly based on Shannon's separation theorem. In some situation Shannon's theorem does not work properly. For optimal performance of a system where Shannon's theorem breaks down there are basically exist a system with the technique of combined or joint source channel coding that can perform better. The theoretical framework for combined or JSCC is to make a communication system more efficient. The transmission of different kinds of signal which are continuous in nature with respect to amplitude such as speech, music or videos, images, are discussed here. For implementation of such kind of system, the representation of the signals is in discrete-time and it is always possible by the methods of appropriate sampling and subsequent processing (Goertz, 2007).

For a perfect means of communication it is always expect that less error should be obtained over a given channel. Up to a certain point of time it is believed that communication in the presence of noise, it was impossible to get low error rates. This conception was disproved by Shannon through his work

"A Mathematical Theory of Communication". Shannon introduced an innovative concept about channel capacity and proved that as the channel's capacity is greater than the transmission rate then small amount of error rate is possible. Furthermore, he depicted a theorem stating that reliable transmission of a source with entropy H over a channel of channel capacity *C* as long as $H \leq C$. A source coder can be applied to a corresponding system which reduces the rate of the source to a minimum limit. At the receiver side without knowing the source type, the channel decoder provides the most likely codeword to the source decoder. Finally despite being unaware of the channel characteristics, the source coder generates the source content. As this method provides an independent mode of operation to the source and channel coder, it is known as separation theorem. By removing the related intricate, it enables independent operation of source and channel for which this is a popular choice among designers. However, the theorem may not hold well in every condition. In situations where it doesn't hold joint source-channel coding is preferred. Here, we focus on the issues related to joint source-channel coding.

System Model

A basic communication system model is given in Figure 1. To transmit the input signal, the source vector x of N-dimension (e.g., N consecutive samples of a speech signal), with K channel to a destination where we can obtain $\tilde{x}$, a reproduction $\tilde{x}$ of effective quality possible.

Due to delay constraints the block length N is limited. The source vector components x_i, where i = *0, 1,...., N-1,* have continuous amplitudes and the mean squared error as a quality measure.

$$d(x,\tilde{x}) = \frac{1}{N}\sum_{i=0}^{N-1}(x_i - \tilde{x}_i) \tag{1}$$

Here the number K of channel-uses and vector dimension N are assumed to be fixed and known. The goal under this assumption is to find a deterministic encoder-decoder pair which reduces the system distortion.

$$D = E_{X,\tilde{z}}\{d(X,\tilde{X}) = \varsigma(\tilde{Z}))\} \tag{2}$$

The probability density function (PDF) of the input vector is assumed to be known and adjacent source vectors are independent. For convenience, consideration is taken for the PDF by $p(x)$ instead of $p_X(x)$ as long as this notation is unique. The random variables are re-introduced if the notation is ambiguous otherwise (Goertz, 2007).

Figure 1. System model of a communication system

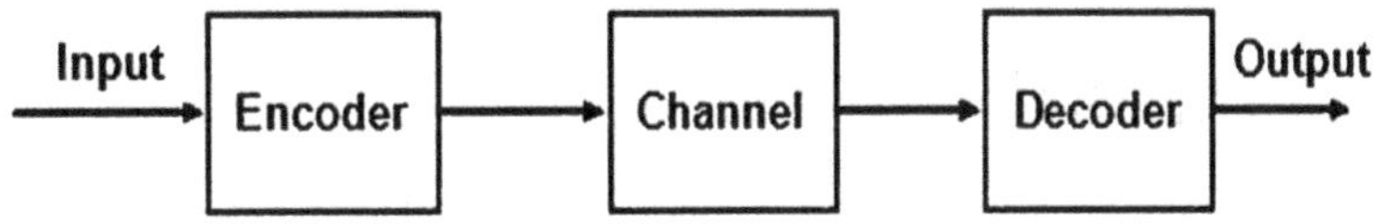

Channel

Here the channel is a discrete-time channel which is used K times for the transmission of the source vector components N. The channel input signal is the K-dimensional vector. The specification of the input alphabet $\mathbb{Z}$ and the output alphabet $\tilde{\mathbb{Z}}$ of the channel are mentioned below. As an example, we may think of a K-bit channel input and a K-dimensional continuous-valued output from an additive white Gaussian noise channel. The model is, however, not restricted to channels with binary input. Moreover, the model can be extended such, that with one channel-use a vector of two real components is transmitted, i.e., the channel has two real dimensions in each of the K uses. Considering the channel to be memory less, stationary and independent of all other signal components, the conditional probability density function $p(\tilde{z} \mid z)$, which describes the probability density of the output vector $\tilde{z}$, given the input vector z, is assumed given as

$$p(\tilde{z} \mid z) = \prod_{l=0}^{k-1} p(\tilde{z}_l \mid z_l) \tag{3}$$

Usually, the input power of the channel is limited when the channel input alphabet is continuous and is expressed as [29]

$$\frac{1}{K} Ex\{\| Z \|^2\} \leq P \tag{4}$$

Power limitation in such a case is quite weak, because it does not limit the peak power of a channel-use (Goertz, 2007).

Encoder

The encoder is described by the unique deterministic mapping expressed as

$$z \in (x) \tag{5}$$

With $x \in \chi = \mathbb{R}^{N}$ and $z \in \mathbb{Z}$, is a device that maps the input source-signal x into the input z of the channel. The encoder output z may have to fulfill a constraint imposed by the channel (e.g., the power constraint mentioned above).

Decoder

The decoder is a deterministic device which maps the channel output $\tilde{z}$ of the channel to the decoder output. The estimation should be proper for perfect reception of the signal at decoder. The decoder mapping for better reception is given by

$$\tilde{x} = \varsigma(\tilde{z}) \tag{6}$$

JSCC in Different Modes of Communication

In this section three system architectures for the problem of lossy transmission of a source over a point to point channel is depicted. A discrete memory less channel is considered in a point to point communication system. In this type of communication, suppose the source $S \sim p(s)$ transmitted over the discrete memory less channel $p(y|x)$ (Minero, Lim & Kim, 2011).

A point to point communication system is depicted in Figure 2. Expected distortion may be obtained by reconstructing the source under some necessary conditions. One of the sufficient conditions for expected distortion is expressed as

$$\frac{1}{n}\sum_{i=1}^{n} E(d(S_i, \hat{S}_i)) \leq D \tag{7}$$

Shannon showed that if channel capacity is greater than the rate distortion function *R (D)* of the source S, then distortion D is achievable is expressed as

$$R(D) < C \tag{8}$$

Where $R(D) = \min_{p(s,\hat{s})\,:\,E(d(S,\hat{S}))\,\leq\,D} I(S;\hat{S})$ is the rate–distortion function for the source S and distortion measure $d(s,\hat{s})$ and

$$C = \max_{p(x)} I(X;Y)$$

is the capacity of the channel $p(y|x)$.

When Separate the Source and Channel Coding

A system architecture of separate source channel coding is depicted in Figure 3. In this architecture the source sequence is mapped into one of 2^{nR} indices *M*. The index coming from the source mapped into a channel codeword X^n, which is transmitted over the channel.

After receiving the channel output Y^n, the decoder finds an output $\hat{M}$ which is an estimate of *M* and reconstructs $\hat{S}^n$ from $\hat{M}$. Between the source code and the channel code a digital interface is provided by the index *M*. According to lossy source coding and channel coding theorem, if the index rate *R* satisfies $R > R(D)$ and $R < C$ then desired distortion is possible (Minero, Lim & Kim, 2011).

Figure 2. Point-to-point communication system

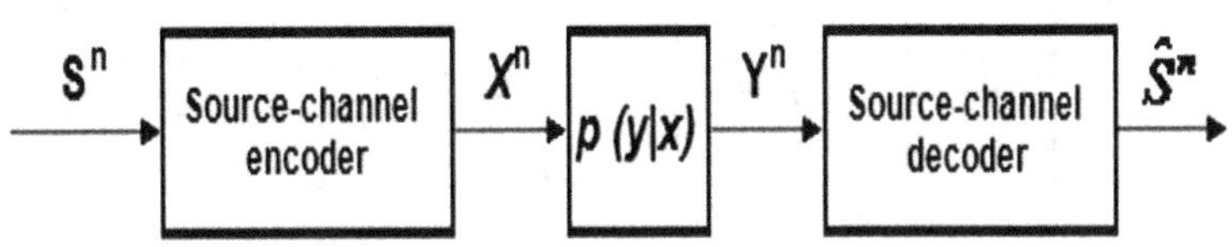

Figure 3. System architecture of separate source and channel coding

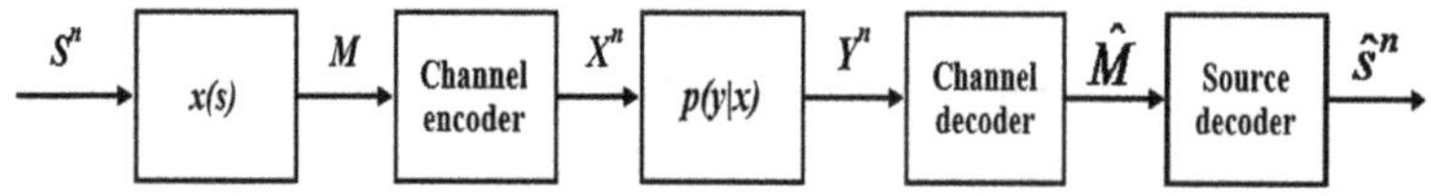

When Uncoded Transmission

Under this uncoded transmission mode of communication, the encoder transmits a symbol-by-symbol mapping of the source sequences as shown in Figure 4.

This mode of communication is optimal in some cases even though it is simple in nature, for example, when communicating a Gaussian source over a Gaussian channel under a squared error (quadratic) distortion measure or communicating a binary source over a binary symmetric channel under a Hamming distortion measure. In these two cases, the desired distortion D can be achieved if $C \geq R(D)$.

A New JSCC System Architecture based on Hybrid Coding

Under this mode of communication the separate source and channel coding and uncoded transmission is combined together by a specific coding, called Hybrid coding as shown in Figure 5.

Source sequence S^n is mapped to one of 2^{nR} sequences $U^n(M)$. This sequence $U^n(M)$ along with S^n is mapped to X^n symbol-by-symbol. This symbol-by-symbol mapping signal is transmitted over the channel. After receiving Y^n, the decoder finds an estimate $U^n(\hat{M})$ of earlier source encoder output signal U^n (M) and reconstructs output $\hat{S}^n$ from U^n (and Y^n) again by a symbol-by-symbol mapping. In this combined mode of communication, the codeword $U^n(M)$ plays the roles of both the source codeword $\hat{S}^n(M)$ and the channel codeword $X^n(M)$ simultaneously. Due to this architecture, source decoder and channel encoder is replaced by the interface of $x(u,s)$ and $\hat{s}(u,y)$. By the lossy source cod-

Figure 4. System architecture of uncoded transmission

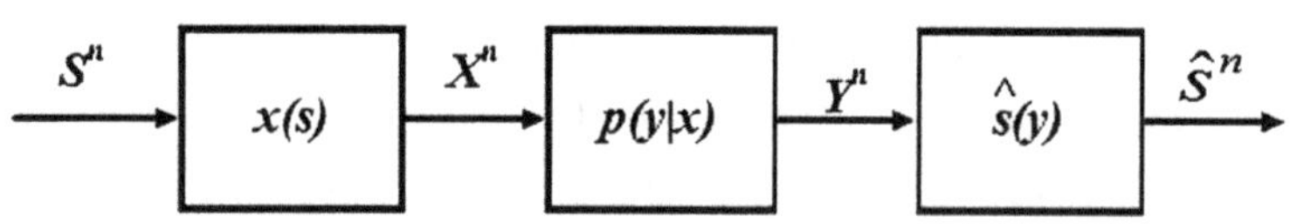

Figure 5. System architecture of a new joint source-channel coding based on hybrid coding

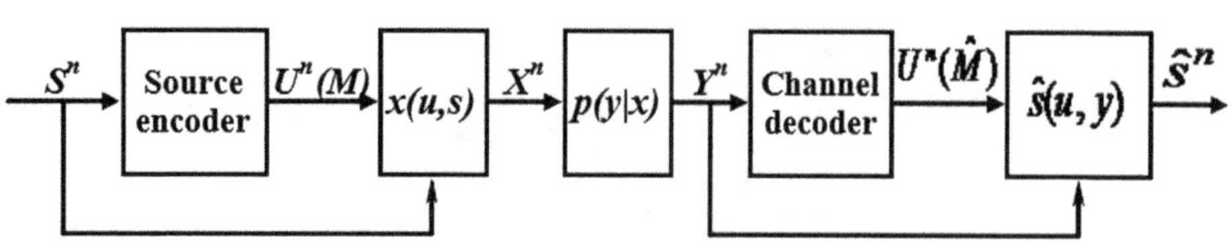

ing theorem, a reliable source encoding is possible if $R > I(U;S)$ and by the channel coding theorem, a reliable channel decoding is possible if $R < I(U;Y)$ which is depicted in the following equation

$$p(y|u) = \sum_{s} p(y|x(u,s))p(s)) \tag{9}$$

Distortion D is achievable if

$$I(S;U) < I(U;Y) \tag{10}$$

for some pmf $p(u|s)$ and functions $x(u,s)$ and $\hat{s}(u,y)$ such that $E(d(S,\hat{S})) \leq D$. By taking $U = (X,\hat{S})$ where $\hat{S} \sim p(\hat{s}|s), X \sim p(x)$ is independent of S and $\hat{S}$ and using the memoryless property of the channel. On the other hand, by taking $U = \phi$, hybrid coding reduces to uncoded transmission (Minero, Lim & Kim, 2011).

This new coding scheme is as simple as the separation scheme. The precise analysis of its performance, however, involves a technical subtlety. In particular, because $U^n(M)$ is used as a source codeword, the index M depends on the entire codebook $C = \{U^n(M) : M \in [1:2^{nR}]\}$. But the conventional random coding proof technique for a channel codeword $U^n(M)$ is developed for situations for which the index M and the (random) codebook C are independent of each other.

MIMO (Multiple Input and Multiple Output)

To increase the overall bitrates MIMO spatial multiplexing is used through transmission of two or more different data streams on two or more different antennas using the same resources in both frequency and time, separated only through. use of different reference signals to be received by two or more antennas as shown in Figure 6. MIMO can increase spectral efficiency and channel capacity. Diversity can be combined with MIMO spatial multiplexing to improve performance. In MIMO the path must be decorrelated. Base station uses MIMO beamforming or precoding technique to transform the signal to match the channel before transmission. Precoding is used to equalize the performance of two streams of data. The base station sends a signal to the handsets, the handsets then estimates the channel and fixed the optimum precoding from the predefined codeword which bought to base station and mobile here. It tends a signal back to base station, the base station then modifies to transform the signal to match the precoded message. In spatial diversity array of antennas are employed at the transmitter and the receiver end with minimum separation of $\lambda / 2$. Between them no data rate are increasing. But special multiplexing is used to increase the data rate because here the data is divided into different streams and these streams are transmitted independently via separate antenna. If this signals arrive at the receiving end which sufficiently different signals ends the receiver and separate this to parallel channels. Thus it increases the data throughput capability. Fading introduces as the signal travel from the transmitter to the receiving end .The attenuation effect a signal over a certain propagation medium. It decreases the signal to noise ratio of the signal and it causes due to reflectors present around the transmitter and receiver.

Figure 6. System architecture of MIMO spatial multiplexing

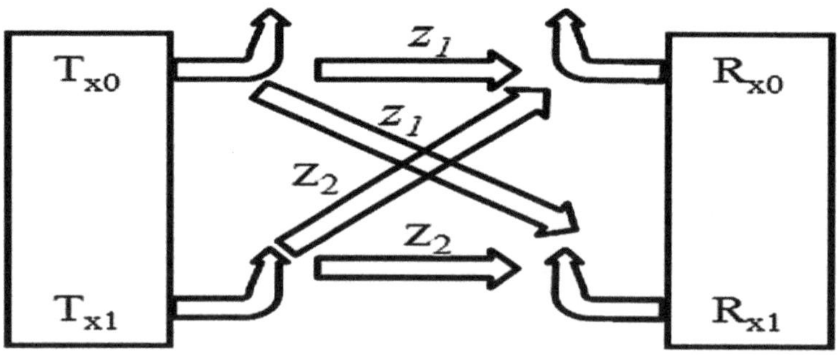

According to Shannon's capacity on MIMO fading channels

$$C = B\log_2(1 + \frac{S}{N}) \tag{11}$$

Where, C = Capacity of Channel

B = Channel bandwidth

S/N = Signal-to-noise ratio (SNR)

Aspects of spatial multiplexing in recent wireless communication system are

1. Different data streams sent simultaneously on different antennas.
2. Higher data rate.
3. No diversity gain.
4. Limitations due to path correlation.
5. Beamforming.

LTE and LTE-A

Long-Term Evolution mainly incorporate with 4G LTE which is a standard form of wireless communication with respect to high-speed data for data terminals and mobile phones. It is based on the network technologies of GSM/EDGE and UMTS/HSPA, increasing the speed and capacity using a different interface of radio together with the improvements of core network. The LTE technology provides uplink peak data rates of 75 Mbps, downlink peak data rates of 300 Mbps and provisions of QoS permitting a transferlatency in the radio access network of less than 5 ms. LTE has the provision to manage fast-moving mobiles and supports broadcast and multi-cast streams. The standardization of LTE technology became one of the most important technology shifts in cellular networks since the introduction of WCDMA. However, it was not until the introduction of LTE - Advanced in Release 10 that the requirements established by the ITU for 4G technologies were finally achieved (Windstorm, 2013).

Introduction of 8x8 MIMO spatial multiplexing in the downlink and 4x4 in the uplink is a major change with respect to LTE-Advanced technology. Higher capacity is also the main focus in LTE-Advanced technology. With respect to high quality radio channel, i.e. high signal to noise ratio, we can used MIMO spatial multiplexing technique. On the other hand situations, where low quality radio channel, i.e. low signal to noise ratio is present it is suitable to use other types of TX-diversity technique. Further development of LTE towards LTE–Advanced was to provide higher bitrates in a cost efficient way and also to completely fulfill the requirements set by ITU for IMT Advanced which is referred to as 4G (Wannstrom, 2013). The main goal of this system is:

1. To increase spectral efficiency from an existing maximum value of 16bps/Hz in R8 to 30 bps/Hz in R10.
2. To increase peak data rate, i.e. for DL 3 Gbps and for UL 1.5 Gbps.
3. At cell edges performance should be improved, e.g. for DL 2x2 MIMO at least 2.40bps/Hz/cell.
4. Increased number of simultaneously active subscribers.

Most of the functionalities which are introduced for the development of LTE-Advanced technology are very much significant. This significant aspect mainly includes Carrier Aggregation (CA), support for Relay Nodes (RN) and enhanced use of multi-antenna techniques (Table 1).

LTE Carrier Aggregation Basics

The 4G LTE-Advanced technology has one of the most suitable distinct features which is called carrier aggregation or sometimes referred to as channel aggregation. By aggregating LTE-carriers it is possible to obtain higher data rates in wireless communication. Under this technique, aggregation is taking place up to five LTE-carriers. Due to this property the transmission bandwidth is extended up to 100 MHz. This improvement in spectral efficiency is not sufficient to provide the required data rates within the channel of maximum 20 MHz. The overall bandwidth must be increased to achieve the required higher data rates. The upper limit for this purpose is set to 100 MHz by IMT Advanced, but for minimum performance it is expected upto 40 MHz. The aggregated carrier also known as component carrier can have different bandwidth of 1.4, 3, 5, 10, 15 or 20 MHz and maximum provision for aggregation is limited to five component carriers. Hence the total bandwidth is 100 MHz. For a communication system the number of aggregated carriers in DL and UL connection can be different, however it is considered that the number of component carriers in uplink (UP) connection is never larger than the number of compo-

Table 1. Comparison among LTE, LTE-Advanced and IMT-Advanced requirements

Properties	LTE 3GPP Release 8	LTE-Advanced 3GPP Release 10	IMT-Advanced InterTelecommunication Union "true 4G"
Peak data rate	DL-300 Mbps UL-75Mbps	DL-1Gbps UL-500Mbps	DL-100Mbps UL-1Gbps
Peak spectrum efficiency	DL-15, UL-3.75	DL-30, UL-15	DL-15, UL-6.75
Tx Bandwidth	Up to 20 MHz	Up to 100MHz	Up to 40MHz

nent carriers in downlink connection (DL). For both the frequency division duplexing (FDD) and time division duplexing (TDD), carrier aggregation can be applied.

In case of wireless communication it is generally considered that spectrum is an essential commodity. With perfect implementation of the transmitted spectrum we can increase the overall system performance. With this regard aggregation of LTE carriers can be done through a number of ways.

Intra-Band Contiguous and Non-Contiguous

Single band is preferred by this form of carrier aggregation technique. As defined for LTE, when a number of component carriers are continuously aggregated within the same operating frequency band then it is called intra-band contiguous carrier aggregation. From the RF viewpoint it is picturised that the terminal considered the component carrier as a single enlarged channel. Only one transceiver is required for this purpose within the terminal or UE, whereas for the non-adjacent channels the requirement is more (Figure 7).

Whereas intra-band non-contiguous carrier aggregation combine channels in the same frequency band in a non-continuous way. Such combinations are more complicated when aggregation is taking place for adjacent carriers. Multi-carrier signal can not be treated as a single signal and for this purpose two transceivers are required which adds complexity to the system where cost and power are prime considerations.

Inter-Band Non-Contiguous

Double band is preferred by this form of carrier aggregation technique. As defined for LTE, when a number of component carriers are aggregated with two different operating frequency bands then it is called inter-band carrier aggregation technique (Figure 8).

Due to the fragmentation of bands this aggregation technique is particularly used in communication system. With this technique, multiple transceivers are required by the UE within the single item, having

Figure 7. Intra-band carrier aggregation

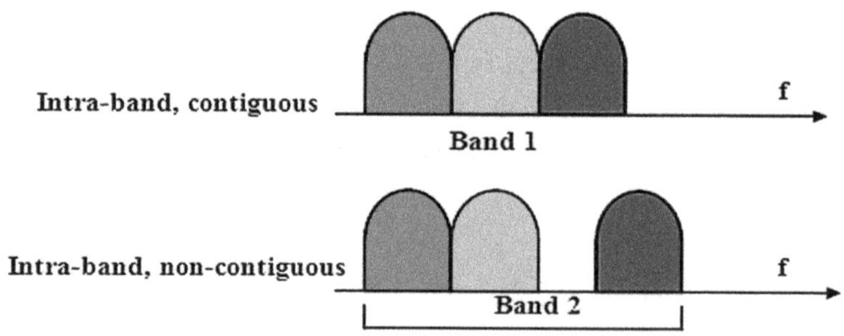

Figure 8. Inter-band non-contiguous carrier aggregation

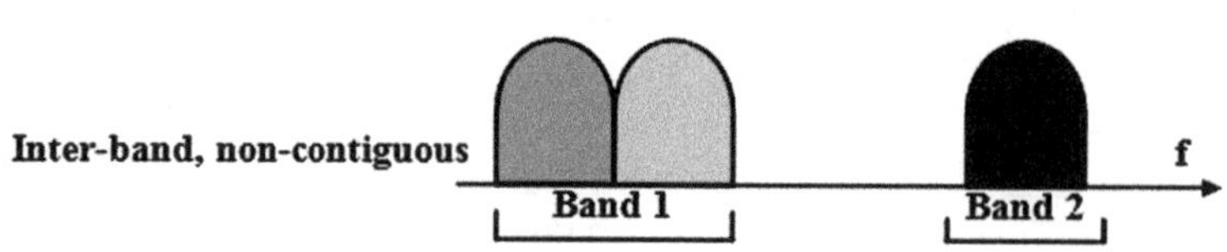

usual impact on performance, power and cost. Additional complexities from the two transceivers are generated with respect to cross modulation and inter modulation (Poole, 2013).

LTE Aggregated Carriers

Each aggregated carrier in carrier aggregation technique is referred to as a component carrier. There are basically two categories

1. **Primary Component Carrier (PCC):** Primary component carrier is the main component carrier in any carrier aggregation mode. There will be a associated primary uplink and downlink component carrier.
2. **Secondary Component Carrier (SCC):** In a LTE aggregated system there may be one or more secondary component carriers.

Different carriers may be use by different terminals. There is no specific definition of using a component carrier as a primary component carrier. The specification of primary component carrier is based on loading of the various carriers which are terminal specific.

In addition to this the cooperation between the uplink primary component carrier and the corresponding downlink primary component carrier is cell specific. When carriers are aggregated a number of serving cells are associated with each of the component carrier. Based on the component carrier frequencies the coverage of the specific serving cells may differ. Suppose for a tree component carrier scenario, the coverage of all the three frequency band can be handled by one cell, which is referred to as Primary serving cell, served by the Primary component carrier (DL and UL PCC). The other component carriers are all referred to as Secondary component carrier (DL and possibly UL SCC), serving the Secondary serving cells. Carrier aggregation on all three component carriers is only possible for the black UE; the white UE is not within the coverage area of the red component carrier (Poole, 2013).

4G LTE CoMP (Coordinated Multipoint)

Coordinated multipoint transmission and reception actually refers to a wide range of techniques that enable dynamic coordination or transmission and reception with multiple geographically separated eNBs. Its aim is to enhance the overall system performance, utilize the resources more effectively and improve the end user service quality. One of the key parameters for LTE as a whole, and in particular 4G LTE Advanced is the high data rates that are achievable. These data rates are relatively easy to maintain close to the base station, but as distances increase they become more difficult to maintain. Obviously the cell edges are the most challenging. Not only is the signal lower in strength because of the distance from the base station (eNB), but also interference levels from neighbouring eNBs are likely to be higher as the UE will be closer to them. 4G LTE CoMP, Coordinated Multipoint requires close coordination between a numbers of geographically separated eNBs. They dynamically coordinate to provide joint scheduling and transmissions as well as proving joint processing of the received signals. In this way a UE at the edge of a cell is able to be served by two or more eNBs to improve signals reception / transmission and increase throughput particularly under cell edge conditions.

In essence, 4G LTE CoMP, Coordinated Multipoint falls into two major categories:

1. **Joint Processing:** The coordination between multiple entities and base stations introduces the concept of joint processing in case of wireless communication. With this technique simultaneous transmission or reception from UEs is possible.
2. **Coordinated Scheduling or Beamforming:** This coordination multipoint (CoMP) technique is a one of the specific form of coordination where a single transmission or reception between a UE and a base station is possible. However among several coordinated entities with an exchange of control the communication is made.

On the channel properties a highly detailed feedback in a fast manner is required to achieve either of these techniques. The other necessity is for very close coordination between the eNBs to facilitate fast switching of the cells or the combination of data. The techniques used for coordinated multipoint, CoMP are very different for the uplink and downlink. This results from the fact that the eNBs are in a network, connected to other eNBs, whereas the handsets or UEs are individual elements (Poole, 2013).

Downlink LTE CoMP

Dynamic coordination is required by the downlink LTE CoMP amongst several geographically separated eNBs which transmitting to the UE. There are basically two formats of CoMP for downlink connection.

1. **Joint Processing for Downlink Transmission**: Using this format of LTE CoMP, different eNBs transmitting data simultaneously to the UE.to improve the received signal strength and quality is the main objective of this format. Cancellation of interference among different transmitted signals which are intended for other UEs is also an aim of this technique. Such a coordinated multipoint creates a high load on the backhaul network. This is because of the fact that the data to be transmitted to the UE is required to be sent to each of the eNB which shall be transmit it to the UE. It can double or triple the amount of data transmitted through the network. It is dependent upon the eNBs that shall send the data. Further joint processing data are also required to be sent between all eNBs present in the CoMP setup.
2. **Coordinated Scheduling or Beamforming:** Using this concept, data to a single UE is transmitted from one eNB. The scheduling decisions as well as any beams are coordinated to control the interference that may be generated. The advantage of this approach is that the requirements for coordination across the backhaul network are considerably reduced for two reasons. UE data does not need to be transmitted from multiple eNBs, and therefore only needs to be directed to one eNB. Only scheduling decisions and details of beams needs to be coordinated between multiple eNBs (Poole, 2013).

Uplink LTE CoMP

1. **Joint Processing and Reception:** Utilization of antennas at different sites is the main goal behind this format. A virtual antenna array can be formed by coordinating between the different eNBs. The signals which are received by the eNBs are then combined and processed to produce the final output signal. This technique is basically allows for weak signals, or in some areas where signals are

masked by interference to be receiving with less number of errors. To operate with this technique in an effective way, it is required to be transferred a large amounts of data between the eNBs.

2. **Coordinated Scheduling:** Amongst the eNBs to operate with a minimum number of interference this scheme coordinating the scheduling decisions in a systematic way. In case of downlink connection, this format has an advantage of much reduced load in the backhaul network. It is because only the scheduling data needs to be transferred between the different coordinating eNBs.

To provide a very low level of latency is one of the key requirements for LTE technology. For multiple site transmission and reception the additional processing required which could add significant amount of delays in processing. To overcome this problem, it is expected that the different sites may be connected together in a form of centralized radio access network (C-RAN) (Poole, 2013).

Advantages of LTE CoMP

Most of the advantages related to the LTE-advanced CoMP technology is remarkably suitable in a communication system although it is a complex set of techniques. This technique mainly includes the following advantages to the users.

1. **Provision for Better Utilization of Network:** By providing connections to several base stations at once, using CoMP, data can be passed through least loaded base stations for better resource utilization.
2. **Reduction of Interference:** By using specialized combining techniques it is possible to utilize the interference constructively rather than destructively, thereby reducing interference levels.
3. **Enhanced Reception Performance:** Using several cell sites for each connection means that overall reception will be improved and the number of dropped calls should be reduced.
4. **Multiple Site Reception Increases Received Power:** The joint reception from multiple base stations or sites using LTE Coordinated Multipoint techniques enables the overall received power at the handset to be increased.

LITERATURE SURVEY OF VARIOUS WORKS DONE RELATED TO OUR TOPIC OF DISCUSSION:

A number of works across the globe have been carried out all over the world in the last few decades that deals with the joint source channel coding (JSCC) for wireless communication system.

Review of Previous Work Done for Joint Source Channel Coding (JSCC)

The following section provides a detailed survey on the study of the literature related to the work that deals with joint source channel coding (JSCC).

1. Holiday, Goldsmith and Poor (2008) in their work have reported an efficient technique of multiple-input multiple-output (MIMO) system for optimal balanced among diversity, multiplexing, and delay which can minimize distortion for a practical source and channel codes. The main objective

is to find the balance in a systematic way between the antenna multiplexing increased data rate and the transmission error reduction provided by the technique of antenna diversity, ARQ and the delay introduced by the method of automatic repeat request (ARQ). They have also extended their analysis to MIMO systems which is incorporated with an ARQ retransmission protocol. Corresponding to the system, at the expense of delay it is possible to provide additional diversity through the automatic repeat request (ARQ) technique.

2. Cheung and Zakhor (2000) have reported an efficient algorithm for bit allocation with respect to a joint source/channel video codec, when transmission takes place over noisy channels. Their main approach of their design is to minimized expected distortion by distributing the available bits of source and channel coding among the existing sub-bands. The constructed distortion curves bound the performance degradation should the channel be estimated incorrectly. The algorithm can be used in other similar distortion minimization problems with two constraints, such as power or complexity.
3. Barrenechea et al. (2003) in their work have described the interaction of transport mechanism and the source coding mechanism for real transmission of data packet through dense networks. They have also shown that use of network capacity in effective mode is not possible where single path routing is used whereas the technique of multiple path routing can do it. They also investigate the multi path routing and multiple description source coding (MDC) which can perform notably better than the existing single description and single path scheme. This improvement becomes more remarkable in accordance with low rates and delay (about 18 dB).
4. Kliewer and Thobaben (2005) have present a posteriori probability (APP) decoder of symbol-based soft-input which can packetized transmitted encoded source indexes of variable-length over wireless channels where after source encoding the residual redundancy is utilized for error protection. The decoding process is almost optimal in combination with reconstructed source data APP estimation or a mean-square estimation technique. Furthermore different solutions for reducing the complexity in the proposed APA decoder are discussed and compared for the best optimal solution. Additionally when channel codes are utilized for protecting the encoded data of variable-length in nature then it is generally believed that an insistent source-channel decoder is obtainable for serially concatenated codes.
5. Sayood and Borkenhagen (1991) have described an innovative technique that led to the development of different data compression schemes. This scheme basically corresponds to the transmission of large amounts of data through a band-limited channel. Many of these techniques try to remove insignificant bits from the data stream. This approach has an unwanted side-effect due to which the information transfer process become more unsafe to channel noise. Involvement of the reinsertion of redundancy and an increase in bandwidth requirements is the main effort to protect their schemes against errors. They have presented a technique to provide protection against errors without the extra overhead required for the method of channel coding. During source coder design they have taken simple assumptions about the source. Most of the times these assumptions are irrelevant with respect to communication systems. They have used the residual redundancy method to provide error protection which is almost the same way as convolutional coding error protection with redundancy insertion. In this paper they develop an approach for utilizing this redundancy.
6. Fitzek et al. (2004) in their work have introduced and advocate a specific coding technique, generally known as multiple description coding (MDC) associated with the advanced 4G wireless communication networks with channel diversity. Due to the advantages of simultaneous support of

heterogeneous terminals and efficient use of spectrum makes multiple description coding (MDC) scheme more convenient in wireless communication systems. In this paper they have investigated a specific type of voice and video traffic in presence of MDC technique. They have also depicts the drawbacks of this coding scheme in terms of generated overhead by the encoding process and the network layer. For the received video they have presented the specific quality measurement. This scheme becomes more suitable to provide scalable and robust video transmission through lossy wireless environments due to the ability of supporting heterogeneous terminals easily.

7. In this paper, Cen (2013) the author has described about the Low-density parity-check (LDPC) codes with the parity-based approach for distributed joint source channel coding (DJSCC) with decoder side information. The parity-based approach is theoretical limit achievable. Different edge degree distributions are used for source variable nodes and parity variable nodes. Particularly, the codeword-averaged density evolution (CADE) is presented for asymmetrically correlated non-uniform sources over the asymmetric memory less transmission channel. Extensive simulations show that the splitting of Variable nodes can improve the coding efficiency of suboptimal codes and lower the error floor.
8. In this paper, Jennequin (2004) the author has described about flexible reconfigurable network architecture based on 4G technology which enables bandwidth optimization and Quality of Service (QoS) management simultaneously. With the co-operative optimization technique it is generally believed that communication system components may be effectively achieved through optimal allocation of system resources and users. This approach commonly known as joint source channel coding and decoding (JSCC/D) that allows for strategies including source-channel coding with modulation techniques and network parameters are combinely determined for the best performance of end-to-end system execution. The main target of JOCO is to enable JSCC/D technique through an innovative scheme. This scheme includes the development of a specifically optimized flexible modulation and channel coding schemes with the adaptation of existing coding schemes for JSCC/D and also to develop an advanced optimal 4G communication network architecture compatible with the joint source channel coding and decoding (JSCC/D) technique.
9. He, Cai, and Chen (2002) in their work have described about the development of an innovative rate-distortion (R-D) model for video coding, i.e. DCT-based video coding which is incorporating with intra refreshing rate of the macro block (MB). With respect to any given intra refreshing rate and bit rate, this model of video coding is capable of estimating corresponding distortion of coding even though a specific video frame is coded before. They then present an efficient paradigm about theoretical analysis of the distortion caused by successive inter-frame propagation and channel error. Based on the analysis paradigm, they have developed a model of statistical in nature to estimate such induced distortion in channel for different encoder settings and channel conditions. They have mathematically described the channel errors of complex behavior in nature for a transmission system with video coding. In case of wireless video applications this analytical model as compared to other reported literature for distortion estimation approaches is very much efficient with respect to computational complexity and implementation cost. They have experimentally shown the results of estimated channel error which induced distortion with respect to delay in processing. Based on the proposed model of source coding R-D and estimation of analytic channel-distortion, they have derived a solution for joint source-channel rate control and adaptive intra mode selection under time-varying conditions of wireless channel. This scheme significantly demonstrates an extensive

experimental result that improves the quality of end-to-end video with respect to transmission of wireless video with coding.

10. Guillemot and Christ (2003) in their work have described about the development of 'Beyond 3G'—B3G—or '4G' networks architecture and their applications corresponding to all-IP based heterogeneous infrastructures. For this design the authors have presents a number of specific challenges beyond the existing applicable techniques conceived so far by the international telecommunication union (ITU) and by the IETF basically for Mobile IP. Signalling, transport and management functions have been evolved from different technologies of different link layers. These technologies have evolved end-to-end and smoothly to the IP-network layer of packet switched. Global mobile network architecture types generally require better capability for smooth and continuous handover between different technology and administrative domains. They have also extending the technique for the present edge concept of mobility to ad-hoc networks and also for mobile networks. In this paper authors incorporate about the technique of joint source and channel coding pattern into a QoS framework associated with the 4G mobile wireless multimedia communications.
11. Camiciotti et al. (2001) in their work have reported innovative schemes for enabling the JSCC/D technique. This incorporated the development of an efficient technique for both the modulation and channel coding schemes, the existing source coding schemes are adapted depending on their ability for the technique of JSCC/D and the optimal development of the new scheme is specifically optimized for the global network architecture design based on JSCC/D for the future 4G mobile communication systems. Development of an efficient JSCC/D controller which will combinely control the source channel coding blocks and also the development of the innovative approach of network transparency which will allow the techniques of JSCC/D, that can be applicable in simple and fully IP-based network and especially for the wireless mobile communication in the 4G ones.
12. In this paper, Huang and Narayanan (2011) have illustrated the problem of joint source-channel coding of transmitting an analog source of discrete-time over a channel of additive white Gaussian noise (AWGN) with known interference at transmitter. They consider the case when the source and the interference are correlated. We first derive an outer bound on the achievable distortion and then, we propose two joint source-channel coding schemes. The first scheme is the superposition of the uncoded signal and a digital part which is the concatenation of a Wyner-Ziv encoder and a dirty paper encoder. In the second scheme, the digital part is replaced by the hybrid digital and analog scheme proposed by Wilson et al. When the channel signal-to-noise ratio (SNR) is perfectly known at the transmitter, both proposed schemes are shown to provide identical performance which is substantially better than that of existing schemes. In the presence of an SNR mismatch, both proposed schemes are shown to be capable of graceful enhancement and graceful degradation. Interestingly, unlike the case when the source and interference are independent, neither of the two schemes outperforms the other universally. As an application of the proposed schemes, they provide both inner and outer bounds on the distortion region for the generalized cognitive radio channel.
13. Garrett and Vetterli (1993) in their work have investigated about the interaction of congestion control. This interaction with source information into components are done by a one kind of special partitioning based on importance of varying quality for variable bit-rate packet video and packet voice. The higher priority transmission for the important signal components in their system shows that objective service quality is increased in substantial way. Using a Markov chain voice source model in the presence of simple PCM speech encoding technique with a priority queue, the practical

simulation results show an improvement of 45 dB signal-to-noise ratio with two priorities through an unprioritized system. Performance is sensitive in the system due to the placement of fraction of traffic in each priority and also the optimal partition mainly depends upon the conditions of network loss. A number of variations are investigated for further partition of information with low-priority into information of multiple priorities. In this paper a comparison is made between timeout which is based on service requirements and dropping packets on overflow of buffer.

14. Akyol, Rose and Ramstad (2010) in this paper have illustrated the source-channel communication system corresponding for a specific finite blocks length coding. The Problem associated with the vector transformations which can optimally map between the source space of m-dimension and the channel space of k-dimention are considered under a given channel power constraint and mean square error distortion measure. The checking of optimal mapping for the encoder and decoder are derived through some necessary conditions of closed form. Such mappings are mainly associated with a practical system of analog joint source channel that transmits an alphabet in continous mode of discrete time source through a noisy channel. A number of numerical results associated with several source channel distributions are presented and from the results it is observed that for both bandwidth expansion and compression the optimal mappings outperform the previous heuristic mappings.
15. In this paper, Wang, Lau and Ng (2000) have analyzed about the performance for a mobile radio channels with feedback corresponding to an optimal multiresolution joint source and channel coding (MR-JSCC). Based on the channel state information (CSI) the transmitter and receiver can synchronously match the specific resolution of each other. By adjusting the resolution thresholds bandwidth requirement could be met. In this paper the authors properly outlined the threshold determination algorithm. With respect to the same bandwidth and power, here the simulation results show that MR-JSCC scheme has much better performances as compared to the baseline-I (No UEP and No feedback) and baseline-I1 (UEP without feedback).
16. Oyman and Foerster (2010) in this paper investigated the performance of MIMO link adaptation with distortion-aware properties. This adaptation is related to realistic system-level simulation (SLS) environments. It is also linked to link level simulation (LLS) using orthogonal frequency division multiplexing (OFDM) under frequency-selective fading in broadband channels. It uses a multicast broadcast service (MBS). The work demonstrated the advantages of MBS over MIMO link adaptation techniques using parameters like peak SNR and end to end distortion.
17. Li, Q, Qian and Wu (2012) in this paper have described cooperative communications which enable efficient utilization of communication resources, by allowing nodes or terminals in a communication network to collaborate with each other in information transmission. In this article, they first survey cooperative communication schemes and discuss their advantages in improving system capacity and diversity. Following that, they examine the applications of cooperative relaying schemes in LTE-advanced systems. Specifically, they investigate two intra-cell coordinated multi-point schemes in LTE-advanced systems, and evaluate the performance of the schemes. It is shown that cooperative relaying leads to both network coverage extension and capacity expansion in LTE-advanced systems. Cooperative communications can significantly improve the system spectrum efficiency and performance.
18. In this paper, Zhang, Qie and Xu (2013) have reported a technique about the optimization techniques of linear precoder for packet retransmissions in multi-input multi-output (MIMO) nonregenerative relay systems. For fully utilization of the time diversity which is provided by automatic repeat

request (ARQ) corresponding to progressive linear precoders are designed for maximizing the mutual information which is delivered by multiple transmissions of the same packet. Firstly, when the channel state information (CSI) is available at the relay and the destination then the relay-only precoder is considered. Furthermore, they analyze about asymptotic performance when the two hops' signal-to-noise ratio (SNR) goes to infinity or zero respectively. Secondly, the joint source and relay precoder is studied assuming the CSI known at all nodes. The ARQ precoder is derived by matrix diagonalization and channel pairing between previous sub channels and current sub channels. Then the power allocation at the source and relay is formulated as a nonconvex multi-variable optimization problem where we employ an iterative algorithm to obtain the suboptimal solution.

19. Lim, Minero and Kim (2010) in this paper have proposed an optimal bit allocation algorithm corresponding to a joint source channel video codec. This assumption is taken over noisy channel. The main approach of this work is to distribute the available bits of the source and channel coding among the sub bands. Here the distribution process is associated with the technique in such a way that the distortion can be minimized. This algorithm can be used to minimize the complexity and power consumption with respect to distortion reduction technique.
20. In this paper, Jiang and Hanzo (2007) have described about efficient channel estimation techniques. Basically they have reported that this specific channel estimation technique is designed for multiuser MIMO-OFDM systems. They have designed the system in the so-called rank-deficient scenarios. In that particular scenario the number of supported users or the number of transmit antennas employed exceeds in accordance with the number of receiver antennas.
21. Khalid and Speidel (2010) in this paper have described the performance aspects of different MIMO systems such as single user MIMO (SU-MIMO) and multi user MIMO (MU-MIMO) system. SU-MIMO systems for both the open-loop and closed-loop are reported in this paper, accordance with the particular emphasis of the MIMO technology based on the maximization aspect of the data rate. This paper also touches upon the topic of MU-MIMO capacity as well as the promising convex optimization approaches to MIMO system design.
22. In this paper, Minero, Lim and Kim (2011) have proposed an innovative analog-digital based hybrid coding. This hybrid coding scheme over the two-user discrete memoryless interference channel, achieves the best known performance and recovers as special cases several previous results on lossless and lossy communication over single hop networks.
23. In this paper, Fresnedo (2015) the author has described about the evaluation of low-complexity analog Joint Source Channel Coding (JSCC) methods for the transmission of discrete-time analog symbols over Multiple-Input Multiple-Output (MIMO) Multiple Access Channels (MAC).

Table 2 summarizes the various reported works that have dealt with joint source channel coding, diversity techniques and related optimal methods that can perform better with a communication system to increase the overall system performance in an efficient way.

Generic System Model

Joint source–channel coding generation technique is basically different from Shannon's since Lapidoth and Tinguely in 2010 in their work have taken assumption for a finite block length for the channel code. With respect to diversity- multiplexing trade-off this assumption is inherent since, the channel introduces insignificant error without finite block length. Therefore, the derivation of diversity gain using channel

Table 2. Summary of works depicting the use of coding and diversity techniques with JSCC

SL No.	Authors	Contribution	Application	Reference and Year
1	Holiday, Goldsmith and Poor	System optimizing the tradeoff between diversity, multiplexing, and delay in MIMO systems that minimize end-to-end distortion.	A delay-sensitive system obtains significant performance gains by this technique	(Holiday, Goldsmith and Poor, 2008)
2	Cheung and Zakhor	An efficient GSG algorithm extends the Shoham–Gersho algorithm to another dimension.	For a JSC video codec	(Cheung and Zakhor, 2010).
3.	Barrenechea, Beferull-Lozano, Verma, Dragotti,Vetterli	Efficient use of MDC	Transmission of real data such as video over a dense network	(Barrenechea, Beferull-Lozano, Verma, Dragotti, Vetterli, 2003)
4	Huang and Narayanan	JSCC with correlated interference	Generalized cognitive radio channel	(Huang and Narayanan, 2011)
5	Cen	Effective use of LDPC with DJSCC	DJSCC for asymmetric transmission channels.	(Cen, 2013)
6	Zhang, Qie and Xu	Precoder optimization	In non-regenerative MIMO relay systems.	(Zhang, Qie and Xu, 2013)
7	Li, Hu, Qian and Wu	Evaluate two intra-cell coordinated multi-point schemes	In LTE-advanced systems	(Li, Hu, Qian and Wu, 2012)
8	Minero, Lim and Kim	A new analog-digital hybrid coding	In joint source-channel coding for wireless communication	(Minero, Lim and Kim, 2011)
9	Oyman and Foerster,	An efficient distortion aware MIMO link adaptation techniques	MIMO link adaptation towards enhanced multimedia communications	(Oyman and Foerster, 2010)
10	Fresnedo	Evaluation of low-complexity analog JSCC	Transmission of discrete-time analog symbols over MIMO Multiple Access Channels (MAC)	(Fresnedo, 2015)

error probability has no meaning. The finite block length ensures that the probability of error in the channel transmission is not trivial. Here, a trade-off between source encoder resolution and channel encoder robustness takes place. Lowering source distortion needs a source coder with high data rate. To achieve such a state, multiple antennas at the source is required to achieve multiplexing. An alternative is to use an encoder with a lower rate but allow more distortion. Then channel error probability can be decreased by increasing diversity gain. In such a case, joint source-channel code should provide the optimal trade-off between these two to reduce end-to-end distortion to the minimum (Figure 9). With retransmission and delay sensitive attribute of the source, further trade-off between channel error reduction and delay due to retransmission can be achieved (Lapidoth & Tinguely, 2010).

1. **Channel Model:** Suppose a wireless channel has M transmits antennas and *N* receives antennas. h_{ij} is the fading coefficient that model the gain from transmit antenna *i* to receive antenna *j* are independent and identically distributed *(i.i.d.)* complex Gaussian with unit variance . It is assumed that the channel gain matrix *H* with elements $H_{ij} = (h_{ij} : i \in \{1,...,M\}, j\{1,...,N\})$ is to be known

Figure 9. Generic system model for joint source channel coding (JSCC)

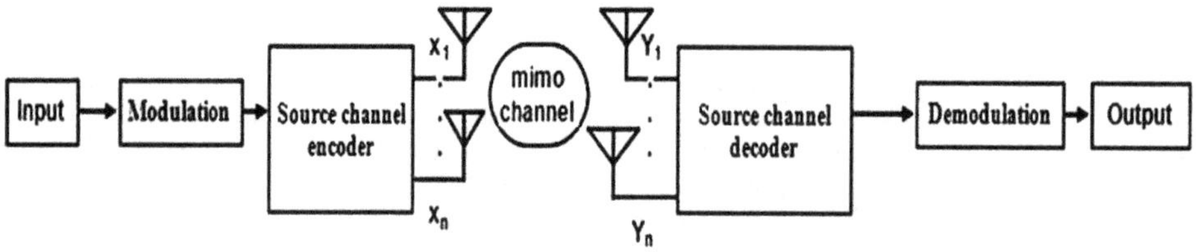

at the receiver and unknown at the transmitter. Over a block of T symbols, it is may also assumed that the channel is constant, while each block is i.i.d. Therefore, the channel can be represented in each block expressed as

$$y = \sqrt{SNR/_{M}}HX + W \tag{12}$$

Here, $X \in c^{M\times N}$ and $y \in c^{N\times T}$ are the transmitted and received signal vectors, respectively. The additive noise vector W is *i.i.d.* complex Gaussian with unit variance.

Under the long-term static channel model, in round $l \in \{1...L\}$ of an ARQ block we can represent the channel as

$$y_l = \sqrt{SNR/_{M}}Hx_l + W \tag{13}$$

Here, $x_l \in c^{M\times N}$ and $y_l \in c^{N\times T}$ are the transmitted and received signal vectors, respectively. The additive noise vector W is *i.i.d.* complex Gaussian with unit variance.

The use of MIMO technique in a wireless relay network has the ability to improve the spectral efficiency and diversity gain. Cooperative relaying is a promising technique due to its ability to extend propagation distance and improve spectral efficiency. The objective of precoder design is to maximize the channel capacity provided by each ARQ transmission (Lapidoth & Tinguely, 2010).

2. **Carrier Aggregation:** It is necessary to add more bandwidth to increase capacity. Since it is important to keep backward compatibility with R8 and R9 mobiles the increase in bandwidth in LTE-Advanced is provided through aggregation of R8/R9 carriers. Both frequency division duplex (FDD) and time division duplex (TDD) can be used by carrier aggregation technique. Each aggregated carrier is referred to as a component carrier. The component carrier can have a bandwidth of 1.4, 3, 5, 10, 15 or 20 MHz and a maximum of five component carriers can be aggregated. Hence the maximum bandwidth is 100 MHz. The number of aggregated carriers can be different in downlink (DL) and uplink (UL), however the number of UL component carriers is never larger than the number of DL component carriers. The individual component carriers can also be of different bandwidths.
3. **Spatial Diversity:** Multiple antennas can be utilized as a means of improving the reliability of the transmission. Multiple antennas at the transmitter transmit precoded versions of the same data stream, while multiple antennas at the receiver in diversity mode may perform different types of

signal combining approaches to improve the received SNR. Diversity approaches supported in Rel-9 LTE include space frequency block coding (SFBC) and frequency shift time diversity (FSTD).

4. **Spatial Multiplexing:** The user throughput can be increased via MIMO communication by simultaneously transmitting several data streams (also called layers). This is known as spatial multiplexing and requires precoding at the transmitter to orthogonalize the streams.
5. **Beamforming:** The directivity of the transmission can be improved by applying beamforming techniques to the array of antennas, thus properly reaching users at the edge of the coverage area.
6. **Diversity–Multiplexing Tradeoff:** Source encoder concatenated with a MIMO channel that is sometimes restricted to transmission over finite block lengths. The channel introduces errors even at transmission rates below capacity. These channel errors give rise to the diversity–multiplexing tradeoff. Under finite block-length channel coding, we may obtain a source and channel coding strategy to minimize end-to-end distortion.
7. **Diversity–Multiplexing Delay Tradeoff:** Through ARQ many wireless system perform error correction instead of accepting decoding errors in the channel. In general, the receiver has some form of error detection code. If a transmission error is detected on a given packet, a feedback path is used to send this error information back to the transmitter, which then resends part or the entire packet to increase the chance of successful decoding. The packet retransmissions, combined with random arrival times of the messages at the transmitter, cause queues to form in front of the source coder and hence each block of data will experience random delays. While ARQ increases the probability of decoding a packet correctly, it also introduces additional delay. The window size of the ARQ protocol determines how many retransmission attempts will be made for a given packet. The larger this window size, the more likely the packet will be successfully received, and the larger the possible delays associated with retransmission will be. ARQ can be viewed as a form of diversity, and hence it complements antenna diversity in MIMO systems. For MIMO systems with ARQ, there are a three-dimensional trade-off between diversity due to multiple antennas and ARQ, multiplexing, and delay. The diversity gain for the ARQ protocol with a maximum of L blocks is

$$D(r,L) = d(r/l) \tag{14}$$

ARQ achieved the diversity gain which is quite remarkable. For any $r \leq \min(M,N)$ we can consider full diversity gain $d=MN$ for sufficiently large L. Thus, for L sufficiently large, there is no reason to utilize spatial diversity since all needed diversity can be obtained through ARQ. For L not sufficiently large, the maximum ARQ window size would still be utilized to minimize the amount of spatial diversity required (Lapidoth & Tinguely, 2010).

CONCLUSION

One of the major problems with the existing joint source channel coding approaches is that consideration of the network between the server and the client as a single transport link. The situation becomes more complicated in the context of multiple available links because involving a low-bandwidth, highly lossy, or long-delay wireless network in the transmission will only degrade the video quality. For this purpose to minimize the expected distortion there is a need to distribute the available channel and source coding bits in a systematic way among the sub bands. In addition, there is also one specific requirement to find the perfect balance between the increased data rate which is provided by antenna multiplexing. By using a retransmission protocol, system performance with a combination of multiple input and multiple output (MIMO) and delay-constrained sources can be increased. In this regard the use of an automatic repeat request (ARQ) retransmission protocol is more convenient because at the expense of delay it provides additional diversity in the system. There is a lot of active work both in industry and in academia to further enhance the capabilities of relays for LTE-Advanced.

It is restricted to transmission over finite block lengths where source encoder is connected with a MIMO channel. Even at transmission rates below capacity, the channel introduces errors with this restriction. These errors which are introduced by the channel give rise to the diversity–multiplexing trade-off. Under this channel coding of finite block-length, a strategy of source and channel coding should be obtained for the communication system to minimize end-to-end distortion.

Nevertheless, both industry and academia have continued improving LTE-Advanced through enhancements in the core technologies of carrier aggregation, MIMO, relaying, and cooperative multipoint communications. In addition, in order to cope with the ever increasing demand for ubiquitous and high-speed data access, the efforts have recently focused on improving the support of heterogeneous networks, as well as device-to device and machine-type communications. In addition to these technologies, the new paradigm of self-organizing networks has the potential of transforming the cellular network into a dynamic entity capable of automatically adjusting itself to guarantee the best possible service by exploiting all the aforementioned technologies. This evolution will continue, not only by improving these technologies, but also by introducing new ones, especially at higher frequency bands capable of satisfying the demand for even faster data access than the ones seen today by introducing new joint source channel coding techniques with MIMO technology.

REFERENCES

Abou-Rjeily, C. (2011). Joint channel-diversity coding for the three-way free-space optical systems through atmospheric-turbulence channels. *International Journal on New Computer Architectures and Their Applications*, *1*(1), 162–175.

Akyol, E., Rose, K., & Ramstad, T. (2010, January). Optimal mapping for joint source channel coding. In *Proceedings of 2010 IEEE Information Theory Workshop (ITW)* (pp. 1-5). doi:10.1109/ITWKSPS.2010.5503179

Azami, S., Duhamel, P., & Rioul, O. (1997). Combined source-channel coding for binary symmetric channels and uniform memoryless sources. In *16° Colloque sur le traitement du signal et des images, FRA, 1997*. GRETSI, Groupe d'Etudes du Traitement du Signal et des Images.

Barrenechea, G., Beferull-Lozano, B., Verma, A., Dragotti, P. L., & Vetterli, M. (2003, April). Multiple description source coding and diversity routing: A joint source channel coding approach to real-time service over dense networks. In *Proceedings of the Packet Video Workshop*.

Camiciotti, L., Lamy, C., Meilhac, L., Oliveieri, S., & Verdi, P. (2001, May). Joint source channel coding for 4G multimedia streaming. In *2nd WWRF meeting, WG3*.

Cen, F. (2013). Distributed Joint Source and Channel Coding with Low-Density Parity-Check Codes. *IEEE Journal on Communications Letters*, *17*(12), 2336–2339. doi:10.1109/LCOMM.2013.101613.131616

Cheung, G., & Zakhor, A. (2000). Bit allocation for joint source/channel coding of scalable video. *IEEE Transactions on Image Processing*, *9*(3), 340–356. doi:10.1109/83.826773 PMID:18255407

Cheung, G., & Zakhor, A. (2000). Bit allocation for joint source/channel coding of scalable video. *IEEE Transactions on Image Processing*, *9*(3), 340–357. doi:10.1109/83.826773 PMID:18255407

Fitzek, F. H., Can, B., Prasad, R., Park, D. S., & Cho, Y. (2004). Application of Multiple Description Coding in 4G Wireless Communication Systems. *WWRF 8 bis.*

Fresnedo, O. (2015). Design of analog joint source-channel coding systems for multiple access channels. *IEEE Transactions on Communications.*

Garrett, M. W., & Vetterli, M. (1993). Joint source/channel coding of statistically multiplexed real-time services on packet networks. *IEEE/ACM Transactions on Networking*, *1*(1), 71–80. doi:10.1109/90.222908

Goertz, N. (2007). *Joint source-channel coding of discrete-time signals with continuous amplitudes.* London: Imperial College Press.

Guillemot, C., & Christ, P. (2003). Joint source channel coding as an element of a QoS framework for '4G' wireless multimedia. *ELSEVIER Journal on Computer Communication*, *10*(5), 54–61.

He, Z., Cai, J., & Chen, C. W. (2002). Joint source channel rate-distortion analysis for adaptive mode selection and rate control in wireless video coding. *IEEE Transactions on Circuits and Systems for Video Technology*, *12*(6), 511–523. doi:10.1109/TCSVT.2002.800313

Holiday, T., Goldsmith, A. J., & Poor, H. V. (2008). Joint source and channel coding for MIMO systems: Is it Better to be robust or quick? *IEEE Transactions on Information Theory*, *54*(4), 1393–1405. doi:10.1109/TIT.2008.917725

Huang, Y. C., & Narayanan, K. R. (2011, August). *Joint source channel coding with correlated interference*. Paper presented at the Conference of IEEE International Symposium on Information Theory (ISIT). St. Petersburg, Russia. doi:10.1109/ISIT.2011.6033709

Jennequin, C. (2004). *JOCO (Joint source-channel Coding-driven digital baseband design for 4G multimedia streaming)*. Paper presented at the Conference of JOCO Project. France.

Jiang, M., & Hanzo, L. (2007). Multiuser MIMO-OFDM for next-generation wireless systems. *Proceedings of the IEEE*, *95*(7), 1430–1469. doi:10.1109/JPROC.2007.898869

Kliewer, J., & Thobaben, R. (2005). Iterative joint source-channel decoding of variable-length codes using residual source redundancy. *IEEE Transactions on Wireless Communications*, *4*(3), 919–929. doi:10.1109/TWC.2005.847032

Lapidoth, A., & Tinguely, S. (2010). Sending a bivariate Gaussian over a Gaussian MAC. *IEEE Transactions on Information Theory*, *56*(6), 2714–2752. doi:10.1109/TIT.2010.2044058

Li, Q., Hu, R. Q., Qian, Y., & Wu, G. (2012). Cooperative communications for wireless networks: techniques and applications in LTE-advanced systems. Wireless Communications, IEEE, 19(2).

Lim, S. H., Minero, P., & Kim, Y. H. (2010, September). Lossy communication of correlated sources over multiple access channels. In *Proceedings of the 48th Annual Allerton Conference on Communications, Control, and Computing*. doi:10.1109/ALLERTON.2010.5706997

Minero, P., Lim, S., & Kim, Y. H. (2011, July). Joint source-channel coding via hybrid coding. In *Information Theory Proceedings (ISIT), 2011 IEEE International Symposium on* (pp. 781-785). IEEE. doi:10.1109/ISIT.2011.6034241

Oyman, O., & Foerster, J. (2010, September). Distortion-aware MIMO link adaptation for enhanced multimedia communications. In *Proceedings of 2010 IEEE 21st International Symposium on Personal, Indoor and Mobile Radio Communications Workshops (PIMRC Workshops)* (pp. 387-392). doi:10.1109/PIMRCW.2010.5670400

Poole, I. (2013). *4G LTE CoMP Coordinated Multiple Tutorials*. Retrieved November, 2013, from http://www.radio-electronics.com/info/cellulartelecomms/lte-long-term-evolution/4g-lte-advanced-comp-coordinated-multipoint.php

Ramchandran, K., Ortega, A., Uz, K. M., & Vetterli, M. (1993). Multiresolution broadcast for digital HDTV using joint source/channel coding. *IEEE Transactions on Selected Areas in Communications*, *11*(1), 6–23. doi:10.1109/49.210540

Sayood, K., & Borkenhagen, J. C. (1991). Use of residual redundancy in the design of joint source/channel coders. *IEEE Transactions on Communications*, *39*(6), 838–846. doi:10.1109/26.87173

Shannon, C. E. (2001). A mathematical theory of communication. *Mobile Computing and Communications Review*, *5*(1), 3–55. doi:10.1145/584091.584093

Wang, X., Lau, V. K. N., & Ng, T. S. (2000). Performance analysis of multiresolution joint source-channel coding for mobile radio channels with feedback. *IEEE International Conference on Personal Wireless Communications* (pp. 157-161). doi:10.1109/ICPWC.2000.905793

Wannstrom, J. (2013). *LTE-Advanced for 3GPP*. Retrieved June, 2013, from http://www.3gpp.org/technologies/keywords-acronyms/97-lte-advanced

Zhang, L., Qie, L., & Xu, J. (2013). Progressive linear precoder optimization for ARQ packet retransmissions in nonregenerative MIMO relay systems. *IEEE Transactions on Signal Processing*, *61*(1), 68–81. doi:10.1109/TSP.2012.2222385

KEY TERMS AND DEFINITIONS

Automatic Repeat Request (ARQ): ARQ also known as Automatic Repeat Query, is an error-control method for data transmission that uses acknowledgements (messages sent by the receiver indicating that it has correctly received a data frame or packet) and timeouts (specified periods of time allowed to elapse before an acknowledgment is to be received) to achieve reliable data transmission over an unreliable service.

Carrier Aggregation (CA): Carrier aggregation is the first step of LTE Advanced. It offers very high peak data rates as well as increased data rates for all users in the cell, and can bring higher capacity for bursty applications. Carrier aggregation continues to evolve, supporting more carriers, aggregation across FDD/TDD, across cells and more.

Coordinated Multipoint (CoMP): Coordinated multipoint transmission and reception actually refers to a wide range of techniques that enable dynamic coordination or transmission and reception with multiple geographically separated eNBs. Its aim is to enhance the overall system performance, utilize the resources more effectively and improve the end user service quality.

Joint Source Channel Coding (JSCC): JSCC design combines techniques for optimizing source codes to match channel noise characteristics with techniques for optimizing channel codes to match source code characteristics into a single joint design aimed at minimizing the end-to-end distortion experienced in the communication system as a whole.

LTE: Long-Term Evolution mainly incorporate with 4G LTE, is a standard for wireless communication of high-speed data for mobile phones and data terminals. It is based on the GSM/EDGE and UMTS/HSPA network technologies, increasing the capacity and speed using a different radio interface together with core network improvements.

LTE-A: Higher capacity is the main focus in LTE-Advanced technology. Introduction of 8x8 MIMO in the downlink and 4x4 in the uplink is a major change in LTE-Advanced.

Multiple Input Multiple Output (MIMO): MIMO (multiple input, multiple output) is an antenna technology for wireless communications in which multiple antennas are used at both the source (transmitter) and the destination (receiver). The antennas at each end of the communications circuit are combined to minimize errors and optimize data speed.

Multiplexing: Multiplexing is the process in which multiple Data Streams, coming from different Sources, are combined and Transmitted over a Single Data Channel or Data Stream.

Quality of Service (QoS): In case of wireless communication the QoS comprises requirements on all the aspects of a connection, such as service response time, loss, signal-to-noise ratio, crosstalk, interrupts, frequency response levels, and so on.

Chapter 9
Direction of Arrival (DOA) and Channel Estimation

Shouman Barua
University of Technology Sydney, Australia

Robin Braun
University of Technology Sydney, Australia

ABSTRACT

Future fit demand combined with a flexible technical solution that is by latest wireless technology stands for. Estimating of the user's location is going to be an integral system with the upcoming mobile technology. This chapter shows some techniques for estimating the direction of arrival (DOA) with mathematical elaboration and simulation results as well. Estimating the DOA in this chapter is regarded to the purpose of using Smart antenna system. It is possible to estimate the location of a user by considering the uplink transmission system of the mobile communication system. Estimating the channel and accurate path delay is also an important task which might be done by using 1D Uniform Linear Array (ULA) or 2D Uniform Rectangular (URA) array antenna system. In this chapter, 1D ULA is considered in order to utilize some popular techniques. The performance of a communication system between two ends is substantially determined by the behaviors of the channel characteristics. It determines signal transformation while propagating through the channel between receivers and transmitters. Accurate channel information is crucial for both the transmitter and receiver ends to perform their best services. The ultimate focus of this chapter is to estimate the channel based on 2D parameter estimation. Uniform Rectangular Array (URA) is used to perform the 2D parameter estimation. It is possible to estimate Azimuth and Elevation of a source by using URA model.

INTRODUCTION

During the last few decades, wireless communication technology was improved enormously. However, the technology did not only perform the communication between two ends. The cutting-edge technologies are not able to meet all the demands required by the people. An interest of recent researchers is to efficiently design a model to estimate the distance, direction, path delay and location of a signal source which might be a mobile user. A proper estimation model technique is very important to identify and

DOI: 10.4018/978-1-5225-0239-5.ch009

perform those tasks efficiently. Therefore, problem of proper channel estimation arises in a large variety of important applications for the researchers.

In order to meet the demand with the increasing data rate, Third Generation Partnership Project (3GPP) standardized the Long Term Evolution (LTE) and LTE-Advance. It is also one of the most usable technologies nowadays because of its higher spectrum efficiency, low latency, and seamless mobility. MIMO (Multiple Input Multiple Output (MIMO) and Orthogonal Frequency Division Multiple Access (OFDM), the two latest technologies in wireless communication system are adopted in LTE system so that it executes better than any other technologies. Lam et al. (2015) discussed about the frequency reused algorithm for LTE-A system to increase the data rates.

In this chapter, techniques for estimating direction of arrival (DOA) have been just investigated. It is expected for the upcoming mobile generation like 5G that direction of arrival will play a vital role to track the user's location exactly. The need for DOA estimation increases enormously to keep pace with the technology advances like finding the geographical location of the User Equipment (UE). Positioning has many applications such as emergency services, location-based billing, hostage rescue, etc. In that case, beam pattern of the antenna array can, therefore, be changed without physically changing the antenna pattern. In order to estimate DOA, channel estimation should be measured exactly.

Propagation signals transmitted from arbitrary sources contain a lot of information about the sources which produce them. This information includes the location of the sources, path delays, angular elevation, etc. Using smart antenna technologies, it is possible to estimate that information as well as the channel between two ends.

The antenna array receives spatial samples of propagating wavefield, which are then processed for DOA estimation. DOA estimation technique performs spatial filtering to separate signals from different spatial locations that have overlapping frequency contents (Van et al., 1988) . But there are many other techniques also available and some of them will be discussed in this chapter.

A ULA antenna configuration is considered at the receiver where all the antennas are placed in a line and the distance between two adjacent antennas are same. Other array configuration options are also possible, e.g., Uniform Circular Array (UCA), Uniform Rectangular Array (URA). The purposes of using smart antenna system, i.e., MIMO is regarded with the respect of

- Direction of Arrival (DOA) estimation,
- Path delays estimation, and
- Accurate Channel Estimation.

Only estimation of DOA is considered in this chapter.

1. SIGNAL MODEL FOR DOA ESTIMATION

Consider a ULA that consists of L elements and the distance between two adjacent antennas is d according to the Figure 1. It is assumed that total number of sources is P. Sources are placed at a distance where near field effect of the receiving antenna array is avoided. Consider a wavefront of a source at an angle θ from the far field region impinges first on the left most antenna element of the receiving antenna array. Each antenna receives the signal from the same source with a time delay and an amount of phase shift. The fact is that the wavefront of the source signal takes extra time to reach the each antenna element relative to the first element. The Figure 2 exactly shows the geometry of how the signal wavefront arrives.

Figure 1. Basic beamforming technique.

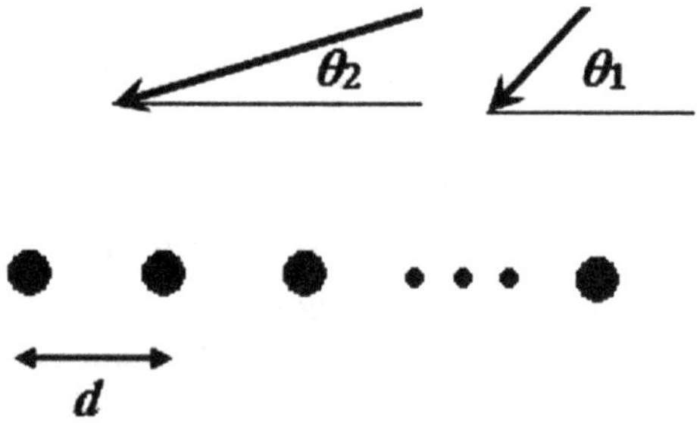

Figure 2. Uniform Linear Array geometry

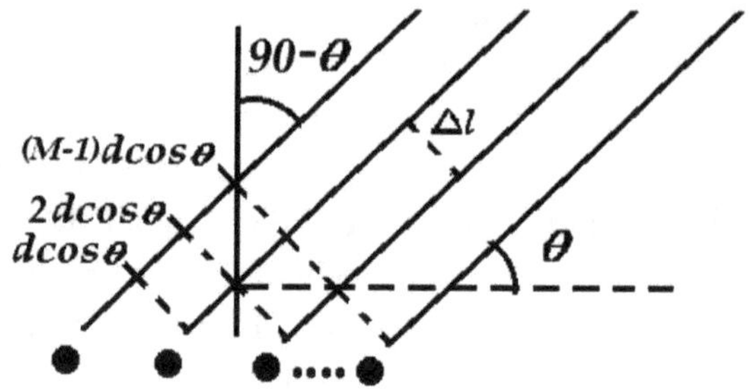

After arriving at the first antenna, the same signal plane then arrives at the second antenna element traveling an additional path Δl . Now using basic trigonometric formula,

$$\Delta l = -d\cos\theta\,. \tag{1.1}$$

Then delay becomes

$\tau = -d\cos\theta\,/\,\mathrm{c}$ where, c is the speed of light, 3×10^{8} m/s. (1.2)

The path difference leads to a phase shift of μ between the signals arriving at the two adjacent antennas. Therefore, the propagation phase shift μ between any consecutive two elements is

$$\mu = \left(\frac{2\pi}{\lambda}\right)\times\Delta l \tag{1.3}$$

where, λ is the wavelength of the signal.
Using equation (1.1),

$$\mu = -\left(\frac{2\pi}{\lambda}\right)\times d\cos\theta \tag{1.4}$$

Now, for L-th element in a ULA consisting of L elements, the phase shift shown in Figure 2 becomes (Rasheed, F., 2008).

$$\mu = -(L-1)\times\frac{2\pi d\cos\theta}{\lambda} \tag{1.5}$$

Using equation (1.4), DOA can be calculated easily as

$$\theta = -\cos^{-1}\left(\mu\Big/2\pi d\right)\times\lambda \tag{1.6}$$

Generally the signal is assumed to be narrowband but for simplicity in practice, a sinusoidal signal is considered which can be represented in a complex form as $e^{j\omega_0 t}$ where, $\omega_0 = 2\pi f_0 t$, f_o is the frequency, t represents time index. This signal is received by the array antenna elements. In order to estimate the delay in next Chapter another reference signal is used for simulation which is called Zadoff-Chu (ZC) sequence. ZC sequences are used as a reference signal in the LTE standard specified by the 3GPP.

Signal received by the first antenna is

$$s_1(t) = e^{j2\pi f_0 t} \tag{1.7}$$

According to the Figure 2, the second antenna will receive the signal after additional plane wave traveling the distance Δl which is the delayed version of signal the received by the first antenna. According to the equation (1.2), the delayed time is τ. Therefore, the received signal by the second antenna is

$$s_2(t) = s_1(t - \tau) \tag{1.8}$$

$$= e^{j2\pi f_0 t} \times e^{-(j2\pi f_0 \tau)} \tag{1.9}$$

$$= e^{j2\pi f_0 t} \times e^{-(j2\pi f_0 d \cos\theta / c)} \tag{1.10}$$

$$= e^{j2\pi f_0 t} \times e^{-(j2\pi d \cos\theta / \lambda)} \tag{1.11}$$

Using equation (1.4),

$$s_2(t) = s_1(t) e^{j\mu} \tag{1.12}$$

Since noise is added by the system, the first antenna will receive the noise in addition to the signal. Therefore,

$$x_1(t) = s_1(t) + n_1(t) \tag{1.13}$$

where, x_1 is the total signal received by the first antenna.

Second antenna will receive

$$x_2(t) = s_1(t - \tau) + n_2(t) \tag{1.14}$$

Continuing and representing it in matrix form,

$$\begin{bmatrix} x_1(t) \\ x_2(t) \\ x_3(t) \\ \vdots \\ x_L(t) \end{bmatrix} = s_1(t) \begin{bmatrix} 1 \\ e^{j\mu} \\ e^{j2\mu} \\ \vdots \\ e^{j(L-1)\mu} \end{bmatrix} + \begin{bmatrix} n_1(t) \\ n_2(t) \\ n_3(t) \\ \vdots \\ n_L(t) \end{bmatrix} \quad (1.15)$$

where,

$\begin{bmatrix} 1 \\ e^{j\mu} \\ e^{j2\mu} \\ \vdots \\ e^{j(L-1)\mu} \end{bmatrix} = \mathbf{a}(\mu)$ is called array steering vector.

Considering superposition of all the signals generated by P sources, s_i, $1 \le i \le P$ and L antenna noises, the output received by the l-th array element at time t can be formulated for l=1, 2, …, L as

$$x_l(t) = s_i(t) \sum_{i=1}^{P} e^{j(l-1)\mu_i} + n_l(t) \quad (1.16)$$

In general, equation is

$$\mathbf{x}(t) = \left[\mathbf{a}(\mu_1)\mathbf{a}(\mu_2)\mathbf{a}(\mu_3)\cdots\mathbf{a}(\mu_P)\right] \begin{bmatrix} s_1(t) \\ s_2(t) \\ s_3(t) \\ \vdots \\ s_P(t) \end{bmatrix} + \mathbf{n}(t) \quad (1.17)$$

where, $\mathbf{x}(t)$ = $[x_1(t)\ x_2(t)\ldots\ x_L(t)]^T$ indicates the data column vector received by the L receiving array antennas, and

$$\mathbf{a}(\mu_i) = \begin{bmatrix} 1 \\ e^{j\mu_i} \\ e^{j2\mu_i} \\ \vdots \\ e^{j(L-1)\mu_i} \end{bmatrix} \quad (1.18)$$

is called array steering column vector.

Finally, signal model equation could be written as

$$x(t)=A(\theta)s(t)+n(t) \tag{1.19}$$

where,

$\mathbf{n}(t) = [n_1(t)\ n_2(t)...n_L(t)]^T$ indicates the zero mean spatially uncorrelated noise.

$\mathbf{s}(t) = [s_1(t)\ s_2(t)...s_P(t)]^T$ indicates the signal column vector generated by P sources, and

$\mathbf{A}(\theta) = \left[\mathbf{a}(\mu_1)\mathbf{a}(\mu_2)\cdots\mathbf{a}(\mu_i)\cdots\mathbf{a}(\mu_P)\right]$ is called the $(L \times P)$ steering matrix.

Referring to the equation (1.18), $A(\theta)$ can be written as

$$\mathbf{A}(\theta) = \begin{bmatrix} 1 & \cdots & 1 & \cdots & 1 \\ e^{j\mu_1} & \cdots & e^{j\mu_i} & \cdots & e^{j\mu_P} \\ \vdots & \vdots & \vdots & \vdots & \vdots \\ e^{j(L-1)\mu_1} & \cdots & e^{j(L-1)\mu_i} & \cdots & e^{j(L-1)\mu_P} \end{bmatrix} \tag{1.20}$$

where, $\mu_i = -\left(\frac{2\pi}{\lambda}\right) \times d\cos\theta_i$ (1.21)is called the spatial frequency for the ith source that generates the signal of the incident angle θ_i (Chen et al., 2010).

Referring to the Figure 3, it is possible to design a weight vector **w** to linearly combine the data received by the array elements to form a single output signal $y(t)$ (Krim et al., 1996)

$$y(t) = \sum_{i=1}^{P} w_i^* n_l(t) \tag{1.22}$$

$$= \mathbf{w}^H \mathbf{x}(t) \tag{1.23}$$

Figure 3. Adding the weight vector to the received signal at each antenna

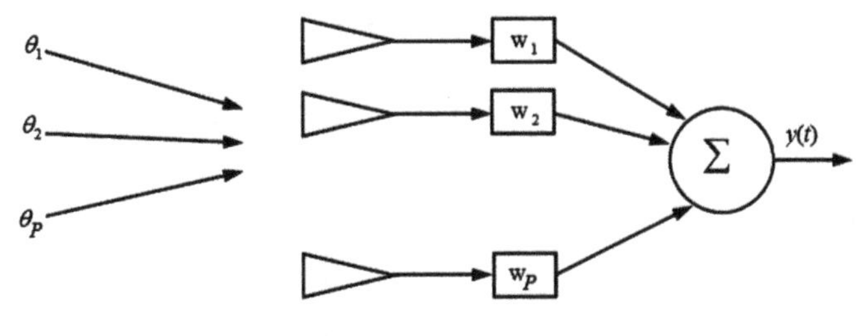

Total averaged power at the output of the array using $y(1), y(2), \ldots, y(T)$ data samples can be expressed as (Krim et al., 1996)

$$\bar{p}(w) = \frac{1}{T}\sum_{t=1}^{T}\left|y(t)\right|^2 \tag{1.24}$$

$$= \frac{1}{T}\sum_{t=1}^{T}\mathbf{w}^H\mathbf{x}(t)\mathbf{x}^H(t)\,\mathbf{w} \tag{1.25}$$

$$= \mathbf{w}^H\hat{\mathbf{R}}_{XX}\mathbf{w} \tag{1.26}$$

where, $\hat{\mathbf{R}}_{XX}$ is a covariance matrix which can be defined as

$$\hat{\mathbf{R}}_{XX} = \frac{1}{T}\sum_{t=1}^{T}\mathbf{x}(t)\mathbf{x}^H(t) \tag{1.27}$$

2. CONVENTIONAL BEAMFORMER TECHNIQUE

Conventional beamformer is also known as Delay and Sum method, Fourier method or Bartlett method. It is one of the simplest methods of estimating DOA. It is computationally attractive and less costly compared with other techniques. It is also robust against array imperfection. When the source signal is computed with interference and noise, this algorithm is still able to give a good estimation of the DOA (Rasheed, F., 2008).

In this type of technique, the array is steered electronically (by changing the phase delays) in one direction at a time and the output power is measured. The DOA is estimated for the incident sources by the directions that give the maximum output power.

Maximizing the output power can be calculated as

$$\bar{P}_{CB}(\theta) = \max_{\mathbf{w}} E\left\{y(t)\right\}^2 \tag{1.28}$$

where, $E\{.\}$ denotes the statistical expectation. Referring to the equation (1.23),

$$\bar{P}_{CB}(\theta) = \max_{\mathbf{w}} E\left\{\mathbf{w}^H\mathbf{x}(t)\mathbf{x}^H(t)\mathbf{w}\right\} \tag{1.29}$$

$$= \max_{\mathbf{w}} \mathbf{w}^H E\left\{\mathbf{x}(t)\mathbf{x}^H(t)\right\}\mathbf{w}. \tag{1.30}$$

Using equation (1.19),

$$\bar{P}_{CB}(\theta) = \max_{\mathbf{w}} \left\{ E\left|\mathbf{s}(t)\right|^2 \left|\mathbf{w}^H \mathbf{A}(\theta)\right|^2 + \sigma^2 \left|\mathbf{w}\right|^2 \right\} \tag{1.31}$$

here, maximization is determined as $\max_{\mathbf{w}} \left|\mathbf{w}^H \mathbf{A}(\theta)\right|^2$ subject to $\mathbf{w}^H\mathbf{w}=1$ which imply that

$$\left\|\mathbf{w}^H \mathbf{A}(\theta)\right\|^2 \le \left\|\mathbf{w}\right\|^2 \left\|\mathbf{A}(\theta)\right\|^2 \tag{1.32}$$

$$= \left\|\mathbf{A}(\theta)\right\|^2. \tag{1.33}$$

According to Krim et al., (1996), the resulting solution can be formulated as

$$\mathbf{w}_{CB} = \frac{\mathrm{A}(\theta)}{\sqrt{\mathrm{A}^H(\theta)\mathbf{A}(\theta)}}. \tag{1.34}$$

Substituting this into equation (1.26)

$$\bar{\mathrm{P}}_{CBF}(\theta) = \frac{\mathrm{A}^H(\theta)\hat{\mathbf{R}}_{\mathrm{XX}}\mathrm{A}(\theta)}{\mathrm{A}^H(\theta)\mathbf{A}(\theta)} \tag{1.35}$$

where, θ is from 0 to 180 degree.

2.1 Some Observations

- Sensor spacing is usually half of the wavelength. Larger spacing results from the performance degradation.
- Smaller spacing usually reduces the total aperture thus results from lower spatial resolution, i.e., the ability to distinguish two closely spaced sources.
- Spatial resolution directly depends on number of sensors, i.e., if the number of sensors with proper spacing are increased spatial resolution is also improved.

2.2 Advantages and Disadvantage of Conventional Beamformer

- For accurate estimation, the array has to be fully calibrated.
- Unless we have a good number of array antennas, spatial resolution determined by the sensors array is not generally very good.
- Multipath is not that much problem since signals are not necessarily to be uncorrelated.

3 CAPON BEAMFORMER TECHNIQUE

The main concept of Capon beamformer is the Minimum Variance Distortionless Response (MVDR) algorithm (Capon et al., 1969) It overcomes the limitation of poor resolution problem experienced by conventional beamformer. "*It proposes to minimize the contribution of undesired signals and noises by minimizing the total output power while maintaining the gain along to the look direction constant*" (N. A. M. Tayem, 2005)

Equation (1.26) can be written as

$$\min_{\mathbf{w}} \bar{P}\left(\mathbf{w}\right) = \min\left(\mathbf{w}^H \hat{\mathbf{R}}_{XX} \mathbf{w}\right), \text{ subject to } \left|\mathbf{w}^H \mathbf{A}\left(\theta\right)\right| = 1 \tag{1.36}$$

The weight factor $\mathbf{w}_{\text{Cap}}$ can be found using the technique of Lagrange multipliers that results as

$$\mathbf{w}_{Capon} = \frac{\hat{\mathbf{R}}_{\text{XX}}^{-1} \text{A}(\theta)}{\text{A}^H(\theta)\hat{\mathbf{R}}_{\text{XX}}^{-1}\mathbf{A}(\theta)} \tag{1.37}$$

Substituting this weight vector to the equation (1.26) leads to

$$\bar{\text{P}}_{Capon}(\theta) = \frac{1}{\text{A}^H(\theta)\hat{\mathbf{R}}_{\text{XX}}^{-1}\text{A}(\theta)} \tag{1.38}$$

4. MULTIPLE SIGNAL CLASSIFICATION METHOD

Schmidt et al., (1986), and Bienvenue et al., (1983) first developed the Multiple Signal Classification (MUSIC) method. It is one of the widely used subspace-based techniques to estimate DOA. It depends on the spatial covariance matrix of the data. It is assumed that noise is spatially and temporarily white and they are uncorrelated (Liberti et al., 1999). Therefore, the spatial correlation RXX for single user can be written as

$$\mathbf{R}_{\text{XX}} = E\left\{\mathbf{x}(t)\mathbf{x}^H(t)\right\} \tag{1.39}$$

$$= \mathbf{A}\left(\theta\right) E\left\{\mathbf{s}\left(t\right)\mathbf{s}^H\left(t\right)\right\} \mathbf{A}^H\left(\theta\right) + E\left\{\mathbf{n}\left(t\right)\mathbf{n}^H\left(t\right)\right\} \tag{1.40}$$

Assuming that $\mathbf{R}_{\text{S}} = E\left\{\mathbf{s}(t)\mathbf{s}^H(t)\right\}$ is the source covariance matrix and $E\left\{\mathbf{n}\left(t\right)\mathbf{n}^H\left(t\right)\right\} = \sigma^2\mathbf{I}$ is the noise covariance matrix. Now Equation (1.40) can be written as

$$\mathbf{R}_{\text{XX}} = \mathbf{A}(\theta)\mathbf{R}_{\text{S}}\mathbf{A}^H(\theta) + \sigma^2\mathbf{I} \tag{1.41}$$

$$= \mathbf{Q}\mathbf{\Lambda}\mathbf{Q}^H \tag{1.42}$$

where, **Q** is a unitary matrix and $\mathbf{\Lambda}$ is a diagonal matrix of real eigenvalues such that $\lambda_1 \geq \lambda_2 \geq \lambda_3 \geq \ldots \geq \lambda_L$.

Any vector that is orthogonal to **a** is an eigenvector of $\hat{\mathbf{R}}_{XX}$ with the eigenvalue σ^2. It is clear that there are (*L–P*) such vectors that are linearly independent. The rest of the eigenvalues are larger than σ^2. So the eigenvalues (and corresponds their eigenvectors) of $\hat{\mathbf{R}}_{XX}$ can be divided into two groups:

Signal eigenvalues: $\lambda_i > \sigma^2; i = 1, 2, 3, \ldots, P$

Noise eigenvalues: $\lambda_i = \sigma^2; i = P + 1, P + 2, P + 3, \ldots, L$

Equation (1.42) can be formulated as

$$\mathrm{R}_{XX} = \mathbf{E}_S \mathbf{\Lambda}_S \mathrm{E}_S^H + \mathbf{E}_N \mathbf{\Lambda}_N \mathrm{E}_N^H \tag{1.43}$$

where,

$\mathbf{\Lambda}_N$ contains the diagonal noise eigenvalues,
$\mathbf{\Lambda}_S$ contains the diagonal signal eigenvalues,
$\mathbf{E}_N$ contains eigenvectors associated with the noise eigenvalues (noise subspace matrix),
$\mathbf{E}_S$ contains eigenvectors associated with the signal eigenvalues (signal subspace matrix).

Since array steering vectors are orthogonal to the noise subspace eigenvectors at the angles θ_1, θ_2, $\theta_3, \ldots, \theta_P$ i.e., $\mathbf{A}^H(\theta)\,\mathbf{E}_N = 0$ can show that Euclidean distance $\mathrm{A}^H(\theta)\hat{\mathbf{E}}_N\hat{\mathbf{E}}_N^H\mathrm{A}(\theta) = 0$ for each and every arrival angel $\theta_1, \theta_2, \theta_3, \ldots, \theta_P$.

In practice, it gets a sharp peak at the angel of arrival by placing this Euclidean distance in the denominator. Therefore, the DOAs can now be estimated from the *P* peaks of the following MUSIC function (Khallaayoun, A., 2010).

$$\bar{\mathrm{P}}_{MUSIC}(\theta) = \frac{1}{\mathrm{A}^H(\theta)\hat{\mathbf{E}}_N\hat{\mathbf{E}}_N^H\mathrm{A}(\theta)} \tag{1.44}$$

4.1 Some Facts about MUSIC

- MUSIC demonstrates high resolution property.
- A total number of sources must be smaller than the total number of sensors in the array.
- MUSIC is very sensitive to antenna position, gain, and phase errors. Therefore, precise calibration is required.
- MUSIC could be computationally expensive because exhaustive search through the field of view (all θ) is required.
- For closely spaced signals and low SNR conditions, MUSIC cannot perform well (Parvazi, P., 2012).
- Accuracy is also dependent on the number of snapshots (Parvazi, P., 2012).
- Multipath is a major problem since MUSIC is unable to estimate DOAs of the correlated signals correctly.

5. ESTIMATION OF SIGNAL PARAMETERS VIA ROTATIONAL INVARIANCE TECHNIQUE

Estimation of Signal Parameters via Rotational Invariance Technique (ESPRIT) was proposed by Paulraj et al., (1986). This is a search-free DOA estimation technique based on the shift-invariance property of the ULA steering matrix (Schmidt et al., 1986) ESPRIT is considered as a computationally efficient technique compared with MUSIC.

Recalling the equation (2) for the steering matrix **A** and let $\overline{\mathbf{A}}$ and $\underline{\mathrm{A}}$ be the matrices with eliminated first and last row respectively as shown in Figure 4.

Therefore, $\overline{\mathbf{A}}\mathbf{D} = \underline{\mathbf{A}}$ *(1.45)*

$$\text{where, } \mathrm{D} = \begin{bmatrix} Z_1 & 0 & \cdots & 0 \\ 0 & Z_2 & \cdots & 0 \\ \vdots & \vdots & \ddots & \vdots \\ 0 & 0 & \cdots & Z_L \end{bmatrix} \text{ here, } Z_i = e^{j\mu_i}. \tag{1.46}$$

Similarly Let $\overline{\mathbf{E}}\mathbf{s}$ and $\underline{\mathrm{E}}_{\mathrm{S}}$ are formed from the signal eigenvector matrix **Es** in the same way as $\overline{\mathrm{A}}$ and $\underline{\mathbf{A}}$ from **A**. A full rank $P \times P$ matrix **C** existed because $\mathbf{E}_{\mathrm{s}}$ and **A** span the same subspace namely signal subspace.

Therefore,

$$\mathbf{E}_{\mathrm{S}} = \mathbf{AC}$$
$$\underline{\mathbf{E}}_{\mathrm{S}} = \underline{\mathbf{A}}\mathbf{C} = \overline{\mathbf{A}}\mathbf{DC}$$
$$\overline{\mathbf{E}}_{\mathrm{S}} = \overline{\mathbf{A}}\mathbf{C}$$

Figure 4. Formation of $\overline{\mathbf{A}}$ *and* $\underline{\mathrm{A}}$ *from* **A** *steering matrix*

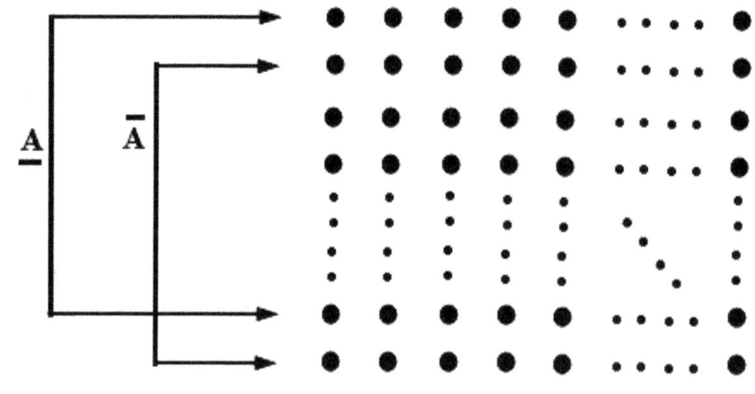

So, $\underline{\mathbf{E}}_S \mathbf{C}^{-1}\mathbf{D}^{-1}\mathbf{C} = \bar{\mathbf{A}}\mathbf{D}\mathbf{C}\mathbf{C}^{-1}\mathbf{D}^{-1}\mathbf{C} = \bar{\mathbf{E}}_S$ (1.47)

and $\underline{\mathbf{E}}_S = \bar{\mathbf{E}}_S \mathbf{C}^{-1}\mathbf{D}\mathbf{C}$ (1.48)

$\Rightarrow \underline{\mathbf{E}}_S = \bar{\mathbf{E}}_S \mathbf{\Psi}$ (1.49)

where, $\mathbf{\Psi} = \mathbf{C}^{-1}\mathbf{D}\mathbf{C}$. (1.50)

Here, **C** and **D** are unknown. Equation (1.50) implies that **D** is a diagonal matrix of eigenvalues of $\mathbf{\Psi}$. Computing the solution for $\mathbf{\Psi}$ from equation (1.49),

$\mathbf{\Psi} = (\bar{E}_S^H \bar{E}_S)^{-1} \bar{E}_S \underline{E}_S$. (1.51)

From (1.46), it is clear that the eigenvalues of $\mathbf{\Psi}$ (contained in D) can be written as

$g_l = -\frac{2\pi}{\lambda} d \cos\theta_l$. *(1.52)*

In practice, the sample covariance matrix $\hat{\mathbf{R}}$ is used. Then, after computing the signal subspace estimation $\hat{\bar{\mathbf{E}}}_S$, the DOAs can be estimated from $\hat{\theta}_l = \cos^{-1}\left(\frac{-\hat{g}_l \times \lambda}{2\pi d}\right)$. Here $\hat{g}_l$ for *l=1, 2, 3, ..., P* are the eigenvalues of $\hat{\mathbf{\Psi}} = (\hat{\bar{E}}_S^H \hat{\bar{E}}_S)^{-1} \hat{\bar{E}}_S \hat{\underline{E}}_S$. (1.53)

6. MULTIPATH CHANNEL ESTIMATION USING 2D PARAMETER ESTIMATION METHOD

A Chanel model equation is introduced which is compared with the equation of 2D parameter estimation. With the help of this technique, it is possible to estimate the user's location (angle) along with the time delay of an incoming signal to the antenna array sensors. In order to estimate this, Unitary 2D ESPRIT method is considered although elaborating 2D ESPRIT is not the aim of this chapter. 1D parameter estimation model that has already been discussed is easy to implement. However, it has some drawbacks as well. After formulating unitary ESPRIT (Zoltowski et al., 1996), 2D parameter estimation technique has become much popular because of its high-resolution estimation ability and computational efficiency. There are several antenna arrays could be implemented by using the unitary ESPRIT, i.e., URA, Triangular Antenna Array. Here, only URA antenna configuration is considered. Nevertheless, this estimation technique could only perform precise result if and only if the array positions are properly spaced with equal distance. A proper signal model is formulated in the next part and finally multipath channel model is compared with that signal model for the 2D rectangular channel estimation method. URA model involves complex geometries (Ullah, H., 2013) (Ren et al., 2013).

7. SIGNAL MODEL FOR CHANNEL ESTIMATION

Consider K and L identical Omni-directional sensors along with x- and y-axis, respectively according to Figure 5. Assume that the distance between two adjacent sensors dx along with x- axis and dy along with y-axis. It is also considered that $dx = dy = \lambda/2$ and P narrowband coherent radiating sources located in the far field region with wavelength λ impinge on the URA at azimuth angle θ_p and elevation φ_p for $p = 1, 2, 3, \ldots, P$.

The received signal at (i, j)th sensor (for $i = 1, 2, 3,\ldots, K$ and $j = 1, 2, 3,\ldots, L$) can be written by using analytical signal representation (Shan et al., 1985)

$$\mathbf{x}(t)=\mathbf{A}\mathbf{s}(t)+\mathbf{n}(t) \tag{1.54}$$

where,

$$\mathbf{x}(t) = [x_{1,1}(t), x_{2,1}(t), \ldots, x_{K\text{-}1,L}(t), \ldots, x_{K,L}(t)]^{\mathrm{T}}$$

and

$$\mathbf{n}(t) = [n_{1,1}(t), n_{2,1}(t), \ldots, n_{K\text{-}1,L}(t), \ldots, n_{K,L}(t)]^{\mathrm{T}}$$

for $t = 1, 2, 3, \ldots, T$ times snapshot. Here $[.]^{\mathrm{T}}$ denotes the transpose of a matrix.

In Equation (1.54), $\mathbf{s}(t)$ represents $P\times 1$ baseband signal vector

$$\mathbf{s}(t) =[s_1(t), s_2(t),\ldots,s_P(t)]^{\mathrm{T}}$$

and $\mathbf{A}$ is a $KL\times P$ steering matrix whose columns have the form as $\boldsymbol{vec}(\mathbf{a}_K(\mu)\ \mathrm{a}_L^T(\nu))$ for

$$\mu = \mu_1, \mu_2, \mu_3, \ldots, \mu_P$$

and $\nu = \nu_1, \nu_2, \nu_3, \ldots, \nu_P$ where, μ_P and ν_P are DOA parameters corresponding to the P-th source along with x-direction and y-direction, respectively, i.e.,

$$\mu = \pi \sin(\theta)\cos(\varphi) \tag{1.55}$$

$$\nu = \pi \sin(\theta)\sin(\varphi) \tag{1.56}$$

for $\theta = \theta_1, \theta_2, \ldots, \theta_P$ and $\varphi = \varphi_1, \varphi_2, \ldots, \varphi_P$.

Here, $\mathbf{a}_K(\mu)$ and $\mathbf{a}_L(\nu)$ denote steering vectors for URA defines as (Ullah, H., 2013)

$$\mathbf{a}_K(\mu) = \left[e^{-j(\mathrm{K}/2)\frac{2\pi}{\lambda}d_x\mu}, \ldots, e^{-j(1/2)\frac{2\pi}{\lambda}d_x\mu}, e^{j(1/2)\frac{2\pi}{\lambda}d_x\mu}, \ldots, e^{j(\mathrm{K}/2)\frac{2\pi}{\lambda}d_x\mu} \right]^T \tag{1.57}$$

Figure 5. Array geometry of the URA

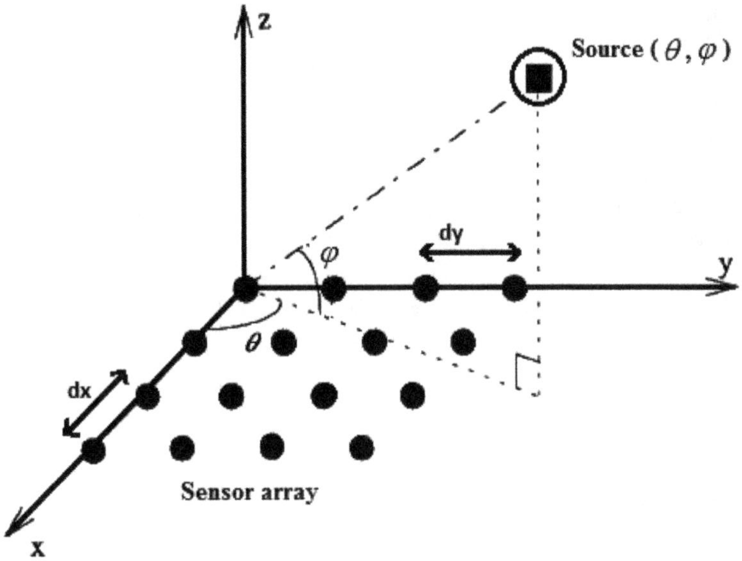

if K is even or

$$\mathbf{a}_K(\mu)=\left[e^{-j((\mathrm{K}-1)/2)\frac{2\pi}{\lambda}d_x\mu},\ldots,e^{-j\frac{2\pi}{\lambda}d_x\mu},1,e^{j(\frac{2\pi}{\lambda}d_x\mu},\ldots,e^{j((\mathrm{K}-1)/2)\frac{2\pi}{\lambda}d_x\mu}\right]^T \tag{1.58}$$

if K is odd, and similarly

$$\mathbf{a}_L(\nu)=\left[e^{-j(\mathrm{L}/2)\frac{2\pi}{\lambda}d_y\nu},\ldots,e^{-j(1/2)\frac{2\pi}{\lambda}d_y\nu},e^{j(1/2)\frac{2\pi}{\lambda}d_y\nu},\ldots,e^{j(\mathrm{L}/2)\frac{2\pi}{\lambda}d_y\nu}\right]^T \tag{1.59}$$

if L is even or

$$\mathbf{a}_L(\nu)=\left[e^{-j((\mathrm{L}-1)/2)\frac{2\pi}{\lambda}d_y\nu},\ldots,e^{-j\frac{2\pi}{\lambda}d_y\nu},1,e^{j(1/2)\frac{2\pi}{\lambda}d_y\nu},\ldots,e^{j((\mathrm{L}-1)/2)\frac{2\pi}{\lambda}d_y\nu}\right]^T \tag{1.60}$$

if L is odd.

The defined steering vectors hold the Conjugate Centro-symmetric property, and it can be expressed mathematically as

$$\mathbf{\Pi}_K\mathbf{A}_K=\mathbf{A}_K^* \tag{1.61}$$

where, $\mathbf{\Pi}_K$ is $K \times K$ exchange matrix with ones on its anti-diagonal and zeros elsewhere.

$$\mathbf{\Pi}_K = \begin{bmatrix} 0 & \cdots & 0 & 1 \\ 0 & \cdots & 1 & 0 \\ \vdots & \cdots & \vdots & \vdots \\ 1 & \cdots & 0 & 0 \end{bmatrix}_{K\times K}$$

From the matrix algebra, $\mathbf{\Pi}_K^2 = \mathbf{I}_K$, where,

$$\mathbf{I}_k = \begin{bmatrix} 1 & & & \\ & 1 & & \\ & & \ddots & \\ & & & 1 \end{bmatrix}$$

Using the property in Equation (2.8),

$$\mathbf{\Pi}_K \mathbf{a}_K = \mathbf{a}_K^* \tag{1.62}$$

A Conjugate Centro Symmetric vector can be found from a simple matrix as in (Ullah, H., 2013) (Ren et al., 2013) (Haardt et al., 1995). If $K = 2v$ is even, then

$$\mathbf{Q}_K = \frac{1}{\sqrt{2}} \begin{bmatrix} \mathbf{I}_K & j\mathbf{I}_K \\ \mathbf{\Pi}_K & -j\mathbf{\Pi}_K \end{bmatrix} \tag{1.63}$$

and if $K = 2v + 1$ is odd, then

$$\mathbf{Q}_K = \frac{1}{\sqrt{2}} \begin{bmatrix} \mathbf{I}_K & 0 & j\mathbf{I}_K \\ 0 & \sqrt{2} & 0 \\ \mathbf{\Pi}_K & 0 & -j\mathbf{\Pi}_K \end{bmatrix} \tag{1.64}$$

A vector $\mathbf{d}_\mathbf{K}(\mu)$ can be defined as (Ullah, H., 2013)

$$\mathbf{d}_K(\mu) \cong \mathbf{Q}_K^H \mathbf{a}_K(\mu)$$

Now, it is seen that $\mathbf{a}_K(\mu)$ is transformed into real-valued steering vector $\mathbf{d}_\mathbf{K}(\mu)$.

The covariance matrix for the array output signal $\mathbf{x}(t)$ can be determined as discussed in the previous chapter. According to the previous chapter, $\hat{\mathrm{R}}_{\mathrm{XX}}$ has $KL - P$ eigenvalues equal to the noise power σ^2. In order to obtain the signal and noise subspaces, Eigen-decomposition is applied to the covariance matrix.

$$\hat{R}_{XX} = \mathbf{E}_S \Lambda_S E_S^H + \mathbf{E}_N \Lambda_N E_N^H \tag{1.65}$$

where, $(KL - P)$ matrix $\mathbf{E}_S$ contains the eigenvectors corresponding to the signal eigenvalues and $KL\times(KL\text{-}P)$ matrix $\mathbf{E}_N$ contains the eigenvectors corresponding to the noise eigenvalues. In (1.65), $P \times P$ diagonal matrix Λ_S and $(KL\text{-}P) \times (KL\text{-}P)$ diagonal matrix Λ_N contain signal and noise eigenvalues, respectively.

ESPRIT algorithm was first proposed for 1D DOA estimation in the ULAs. Later on 1D ESPRIT technique was generalized into unitary 2D- and multi-dimensional (MD) ESPRIT (Liberti et al., 1999) (Khallaayoun et al., 2010) for the URAs.

8. MULTIPATH CHANNEL ESTIMATION

Consider a MIMO system in which the signal is assumed to travel from the transmitter to the receivers. Consider a multipath environment where P discrete propagation paths existed (each path is considered as a source of the signal). Assume that each path is involved with different parameters like DOA (θ), propagation delay (τ_p). To fit with standard data transmission, first perform band-limiting by transmitting the data through a low pass filter and after that sampling is done at sampling period *Ts*.

In order to make it simple, only 2D case in detailed was assumed. Therefore, only two sources and an array receive antennas are only considered. The same signal was transmitted from the two different sources but with a known delay was imposed in the second source. Now, parameters of interest are propagation delays τ_p and DOA θ_p for $p = 1, 2, 3, \ldots, P$. The goal is to estimate P pairs of exponential (a_p, b_p) which can be seen as a 2D harmonic problem (Pesavento et al., 2004)

Considering $k = 1, 2, \ldots, K$ is the frequency subcarrier index along one dimension of the array output data and $l = 1,2, \ldots, L$ denotes the antenna element index along the other dimension of the array output data.

Now, from Pesavento et al. (2004), formulate an equation for 2D as

$$y_{k,l}(\mathrm{i}) = \sum_{p=1}^{P} z_p(\mathrm{i})\, \mathrm{a}_p^k\, b_p^l + n_{k,l}(\mathrm{i}) \tag{1.66}$$

where,

$$a_p = e^{-j(2\pi/\mathrm{K})\tau_p} \tag{1.67}$$

And

$$b_p = e^{-j(2\pi d_R/\lambda)\cos\theta_p} \tag{1.68}$$

Our two parameter of interest is now a_p and b_p. d_R represents the elemental spacing on the receive side. Let us define steering vector $\mathbf{a}_\mathrm{p}$ for p-th path $\mathbf{a}_\mathbf{p} = [a_p^1, a_p^2, \ldots, a_p^K]^T$ and vector $\mathbf{b}_\mathrm{p}$ for the p-th path $\mathrm{b}_\mathrm{p} = [b_p^1, b_p^2, \ldots, b_p^L]^T$.

Steering vectors $\mathbf{a}_\mathrm{P}$ form the column of a $K{\times}P$ matrix $\mathbf{A}$ defined as $\mathbf{A} = [\mathbf{a}_1, \mathbf{a}_2, \ldots, \mathbf{a}_\mathrm{P}]$. Vectors $\mathbf{b}_\mathrm{p}$ form the column of a $L \times P$ matrix defined as $\mathbf{B} = [\mathbf{b}_1, \mathbf{b}_2, \ldots, \mathbf{b}_\mathrm{P}]$. Introducing the Khatri-Rao product between B and A, from the equation as

$$\mathbf{B} \circ \mathbf{A} = [\mathbf{b}_1 \otimes \mathbf{a}_1, \mathbf{b}_2 \otimes \mathbf{a}_2, \ldots, \mathbf{b}_\mathrm{p} \otimes \mathbf{a}_\mathrm{P}] \tag{1.69}$$

Introducing a new vector

$$\Omega = [(\mathbf{a}_1, \mathbf{b}_1), \ldots, (\mathbf{a}_\mathrm{P}, \mathbf{b}_\mathrm{P})]$$

Finally, 2D harmonic retrieval problem could be stated as (Pesavento et al., 2004)

$$\mathbf{y}(i) = \mathbf{H}(\Omega)\, \mathrm{z}\,(i) + \mathbf{n}\,(i) \tag{1.70}$$

where, $\mathbf{y}(i)$ denotes the measurement vector and 2D Signal matrix $\mathbf{H}(\Omega)$ is calculated by the Khatri-Rao product of matrices $\mathbf{B}$ and $\mathbf{A}$.

$$\mathbf{H}(\Omega) = \mathbf{B} \circ \mathbf{A} \tag{1.71}$$

In Equation (1.70), $\mathbf{z}\,(i) = [z_I, z_I, \ldots, z_I]$ is the complex envelope of the harmonics, which is considered as a nuisance parameter and $\mathbf{n}(i)$ is a vector stands for additive zero-mean complex Gaussian noise with covariance matrix $\sigma^2\mathbf{I}_{KL}$.

It is to be noted that Equation (1.70) has the same structure as the signal model introduced in Equation (1.54). Matrix $\mathbf{H}$ has the parameters of interest as in structure $\mathbf{A}$ and $\mathbf{z}(i)$ has a similar structure as the complex vector $\mathbf{s}(\mathrm{t})$ in Equation (1.54). Therefore, finally apply the unitary 2D ESPRIT to estimate the DOA and path delays of the channel from data vector $\mathbf{y}(i)$ in Equation (1.70). Here, $\mathbf{H}(\Omega)$ corresponds to the channel in the frequency domain (Barua et al., 2015).

9. SIMULATED RESULTS

Figure 6 shows the simulation results of various estimation techniques for the angles 40, 55 and 120 degree where conventional beamformer technique is not able to recognize the direction of arrival for low resolution.

Two paths were considered from the transmitter to receiver. A multipath model environment was developed with two paths by transmitting the same source signal via two antennas while the signal at the second antenna is delayed by a small period. The basic parameter setup is depicted in Table 1. Table 2 shows the comparison of delays using 2D ESPRIT technique and Spatial Channel Model for several experimental setups.

Figure 6. DOA estimation using various techniques

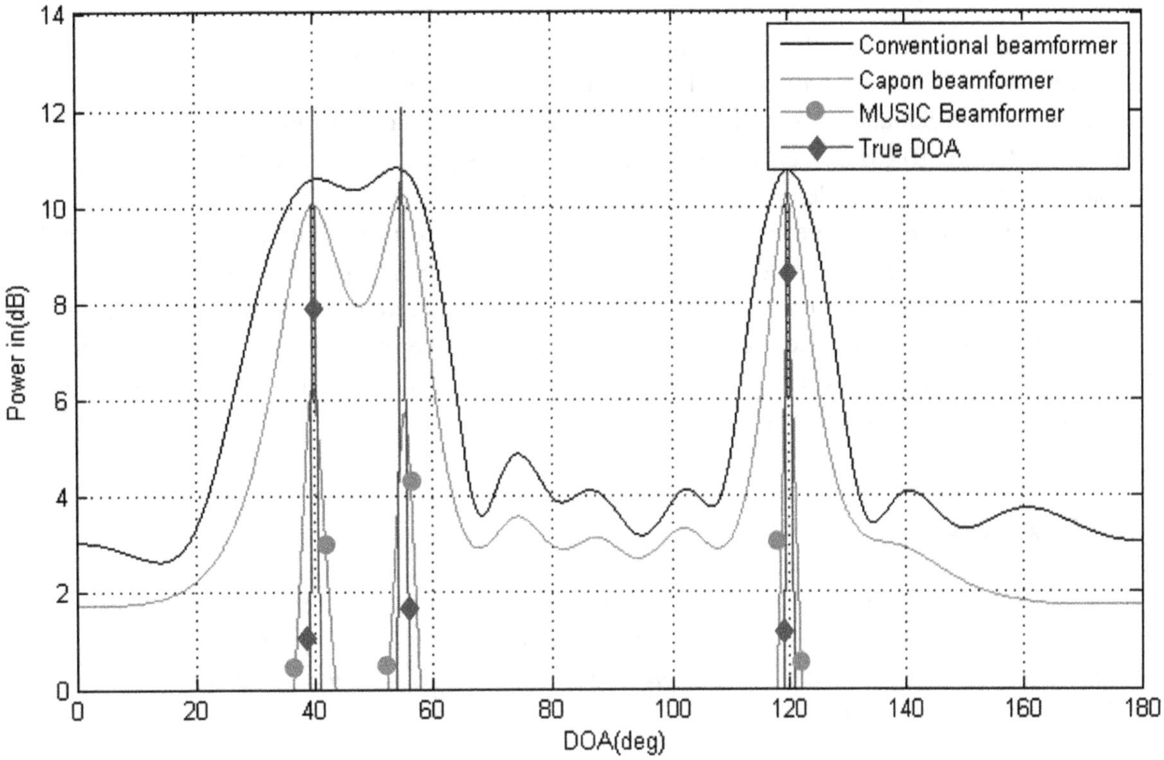

Table 1. Basic parameter setup

Attribute	Values	Attribute	Values
Sources	2	Carrier Frequency	2.4 GHz
Sensors	8	Distance between two adjacent antennas	0.0625
Time samples	100		
Power (w)	[1 1 1]	Noise Variance	2
True Angel	[40 55 120]		

Table 2. Comparison of the direction of arrival and multipath delay estimation using 2D ESPRIT technique and SCME model between true and estimated data.

Text	True	Estimated
Direction of Arrival (1)	[20 60] Degree	[20.156 60.001] Degree
Delay (second)(1)	[1.0e-06 * 0 1.0e-06 * 0.2604]	[1.0e-06 * 0 1.0e-06 * 0.2644]
Direction of Arrival (2)	[60 120] Degree	[60.602 120.221] Degree
Delay(second) (2)	[1.0e-06 * 0 1.0e-06 * 0.2604]	[1.0e-06 * 0 1.0e-06 * 0.2504]
Direction of Arrival (3)	[30 90] Degree	[30.156 90.031] Degree
Delay(second) (3)	[1.0e-06 * 0 1.0e-06 * 0.2604]	[1.0e-06 * 0 1.0e-06 * 0.2515]

10. CONCLUSION

DOA estimation using basic Beamforming techniques has been discussed in this chapter with the simulation results. Simulation for the ESPRIT technique cannot be plotted on the graph since it is not the search-based technique. It is seen from the plot MUSIC technique is the best among other earlier techniques. But ESPRIT is also promising technique which is going to be the best technique in order to estimate DOA in future mobile technology because of its preciseness. In this paper, a channel estimation technique using LTE reference signals on SCME model has been simulated. In order to find the channel, LTE standard provided by the 3GPP were followed. The results were compared with known parameters and found almost similar results. Since, only two paths were considered for channel estimation; it is also possible to consider more paths, which might be the future work. It is expected that with this model it is possible to explore the new era of communication system by extending the work that has already been done in this paper.

REFERENCES

Barua, S., Lam, S. C., Ghosa, P., Xing, S., & Sandrasegaran, K. (2015, June). A survey of Direction of Arrival estimation techniques and implementation of channel estimation based on SCME. In *Electrical Engineering/Electronics, Computer, Telecommunications and Information Technology (ECTI-CON), 2015 12th International Conference on* (pp. 1-5). IEEE. doi:10.1109/ECTICon.2015.7206986

Bienvenu, G., & Kopp, L. (1983). Optimality of high resolution array processing using the eigensystem approach. *Acoustics, Speech and Signal Processing. IEEE Transactions on*, *31*(5), 1235–1248.

Capon, J. (1969). High-resolution frequency-wavenumber spectrum analysis. *Proceedings of the IEEE*, *57*(8), 1408–1418. doi:10.1109/PROC.1969.7278

Chen, Z., Gokeda, G., & Yu, Y. (2010). *Introduction to Direction-of-arrival Estimation*. Artech House.

Haardt, M., & Nossek, J. A. (1995). Unitary ESPRIT: How to obtain increased estimation accuracy with a reduced computational burden. *Signal Processing, IEEE Transactions on*, *43*(5), 1232–1242. doi:10.1109/78.382406

Khallaayoun, A. (2010). *High resolution direction of arrival estimation analysis and implementation in a smart antenna system*. (Doctoral dissertation). Montana State University-Bozeman, College of Engineering. Available at http://scholarworks.montana.edu/xmlui/handle/1/1632

Krim, H., & Viberg, M. (1996). Two decades of array signal processing research: The parametric approach. *Signal Processing Magazine, IEEE*, *13*(4), 67–94. doi:10.1109/79.526899

Lam, S. C., Subramanian, R., Sandrasegaran, K., Ghosal, P., & Barua, S. (n.d.). *Performance of well-known frequency reuse algorithms in LTE downlink 3GPP LTE systems*. Academic Press.

Liberti, J. C., & Rappaport, T. S. (1999). *Smart antennas for wireless communications: IS-95 and third generation CDMA applications*. Prentice Hall PTR.

Parvazi, P. (2012). *Sensor Array Processing In Difficult And Non-Idealistic Conditions*. (PhD Thesis). TU Darmstadt. Available at http://tuprints.ulb.tu-darmstadt.de/view/person/Parvazi=3APouyan=3A=3A.html

Paulraj, A., Roy, R., & Kailath, T. (1986). A subspace rotation approach to signal parameter estimation. *Proceedings of the IEEE*, *74*(7), 1044–1046. doi:10.1109/PROC.1986.13583

Pesavento, M., Mecklenbräuker, C. F., & Böhme, J. F. (2004). Multidimensional rank reduction estimator for parametric MIMO channel models. *EURASIP Journal on Applied Signal Processing*, *2004*(9), 1354–1363. doi:10.1155/S1110865704401036

Rasheed, F. (n.d.). *Beamforming doa estimation for a uniform linear array*. Tech. Rep. Electrical and Computer Engineering Department, Drexel University.

Ren, S., Ma, X., Yan, S., & Hao, C. (2013). 2-D unitary ESPRIT-like direction-of-arrival (DOA) estimation for coherent signals with a uniform rectangular array. *Sensors (Basel, Switzerland)*, *13*(4), 4272–4288. doi:10.3390/s130404272 PMID:23539031

Schmidt, R. O. (1986). Multiple emitter location and signal parameter estimation. *Antennas and Propagation. IEEE Transactions on*, *34*(3), 276–280. doi:10.1109/TAP.1986.1143830

Shan, T. J., Wax, M., & Kailath, T. (1985). On spatial smoothing for direction-of-arrival estimation of coherent signals. *IEEE Transactions on Acoustics, Speech, and Signal Processing*, *33*(4), 806–811. doi:10.1109/TASSP.1985.1164649

Tayem, N. A. M. (2005). *Direction of arrival angel estimation schemes for wireless communication system*. (Master's thesis). Wichita State University. Available at http://soar-dev.wichita.edu/bitstream/handle/10057/688/d05020.pdf?sequence=1

Ullah, H. (2013). *Multi-antenna aided positioning in LTE networks*. (Master's thesis). Technische Universitaet Darmstadt. Available at http://www.academia.edu/4256293/Theis_report

Van Veen, B. D., & Buckley, K. M. (1988). Beamforming: A versatile approach to spatial filtering. *IEEE ASSP Magazine*, *5*(2), 4–24. doi:10.1109/53.665

Zoltowski, M. D., Haardt, M., & Mathews, C. P. (1996). Closed-form 2-D angle estimation with rectangular arrays in element space or beamspace via unitary ESPRIT. *Signal Processing. IEEE Transactions on*, *44*(2), 316–328. doi:10.1109/78.485927

Section 3
Network Design, Control, and Performance

Chapter 10
A Key to 5G Networks of Tomorrow

Aqeel ur Rehman
Hamdard University, Pakistan

Ahmar Murtaza
Hamdard University, Pakistan

Syed Muhammad Kashif Alam
Hamdard University, Pakistan

Iqbal Uddin Khan
Hamdard University, Pakistan

ABSTRACT

The emergence of mobile telephony increases the options of its deployment and utilization. As the research goes further, it also delivered Generations of Mobile Communication to us after rapid intervals of time. Each advancement is considers as the generation. Till now Forth Generation (4G) is in its maturity phase and researchers are planning to see 5G commercially around 2020. In upcoming Fifth Generation (5G), vision is to reduce latency typically it will be of less than 1ms, as the latency decrease one more thing will be achieved which can be named as Virtually Zero Distance. As the world is becoming the connected world, the availability of data on global network is going to be huge in size. In addition to that the new emerging concepts like Internet of Things (IoT) are giving boost to the requirement of big data storage and analysis. Internet of Things (IoT) is a concept of providing uniquely identifiable objects connectivity to Internet. This chapter is proposed to highlight the IoT concept and its requirements that are directly linked with the 5G Network technology services.

INTRODUCTION

What is IoT?

Nowadays, “Internet of Thing” (IoT) is one of the most famous topic in researcher’s, business developers and IT industries because they are expecting more than 20 billion devices will be on internet by 2020 and their aim is to enable the interconnection between peoples as well as their environment things based on standardized communication protocols under the name of “Internet of Thing” (IoT). In other word, the monitoring of critical situations, remote management and the wish of keep in touch pushes multiple fields to adopt the terminology called IoT. The Internet of Things (IoT) is the emerging area for the interest of researchers which provides a unique communication and remote control capabilities of monitoring and

DOI: 10.4018/978-1-5225-0239-5.ch010

controlling to any physical or virtual objects having any shape and any size, anywhere, anytime and by any one via secured infrastructure of internet with evolving of existing and future technologies. The key of the IoT is that each and every thing which is laid under the term Electronics and can controlled may be connected to the cloud of network and can or may continuously transmit its status or data or receive the date to store or any instruction to be followed. As the above said lines reflects not only the vast area to be considered for development and advancement.

As the technology advances, many different platforms started to evolve with communication features. As we can observed on graph, that in past decade internet started to grow like an impulse. Integration of device sizes, appreciation of Wireless technologies, and growth in demand of palm top or hand held devices led us towards a though that we can have control of each and every electronic device if it can be connected to internet. Evolution of the Internet has occurred five tiers. As the internet is the network of networks, it begins from the connection of two computers and then with help of research and development advanced towards World Wide Web, i.e. huge numbers of computers are connected all together. Mobile-Internet is also a thick branch of internet providing platform to mobile devices to become the part of Internet. Then, via internet social networks emerged. And now, it is rapidly moving towards IoT, means connecting each and everyday objects to the Internet (Perera, Zaslavsky, Christen, & Georgakopoulos, 2014).

Available Wireless Technologies

As every field has its own requirements which may be of high data rate, every field has its own accessing techniques over RF, which may be WiFi, WiMax or any other guided and unguided standard. Similarly, there are numerous technologies and protocol standards will be unified in macro IoT and maximum adoption will be done via unified infrastructure of internet mobility, sensing network and medical field and many more, as all of them could has to transmit bulk amount of data in real time.

Furthermore, the emergence of mobile telephony increases the options of its deployment and utilizations, which pushes researches to increase the options and facilitates over the platform of communication and under the word mobility. As the research goes further, it also delivered Generations of Mobile Communications to us after rapid shit of time intervals. Starting from few character texting services towards the sharing and downloading of bulk data in Gigabytes has come to us. Each advancement is considers as the generation. Till now Forth Generation (4G) is in its maturity phase and researchers are planning to see 5G commercially around 2020.

Evolution of 5G

The sharing of information exhibits more constrains as the distance between communicating entities changes from far too far-off. These constrains includes the increase of latency per hop using single technology as well as the use of multiple technologies per transfer over RF platform.

5G technology, known as the 5th generation mobile technology, is the upcoming mobile communication technology standard. 5G is expected to be rollout in 2020. In upcoming Fifth generation (5G), vision is to reduce latency typically it will be of less than 1mSec, as the latency decrease one more thing will be achieved which can be named as Virtually Zero Distance. Along with the before points, high data rate will also available with RF platform independency. RF platform independency can be elaborated as the higher data rate will be available over any spectrum and any access technology.

Flat IP network will be the key for 5G network to be acceptable for all kind of technologies. Flat IP architecture provides a way to identify the devices using symbolic name instead of hierarchal addressing scheme used in traditional IP networks.

5G Network Architecture

5G Network is known as the Nanocore architecture (Patil, Patil, & Bhat, 2012) that is the convergence of i) Nano Technology, ii) Cloud Computing, and iii) All IP network. Figure 1 is showing the 5G network architecture.

5G mobile communication technology will enable mobile phones to access different wireless technology at the same time and the terminal will be able to combine different data flows from different technologies. Figure 2 is mentioning the 5G mobile network architecture.

Benefits of 5G Technology

Following are the key benefits of 5G technologies (VIPIN, July 20, 2014):

- High speed, high capacity, and low cost per bit
- Support interactive multimedia, voice, streaming video, Internet, and other broadband services

Figure 1. 5G Architecture: the nano core

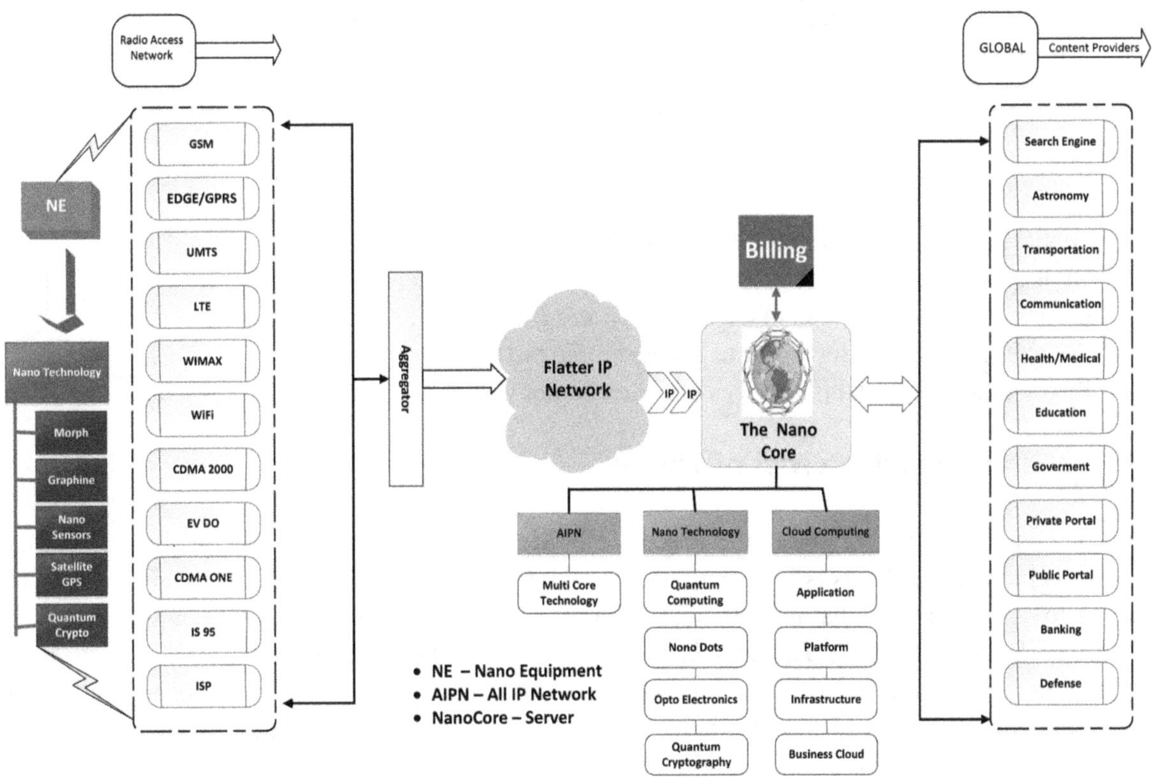

Figure 2. 5G Mobile Network Architecture

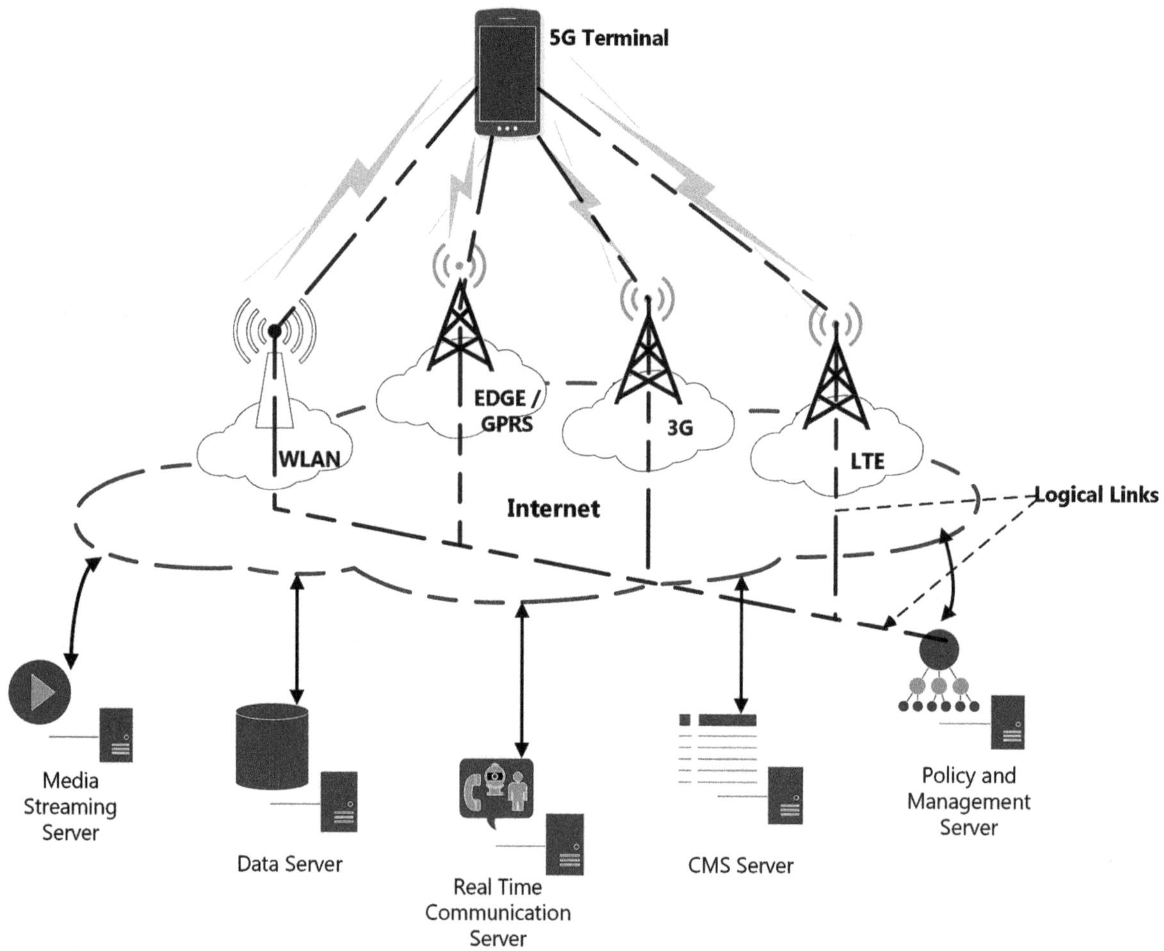

- Global access, service portability, and scalable mobile services
- Bi directional, accurate traffic statistics.
- Large broadcasting of data in Gigabit, supporting almost 65,000 connections
- Remote management for a user to get better and fast solution
- Policy based high quality services of 5G technology to avoid error

5th Generation and Internet of Things

As 5G is in the discussion phase, the concept of each and every device connected to internet, also known as Internet of Things is in research phase and will be granted a standard very soon. Under the roof of IoT, it is predicted that more than 20 billion devices will be the subset of the internet up to the year 2020 (Bassi et al., 2013). The internet enabled devices will be mostly from the field of medical and sensing networks as they are deployed in critical situations and areas. Just imagine the amount of traffic which will be generated by the predicted amount of devices connected to internet, which increases the importance of data collection in minimum latency, as data could be sensitive as well critical if it is transmitted from and patient or sensing network. The size of data bunch generated by any medical or sensing network node can vary and can be of large size with the condition of real time delivery. So all these predicted situations are promised to be solved and will be well entertained in 5G.

IoT VISION AND ARCHITECTURES

IoT Vision

In 1998 Kevin Ashton names the concept of connecting everything to internet, The Internet of Things. The scope of internet of things can be as vaster as much as we think about it (Bassi et al., 2013). As it is understood that the wider the scope of technology the more sophisticated and complex solutions required to heading problems, Which elaborated that each and every thing can and will be connected to internet . Internet of Things (IoT) is a huge domain, which connects any smart device with Internet using a dependable protocol and numerous information sensing equipment such as simple wireless senor may be including RFID, other infrared sensors, global positioning system (GPS), mobility and a huge data center or a web server or anything that can come under any processing limit and size that can be communicated to each other or to a specific device/thing or user by using any path/network or service (Zhang & Zhu, 2011). The larger the scale of Internet of thing the more vulnerable it becomes. The equipment, nodes or things in *IoT* may have many features to be considered and many core requirements shall be considered to develop a mature architecture of *IoT* platform (Bassi et al., 2013). The general view of Internet of things is illustrated in Figure 3.

We have a vision of IoT, referring ubiquitous computing that how and where it can be deployed. In the Figure 4 we have illustrated three broad areas under IoT Vision.

Wireless Sensor Network

Wireless Sensor Network (WSN) is a network comprising over miniaturized smart sensor devices to acquire the natural phenomena. Effective use of WSN may provide good monitoring and control solutions in almost every field of life. Following are some of the areas where WSN technology is in effective use:

- ***Agriculture and Breeding:*** To keep an eye to all physical and environmental condition to vast cultivated lands and on large number of animals is fairly impossible. To monitor all said conditions technology is deployed to collect our desired stats and to react in a reasonable time to avoid any loss or unwanted situation (Saint-Exupery & Antoine, 2009). The situation can be of extreme dryness, attack of insects, and interaction with viral disease of live stocks and critical health situation of animals and will lead to ward the catastrophe loss and increase of demand of relevant consumable.
- ***Oil and Gas:*** When talk about fuel and remote locations, one domain that must come in our mind is hunt of fossil fuel. To search and dig the deposit of fossil fuel required enormous amount of assets, human power and most important the finance. Any anomaly regarding aforesaid resources will not be easily bearable. To keep updated information of every instant and mobility, it required a platform which should be highly deliverable. Oil and Gas fields encompass various points to be monitored according to the maturity of the well. These monitoring points have sensors mounted

Figure 3. A General View of Internet of Things

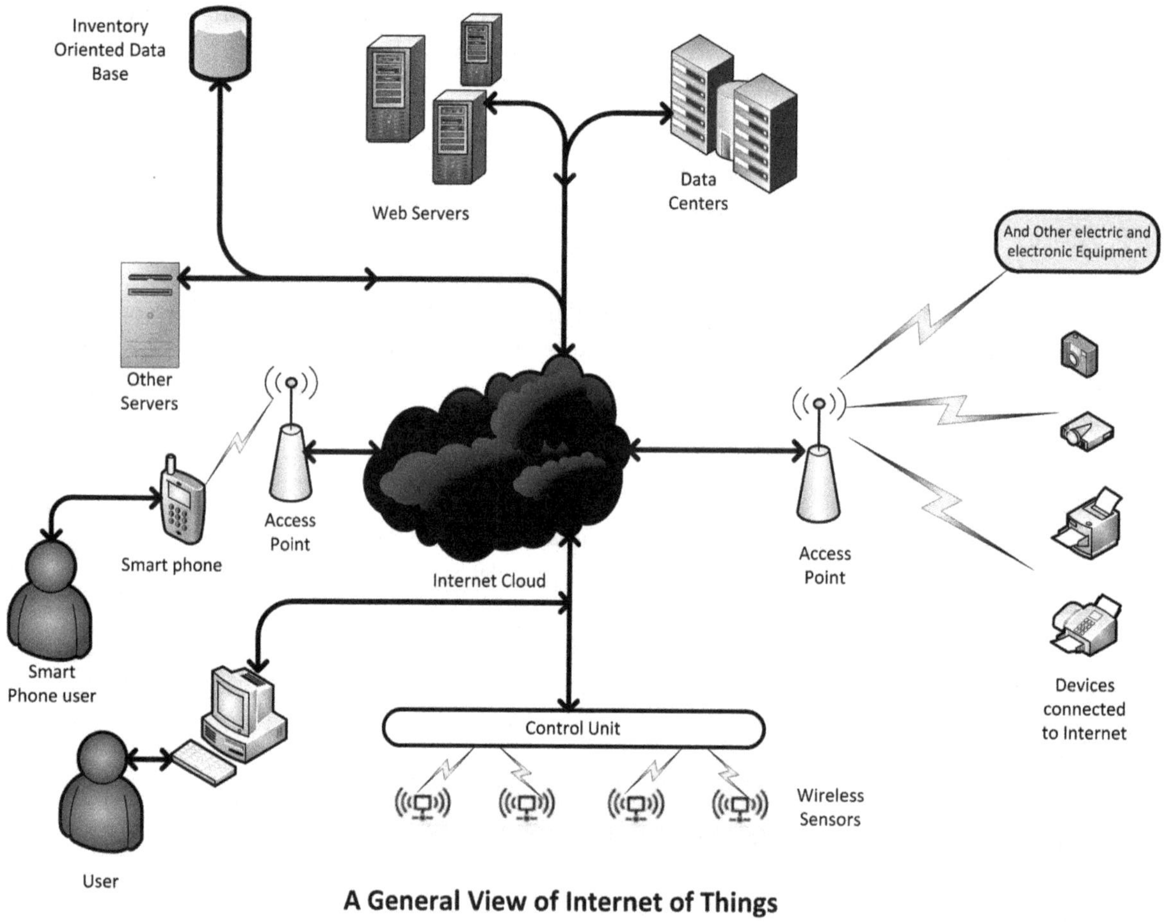

like wire line tension force seniors, load cells, pressure transducers and various temperature and Hydro-carbon /gas sensors (Saint-Exupery & Antoine, 2009). It also includes the remote assets management to track down all assets on a particular site as well as keep update information of their condition, required maintenance of any section of any portable testing and digging units. So that can be reformed and required components and manpower could be delivered at site with in due time.

- ***Logistics and Supply Chain:*** Using RFID tags on each product, or on a bunch of each product (if product is cheaper than the cost of RFID tag) can make task of retail counter easier as well as it can also protect the item from un-fair mobility. Even the collection of data at retail counter can be exchanged to relevant manufactures to maintain the supply as per demand. Using low cost active/ Passive tags and sensors along or within various parcels and logistics can help owner or vendor to keep track of their goods (Saint-Exupery & Antoine, 2009). For product tracking from manufacturer, whole sale to retailer's storage can be done for sake of inventory and stock management.

Figure 4. Application Domains of IoT

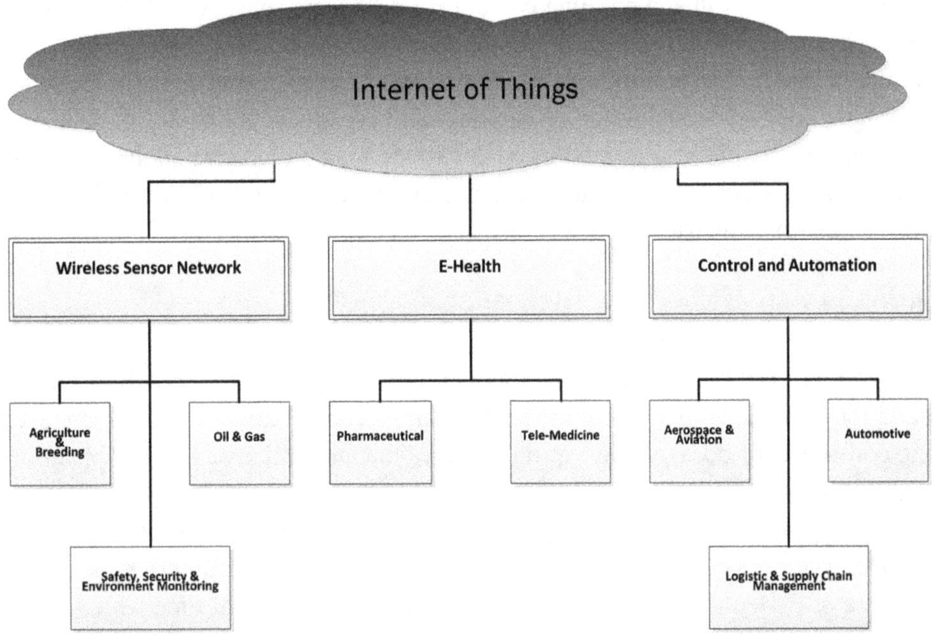

Control and Automation

- ***Aerospace and Aviation:*** As in avionics, the word safety is far more prior than the word trust among avionic giants and costumers. An Air craft is in real a complete pack of Science, technology and engineering, which are molded together in form of modules. Each module is maintained in regular frequency but missing it caused disasters (Saint-Exupery & Antoine, 2009). By applying smart sensors in different modules of aircrafts a centralized server can keep update of their service logs and only permitted air craft will be allowed to take-off or to leave hanger. In any aircraft, there are various modules which can be mounted and un-mounted as an independent module. The tasks for those modules are to transmit specific data to air or to concern department, for example location transponder. In some unlike incidents, the flight recorded (black box) was unable to locate near site, but by applying IoT, a self-powered location transmitter can be added to it for rapid tracking.
- ***Automotive:*** The concepts of Vehicle to Vehicle (V2V) interaction and Vehicle to Infrastructure (V2I) interaction are the concept of safety and reliability which comes under the umbrella of IoT. V2V discusses the smart interaction of vehicles to avoid accidents and V2I encompasses the interaction of city infrastructure to vehicle interaction (Luigi, 2011). Utilizing the said concepts, the condition and functional report can be obtained by security and insurance companies. From day of manufacturing to date of sale and to the recent day every record of performance can be collected. As discussed earlier, V2V communication cannot only manage the flow of traffic but can also help any human, in case of any emergency.
- ***Safety, Security and Environment Monitoring:*** Identifiable devices can be used to mature the security of any apartment, factory or office. These devices may be implemented in lockers or bank

safes and at doors of homes and apartments to ensure the access of right person. The environment monitoring system within any building can collect data from sensors outside that building to adjust the environment of that specific building using chillers and air conditioners (Saint-Exupery & Antoine, 2009). End to end transfer can be tracked in case of transfer of cash, weapons or any other valuable goods. Some wildlife researches can keep track and record of species they are researching on. Even explorers can be tracked in case of any emergency and the logs of temperature, height and other environmental quantities can be automatically transmitted while they are on their adventure or exploration trip.

E-Health

- ***Pharmaceutical:*** Talking about identifiable devices or tags can be placed on the drugs can be beneficial for patients, while storing the package in its cabinet and the smart cabinet can keep informing consumers of dosages and expiration date, and authenticity as well. This information can be shared by the manufactures and vendors of related drug(s).
- ***Tele-Medicine:*** Monitoring of critical patients is now easier as wearable and implantable sensors arrived. These sensors can continue transmit data of any patient over internet via control unit. The telemedicine or E-Health comes with issues of bulk data and in real time which can be available in upcoming generations of communication platforms. Some patients need to be observed critically, so doctor can keep himself updated by observing all graphs and reading on any android application.

IoT Architecture

As Internet of things is still in discussion phase, no mature architecture is proposed or standardized. For the reference we will consider the most resent published architecture for Internet of things (Bassi et al., 2013).

It is not standardized yet and still open for suggestions, discussion and for proposals from researchers. As a generic architecture of a communication model there are three basic layers as illustrated in Figure 5. Layer – 1 is Device layer, Layer – 2 is Functional layer and Layer – 3 is Application layer. The middle layer is responsible for distinguishing among other and Internet of things architecture, the actual issues resides inside layer – 2.

IoT – ARM 3.0

The IoT – ARM, consists of 3 basic blocks; i) The Reference Model, ii) The Architecture, and iii) The Guidelines. Figure 6 illustrates all constitute parts of ARM. Each block is discussed below in detail.

- ***IoT Reference Model:*** The IoT Reference Model is conforming to the OASIS reference model definition. It generates a common understanding of the IoT domain. Its description includes a general exchange of knowledge related to the IoT domain, including an IoT Domain Model for the top-level description, an IoT Information Model which explains that how IoT knowledge is going

Figure 5. Generic Architecture of a Communication Model

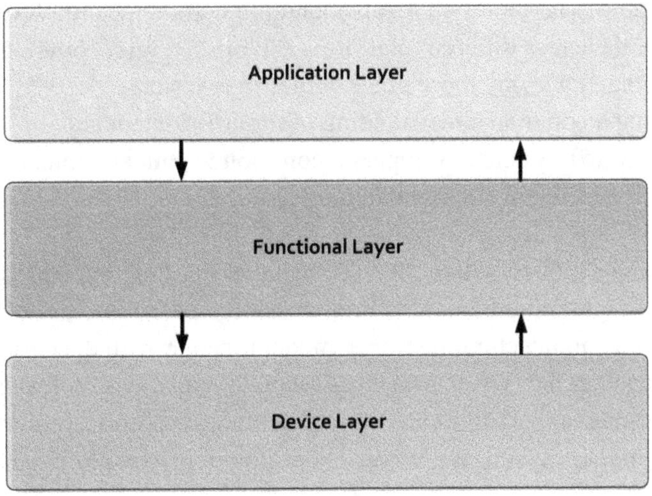

Figure 6. IoT-A Architectural Reference Model

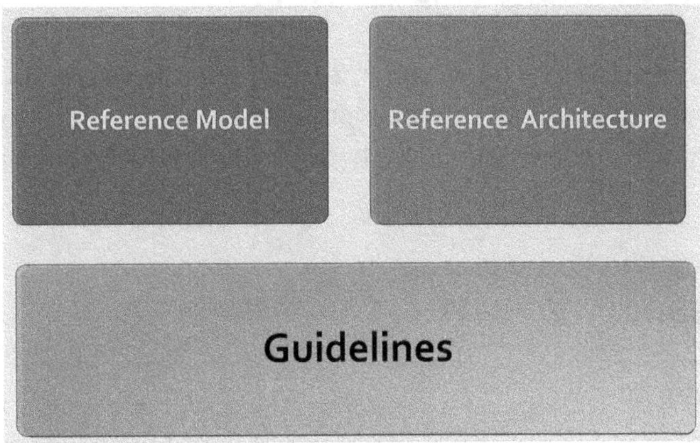

to be molded, and an IoT Communication Model which provides a specific understanding about communication between many heterogeneous IoT devices and the Internet (Bassi et al., 2013).

- ***IoT Reference Architecture:*** It converge the focus on abstracted mechanisms rather than a consolidated applications, according to the concern of IoT stakeholders. All the abstracted perspectives are used according to the general literature and standards. The reference model can be used for deciding the working boundary for the researchers and analyzers. So that can limit the work and to give proper shape in their proposed schemes and architectures, in other words by reference model organization of IoT standard can easily be done.
- ***The Guidelines:*** The guidelines discuss how purposed Models, within the borders of reference model, views and perspectives can be concretely used. It aims at explaining the usage of the IoT ARM. One of the focused focus area of this guidance is to derive out domain-specific architectures from the ARM .

- ***IoT – A and Functional Model:*** IoT – A Architectural Reference Model wants to promote high level of interoperability needs for all levels including communication level, the service level and the knowledge levels across different platforms established on a common grounding. The IoT-A project considers that achieving those goals comes in two steps:
    - In establishing a common understanding of the IoT domain and
    - In providing to IoT system developers a common technical foundation and set of guidelines for deriving a concrete IoT system architecture

Both aspects are being captured within the IoT Architectural Reference Model.

IoT-A, the European Lighthouse Integrated Project, addresses the Internet-of-Things Architecture. It proposes the creation of an architectural reference model together with the definition of an initial set of key building blocks. Together they are envisaged as crucial foundations for fostering a future Internet of Things. Using an experimental paradigm, IoT-A will combine top-down reasoning about architectural principles and design guidelines with simulation and prototyping to explore the technical consequences of architectural design choices (Bassi et al., 2013).

COMMUNICATION TECHNOLOGIES FOR IoT

In Table 1 (Aqeel-ur-Rehman, Mehmood, & Baksh, 2013; Brunnenmeyer, Mills, Patel, Suarez, & Kung, 2012; iDirect, 2014) four tiers of communication is presented which are currently available and possibly can be used in IoT to certain extent. Figure 7 illustrates that how possibly multiple technologies can be the part of IoT.

CHALLENGES ASSOCIATED WITH IoT

Multiple Environments

IoT primarily focuses on the sensing networks, electronics health and control and automation areas. The time will force the single sensing environments to hybrid environments. For understand we can say that hybrid environments can be partially health and partially control and automation. As to maintain the room temperature according to the need of the admitted patient in a hospital and will vary from patient to patient, more over for an example an intelligent building management system can adjust the inner condition of building including light and temperature according or in correspondence of sensors providing data form outside the buildings. So in these types of deployments temperature sensors, humidity sensors and air flow sensors will vary according to the place of installation as a single system can acquire multiple standards of sensors for inside and outside installations.

Connectivity

Connectivity of devices among themselves as well as with the infrastructure will be a great challenge as the devices will be of different types and surly will be at different locations (Chase, 2013). The different types of devices will present heterogeneity in IoT domain and the demand for a platform to resolve

Table 1. Communication Technologies for IoT

	Wireless Personal Area Network (WPAN)				Wireless Local Area Network (WLAN)	Wireless Metropolitan Area Network (WMAN)	Wireless Wide Area Network (WWAN)	
	Bluetooth	**Zigbee**	**RF Link (RFID)**	**Z-Wave**	**WiFi**	**WiMax**	**Satellite**	**Cellular**
IEEE Std.	IEEE 802.15.1-3	IEEE 802.15.4	IEEE C95.1	Z-Wave Alliance	IEEE 802.11a/b/g/n	IEEE 802.16	IEEE Std. 521-2002	IEEE 802.21
Topology	Star	Mesh, Star, Tree	--	Mesh	Star	Star, Mesh	Star, Mesh and Hybrid	Star
Bandwidth	1 Mbps	250 Kpbs	18 MHz	900 MHz	Up to 54 Mbps	134 Mbps	27 MHz	200 KHz
Channel Bandwidth	1-2480 MHz	0.3- 2 MHz	--	868 MHz	<20 MHz	<20 MHz	6 MHz	25 KHz
Max. data Rate	0.72- 10 M bit/s	0.25 M bit/s	1 M bit/s	9600 bits or 40 kbits/s	54 M bit/s	1 G bit/s	1 M bit/s	200Kbps
Spectrum	2.4 GHz	2.4 GHz	2.4 GHz	2.4 GHz	2.4-5 GHz	3.5-10 GHz	Ka / Ku band	900 – 1800 MHz
Range	5-30 m	10-300 m	<3 m	30 m	4-20 m	20 Km	N/A	400 meters
Power Consumption	Very Low	Very Low	Very Low	Very Low	Low	High	Medium	Medium
Performance	Better	Good	Better	Better	Excellent	Excellent	Good	Better
Application Area	Ad-Hoc Network	E-Health	WSN	Home Automation	Building Automation	Residential and enterprise environments	Global	Mobility

heterogeneity will increase. Although the there are various solutions are available but the latency and connectivity issues may be resolve in upcoming 5G as promised.

Power Critical

As disused in earlier heading that heterogeneity is challenge to IoT, so heterogeneity comes as devices implanted or deployed in different areas and the power supply to them is critical, as researchers and engineers focus on the low power issues they have to compromise over encryption schemes, processing powers and connectivity techniques. In this regards to have all optimized the power consumption is very critical if sensors are planted within human body or deployed in some remote areas to work for their assessed time (Chase, 2013).

Security

As IoT is a platform which mostly transmit its data over RF, so it is vulnerable in at-least two ways, one is as it is RF so it can be spoofed very easily, secondly due to low processing powers of devises large and complex encryption schemes cannot utilized efficiently.

Figure 7. Concept of Multiple Technologies in IoT

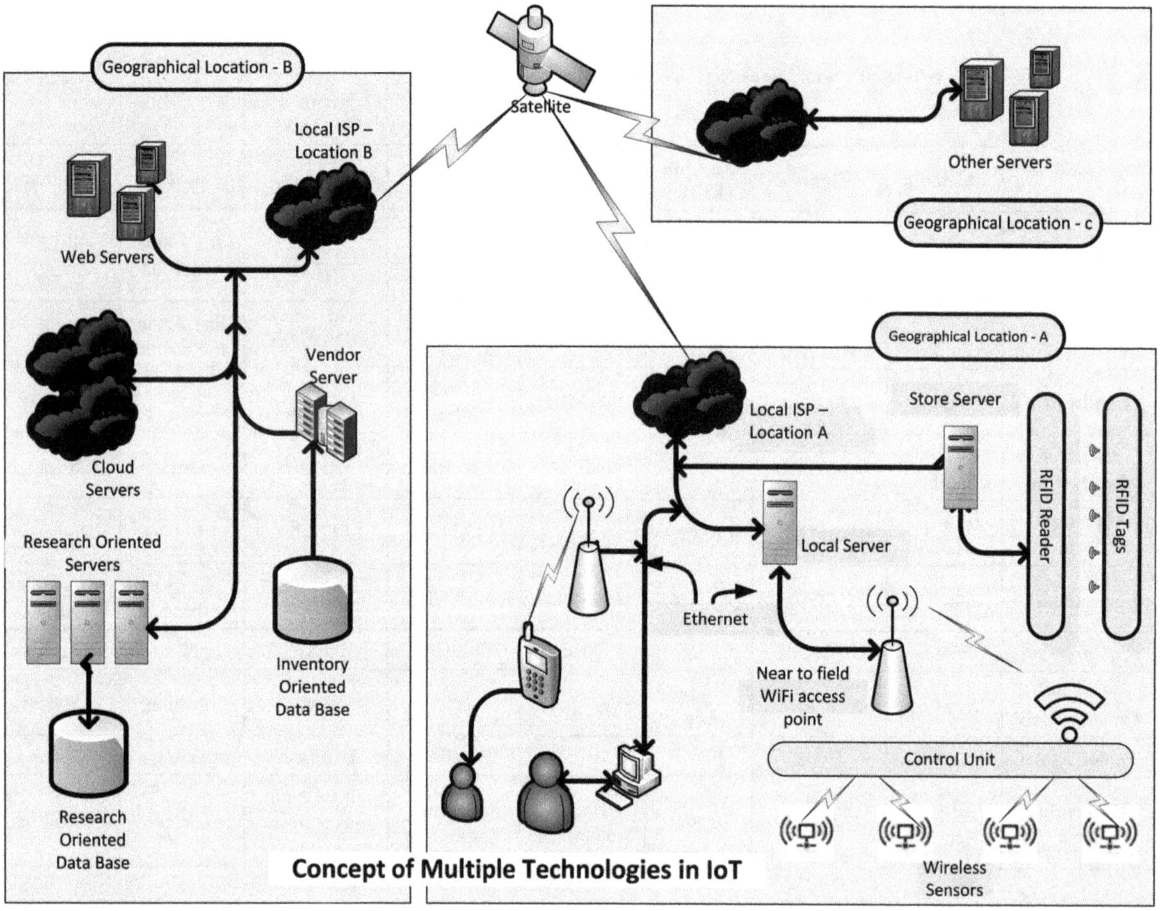

Complexity

Multiple layers of communications and platforms overlap each other in IoT. It involves number of communication platforms, hardware architectures, programming languages and operating systems. So its complexity is not only in their deployment phase but as well as in development phase as a generic application platform is not easy at present time so as common fault tracing techniques cannot be applied in IoT.

Cloud and IoT

Cloud is so far is understood as some storage on web, despite of this general understanding cloud has its own vast visions and applications. Few terminologies are right now very famous for cloud computing like Metal as a Service (MaaS), Platform as a Service (PaaS), Software as a Service (SaaS) and Infrastructure as a Service (IaaS). Available platforms, discussed earlier can be utilized not only for data storage but analyzing data, even visualizing data any time anywhere. As Data can be collected from remote locations no mater from sensor networks or transmitted / uploaded from any smart phone will be captured on cloud for further processing and analyzing.

Data Flood

As we can understand that the huge amount of smart devices will come under umbrella of IoT, So as we must prepare for the flood of data on internet which will not only be generated by the cellular or smart phone uses but also but the sensors deployed as wireless sensors or in Tele-medicine field. As the flow of data increases it will bottle neck the systems as well as may no full fill the requirement of Just in Time. In the currently deployed infrastructure it is not possible but 5G has promise to handle it too.

APPLICATIONS OF IoT IN DIFFERENT SECTORS OF LIFE

Some of possible applications in different sectors of life are as follows:

E-Health

Smart sensors may be worn by people to for location tracking, health, or for fitness record and may be by sports men on their athletic activities to monitor them (Anderson & Rainie, 2014). The sensors may be implanted, wearable or deployed to nay physical location will be used to monitor the patients or employees where special assistance needed. The technologies are getting mature with the passage of time we are having more advanced sensors and their control unit much smaller and cheaper.

Smart Homes

Smart homes encompass all type of monitoring what a human can do. Remote management of household facilities including lighting, air conditioning, security and energy management will be adopted by most homes on a series of intricate networks (Anderson & Rainie, 2014).

The embedded and smart appliances will assist the keeper to maintain garden and kitchen. Most of the appliances are set on their timer these days, but will not be in near future all will be managed according to your setting also known as *Profile.*

Smart Transportation

Smart Transportation is a concept under umbrella of Internet of things. It is also discussed as IoV, The Internet of Vehicles. The concept of IoV, is an advanced form of Telematics. Telematics was introduced as fleet management technique used for transportation purpose and in Formula one cars (Gerla, Lee, Pau, & Lee, 2014). This will be the part of a complex adopted and integrated network system of any smart city. This will be able to connect multiple persons within their transport or different locomotives with each other to share information of multiple locations within any city. The merging of technology includes statistics communications, environmental protection, energy conservation, and safety. To gain prospective ratio in this emerging market, achievement of core technologies and proposed standards will be vital to securing a calculated advantage. However, the integration of the IoV with other infrastructures should be as important as the building of the IoV technologies themselves. As an outcome of these thoughts, the IoV will become an integral part of the largest Internet of Things (IoT) infrastructure by its completion as per recommendation. It must be emphasized as prime, that collaboration and interconnection between the transportation sector and other sectors (such as energy, health-care, environment, manufacturing, and agriculture) will be the next step in IoV development.

The Internet of Vehicles will encompasses, establishment of Ad Hoc networks for stats sharing and communication among multiple vehicles or drivers, communication could be via phones within any vehicle to another vehicle. IoV will also obtain several services through Ad-Hoc group, which may involve multiple peoples and smart computing and various environments and emergencies. As a whole, connectivity among multiple vehicles will allow every vehicle (node) in that specific network to share their data of electronic sensors, another communication platform like cellular network (any node can allow hop of data from other node within any established network), share traffic's status, inform or ask about current location, in emergency vehicles like ambulances and fire tenders situation can be monitored, environment and climate conditions can also be shared among all (Shengguang, Lin, Yuanshuo, & Rucai, 2013). Furthermore, any vehicle can transmit its specifications like License plate number, Identification specs and driving license status of a driver, over internet or any smart check point or Toll stops. Applications of IoV in some practical and beneficial areas are discussed below:

Practical Areas of IoV

There is variety of areas and modes of IoV deployment in vehicle based environment. The solicitations of IoV or vehicle based communications speeded from safety to comfort and from commercial to entertainment purpose.

Below is the discussion on the three broad areas observed in emerging vehicle oriented, internet based applications and offers a broad vision on drifts and developments toward an intelligent vehicle network and create impression on the autonomous vehicle theory and its upcoming industry (Gerla, Lee, Pau, & Lee, 2014; Shengguang, Lin, Yuanshuo, & Rucai, 2013).

Presented Systems of IoV

Development towards ubiquitous IoV schemes will need to be conducted in stages, beginning with low-risk, simple applications, and learning from the outcomes of those to plans and design widespread systemic organizations, while allowing significant time for population adaptation and regulatory management systems to be developed. As an example, computer-calibrated control of vehicle movements and collision avoidance systems would be verified and improved in closed or control environments, such as warehouses, then implemented more extensively between specialized driverless vehicles on designated roads, before wide deployment to public and private transport for intact cities. Data linkages would start with basic information interchange, such as traffic monitoring or reporting on vehicular emissions, passenger numbers, cargo loadings, locations or available travel routes, before progressing to two-way telemetry, active traffic management and external control of vehicle functions.

Accessible IoV Data as a Source to Permit Comprehensive Research

The data groups produced by functioning IoV systems will be ironic and varied, and will establish a valuable resource in their own right. For this, the data has to be measured not as a 'replaceable' or 'throwaway' product to meet immediate needs of IoV users, but as a gathering economic and scientific resource, with many possible future users. Such massive data groups (occasionally labeled 'Big Data') can provide a foundation for researchers in many other disciplines, not only for the advancements and development of IoV systems, or the observing and supervision of vehicles, traffic, road systems and their economic impacts or industrial progress and development.

Through their nature, productivity and steadiness, the emergent of IoV data groups will similarly update research into areas as dissimilar as human conduct and social sciences, urban policy, state security, medication and epidemiology, inhabitants dynamics, geo-political capital distribution and economic development, weathercasting, market retorts to publicity and price setting, resource and utilities managing, food trading, exhibiting the spread of aggressive plants, pathogens and pests, freight logistics, tourism movements, planning of education systems, analysis of media feeding and propagation, agricultural development, and the fundamental mathematics of complex dynamic structures.

The cumulative remunerations of access to IoV data in these other economically important phases of research may be analogous to, or even exceed the direct economic benefits of IoV solicitation.

Promotion of Integration of IoV plus Vehicles

By means of joint mutual laboratory, values of IoV in safety and economic performance can be further employed. Multinational industry-university-research collaboration should be boosted to speed up the revolutionized and upgrading of automobile industry.

Smart Cities

Smart city is well-defined by IBM as the use of statistics and communication technology to intellect, examine and fit in the key information of core systems in running cities. At the same time, smart city can make intellectual reaction to different kinds of needs, together with daily occupation, ecological protection, public protection and city facilities, industrial and commercial accomplishments (Su, Li, & Fu, 2011).

Smart cities have been the chat of the town in residential executive institutions and associations of cities. Unluckily, the span is being used in an abundant approach by city officials and technology dealers, clouding the vision on what it actually takes to develop a smart city. This research plans to go beyond debates on the definition of the smart city and the propaganda connected to the idea of smart city. It is converging explicitly on the organizational challenges for a metropolitan management that originate with a smart city ambition – a much needed insight for cities that want to make the most of the smart city concept and alter the fabric of the metropolitan management accordingly. Present research on premeditated organizational transformation summaries that authority will have to balance control with empowerment, teamwork and collaboration will have to take on more open forms, business processes will be co-constructed with an endorsing community outside of the organization and will, very likely, involve unceasing improvement and conducting tests, and finally the general cultural stance towards the acceptance and adoption of technology will necessarily have to be one of 'want to' use and capitalize in the technology use instead of 'have to', meaning that experimentation is required. The concluding puts motivational issues of all the parties that are involved in the smart city ecosystem central to the success of that ecosystem. In sum, the supposition is that the actors that design and carry out the organizational transformation in smart cities, will have to vigorously adjust their design, planning, and control of organizational change to reflect the environment perspective and its key success factors (Van den Bergh & Viaene, 2015).

There might be a number of topics of interest that could be able to serve to well understand the influence of the smart city motivation, and in what way to turn it into truth that are discussed below. The subsequent observations are noteworthy for any smart city participant. In particular, focuses are put on the effects from a city administration standpoint, as no architecture is specifically for smart city concept.

These years, by way of the impression of "smart planet" being drove forward, smart city, smart grid and smart enterprises have been projected as important parts of smart planet in succession. Smart city, as not only a distinctive application of smart planet, but also one of the most popular topics and the most cutting-edge issues, has caused extensive concern. In most recent years, from London to Taipei, from New York to Singapore, one by one, the construction-project of smart city (originally known as the wireless digital city or wireless city) is the same as the inspiration, increasing around the globe.

Smart city, the essential strategy of IBM, primarily emphases on applying the next-generation information technology to all walks of life, set in sensors and equipment to hospitals, power grids, railways, bridges, tunnels, roads, buildings, water systems, dams, oil and gas pipelines and other objects in every corner of the world, and forming the "Internet of Things" via the Internet. Then the Internet of Things can be integrated through super computers and cloud computing. In this case, people can accomplish production and life in a more meticulous and dynamic way, achieving the state of worldwide intelligence and in the end reach;

"Internet + Internet of Things = Smart Planet" (Su, Li, & Fu, 2011)

In brief, "Smart City" is the concrete approach of "Smart Planet" applying to detailed region, succeeding the informational and integrated administration of metropolises. It can also be said to be an operative combination of smart organization concepts, smart construction methods, smart administration methods, and smart expansion methodologies. Through the digital grid management of municipal layout, resources, environment, economic, social and other systems, as well as the digital and informational handling and application of urban frame and elementary environment, we can achieve intellectual urban administration and facilities, in this manner promote the more proficient, more convenient and pleasant operation of modern metropolises.

City and metropolitan surroundings represent puzzling social environments which encompass local government, citizens and organization units (Clohessy, Acton, & Morgan, 2014). Currently these units are experiencing explicit requirements concerning to striking themes such as business and employment formation, sustainable development, energy and water, public welfare, environment, healthcare, education and public services, all of which are, in some form or another, are being progressively assisted and supported by ICT (Information Communication Technologies). Parallel to these necessities, the latest unsettled global economic decline is increasingly employing pressure on metropolises to cut budgets resulting in damaging effects not only on the maintenance and advancement of existing ICT infrastructure and facilities but also on future modernization polices. Nevertheless, the perception of a smart city, also known in the appearances of intelligent city, information city, digital city, e-city and virtual city, has been acknowledged as being a classic example of a response to discourse the current and future multifaceted challenges of aggregate resource proficiency, reducing releases, sustainable health care facilities for ageing residents, empowering youth and join in basic elements (Clohessy, Acton, & Morgan, 2014).

The IoT's rooted in the smart homes yield circumstantial information, which is logically managed by the smart homes. However, taking the next step by uniting the smart homes into one large organic unit arrange for basis for evolving fresh services and new infrastructures where the smart homes are the information providers handled in a new communication theory.

By means of Cloud of Things (CoT) in this situation to incorporate the smart homes into a smart city concept produces new bearable service opportunities, creates a foundation for refining the quality of life for populates, and it forms different means of implementing municipal control (Skouby & Lynggaard, 2014). In this revelation the Cloud of Things technology is an animated player because it handles the massive sum of information produces by the IoT based entities. In simple expressions a Cloud of Things is a group of possessions and calculation know-hows interconnected by the Internet. For smart cities uniting IoT and Cloud of Things (CoT) is critical, as this consent IoT data to be processed and deposited. This radical smart city frame presents a numeral amount of experiments spaced out from the growth of a 5G system. The IoT devices exhibits challenges as they require interconnecting with each other (machine to machine communication); they need to interconnect with the cloud services which assemble big-data, and they need to manipulate complex data using artificial intelligence. To handle these encounters with the massive amount of data involved, 5G technologies are required. Initially, the M2M interconnection have need of scalable and resourceful radio accommodations like cognitive radios (5G) which are able to assign bandwidth as per requirement, capable to handle high interference points, and able to conserve power by adjusting resources dynamically. Furthermore, uploading big-data to the cloud services involves considerable bandwidth with an inconsequential delay, i.e. it must be possible to offer mined big-data services in real time. 5G technologies are predicted to support this. Lastly, processing compound data using Artificial Intelligence is stimulating because it is allocated on the source constrained IoT devices. Hence, in order to handle complex IoT outlines an Advanced Artificial Intelligence (AAI) system is desired. This Advanced Artificial Intelligence system can be implemented as a disseminated interface between the IoTs, the smart home artificial intelligence systems, and the artificial intelligence systems obtainable as cloud services, where Cloud of Things is the integrating technology (refer to Figure 8). An IoT device comprises of an intelligent core and a M2M communication interface. The intelligent core is a source reserved device which offers Artificial Intelligence processing capability and a M2M interface which handles 5G based IoT communication (refer to Figure 9). In future, the communicating IoTs, the smart home servers, and the cloud services establish a Ubiquitous Network which proposition Ubiquitous connectivity's (Skouby & Lynggaard, 2014).

Figure 8. Advanced ICT Based Infrastructure for Future Smart Cities

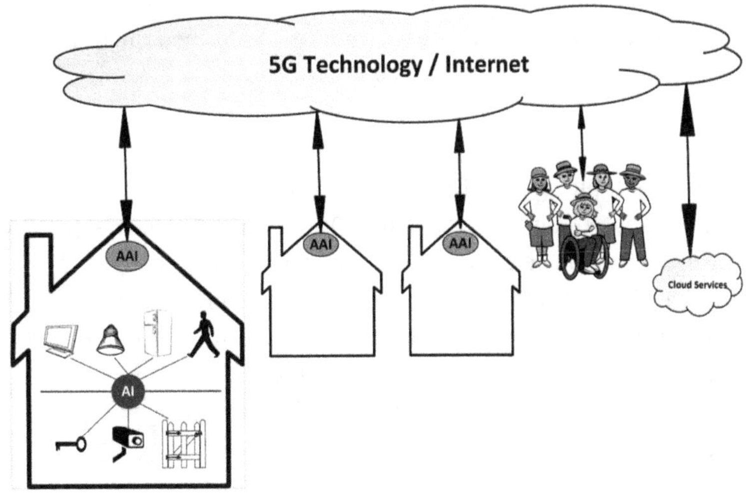

Figure 9. 5G Communication Concept

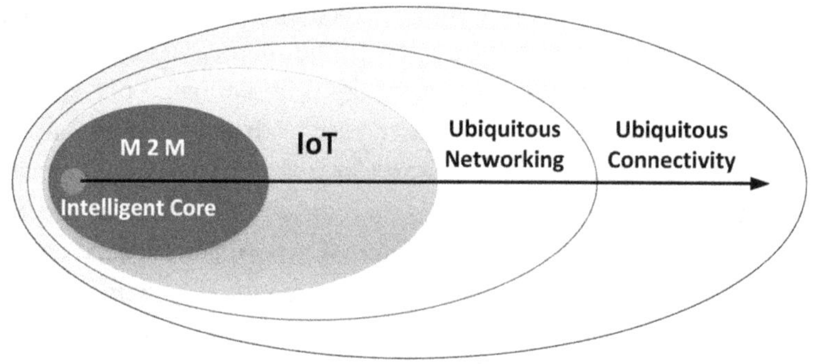

Communities

GPS management system will be more advanced along with proper deployment of IoT, Communities and their surroundings their neighborhoods and associated entities will be connected via smart planning. The surveillance within communities will be done more carefully to provide and promote a safer environment for families. Transportation infrastructure can become more deliverable (in time) along with better safety measures (Anderson & Rainie, 2014). Even more facilities given in any area of city like water distribution and its management will be done through a smart system and can simultaneously it will approximately eliminates the possible wasting of the natural resources.

Business and Commerce

Corporate sectors, businesses, organizations and other financial sectors rely on multiple systems to maintain and monitor their sales, records, invoices and many others as well as do the risk analysis over a defined period of time. If the system of any company fails, it must suffer a financial setback at every time re-occurrence. Along with an IoT connected system, a fusion of systems can be created with the involvement of the micro or subsystems, can potentially create a stronger system (Anderson & Rainie, 2014). Still there will be the risk; the insubstantial profit can hit the roof. Time and energy will be less consumed so the cost can be significantly suppressed if all factors of a company are regularly monitored via verbal communication or statistical collection either by predefined interval or on a requirement.

Environment

It is always the controversial topic, many of the valued researchers and analysts are in the favor to possible restoring and preservation of our mother planet's environment (Anderson & Rainie, 2014). A live stream of natural measurements can be collected to predict upcoming changes or updates on pre-existing reforms. Researchers have vision that there will be real-time feeds from sensors in remote fields, oceans, forests, and other urban areas about the pollution (environmental, sound and light) levels, harvesting field's information, and extraction of resource that will allow the closer monitoring of issues and problems (Anderson & Rainie, 2014). Atypically, a technological advancement may be just the right thing the environmental experts could use to produce more accurate results.

SUMMARY

In the above chapter, we have discussed the term of internet of thing and its available platform along with the currently available wireless communication technologies. Furthermore, we have discussed the limits of currently available and recently deployed generations of wireless technologies. Moving toward 5G we have enlightened the key features available in it and discussed the necessity of 5G for proper deployment of IoT as its vision, also discussed the areas where IoT can be deployed with the requirement of high speed data link for real time or just in time response.

REFERENCES

Anderson, J., & Rainie, L. (2014). *Digital Life in 2025: The Internet of Things Will Thrive By 2025.* Pew Research Center.

Aqeel-ur-Rehman, Mehmood, K., & Baksh, A. (2013). Communication Technology That Suits IoT-A Critical Review. In Wireless Sensor Networks for Developing Countries (pp. 14-25). Springer.

Bassi, A., Bauer, M., Fiedler, M., Kramp, T., Kranenburg, R., Lange, S., & Meissner, S. (2013). *Enabling Things to Talk: Designing IoT Solutions with the IoT Architectural Reference Model.* Springer. doi:10.1007/978-3-642-40403-0

Brunnenmeyer, D., Mills, S., Patel, S., Suarez, C., & Kung, L.-B. (2012). *Ka and Ku Operational Considerations for Military SATCOM Applications.* Paper presented at the The 2012 Military Communications Conference – Track 4 – System Perspectives, Milcom. doi:10.1109/MILCOM.2012.6415563

Chase, J. (2013). *The Evolution of the Internet of Things.* Dallas, TX: Texas Instruments Incorporated.

Clohessy, T., Acton, T., & Morgan, L. (2014). Smart City as a Service (SCaaS): A Future Roadmap for E-Government Smart City Cloud Computing Initiatives. In *Proceedings of the 2014 IEEE/ACM 7th International Conference on Utility and Cloud Computing.* doi:10.1109/UCC.2014.136

Gerla, M., Lee, E.-K., Pau, G., & Lee, U. (2014). *Internet of vehicles: From intelligent grid to autonomous cars and vehicular clouds.* Paper presented at the IEEE World Forum on Internet of Things (WF-IoT). Available at http://www.idirect.net/Company/Resource-Center/Satellite-Basics.aspx

Luigi, G. (2011). Vehicle-to-Vehicle/Vehicle-to-Infrastructure Control. In T. Samad & A. M. Annaswamy (Eds.), The Impact of Control Technology (p. 211). IEEE Control Systems Society. Available at http://ieeecss.org/sites/ieeecss.org/files/documents/IoCT-Part4-13VehicleToVehicle-LR.pdf

Patil, S., Patil, V., & Bhat, P. (2012). A review on 5G technology. *International Journal of Engineering and Innovative Technology, 1*(1), 26–30.

Perera, C., Zaslavsky, A., Christen, P., & Georgakopoulos, D. (2014). Context aware computing for the internet of things: A survey. *IEEE Communications Surveys and Tutorials, 16*(1), 414–454. doi:10.1109/SURV.2013.042313.00197

Saint-Exupery, D., & Antoine. (2009). Internet of things, strategic research roadmap. *Surrey: Internet of Things Initiative.*

Shengguang, L., Lin, T., Yuanshuo, Z., & Rucai, Z. (2013). *Internet of Things for special materials transportation vehicles*. Paper presented at the IEEE International Conference on Green Computing and Communications (GreenCom), and IEEE Internet of Things (iThings/CPSCom) and IEEE Cyber, Physical and Social Computing. doi:10.1109/GreenCom-iThings-CPSCom.2013.351

Skouby, K. E., & Lynggaard, P. (2014). *Smart home and smart city solutions enabled by 5G, IoT, AAI and CoT services*. Paper presented at the International Conference on Contemporary Computing and Informatics (IC3I).

Su, K., Li, J., & Fu, H. (2011). *Smart city and the applications.* Paper presented at the International Conference on Electronics, Communications and Control (ICECC). doi:10.1109/ICECC.2011.6066743

Van den Bergh, J., & Viaene, S. (2015). *Key Challenges for the Smart City: Turning Ambition into Reality.* Paper presented at the 48th Hawaii International Conference on System Sciences (HICSS).

VIPIN. (2014). *5G Mobile Technology wiki - wikisedia.* Retrieved May 2015, from http://wikisedia.com/5g-mobile-technology-wiki/

Zhang, H., & Zhu, L. (2011). *Internet of Things: Key technology, architecture and challenging problems.* Paper presented at the IEEE International Conference on Computer Science and Automation Engineering (CSAE).

ADDITIONAL READING

Aggarwal, C. C., Ashish, N., & Sheth, A. (2013). *The internet of things: A survey from the data-centric perspective Managing and mining sensor data* (pp. 383–428). Springer.

Ahn, S.-H., Kim, N.-U., & Chung, T.-M. (2014). *Big data analysis system concept for detecting unknown attacks.* Paper presented at the 2014 16th International Conference on Advanced Communication Technology (ICACT). doi:10.1109/ICACT.2014.6778962

Athreya, A. P., & Tague, P. (2013). *Network self-organization in the internet of things.* Paper presented at the 2013 10th Annual IEEE Communications Society Conference on Sensor, Mesh and Ad Hoc Communications and Networks (SECON).

Atzori, L., Iera, A., & Morabito, G. (2010). The internet of things: A survey. *Computer Networks, 54*(15), 2787–2805. doi:10.1016/j.comnet.2010.05.010

Bandyopadhyay, D., & Sen, J. (2011). Internet of things: Applications and challenges in technology and standardization. *Wireless Personal Communications, 58*(1), 49–69. doi:10.1007/s11277-011-0288-5

Barnaghi, P., Wang, W., Henson, C., & Taylor, K. (2012). Semantics for the Internet of Things: Early progress and back to the future. [IJSWIS]. *International Journal on Semantic Web and Information Systems, 8*(1), 1–21. doi:10.4018/jswis.2012010101

Bin, S., Yuan, L., & Xiaoyi, W. (2010). *Research on data mining models for the internet of things.* Paper presented at the 2010 International Conference on Image Analysis and Signal Processing (IASP).

Chun-Wei, T., Chin-Feng, L., Ming-Chao, C., & Yang, L. T. (2014). Data Mining for Internet of Things: A Survey. *IEEE Communications Surveys and Tutorials, 16*(1), 77–97. doi:10.1109/SURV.2013.103013.00206

Costa-Pérez, X., Festag, A., Kolbe, H.-J., Quittek, J., Schmid, S., Stiemerling, M., & Van Der Veen, H. (2013). Latest trends in telecommunication standards. *Computer Communication Review, 43*(2), 64–71. doi:10.1145/2479957.2479968

Gubbi, J., Buyya, R., Marusic, S., & Palaniswami, M. (2013). Internet of Things (IoT): A vision, architectural elements, and future directions. *Future Generation Computer Systems, 29*(7), 1645–1660. doi:10.1016/j.future.2013.01.010

Koch, C. (2013). *Compilation and synthesis in big data analytics Big Data* (pp. 6–6). Springer.

Labrinidis, A., & Jagadish, H. (2012). Challenges and opportunities with big data. *Proceedings of the VLDB Endowment, 5*(12), 2032–2033. doi:10.14778/2367502.2367572

Madden, S. (2012). From databases to big data. *IEEE Internet Computing, 16*(3), 0004-0006.

Mainetti, L., Patrono, L., & Vilei, A. (2011). *Evolution of wireless sensor networks towards the internet of things: A survey.* Paper presented at the 2011 19th International Conference on Software, Telecommunications and Computer Networks (SoftCOM).

Mayer-Schönberger, V., & Cukier, K. (2013). *Big data: A revolution that will transform how we live, work, and think.* Houghton Mifflin Harcourt.

Patil, P. S., Rao, S., & Patil, S. B. (2011). *Optimization of data warehousing system: Simplification in reporting and analysis.* Paper presented at the IJCA Proceedings on International Conference and workshop on Emerging Trends in Technology (ICWET).

Sagiroglu, S., & Sinanc, D. (2013). *Big data: A review.* Paper presented at the 2013 International Conference on Collaboration Technologies and Systems (CTS). doi:10.1109/CTS.2013.6567202

Said, O., & Masud, M. (2013). Towards internet of things: Survey and future vision. *International Journal of Computer Networks, 5*(1), 1–17.

Schroeder, R., & Meyer, E. (2012). *Big data: what's new. Internet, Politics, Policy 2012: Big Data.* Big Challenges.

Shi, W., & Liu, M. (2011). *Tactics of handling data in internet of things.* Paper presented at the 2011 IEEE International Conference on Cloud Computing and Intelligence Systems (CCIS). doi:10.1109/CCIS.2011.6045121

Stankovic, J. A. (2014). Research Directions for the Internet of Things. *Internet of Things Journal, IEEE, 1*(1), 3–9. doi:10.1109/JIOT.2014.2312291

Villars, R. L., Olofson, C. W., & Eastwood, M. (2011). Big data: What it is and why you should care. White Paper, IDC.

Whitmore, A., Agarwal, A., & Da Xu, L. (2014). The Internet of Things—A survey of topics and trends. *Information Systems Frontiers*, 1–14.

KEY TERMS AND DEFINITIONS

Big Data: Huge Quantity of Data generated by Smart Devices.

Cloud Computing: The practice of using a network of remote servers hosted on the Internet to store, manage, and process data.

Cognitive Radio: an adaptive, intelligent radio and network technology that can automatically detect available channels in a wireless spectrum and its attributes.

CoT: Cloud of Things.

5G: 5th generation mobile networks.

IBM: The International Business Machines Corporation, an American multinational technology and consulting corporation.

Internet of Things: Global Network of Smart Devices.

IoT: Standard Abbreviation of Internet of Things.

IoV: Standard Abbreviation of Internet of Vehicles.

M2M: Machine to Machine.

V2I: Vehicle to Infrastructure.

V2V Communication: the wireless transmission of data between motor vehicles.

Wireless Sensor Network: wireless network consisting of spatially distributed autonomous devices using sensors.

WSN: Standard Abbreviation of Wireless Sensor Networks.

Chapter 11
Synthetic Modeling of Human Mobility Patterns

Ali Diab
Al-Baath University, Syria & Ilmenau University of Technology, Germany

Andreas Mitschele-Thiel
Ilmenau University of Technology, Germany

ABSTRACT

One of the wide used modeling techniques applied to model human mobility patterns is synthetic mobility modeling. Synthetic models are largely preferred and widely applied to simulate mobile communication networks. They try to capture the patterns of human movements by means of a set of equations. These models are traceable, however, not capable until now of generating realistic mobility models. The chapter highlights the state of art and provides a comprehensive investigation of current research efforts in the field of synthetic mobility modeling. It reviews well-known approaches along with their pros and cons and finalizes with a qualitative comparison between them.

1. INTRODUCTION

Ubiquitous access to information anywhere, anytime and anyhow is a main feature of future communication networks which experience a tremendous development clearly evident from the wide variety of new services and broadband applications way beyond the classical application, i.e. telephony (Diab & Mitschele-Thiel, 2014). The Internet itself emerges towards what is today termed "the Internet of things", which assumes that users of future communication networks are objects, not only humans. Each object is embedded with sensors and is able to communicate. In other words, the physical world itself including communication networks, humans and objects is becoming a type of huge information system. To satisfy always increasing demands of users within such communication and information systems, the better support of mobility, the provision of services at low costs, etc. are the drivers standing behind the standardization of 4th Generation (4G) mobile communication networks, termed Long Term Evolution (LTE). Researchers expect that the number of LTE subscribers will reach the base of 3G/UMTS networks (1.087 million subscribers) by the year 2015 (Garza, Ashai, Monturus & Syputa, 2010).

DOI: 10.4018/978-1-5225-0239-5.ch011

One of the major aspects in this context and also the focus of this chapter is the development of novel methods to model human mobility patterns, which have myriad uses in crucial fields such as mobile communication research, urban planning, ecology, epidemiology, etc. The authors in (Gonzalez, Hidalgo & Barabasi, 2008) state that human mobility patterns are far from being random. They show a high degree of temporal and spatial regularity, thus, follow simple reproducible patterns. So, faithfully reproduction of the movements of real people significantly helps in answering crucial questions necessary to improve the experience of individuals, organizations, etc. For instance, an accurate reproduction of how people move around a city can help to evaluate if the installation of a specific sensing application on mobile devices would be able to reach the desired coverage, see (Isaacman, et al., 2012).

The modeling of how large as well as small populations move within small, medium and metropolitan areas is the goal that various human mobility modeling techniques tried to achieve. The modeling techniques tried to capture how humans move between places that have special importance in their life, e.g. work places, homes, etc. Many studies have shown that individuals tend to spend the majority of their time at few such places, see (Gonzalez, Hidalgo & Barabasi, 2008), (Isaacman, et al. 2011) and (Song, Qu, Blumm & Barabási, 2010). Keep in mind that the reproduction of human density over time at various geographic scales helps in addressing essential societal issues like the environmental impact of home-to-work commutes for instance. Furthermore, different geographic distributions of communities, commercial centers, companies, etc. heavily affect human mobility patterns, which are affected by and also must be reflected in various factors such as transportation infrastructures, plans for cities growth, etc. Previous efforts have reported significant differences between cities in terms of metrics such as commute distances for instance, see (Isaacman, et al. 2011a), (Isaacman, Becker, Cáceres, Kobourov, Rowland & Varshavsky, 2010) and (Noulas, Scellato, Lambiotte, Pontil & Mascolo 2012).

Models of human mobility patterns are categorized into three categories, namely synthetic, trace-based and community-based models, see (Musolesi & Mascolo, 2009), (Camp, Boleng & Davies, 2002) and (Ranjan Roy, 2011). Synthetic models are mathematical ones trying to capture the patterns of human movements by means of a set of equations. The key idea of trace-based models is the exploitation of available measurements and traces achieved in deployed systems, e.g. connectivity logs, location information, etc., to reproduce synthetic traces characterized by the same statistical properties of real traces. Community-based modeling of human movements depends on the fact stating that mobile devices are usually carried by humans, which implies that movement patterns of such devices are necessarily related to human decisions and socialization behaviors. So, human movement routines heavily affect the overall movement patterns resulting, e.g. people move daily morning from their houses to their offices and come back at the afternoon for instance. One of the major contributions in this context is the application of social network theory to generate more realistic human movement patterns, see (Musolesi & Mascolo, 2009).

One of the critical issues concerning human mobility patterns additional to the modeling of them is the prediction of future locations of users. The performance of a wide range of protocols, services and applications in mobile communication networks will be optimized, and even better decisions for other concerns in other fields can be taken (e.g. economy, planning of metropolitan areas, etc.) if an accurate prediction of future users locations is achievable (Alnasouri, Mitschele-Thiel & Diab, 2009). Currently, new users locations are determined either statically (using the neighbor graph (Mishra, Shin & Arbaugh, 2004) for instance) or dynamically with the help of history caches and mechanisms like data mining to predict new mobility patterns based on old ones, see (Franois, 2007) and (Su, Lee & Gerla, 2001).

Synthetic models are categorized into two categories, namely synthetic single node mobility models and synthetic group mobility models (Camp, Boleng & Davies, 2002). The first category assumes that each user is allowed to move freely without caring of other users, while the second assumes that movements of each user relate to the movements of a group of users, e.g. a group of students, tourists, vehicles, solders, etc. Users, thus, are not allowed to move freely.

The chapter highlights the state of art and provides a comprehensive investigation of current research efforts in the field of synthetic modeling of human movement patterns. It provides an insight into synthetic single-node and group mobility models and overviews well-known approaches along with their strengths and drawbacks. The chapter summarizes with a qualitative comparison between the discussed models and concludes with the main results.

2. SYNTHETIC SINGLE NODE MOBILITY MODELS

These models are also referred to as synthetic individual/entity mobility models in the literature. The movement of each individual Mobile Node (MN) is considered based on the MN's speed, MN's direction, MN's transition length/time between two consecutive positions and the distribution of MNs locations. Keep in mind that the movements of individuals do not relate to each other.

The following overviews well-known synthetic single node mobility models along with a discussion of their strengths and drawbacks. This section finalizes with a qualitative comparison between the described models with respect to the temporal and spatial dependency of the velocity as well as the direction, the existence of geographic restrictions on movements and pause times between movements.

2.1. Random Walk (RW) Mobility Model

The RW mobility model relates to the Brownian motion first described by Einstein in 1926 (Sanchez & Manzoni, 2001). This model simply mimics the statement: many individuals move in nature in extremely unpredictable ways (Davies, 2000). It is simple from the mathematical point of view and, therefore, widely applied to generate movement patterns for simulation purposes in the field of mobile communication networks.

Description: the operation of the RW mobility model can be described as follows: MNs are first distributed in the area based on a specific distribution model. Each MN moves, following that, from its current location to a new location, selected randomly applying a uniform distribution from the range $\left[0 \rightarrow 2\pi\right]$, with a speed also selected randomly from a specific speeds range applying a uniform or Gaussian distribution. Note that each movement occurs, in principle, after either a constant time interval (Δt) or a constant distance (d) traveled. Note also that there are no pause times between movements. When a MN reaches an area boundary, it bounces off the boundary with a specific angel determined by the incoming direction. The MN then continues moving along this path.

There exist many derivatives of the RW mobility model including the 1-Dimentional (1-D), 2-D, 3-D, and d-D RW models, see (Camp, Boleng & Davies, 2002). In the 1-D RW model (Davies, 2000) one imagines a gymnast standing in the middle of a balance beam. The gymnast moves on the beam with a random selected speed towards one of its end points for a certain time period (Δt) or distance (d) based on the results of a coin flip. In the 2-D RW model, one imagines the gymnast moving on a planar surface instead of a balance beam, see Figure 1 which shows an example movement pattern generated

Figure 1. An example movement pattern generated by the 2-D RW mobility model (Camp, Boleng & Davies, 2002)[1]

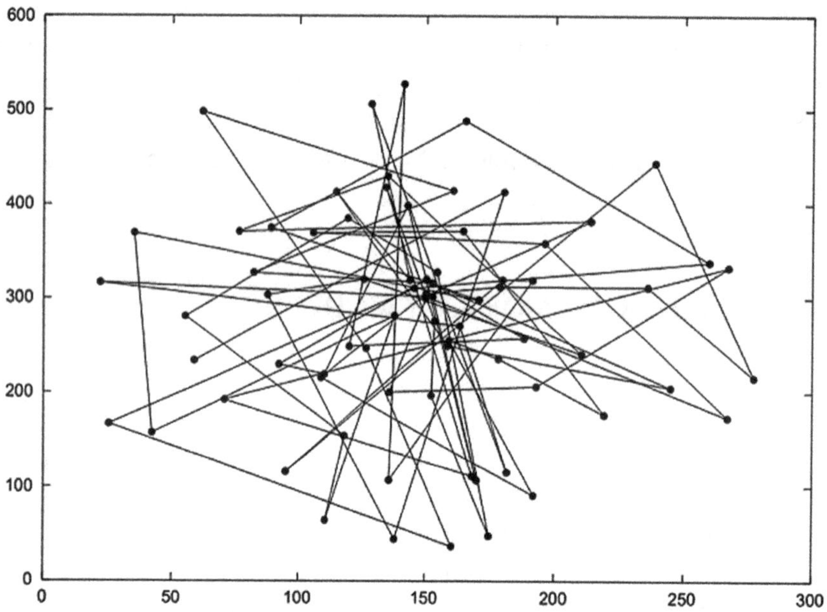

from the 2-D RW model. This model is widely applied to simulate mobility of MNs since the earth surface is modeled using a 2-D representation, see (Camp, Boleng & Davies, 2002). In the 3-D RW model (Kim, Chung & Sung, 1998), (Weisstein, 1998), the vertical motion is additionally taken into account. Such scenarios are adequate to simulate flying devices for instance.

Discussion: the RW mobility model is simple from the mathematical point of view. The direction and speed of MNs at any given time are random and are completely independent of past directions and speeds. In other words, the RW mobility model generates memory-less movement patterns (Basu, Redi, & Shurbanov, 2004). This in turn may lead to an oscillatory-type trajectory (moving forward and backward), see (Zonoozi & Dassanayake, 1997). The randomness in selecting the next direction as well as speed renders this type of motion completely unrealistic and unpredictable. In brief, the characteristics of the RW mobility model generate unrealistic movement patterns with sudden stops, sudden accelerations, sharp turns, etc.

It has been proven in 1921 that any MN operating the 1-D or 2-D RW mobility model will return back to the starting point with a complete certainty (Giner, 2004). This means that the RW mobility model is adequate to simulate MNs moving around their starting points without worrying about never returning back.

The studies achieved have shown that the roaming patterns of MNs operating the RW mobility model will be restricted to a small portion of the simulation area if (Δt) or d is selected short, see (Camp, Boleng & Davies, 2002). The simulation portion increases as (Δt) or d goes up. For an ad hoc network for instance, this means that short values for Δt or d are adequate to simulate semi-static networks. Otherwise, larger values must be used.

Limitations: although the RW model is widely used to generate mobility patterns to simulate mobile communication networks, it is insufficient to capturing following mobility characteristics:

- **Temporal Dependency of the Velocity:** it is not possible in the RW model to express relations between the current and future speeds. In reality and due to the physical constraints of MNs, they do not move abruptly. The new speed relates somehow to previous ones.
- **Temporal Dependency of the Direction:** same as by the velocity, it is not possible to express relations between current and future directions. In fact, the new direction also relates to previous ones and depends on the scenario defined.
- **Spatial Dependency of the Velocity:** due to the physical constraints of MNs and the surrounding environment, the speed of a MN may be influenced by and/or correlated with the speeds of neighbor MNs, e.g. the speed of a car is affected by the speeds of other cars traveling in front of it for instance.
- **Spatial Dependency of the Direction:** the location of a MN may correlate in real-life scenarios with the locations of specific or neighbor MNs, e.g. the location of the members of a tourists group relates to the location of the group's leader.
- **Geographic Restrictions of Movements:** MNs, in reality, are not always allowed to move in arbitrary directions. In cities for instance, MNs move on streets, freeways, etc.
- **Pause Times Restrictions:** the RW model does not allow MNs to pause between movements. This does not reflect real-life scenarios, where MNs stop in specific places such as markets, campus, etc., and move again to other places.

Variations of RW Mobility Model: to solve particular real-life problems, the following variations of the RW mobility model have been proposed:

- **Random Waypoint (RWp) Mobility Model:** this model (Camp, Boleng & Davies, 2002) extends the standard RW mobility model by introducing pause times between movements. An insight into this model will be provided in the following section.
- **Constant Velocity RW Direction (CV-RWd) Mobility Model:** this variation of RW model simply assigns a constant speed to the MNs, while movements' directions remain random, see (Davies, 2000).
- **Directional RW (DRW) Mobility Model:** this model assumes that the probability of moving on the same direction and away from it (the opposite direction) is higher than moving on other directions, see (Davies, 2000) and (Giner, 2004).
- **Ping-pong RW (PpRW) Mobility Model:** this model is applied to simulate ping-pong movements where MNs return back to the previous location with a high probability (Giner, 2004).
- **Turning RW (TRW) Mobility Model:** this model guarantees that the future direction will be different from the current one by setting the probability of moving on the same direction (forward or backward) to zero (Ranjan Roy, 2011).
- **H-Spin RW (H-SRW) Mobility Model:** this model assigns a high probability to the movement in the direction that makes the MN moving in a clockwise triangular pattern, see (Ranjan Roy, 2011).
- **T-Spin RW (T-SRW) Mobility Model:** this model is similar to the previous one. However, it guarantees that the movement pattern will be clockwise (Ranjan Roy, 2011).

- **RW with Drift (RW-D) Mobility Model:** this model is also known as Brownian motion with drift mobility and applied to mimic the movements types, where MNs move randomly within conveys that move in particular directions, see (Turgut, Das & Chatterjee, 2001) and (Li & Rus, 2003). When all MNs in the scenario have the same drift in direction and speed, the model becomes the ordinary Brownian process since all MNs have zero relative drift velocity with respect to each other. Note that the value of the drift heavily affects the resulting pattern, e.g. small drifts in directions and velocities result in semi-static ad hoc network, see (Ranjan Roy, 2011).
- **Markovian RW (MRW) Mobility Model:** this model is known as the probabilistic version of the RW model (Camp, Boleng & Davies, 2002), (Chiang, 1998). It introduces a memory and uses the Markov chain to model the movements. The model will be studied in more details in section 2.1.5.

2.2. Random Waypoint (RWp) Mobility Model

The RWp mobility model (Camp, Boleng & Davies, 2002) is widely used to generate mobility patterns for MNs, especially for ad hoc networks. The model is simple and straightforward stochastic, see (Bettstetter, Hartenstein & P´erez–Costa, 2003). As mentioned above, this model extends the RW mobility model by introducing pause times between movements.

Description: the operation of the RWp model is similar to the RW model. Each MN moves from its current location to a new one by randomly selecting a new location and speed. Different from the RW mobility model, the MN additionally and also randomly selects a pause time duration. Upon reaching the new location, the MN waits for the selected pause time duration and repeats, following that, the process again. The time a MN takes to move from a starting location to a new one is called a movement epoch, termed a movement period and a transition time as well. The distance traveled during a movement epoch is referred to as a transition length. An example movement pattern employing the RWp mobility model

Figure 2. An example movement pattern generated by the RWp mobility model (Camp, Boleng & Davies, 2002)[2]

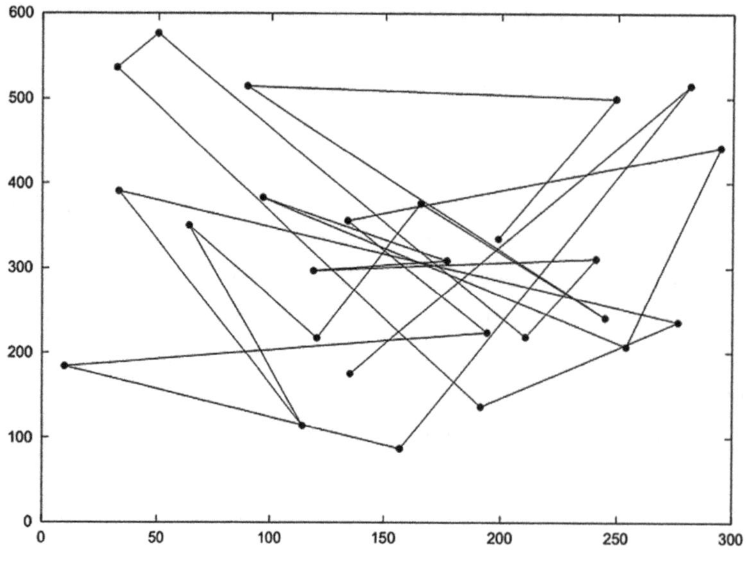

is depicted in Figure 2. Note that the pattern looks similar to that observed when employing the RW model, such similarity is also expected.

Discussion: similar to the RW model, MNs move in completely unpredictable ways. That is, MNs directions as well as speeds at any given time are random and independent of past directions and speeds. Pause times between movements are random as well, without dependency on previous pause times. This implies that the RWp model is also a memory-less model. The randomness results, on one side, in an oscillatory-type trajectory (same as by the RW model) and, on the other side, in unrealistic movement patterns with sudden stops, sharp turns, etc.

There were several studies that analyzed the RWp model. The analysis done in (Navidi & Camp, 2004) has shown that due to the initial random distribution of MNs within the simulation area, it takes some time until MNs reach a stationary distribution. This implies that the initial simulation period must be discarded. How long is this period is a matter of the scenario itself. In (Yoon, Liu & Noble, 2003), it has been proven that the average velocity decreases over time if the allowed minimum value of the speed is equal to 0. The analysis achieved in (Camp, Boleng & Davies, 2002) has shown that there exist a complex relationship between the MN speed and the pause time when operating the RWp mobility model. For instance, an ad hoc network with fast MNs that have long pause times is actually more stable (in terms of links changes per a MN) than an ad hoc network with slow MNs and short pause times. Thus, it is very important to select adequate parameters (speed range, mean speed value, pause times range, etc.) for the RWp model. Figure 2 provides the MNs links breakage rate as a function of the speed and pause time for a sample scenario. The figure shows that a stable network (few links changes per a MN) results when pause times are selected long (over 20 s in the figure), even when MNs move at high speeds. For more details, the readers are directed to (Camp, Boleng & Davies, 2002) and (Karp, 2000).

The study of the stochastic parameters of the RWp model has shown that MNs in the RWp model tend to move back to the middle of the simulation area, see (Ranjan Roy, 2011), (Aschenbruck, Gerhards-Padilla & Martini, 2008) and (Bettstetter & Wagner, 2002). This effect is known as the density wave effect and results in the clustering of MNs in one part of the simulation area. MNs, therefore, appear to converge, disappear and converge again, and so on. Figure 3 depicts the discussed density wave effect resulting from the RWp model.

Limitations: in spite of the wide usage of the RWp model, it is incapable of capturing following mobility characteristics (Bai, Sadagopan & Helmy, 2003):

- **Temporal Dependency of the Velocity:** There is no relation between the current and future speeds of MNs.
- **Temporal Dependency of the Direction:** There exist also no dependencies between the current and future directions of MNs.
- **Spatial Dependency of the Velocity:** The speed of any MN does not relate to the speeds of other MNs in the scenario.
- **Spatial Dependency of the Direction:** There is no relation between the location of any MN and the locations of other MNs moving around.
- **Geographic Restrictions of Movements:** The RWp model does not enable researchers to define obstacles, streets, etc.

Figure 3. The MNs links breakage rate as a function of the speed and the pause time while operating the RWp mobility model (Camp, Boleng & Davies, 2002)

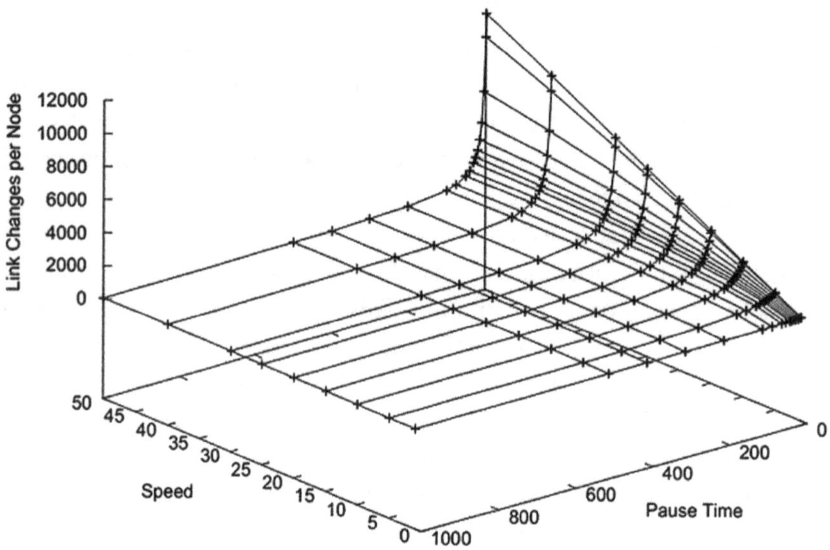

Figure 4. The density of the RWp mobility model (Aschenbruck, Gerhards-Padilla & Martini, 2008)

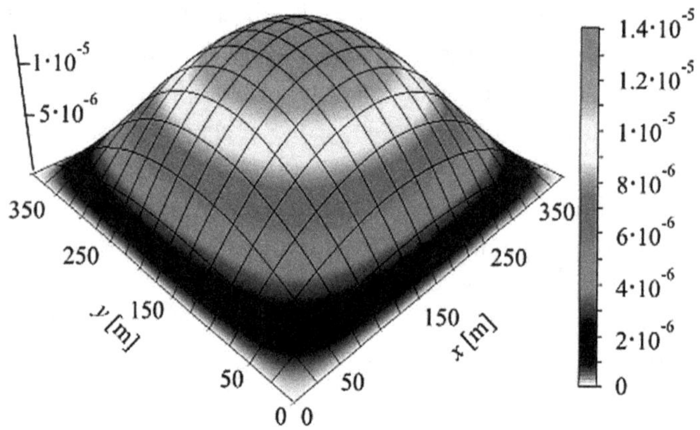

Variations of the RWp Mobility Model: To improve the RWp mobility model, the following variations have been proposed (Boudec & Vojnovic, 2005):

- **RWp on a General Connected Domain (RWp-GCD) Mobility Model:** the model extends the classical RWp model by relaxing the assumption that the simulation area (A) is a convex (rectangular, region, etc.). A is defined as a connected domain, over which a uniform distribution of waypoints is defined. In other words, the waypoints are first defined. Following that, MNs are distributed in the simulation area. Each MN selects a next direction (one of the waypoints) randomly and moves towards this waypoint using the shortest path. Figure 5 shows an example of the RWp-GCD applied on a non-convex domain (termed the City Section (CS) model and will be handled in detail later). The travel between the waypoints simulates movements on streets. Arbitrary numeric speeds are allowed to be assigned to the waypoints.

- **Restricted RWp Mobility Model:** this model also assumes that the simulation area is not a convex (Blazevic, Le Boudec & Giordano 2005). The simulation area contains sub-domains, within MNs move applying the classical RWp model for random selected number of visits. Following that, a new sub-domain is selected randomly and the speed on the trip towards the new sub-domain is chosen according to a distribution that may depend on the characteristics of the origin and destination sub-domains. Note that this type of mobility models is adequate to model large ad hoc networks, or to model the mobility between different administrative mobile communication networks. Figure 6 shows an example mobility pattern generated by the restricted RWp model.
- **RWp on Sphere Mobility Model:** this model is a special case of the RWp-GCD model, where the simulation area is a unit of a sphere. This model is especially adequate to generate mobility patterns for airplanes, where the waypoints represent airports and MNs stand for airplanes. MNs start from a certain waypoint, pick a new destination, speed and pause time randomly and travel on the shortest path towards the destination. The shortest path is the shortest arc on the great circle that includes the start as well as destination waypoint. Figure 7 provides an example mobility pattern generated by the RWp on sphere model.
- **RWp with Wrapping Mobility Model:** this model is also termed the RWp on torus or the RW with wrapping. The main feature of this model is the wrapping when a MN hits a boundary of the simulation area. For instance, when a MN reaches a simulation area boundary at (x,y) it is wrapped to the other side (either $(x,0)$ or $(0,x)$) and continues traveling from there, see Figure 8.
- **RWp with Reflection Mobility Model:** this model is similar to the RWp with wrapping mobility model, however, with a billiard-like reflections rather than wrapping.
- **Random Direction (RD) Mobility Model:** this model (Royer, Melliar-Smith & Moser, 2001) has been developed to overcome the density waves effect observed with the RWp mobility model. Each MN selects a random direction and speed. However, different from the RWp model, the MN travels towards the simulation area border in the selected direction. When the MN reaches the border, it pauses for a random selected pause time. Following that, the MN selects another direction and repeats the process again, and so on. An example movement pattern employing the RD mobility model is shown in Figure 9.
- **RWp with Attraction Points (RWp-APs) Mobility Model:** The model (Bettstetter & Wagner, 2002) extends the classical RWp model by introducing attraction areas and points to generate more realistic non-equally distributed mobility patterns. The probability of selecting an attraction point or a point in an attraction area is greater than the probability of selecting any other waypoint as a destination. It is observed in the RWp-APs model that MNs may visit some points more frequently than other points. Moreover, the density wave effect disappears in the model.
- **Clustered-Mobility (C-M) Mobility Model:** This model (Lim, Yu & Da, 2006) is motivated from the behavior of humans in disaster areas. It is similar to the RWp-APs, since it introduces attraction points and areas. However, the attraction of a point depends on the number of MNs locating nearby.
- **Weighted Waypoint (WWp) Mobility Model:** this model (Hsu, Merchant, Shu, Hsu & Helmy, 2004) is a variation of the RWp model, where the destination is not selected randomly. This model is adequate to model movement patterns for pedestrians on a campus, drivers in a city, etc.

Locations are defined, first, as waypoints and assigned different weights related to the probability of choosing them. Of course, the higher the probability, the greater the weight and the more popular the location. The weight of selecting each destination is time-dependent. Therefore, time-variant Markov model is applied to capture this time-dependent weight assignment. Moreover, the distribution of pause times at each location is a property of that location. The simulations done in (Hsu, Merchant, Shu, Hsu & Helmy, 2004) have shown that an uneven spatial distribution (clustering) arises when using the WWp model. That is, the density of MNs at the more popular locations will be more than at other locations. Furthermore, MNs in the WWp model are less mobile than in the classic RWp model.

Figure 5. An example scenario generated by the RWp-GCD mobility model on a non-convex domain (the city section model)

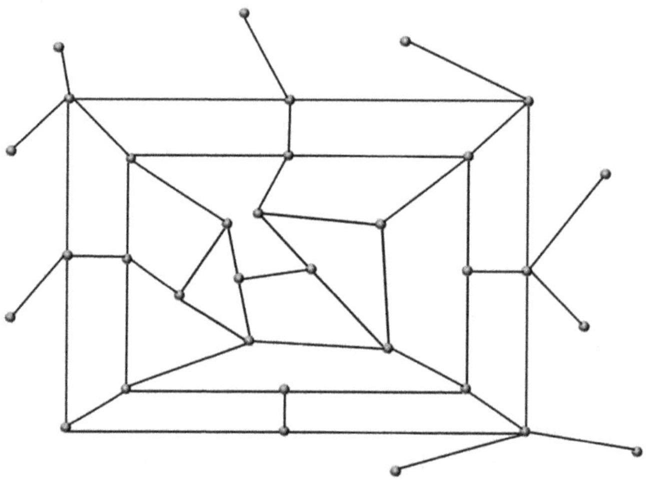

Figure 6. An example mobility pattern generated by the restricted RWp mobility model

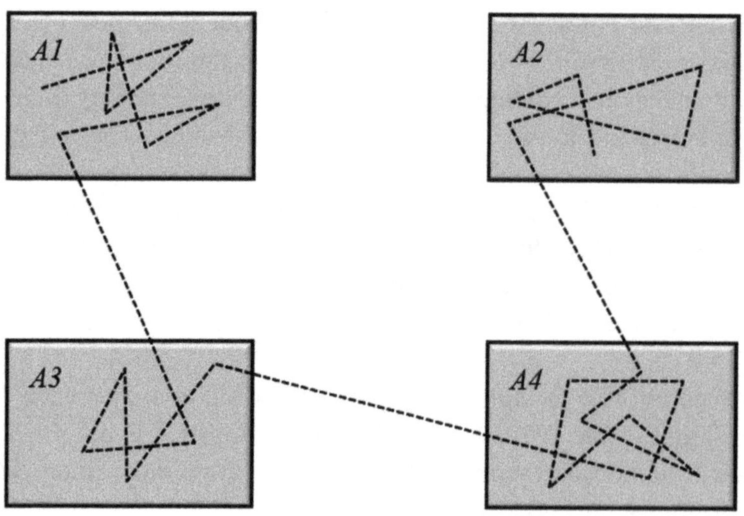

Figure 7. An example mobility pattern generated by the RWp on sphere mobility model

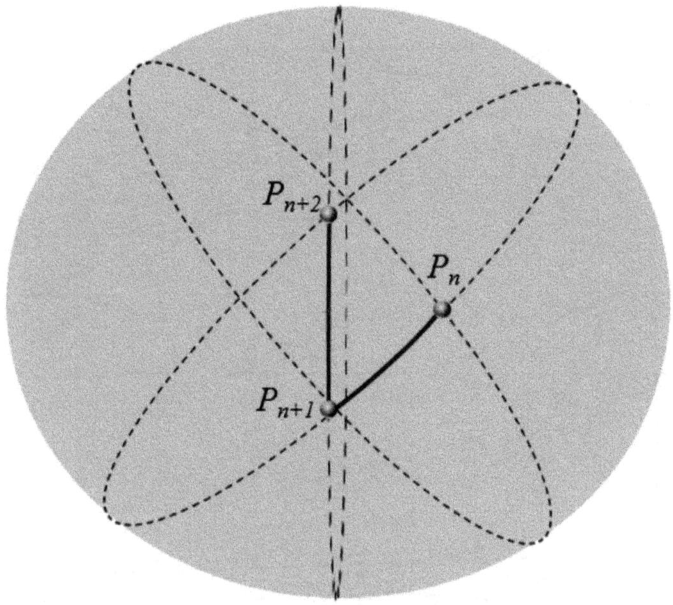

Figure 8. An example mobility pattern generated by the RWp with wrapping mobility model

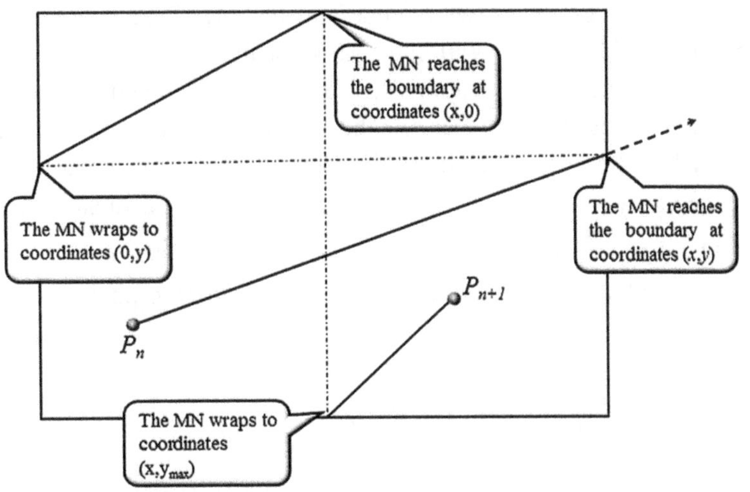

2.3. Probabilistic Mobility Model

The probabilistic mobility model (Diab, 2010) is similar to the WWp model discussed above. It is adequate for generating movement patterns for MNs moving within the coverage area of cellular communication networks. This model relies on the statement saying: MNs' movement patterns are, in fact, far from being random. A high degree of temporal as well as spatial regularity is observed by these patterns. So, they follow simple reproducible patterns [2]. In reality, MNs restrict their movements from a specific location to a location in their geographical neighborhood.

Figure 9. An example movement pattern generated by the RD mobility model (Camp, Boleng & Davies, 2002)[3]

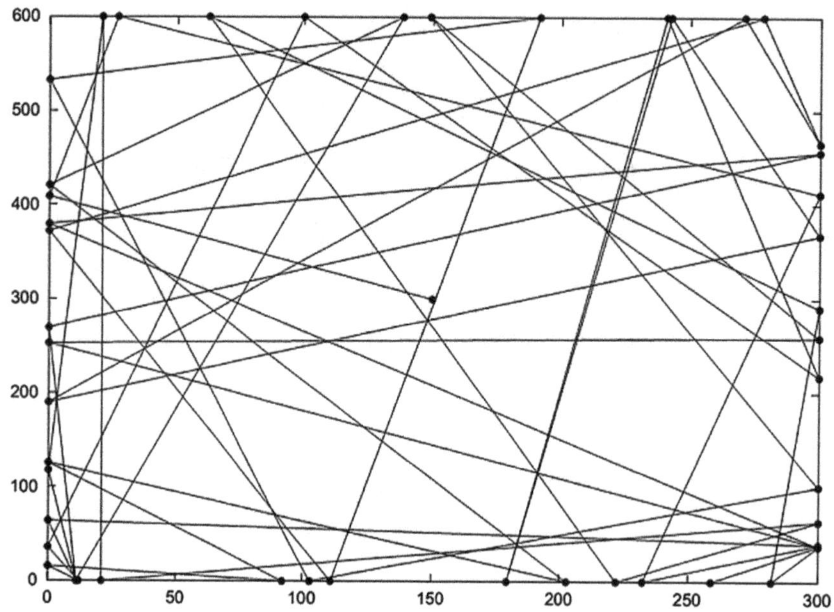

Description: The probabilistic mobility model defines the term "movement" as the transition from a location to a neighbor one. Movements within the same location are not of interest from the proposed probabilistic model point of view. The locations (Z) are first determined. These locations express, in principle, the locations of the base stations serving the cellular cells of the selected mobile communication network. The model assumes two known probabilities, the first is the probability that a MN locates initially in a specific location (q_i , where $i = 1 \rightarrow Z$). The second are the transition probabilities between locations ($p_{i,j}$, where $i, j = 1 \rightarrow Z$). Note that $p_{i,i}$ is equal to zero, since the model does not consider movements inside the same location, i.e. the same cell. The transition probabilities are described by the matrix *P*.

$$P = \begin{bmatrix} P(1,1) & \cdots & P(1,Z) \\ \vdots & \ddots & \vdots \\ P(Z,1) & \cdots & P(Z,Z) \end{bmatrix} = \begin{bmatrix} 0 & \cdots & P(1,Z) \\ \vdots & \ddots & \vdots \\ P(Z,1) & \cdots & 0 \end{bmatrix}$$

So as to derive the probabilities P_i that a MN locates at the location *i* in the steady-state of the system, a start vector $Q^0 = [q_1 \ \ldots \ q_Z]$ is defined. The vector *Q* is calculated, following that, using the recursive equation $Q^{n+1} = Q^n * P$ as follows:

$$Q = \lim_{n \to \infty} Q^n = \left[P_1 \ \ldots \ P_Z\right] \tag{1}$$

Discussion: The model is simple from a mathematical point of view. Furthermore, it significantly helps in analyzing the performance of mobile communication networks, since it provides the probabilities of residing at the locations defined in the steady-state of the system. This in turn simplifies the mathematical analysis of mobile communication networks aspects studied. The transition probabilities defined are not time-depended in the model. They can, however, be extended to adapt to changes of day times. Similar to the WWp model, the density of MNs at the locations with higher residing probabilities will be higher than at other locations having less residing probabilities. That the movements inside the same location are not considered is a drawback of the model. Furthermore, no pause times are defined at the locations. Note that the speed of the MN is not of interest for the model, since the important is where the MN moves, not with which speed. Therefore, this model is not adequate for ad hoc networks.

Limitations: The probabilistic mobility model is incapable of expressing the following:

- *Temporal and spatial dependency of the velocity*: in general, the speed is not a part of the model.
- *Spatial dependency of the location*: it is obvious from the description of the model that that the location of any given MN does not relate to the location of others.
- *Geographic restrictions of movements.*
- *Pause times restrictions.*

2.4. Markovian Random Walk (MRW) Mobility Model

The MRW mobility model (Camp, Boleng & Davies, 2002), (Chiang, 1998) aims at being more realistic than other mobility models. It mimics the following statement: people tend to continue moving in a semi-constant forward direction. They rarely turn suddenly back and retrace their steps. Moreover, they never move randomly hoping that they wind up somewhere toward their destinations.

Description: The MRW model introduces a probabilities matrix to determine the next position (x, y) of a MN at the next time step. Three states are used for the coordinate x and three for the coordinate y, see Figure 10 which shows the flow chart of the MRW mobility model. Note that the state 0 represents the current location (for x or y), the state 1 stands for the previous location of the MN and the state 2 expresses the next MN's location if the MN continues moving in the same direction.

The probabilities matrix used looks as follows, where $p(a,b)$ is the probability that the MN moves from the state (a) to state (b).

$$P = \begin{bmatrix} p(0,0) & p(0,1) & p(0,2) \\ p(1,0) & p(1,1) & p(1,2) \\ p(2,0) & p(2,1) & p(2,2) \end{bmatrix}$$

MNs are initially distributed within the simulation area. After that, each MN determines its new coordinates at the next time step based on the probabilities matrix we introduced. The MN's speed is fixed during the simulation. Note also that there is no pause times between movements. Figure 11 depicts an example movement pattern of a MN applying the MRW mobility model.

Discussion: The principles, the MRW model follows, produce probabilistic rather than purely random movement patterns. This may produce more realistic scenarios than those generated by other mobility

Figure 10. Flow chart of the MRW mobility model

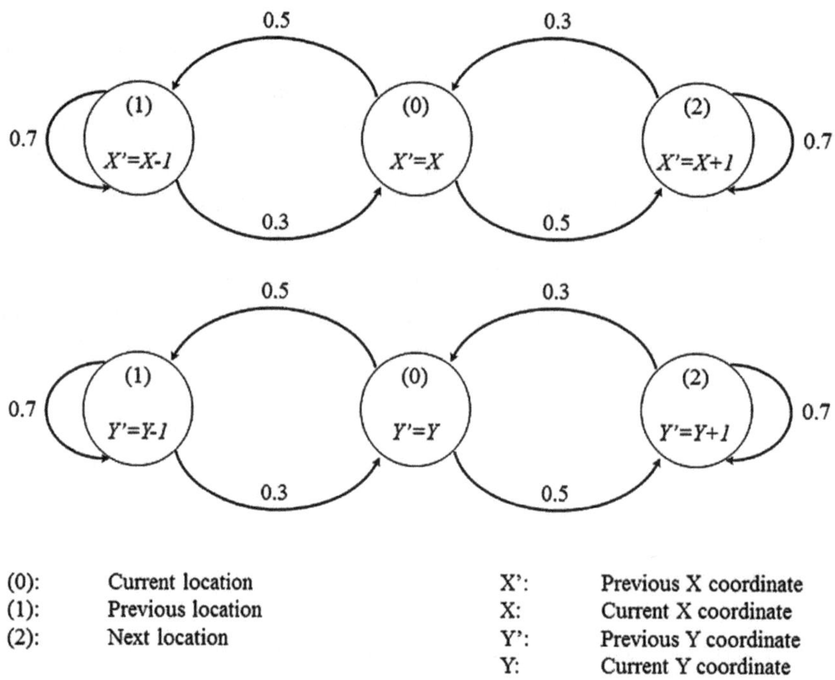

Figure 11. An example movement pattern generated by the MRW mobility model (Camp, Boleng & Davies, 2002)[4]

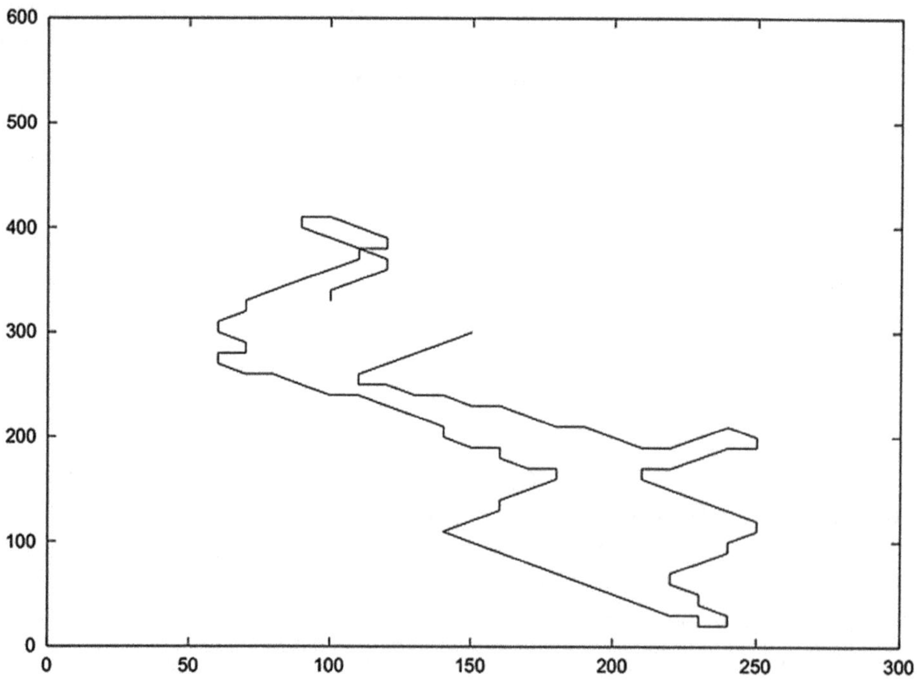

models. In fact, an accurate definition of the probabilities matrix heavily influences the mobility patterns produced. Therefore, the definition of this matrix is done normally based on traces from given real mobility scenarios. The figure above shows that MNs move in straight lines for time periods. The figure does not show high variations in the direction as those produced by the RW mobility model.

Limitations: the discussion provided above shows that the MRW model cannot express the following:

- **Temporal and Spatial Dependency of the Velocity:** The model assumes, in fact, a constant velocity.
- **Spatial Dependency of the Direction:** There are no dependencies between the location of the MN and the locations of other MNs in the scenario.
- **Geographic Restrictions of Movements:** There exist no mechanisms to determine obstacles, streets, certain places with specific characteristics, etc.
- **Pause Times Restrictions:** The model also does not assume pause times between movements.

Concerning the temporal dependency of the direction, one notes the existence of such dependency between future and current locations, which is a result of to the probabilistic nature of the model.

2.5. Gauss-Markov (G-M) Mobility Model

The G-M mobility model (Liang & Haas, 1999), (Tolety, 1999) agrees with the RW mobility model concerning the extremely unpredictable paths MNs move on. However, it does not agree with the way the RW model handles MNs directions and speeds, since the directions as well as speeds will, in general, not change totally randomly. So, there should be a relation between current and past locations as well as speeds.

Description: Same as other mobility models, MNs will have an initial distribution within the simulation area. During this step, each MN is assigned a speed and a direction. Following that, MNs start moving to their defined directions at their defined speeds. After a constant time interval (Δt), movements occur by updating the direction and speed of each MN. More concrete, the direction as well as the speed of a MN at the nth movement is calculated based on the direction and the speed of the MN at the $\left(n-1\right)^{th}$ movement by applying the following equations, where S_n and d_n are the speed and the direction at the nth movement, α is a randomness tuning parameter ($0 < \alpha < 1$), $\overline{S}$ and $\overline{d}$ are the mean values of the speed and the direction when $n \to \infty$, $S_{x_{n-1}}$ and $d_{x_{n-1}}$ are random variables from Gaussian distribution. Note that setting the parameter α to 0 results in the Brownian motion, while setting α to 1 results in a linear motion.

$$S_n = \alpha * S_{n-1} + \left(1-\alpha\right)\overline{S} + \sqrt{\left(1-\alpha^2\right)} * S_{x_{n-1}} \tag{2}$$

$$d_n = \alpha * d_{n-1} + \left(1-\alpha\right)\overline{d} + \sqrt{\left(1-\alpha^2\right)} * d_{x_{n-1}} \tag{3}$$

The location of a MN at the nth movement is given by the following equations, where x_n and y_n are the coordinates at the nth movement, x_{n-1} and y_{n-1} are the coordinates at the $\left(n-1\right)^{th}$ movement.

$$x_n = x_{n-1} + S_{n-1} * \cos\left(d_{n-1}\right) \tag{4}$$

$$y_n = y_{n-1} + S_{n-1} * \sin(d_{n-1}) \tag{5}$$

To ensure that the MN will move away from the simulation area borders, the mean direction variable ($\bar{d}$) is modified when the MN moves within a certain distance of a border, see Figure 12 which provides the mean direction variable ($\bar{d}$) for different locations within the simulation area, and Figure 13 which illustrates an example movement pattern generated using the G-M mobility model.

Discussion: Directions and speeds of MNs when employing the G-M mobility model are not totally random. Moreover, they depend on past directions as well as speeds. In other words, the G-M mobility model introduces a memory to remember last values of speeds and directions. This in turn eliminates the sudden stops and sharp turns. However, directions and also speeds remain unpredictable. So, no pre-defined streets, freeways, etc. with specific mobility characteristics can be defined applying the G-M model. Furthermore, regions where MNs move totally randomly, e.g. within gardens for instance, are not possible as well.

Limitations: Based on the discussion above, following constraints are observed in the G-M model:

- **Spatial Dependency of the Velocity and Direction:** The G-M model does not consider relations between the speed and the direction of a given MN with the speeds and the directions of other MNs.

Figure 12. The mean direction variable ($\bar{d}$ *) for different locations within the simulation area*

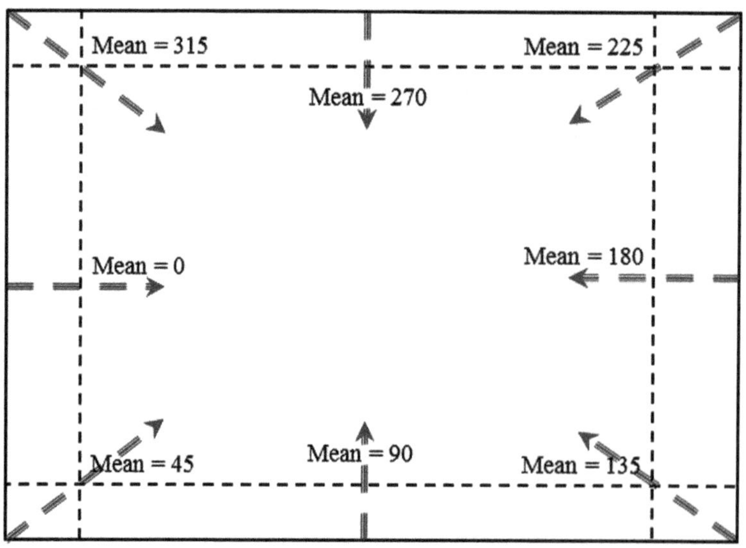

Figure 13. An example movement pattern generated by the G-M mobility model (Camp, Boleng & Davies, 2002)[5]

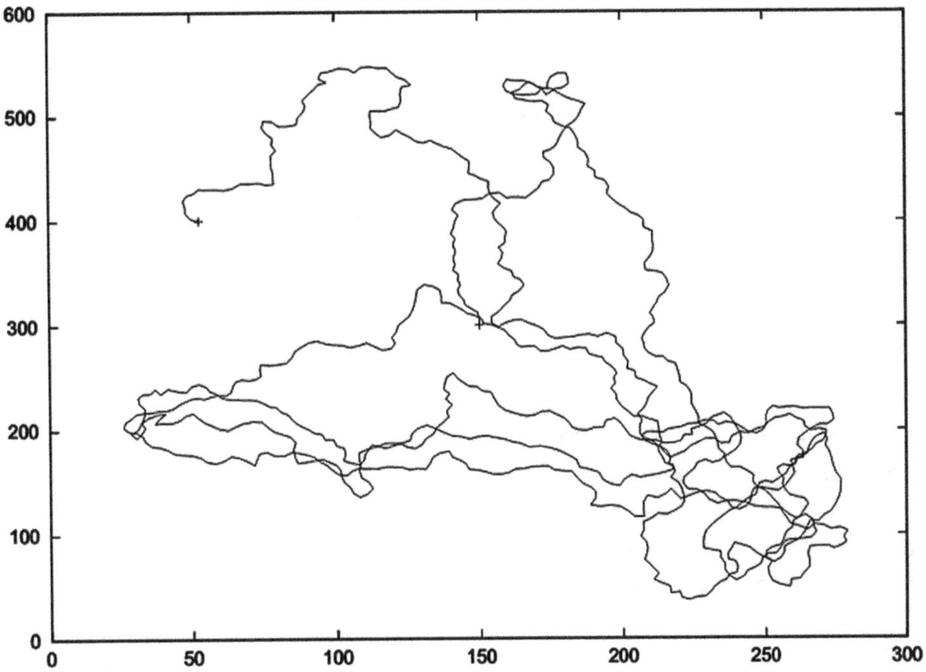

- **Geographic Restrictions of Movements:** As written above, no pre-defined streets with specific mobility characteristics, freeways, etc. are possible.
- **Pause Times Restrictions:** The G-M model does not include pause times between movements.

In terms of the temporal dependency of the velocity and direction, one notes the existence of such dependency since future speeds and directions are calculated based on previous ones.

2.6. A Boundless Simulation Area (BSA) Mobility Model

The BSA mobility model aims at generating mobility patterns that are more realistic than those produced by other mobility models already mentioned. It relies on the fact that neither the locations nor the speeds of MNs are random. Future velocities and locations relate somehow to current ones (Haas, 1997).

Description: MNs are initially distributed in the simulation area. Initial directions and velocities are selected randomly, same as by the RW mobility model. Each MN defines, following that, a velocity vector $V = (v, q)$, where v is the current MN's speed and q is its current location (represented by (x, y) coordinates). The speed vector is updated after time steps using the equations written below.

$$v(t + \Delta t) = min\left[\max\left(v(t) + \Delta v, 0\right), V_{max}\right] \quad (6)$$

$$\theta(t + \Delta t) = \theta + \Delta\theta \quad (7)$$

$$x(t+\Delta t) = x(t) + v(t) * \cos\theta(t) \tag{8}$$

$$y(t+\Delta t) = y(t) + v(t) * \sin\theta(t) \tag{9}$$

V_{max} is the maximum velocity defined, while Δv is the change in the MN's velocity. Δv is uniformly distributed between $-A_{max} * \Delta t$ and $A_{max} * \Delta t$, here A_{max} is the maximum acceleration of a specific MN. $\Delta\theta$ is the change in the MN's direction and is also uniformly distributed between $-\alpha * \Delta t$ and $\alpha * \Delta t$, where α is the maximum angular change in the direction of a specific MN.

Different from other mobility models, the MN that reaches an area boundary continues moving and appears on the opposite side of the simulation area. This creates a torus-shaped simulation area, see Figure 14 which illustrates this concept. An example mobility pattern generated using the BSA mobility model is also provided in Figure 15.

Discussion: The BSA calculates future velocities and locations based on previous ones. So, the model is not a memory-less model, which prohibits sudden stops and sharp turns. However, speeds and also locations are not predictable. This makes it impossible to pre-define streets or regions with specific mobility characteristics as those we see in cities for example.

Limitations: Depending on the discussion above, following constraints are observed in the BSA model:

- *Spatial Dependency of The Velocity and the Direction*: The locations of MNs are not affected by each other. The same is true for the velocity.
- *Geographic Restrictions of Movements*: We explained that the definition of streets with specific characteristics, freeways, etc. is not a part of the BSA model.
- *Pause Times Restrictions*: Pause times between movements are also not a part of the BSA model.

Because future speeds and directions are calculated based on previous ones, there exists a temporal dependency of the velocity and the direction.

Figure 14. The torus-shaped simulation area

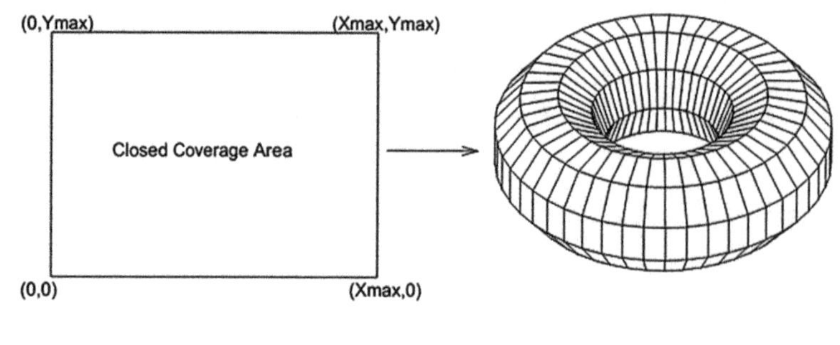

Figure 15. An example movement pattern generated by the BSA mobility model (Camp, Boleng & Davies, 2002)[6]

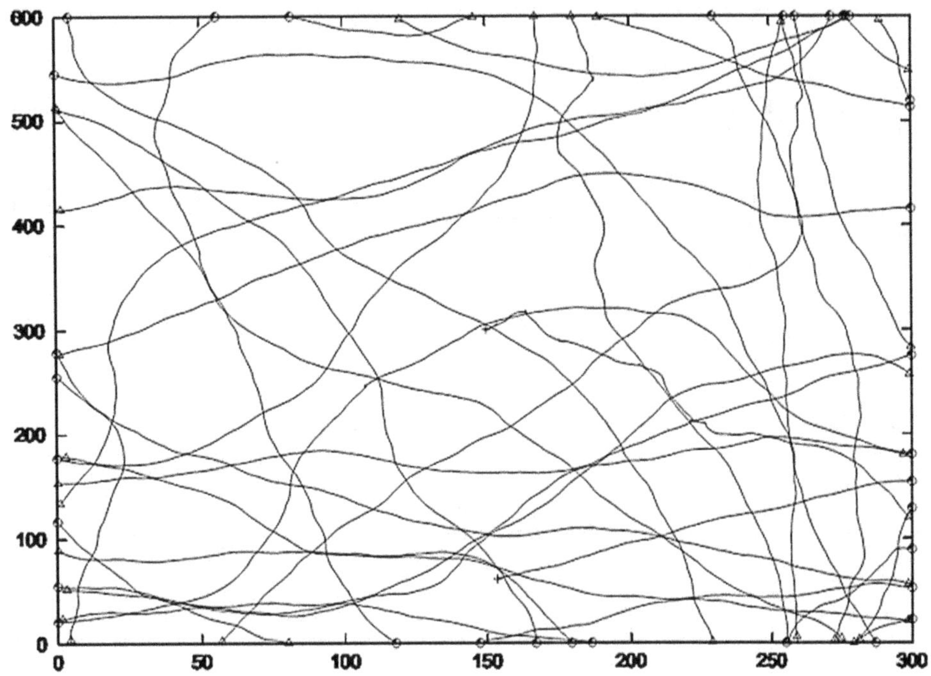

2.7. City Section (CS) Mobility Model

The CS mobility model mimics the mobility of MNs inside cities. The model relies on the fact that users (pedestrians, autos, etc.) move inside cities on streets that also have specific mobility characteristics including speeds limits (Davies, 2000). Users are not always allowed to move randomly in real-life scenarios.

Description: Researchers define first the part of the city, where they want to simulate MNs movements inside. Concrete, they define the streets and the speeds limits on each street. MNs are then distributed on the streets defined. Each MN selects randomly a new destination which will be a point on a certain street. The movement algorithm determines then the shortest path between the current and new destination. Sure, the MN is allowed to move only on the streets. Note that the characteristics of the streets are considered while moving. Upon reaching the destination, the MN repeats the process again after a pause time and so on. Figure 16 presents an example movement patterns generated by the CS mobility model.

Discussion: The CS mobility model provides realistic movement patterns inside cities for nomadic as well as vehicles by restricting the MNs traveling behavior (all MNs must follow pre-defined streets with pre-defined characteristics). Obstacles are implicitly included in the CS mobility model since MNs are allowed to move only on pre-defined streets. This implies that all parts of the simulation area not containing streets are in general obstacles. This, however, does not absolutely match the reality, since there are parts of cities where MNs are allowed to move freely, e.g. inside gardens, houses, etc. In addition, there are not only speed limits on the streets. There may be accelerations and/or decelerations as well. Such behaviors are not possible when applying the CS mobility model.

Figure 16. The An example movement pattern generated by the CS mobility model (Camp, Boleng & Davies, 2002)[7]

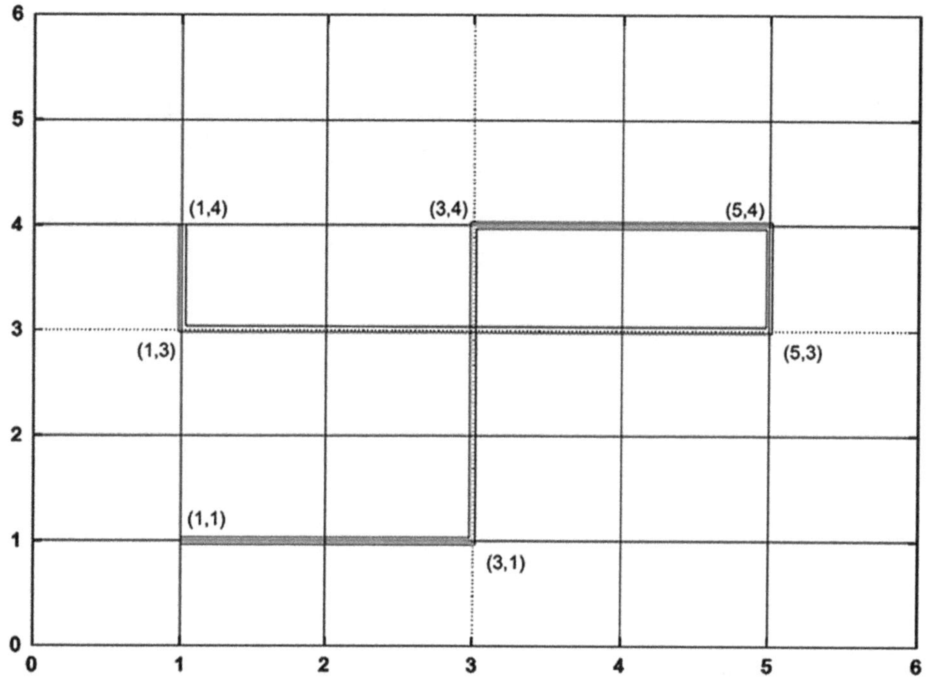

Limitations: Based on the discussion induced above, following constraints are noted in the CS model:

- **Temporal Dependency of the Velocity and the Direction:** neither the future location nor the future velocity of any MN depends on previous values. So, a temporal dependency does not exist.
- **Spatial Dependency of the Velocity and the Direction:** locations and velocities of MNs are not influenced by the locations and velocities of any specific/neighbor MNs.

There exist geographic restrictions of movements, since movements are restricted to streets. There exist pause times between movements as well.

Variations of the CS mobility model: so as to provide more realistic scenarios inside cities, some variations of the CS mobility model have been proposed. These proposals include introducing extra pause times at certain intersections, destinations, etc., accelerations and/or decelerations on the streets, higher/lower density of MNs based on the time of day, etc.

2.8. Behavioral Mobility Model

Different from other mobility models such as the RW, RWp, RD, etc., which describe the mobility behavior of MNs capable of communicating using high radio ranges (from meters to a multiple of kilometers), the behavioral mobility model is intended to model the mobility behavior of MNs at small scales (Legendre, Borrel, Dias de Amorim & Fdida, 2006), e.g. ad hoc networks that use the infrared technology, humans who communicate directly with each others, etc., using Newton's law of forces. The mentioned model decomposes mobility into simple atomic individual behaviors in both space and time (termed microscopic mobility as well) and, following that, combines the behaviors observed to reproduce the overall mobility pattern. The model is adequate to generate mobility patterns for a wide range of applications, e.g. escape panic scenarios (where humans escape in case of emergency), wireless sensor network, ad hoc networks for vehicles, delay-tolerant networks, etc.

Description: the objectives of the behavioral mobility model are the

- Definition of behavior rules governing the mobility of a given physical or biological instance/phenomena, where interactions significantly affect the mobility pattern resulting and
- Precise modeling of individuals whose nature is known.

The behavioral rules describe the expectations of how individuals will react on and how are they influenced by the environment, e.g. their destinations, nearby peers and obstacles, etc. Example behavior rules may look like: select the shortest path towards the destination, avoid bumping into an obstacle by a reflex action when being too closed to the obstacle, stop at a red light, etc. The application of a behavior rule on a specific individual results in what we call "atomic behavior/motion". The combination of these atomic behaviors/motions results in the overall mobility pattern.

Rules in the behavioral mobility model are represented by attractive or repulsive forces, e.g. a MN is attracted by its destination and repulsed by an obstacle. The sum of all forces affected a MN results in an acceleration vector a of the MN. This is the way how the behavioral mobility model applies the Newton's law of forces. The modeling of individuals' behaviors via rules enables flexibility in building scenarios adequate for a wide range of applications, especially those applied to analyze emergency situations. MNs in the scenarios are autonomous in the sense that each MN is allowed to have its own behavior rules. Moreover, each MN/set of MNs is allowed to change in a dynamic way own behavior rules independently or even cooperatively depending on the context they are placed in. Note that the surrounding environment is made up with obstacles.

Now, to precisely define the characteristics of MNs, each MN is geometrically represented by a circle of a radius r, a velocity v_0, a maximum velocity v_{max} and a maximum acceleration a_{max}. The mentioned metrics may be picked up from empirical studies for more accurate modeling, e.g. a normal distribution N to model walking speed of pedestrians (Ranjan Roy, 2011), constant speed for pedestrians (McNett & Voelker, 2005), etc.

Each behavioral rule affecting a specific MN generates an independent acceleration vector (Helbing, Farkas & Vicsek, 2000), (Helbing, 2001) and (Krbalek & Helbing, 2004). All acceleration vectors affecting a specific MN are accumulated to form the final acceleration vector a (also termed acceleration request, see (Legendre, Borrel, Dias de Amorim & Fdida, 2006)), whose direction is followed by the MN. After the calculation of the vector a, the speed of the MN is updated using the following equation, where v_t is the MN's speed at time t and Δt is the simulation time step.

$$v_{t+\Delta t} = v_t + a * \Delta t \tag{10}$$

For clarity and due to the importance of the behavioral mobility model, we will discuss two rules for pedestrians, namely the path following and the wall avoidance rule, see (Ranjan Roy, 2011). The path following rule defines a straight path from the start point to a specific destination. Let us assume that i is the index of the MN under consideration, $S_i(t)$ is the MN's position at time t and $d_i(t)$ are the coordinates of the destination. The path following acceleration vector $a_{pf}\left(t\right)$ at time t is calculated using the equation written below, where β is a normalization scalar of the dimension (1 s) and $v_i\left(t\right)$ is the speed of MN_i at time t.

$$a_{pf}\left(t\right) = \beta[v_0\left\{\frac{d_i\left(t\right) - S_i\left(t\right)}{\left|d_i\left(t\right) - S_i\left(t\right)\right|}\right\} - v_i\left(t\right)] \quad (11)$$

The wall avoidance rule enables pedestrians to avoide coming up against walls. Of course, this rule results in a repulsive force pushing the perestrian away from the wall. The magnitude of the force resulting from this rule is inversely proportional to the square distance that separates both the pedestrian and the wall. Moreover, the force magnitude is bounded by a given distance (1 m for instance), above which the magnitude is set to 0. So, the resulting wall avoidance acceleration vector $a_{wa}\left(t\right)$ at time t is calculated using the following equation, where γ is a normalization scalar of the dimension (1 s^2) and $W_i\left(t\right)$ is the closest point on the wall from MN_i at time t.

$$a_{wa}\left(t\right) = \gamma\left(S_i\left(t\right) - W_i\left(t\right)\right) \quad (12)$$

Figure 17 illustrates an example MN moving while applying the behavioral mobility model. Note that the MN implements the two discussed behavior rules.

Figure 17. An example movement pattern generated by the behavioral mobility model

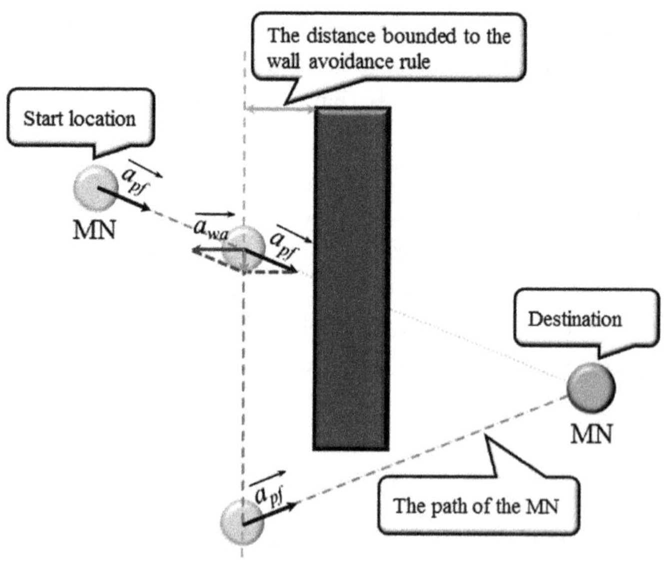

Discussion: The behavioral mobility model is complex from the mathematical point of view and also not simple to implement. However, it is capable of generating more realistic mobility patterns than other studied mobility models. The key issue is, in fact, the definition of the behavior rules, which control the overall mobility pattern. The relaying on rules to control the movements of MNs makes the model very flexible and enables modeling scenarios adequate for a wide range of applications. A major feature of the model is the decomposition of mobility into simple atomic individual behaviors in both space and time domains. The overall mobility pattern results then from the execution of all behavior rules, which in practice work in parallel. This way of work is adequate for self-organizing systems, which are distributed and depend on interactions in the microscopic level between the components of the systems. The overall behavior of a self-organizing system emerges then from the microscopic behaviors, see (Holzer, 2009). The behavior of the model described and also the use of self-organization principles make it possible to obtain overall patterns that are not interpretable via the applied microscopic behaviors. As known, this is a feature of self-organizing systems and called emergency, see (Holzer, 2009).

Limitations: In principle, all temporal and spatial dependencies of velocities and directions are possible. Geometric restrictions of movements and also an integration of pause times are possible as well. In fact, the question is the behavior rules defined, which can be very complicated.

2.9. Knowledge-Driven (KD) Mobility Model

The KD mobility model (Dobson, et al., 2006), (Handorean, Gill & Roman, 2004), (Sen, Handorean, Roman & Hackmann, 2004) follows principles somewhat different from those we have already seen. It is a new trend in mobility modeling that depends on exploiting the knowledge of movement patterns and allows for a total freedom of MNs mobility. In fact, the KD mobility model is capable of accommodating any arbitrary mobility model including synthetic, empirical and also community-based models. From the research point of view, the utilization of the knowledge about mobility patterns to adapt the network to environment changes is considered as a part of recent research trends of autonomic solutions[8].

The basic idea is to exchange information among MNs about their intended movements and also services. So, the KD model deals with the mobility from two points of view, namely the mobility of users and services, see (Ranjan Roy, 2011). Therefore, this model is expected to contribute to the optimization of the performance of service placement mechanisms (Khan, Diab, Brosch & Ahmad, 2012). The following provides more insight into the model.

Description: the KD mobility model assumes a network of a finite set of MNs/hosts (H), each operates a finite set of agents. Each agent is, in principle, a piece of software implemented to offer some functionalities. Note that agents are capable of moving (Sen, et al., 2004). That is, they may move from a MN/host to another. Note that an important feature of the network we consider is the existence of a global clock synchronization.

The global knowledge (K) in the network is written formally as follows, where $K(h)$ denotes the knowledge available at the MN/host h.

$$K = \prod_{h} K\left(h\right) \tag{13}$$

The model assumes that a MN/host h is characterized by a motion profile $\mu\left(h\right)$ and has, as mentioned, knowledge $K(h)$. The motion profile describes the MN/host movement pattern, mostly as a function of the time. Formally, one can write the following, where T is the time domain and L is the set of locations the MN/host h will visit.

$$\mu : H \rightarrow (T \rightarrow L) \tag{14}$$

This simply means that this function takes a time as an input, and delivers the location at that time as an output. The function included in the motion profile determines the mobility model the MN/host h follows. See the following simple examples for clarity. Note that $\mu(h,t)$ expresses the location of a MN/host h at time t. $\mu(h,t)$

$\forall t : \mu(h,t) = l$, where $(l \in L)$. This rule results in a stationary MN/host. (15)

$\mu(h,t) = \mu(h,t) + (t - t_0) * v_h$, where v_h is the velocity of the MN/host. This rule indicates that the MN/host moves in a straight line. (16)

It should be clear now, why the KD mobility model is capable of accommodating any other mobility model including synthetic, empirical and also community-based models. This depends on the motion profile that researchers, in principle, are allowed to construct as they prefer.

So as to obtain a global knowledge of the network concerning motion profiles, these profiles are exchanged between MNs/hosts using a gossiping protocol (Chandra, Ramasubramanian & Birman, 2001), which results in an epidemic flooding of motion profiles in the network.

Concerning the services, the KD mobility model characterizes each service by a unique identifier ($s \in S$, where S is the set of available services), service capabilities ($\aleph(s)$), service performance attributes ($\varphi(s)$), service external dependencies ($\partial(s)$) and a service allocation profile ($\beth(s)$) The term $\aleph(s)$ describes the functionalities of the service, e.g. printer, fax, search engine, etc. The term $\varphi(s)$ reflects the attributes of the service being operated, e.g. for a printer service, the attributes can be the resolution, number of copies, colors system, etc. The external dependencies ($\partial(s)$) include other services necessary to operate the service. The term $\beth(s)$ includes the location of the service instance as a function of the time. This means, at which MN/host the service will reside at a given time. So, one is allowed formally to write:

$$\beth : S \rightarrow (T \rightarrow H) \tag{17}$$

Based on the discussion induced above, a motion profile $\mu(s)$ is defined for each service instance. So, one formally can write the following, where $\beth(s,t)$ returns the MN/host, on which the service instance resides at time t.

$$\mu(s,t) = \mu(\beth(s,t),t) \tag{18}$$

However, unlike the motion profiles of MNs/hosts, motion profiles of services cannot be defined arbitrary. Rather, they are derived from services allocation profiles. Keep in mind that both services motion and allocation profiles are constructed based on the service placement algorithm being operated, e.g. REplication in Dense MANet (REDMAN) (Bellavista, Corradi & Magistretti, 2005), Self-Organizing Network Density (SONDe) (Gramoli, Kermarrec & Le Merrer, 2008), COCOA (Coppens, Wauters, De Turck, Dhoedt & Demeester, 2005), etc. For more information about service placement mechanisms, the readers are directed to (Ali, Mitschele-Thiel, Diab & Rasheed, 2010). Figure 18 illustrates an example network containing MNs and servers hosting two types of services (s_1 and s_2). The MNs are the clients that request the services. Note that a new instance of the service s_1 is created to serve a part of the clients, thus, to enhance the performance.

Note that the movement of each service instance $s \in S$ is, in principle, composed of two components. It has its own motion, derived from its motion profile, in addition to the motion of the MN/host, on which the service instance resides at the moment. Again, services motion profiles are exchanged between MNs/hosts using a gossiping protocol.

Now, let us discuss the provision of services to clients. As known, each client requests the service it needs. The KD model characterizes each request by the client that initiated the request, the capabilities required from the service, the performance level desired, the start time, at which the client aims at utilizing the service and the ending time, after which the service is no more desired. Performance metrics in this context include metrics like service reachability, availability, extensibility, etc. Remember that the KD model exploits the knowledge gathered from each MN/host and service instance. Thus, a pre-planning and proactively management tasks for the purpose of performance optimization are possible,

Figure 18. An example network with MNs and multiple services, an instance of the service s_1 moves to another host

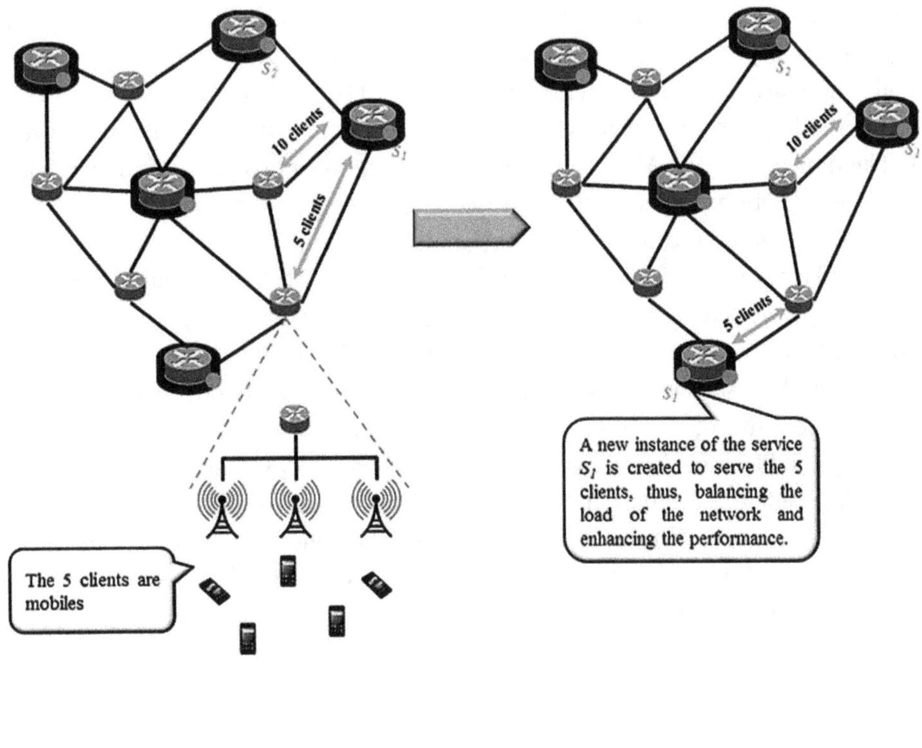

see (Lopez-Nores, García-Duque, Pazos-Aria, & Díaz Redondo, 2008). Moreover, the selection of the service provider best suited to the client gets simpler due to the existence of the global knowledge.

Discussion: As mentioned above, the KD mobility model exploits available knowledge of the mobility of MNs/hosts and service instances. Thus, a pre-planning is possible, which is a great advantage. The KD mobility model is capable of accommodating any arbitrary mobility model, which is also a notable strength. Moreover, the deal with application-level services mobility in addition to the mobility of users is a main feature of the KD model that does not exist in another mobility model. This simplifies studies in a new research field, namely the service placement.

The KD model considers the existence of a global clock synchronization and accurate positioning capabilities. These make the KD mobility model technology-depended and only applicable when all MNs/hosts have such capabilities. A global accurate synchronization in dynamic changing networks is a major problem and hard to be properly realized. This is a weakness of the KD model and may prohibit a practical application. The exchange of local knowledge among network nodes to build the global knowledge is a problem as well, since such tasks will consume network resources, and are not always possible in networks like ad hoc networks for instance, especially when we consider merging and partitioning of these networks. Keep also in mind that researches tend currently to develop distributed systems and approaches without global knowledge.

Limitations: Because the KD mobility model does not define a specific mobility model, but accommodates an existing one by properly constructing the motion profiles of MNs/hosts and also service instances, it inherits the limitations of accommodated mobility models. However, this will not be done primitively, since some features and limitations of accommodated mobility models may disappear, others may emerge. This heavily depends on the relation between or even the behavior resulting from the mobility of service instances and MNs/hosts carrying these instances.

2.10. Other Synthetic Single Node Mobility Models

In addition to the described mobility models, there exist other models aiming at being more realistic and easily mathematically tractable. In this section, we will briefly go through some of these models.

The Smooth Random (SR) mobility model (Bettstetter, 2001), (Bettstetter, 2001a) generates mobility patterns without sudden stops and sharp turns, i.e. smoother movement patterns concerning direction and velocity. The model classifies MNs based on their maximum velocity, preferred velocity, maximum acceleration and deceleration. MNs calculate their new velocities and directions based on the parameters we already mentioned and on previous values of the direction and velocity.

The graph-based mobility model proposed in (Tian, Haehner, Becker, Stepanov & Rothermel, 2002) uses a graph, whose vertices are possible destinations and whose edges represent the paths MNs are allowed to move on. MNs move on the graph by operating the RWp model. They are initially distributed in the scenario. Each MN selects then one of the graph vertices as its destination and moves to this destination. Of course, MNs move on the allowed paths only.

An approach applied to model movements while considering obstacles includes the determination of movement paths inside the simulation area using Voronoi-diagrams. This approach is termed then a Voronoi-based mobility model (Zimmermann & Gruber, 2005) and is first introduced with the obstacle mobility model (Jardosh, Belding-Royer, Almeroth & Suri, 2003). The Voronoi-based mobility model uses the edges of buildings as an input to calculate the Voronoi-diagram. Thus, the graph constructed comprises the Voronoi-diagram and additional vertices that work as intersections for the edges of the

Voronoi-diagram as well as obstacles. In fact, these additional vertices represent the entrances to the obstacles present in the scenario. Movements on the constructed graph are modeled similarly to the graph-based mobility model. Note that the paths in the Voronoi-based mobility model are modeled equidistant from all obstacles. It is worth mentioning that there is an important weakness observed in the Voronoi-based mobility model, namely paths traveled on between start and destination waypoints are not optimal.

Another principle to generate more realistic mobility models is to divide the simulation area into various sub-areas, within arbitrary mobility models are used. Examples of models following this principle are the Area Graph-based (AGb) (Bittner, Raffel & Scholz, 2005) and CosMos (Guenes & Siekermann, 2005) mobility models. The AGb mobility model realizes clusters, i.e. sub-areas, with higher node densities in addition to paths in between. The clusters are considered vertices and the paths are edges in between. Each edge is weighted by a weight that represents the probability of selecting this path when moving outside the current cluster. Inside clusters, MNs operate the RWp mobility model for a randomly chosen time and move, following that, towards another cluster. To do this, one of the paths interconnecting the cluster with others is selected, based on the weights already mentioned. The CosMos model works in a similar manner. The simulation area is divided into non-overlapping zones, inside each an arbitrary mobility model applied. Note that mobility models inside different zones may vary from each other. Transitions between the zones are defined similarly to the AGb model.

There is another principle followed by some approaches to generate realistic mobility models, namely retrieving information about road maps. One of the well-known mobility models developed in the context of the UMTS standardization is the so-called Manhattan-Grid model (ETSI TR 101 112, 1998). The simulation area is divided into squared blocks, whose vertices express streets. The model is used to model the movement of pedestrians, who move on the defined streets. The pedestrians are distributed first in the Grid. Following that, each pedestrian randomly chooses a direction and a velocity. Once a pedestrian reaches a corner, he changes the direction with a certain probability and selects a new velocity. Another model that tries to retrieve information about roads, including speed information and crossroads, is the Random Waypoint City (RWpC) mobility model (Kraaier & Killat, 2005). This model is applied to generate movement patterns for vehicles in urban environments. MNs are thrown initially on the streets following a uniform distribution. Following that, each MN selects a new waypoint as a destination and moves towards its destination. Movements themselves are modeled similar to the RWp model, however, they are restricted to streets.

The freeway mobility model (Markoulidakis, Lyberopoulos, Tsirkas & Sykas, 1997) is another known mobility model applied to realize vehicle mobility on roads. MNs change their velocities in dependence to their current speeds. Moreover, the speed of any MN is affected by the speeds of other MNs moving on the same street inside a certain radius. The STreet RAndom Waypoint model (STRAW) (Choffnes & Bustamante, 2005) uses also information from maps, similar to the RWpC model. However, movements of vehicles are realized based on specific vehicular congestion and simplified traffic control mechanisms. Future velocities of vehicles relate to current ones and are affected by other vehicles, see (Choffnes & Bustamante, 2005) for more details.

A general mobility model capable of modeling complex scenarios is discussed in (Stepanov, Haehner, Becker, Tian & Rothermel, 2003). The model is called User-oriented Mobility Meta (UoMM) mobility model. First, the simulation area is modeled by defining restrictions inside the area in addition to attraction points. The movements of users are, in principle, sequences of attraction points. To be more realistic, the model realizes a temporal as well as spatial dependency of the velocity of users. In fact, the

UoMM is a generic mobility model that can be parameterized to realize other known mobility models. In (Stepanov, Marron & Rothermel, 2005), this model has been applied to generate the typical user movements during a day.

A model to represent movements in disaster areas, named disaster-area model, is provided in (Aschenbruck, et al, 2007). The model supports movements on optimal paths, includes obstacles and enables users to join and/or leave the scenario assumed. Group mobility can optionally be supported as well. The simulation area is divided into polygonal tactical sub-areas, which are further classified based on the civil-protection concept "separation of room", see (Aschenbruck, et al, 2007). Each Node joining the scenario is assigned to one of these tactical sub-areas. The disaster-area model allows different tactical sub-areas to have different speed restrictions, and enables the existence of two types of MNs, namely stationary and transport MNs. Stationary MNs move following the RW model inside their tactical sub-areas, while transport MNs move between sub-areas to carry patients.

So as to define the optimal paths between tactical sub-areas, robot motion planning algorithms are used. Visibility graphs are also utilized to find shortest paths and to avoid obstacles. The vertices of the visibility graph are the borders of the tactical sub-areas. Edges between these vertices represent the paths that must be followed. Note that the term visibility means that there exist no obstacles between any two vertices interconnecting with each other. Transport MNs typically leave the disaster area to transport patients to hospitals, safe places, etc. After a specific time duration, they come back to the disaster area to do their job again. This behavior is realized by introducing entry and exit points. From entry points, MNs join the scenario, while they go out when reaching an exit point.

2.11. Summary

Table 1 sums up the survey we introduced in this section, and provides a qualitative comparison between the reviewed models. Concerning the temporal dependency, one notes that only the G-M model, BSA model, behavioral model, KD model, SR model and UoMM model are capable of providing a temporal dependency of the velocity and the direction. Remaining mobility models either provide a temporal dependency of the velocity or the direction or provide no temporal dependency at all. The provision of a spatial dependency of the velocity as well as the direction was not a focus of synthetic mobility models. There are only three mobility models that considered such dependency, namely the RW-D model, behavioral model and KD model.

For the geographical restrictions, one clearly notices that streets, freeways, etc. are considered in mobility models developed especially to model movements in cities, e.g. the CS model, Graph-based model, freeway model, etc. The definitions of obstacles is seen in few models only, e.g. the obstacle model, Voronoi-based model, disaster-area model, etc.

3. SYNTHETIC GROUP MOBILITY MODELS

Different from the synthetic single node mobility models that study multiple MNs whose actions are completely independent of each other, synthetic group mobility models are intended to investigate the motions of MNs where the communication is done among teams to cooperatively achieve a common goal. In other words, movement of MNs depend on each other. Examples include movements of tourist groups, students groups, solders, etc.

Table 1. A qualitative comparison between the synthetic single node mobility models investigated

Model	Temporal dependency		Spatial dependency		Geographical restrictions		Pause time
	Velocity	Direction	Velocity	Direction	Paths, streets, etc.	Obstacles	
RW model	N	N	N	N	N	N	N
RWp model	N	N	N	N	N	N	Y
CV-RWd model	N[9]	N	N	N	N	N	N
DRW model	N	Y	N	N	N	N	N
PpRW model	N	Y	N	N	N	N	N
TRW model	N	Y	N	N	N	N	N
H-SRW	N	Y	N	N	N	N	N
T-SRW	N	Y	N	N	N	N	N
RW-D model	N	N	Y	Y	N	N	N
Restricted RWp model	For speeds between sub-domains	N	N	N	N	N	Y
RWp on Sphere model	N	N	N	N	N	N	Y
RWp with wrapping model	N	N	N	N	N	N	Y
RWp with reflection model	N	N	N	N	N	N	Y
RD model	N	N	N	N	N	N	Y
RWp-APs model	N	N[10]	N	N	N	N	Y
C-M model	N	N[11]	N	N	N	N	Y
WWp model	N	Y	N	N	N	N	Y
Probabilistic model	N	Y	N	N	N	N	N
MRW model	N[12]	Y	N	N	N	N	N
G-M model	Y	Y	N	N	N	N	N
BSA model	Y	Y	N	N	N	N	N
CS model	N	N	N	N	Y	N	Y
Behavioral model	Rules-based	Rules-based	Rules-based	Rules-based	N[13]	Y	Rules-based
KD model	Motion profiles-dependent	Motion profiles-dependent	Motion profiles-dependent	Motion profiles-dependent	Motion profiles-dependent	Motion profiles-dependent	Motion profiles-dependent
SR model	Y	Y	N	N	N	N	N
Graph-based model	N	N	N	N	Y	N	Y
Obstacle model	N	N	N	N	Y	Y	Y
Voronoi-based model	N	N	N	N	Y	Y	Y
AGb model	N	N[14]	N	N	Y	N	Y

continued on following page

Table 1. Continued

Model	Temporal dependency		Spatial dependency		Geographical restrictions		Pause time
	Velocity	Direction	Velocity	Direction	Paths, streets, etc.	Obstacles	
CosMos	Mobility model-dependent[15]	Mobility model-dependent[16]	Mobility model-dependent	Mobility model-dependent	Y	N	Mobility model-dependent
Manhattan-Grid model	N	N	N	N	Y	N	N
RWpC model	N	N	N	N	Y	N	Y
Freeway model	Y	N	Y	N	Y	N	N
STRAW model	Y	N	Y	N	Y	N	Y
UoMM model	Y	UoMM-dependent[17]	Y	UoMM-dependent	UoMM-dependent	Y	UoMM-dependent
Disaster-area model	N	N	Y	Y	Y	Y	Y

In group mobility models, there exist either a function that describes the behavior of the MNs' group, or MNs coordinate via a leader or a common target. This section overviews well-known synthetic group mobility models going through their strengths and drawbacks. The section sums up with a qualitative comparison between the described models in terms of the temporal and spatial dependency of the velocity as well as the direction of the groups and their members, the existence of geographic restrictions on movements and pause times between movements.

3.1. Reference Point Group (RPG) Mobility Model

The RPG mobility model (Hong, Gerla, Pei & Chiang, 1999) organizes MNs into groups based on their logical relationships. The model is widely used to simulate many existing applications, e.g. routing protocols for ad hoc networks, clustering protocols for ad hoc networks, etc. Moreover, by properly selecting the parameters of the RPG mobility model, several mobility models can be generated as we will see later on.

Description: The RPG model represents a logical relationships between MNs that move together and, thus, form a group. Each group is assigned a logical center termed "reference point", whose movement is followed by all members of the group. This implies that the reference point defines the group's motion behavior including its velocity, location, direction and acceleration. Members of the group are randomly/uniformly distributed in the neighborhood of the reference point, i.e. the geographic scope of the group. Note that such scheme allows, in principle, for an independent random motion behavior of the group's members (inside the group's geographic scope) in addition to the motion of the group itself. Figure 19 illustrates how MNs move employing the RPG mobility model. As the figure shows, the scenario contains two groups, each has 5 members. First, the reference point of a group i at time t_0 selects randomly a new location with a motion vector $W_i\left(t\right)$. Now, the location of a member j of the group i at the new location is defined by adding a randomly selected motion vector deviation $M_{ij}\left(t\right)$

to the reference point, see the figure. Note that the vector $M_{ij}(t)$ is an identically distributed random vector, whose length is independent of the previous location of the MN. $M_{ij}(t)$ is also and uniformly distributed within a certain radius $[0 \rightarrow r_{max}]$, where r_{max} is the maximum allowed distance deviation. Note also that the direction of r_{max} is uniformly distributed in the range $[0 \rightarrow 2\pi]$.

$$V_{ij} = W_i(t) + M_{ij}(t) \quad (19)$$

An example movement pattern of five groups with different number of members (between 2 and 6 members) is shown in Figure 20. Both the movement of the reference points and the members within the groups are implemented using the RWp mobility model. Note that pause times for the MNs within the groups are set to zero, while the groups themselves wait after each movement for the same time duration.

Discussion: As shown, the RPG model allows for a random motion of the groups on one side and of their members inside their geographical scopes on the other side. In fact, any synthetic single node mobility model from the models discussed in section 2.1 can be applied to model the movements of the groups' reference points. This enables generating a wide range of mobility patterns. Sure, researchers should carefully select the patterns more adequate for their studies. Even for a given synthetic single node mobility model, a proper selection of the RPG model parameters, e.g. initial groups locations, reference points paths, etc., enables diverse applications to be modeled (Hong, Gerla, Pei & Chiang, 1999), for instance:

- **In-Place Mobility Model:** Used to model applications like a battlefield situation, where different army battalions are ordered to do specific operations in different areas to achieve the mission

Figure 19. Motion vectors of two groups' members

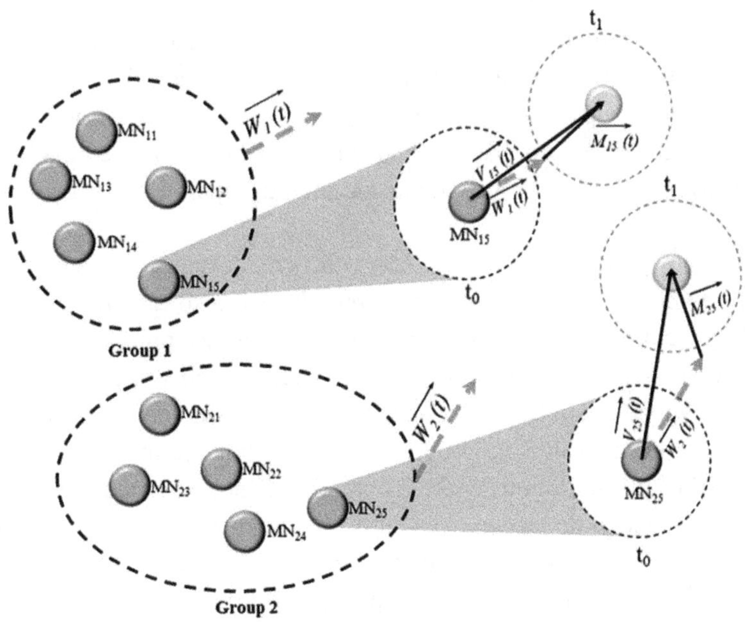

Figure 20. An example movement pattern of five groups of MNs generated from the RPG mobility model (Camp, Boleng & Davies, 2002)

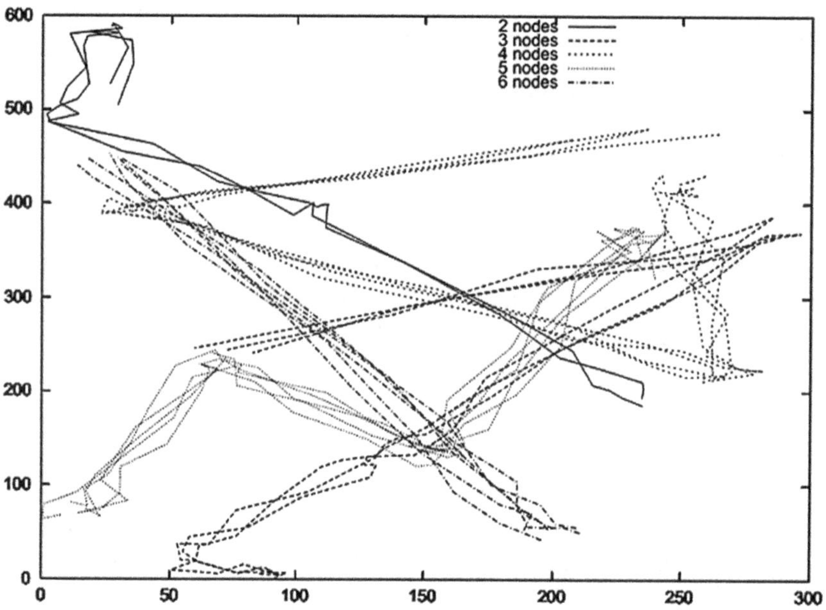

objective. In this model, the simulation area is divided into several adjacent regions with a group or even many in each region. All groups have the same mobility characteristics.

- **Overlap Mobility Model:** Similar to the in-place mobility model. However, there are different groups achieving different tasks and, thus, having different mobility characteristics. This model is adequate for generating mobility patterns for applications like disaster recovery, where different rescue teams exist, e.g. police, the medical assistant team, psychologist team, etc.
- **Convention Mobility Model:** This model is used to generate patterns similar to those observed in conventions and conferences. There are groups that have fixed locations (represent exhibitors in conventions) and groups that move around with different mobility characteristic and visit the locations of the fixed groups (express the attendees groups). By specifying certain mobility characteristics to the fixed groups, so that they move in cooperative/planned manner, applications like a military maneuver with joint aircraft, tank, and infantry operations can be modeled.

Limitations: Because any synthetic single node mobility model can be used to model the movements of groups' reference points, the RPG inherits the strengths and also drawbacks of the selected single node mobility model (of course, only the movements of the reference points are considered here). Note, however, that because of the spatial dependency between each group's reference point and members, it is expected that the RPG model will behave somehow different from the synthetic single node mobility model applied. For the movements of the groups' members, they are modeled, in principle, using the RW model. In spite of this, the limitations of the RW model are not inherited one to one. Concrete, there is no temporal as well as spatial dependency of the velocity and the direction among the members of each group. However, there is a velocity as well as a location spatial dependency between the members of a group and their reference point, known hypothetically as a group leader. Geographic restrictions of movements are possible neither among members nor between the members and leaders. Note that groups' members do not pause between movements, although groups leaders may do. This is a drawback of the model since members, in fact, may stop moving for the pause time durations the leaders stop for and may move around.

Variations of the RPG Mobility Model: many variations of the RPG mobility model are studied in the literature. The following provides more insight into these extensions.

- **Modified RPG Mobility Model:** This model (Bai, Sadagopan & Helmy, 2003) extends the classical RPG model by the derivation of the velocity as well as the direction of groups members in a different way. Concrete, a speed and an angle deviation ratio are assumed to control the deviation of the velocity as well as direction of the group's members from that of the group's leader.
- **Column Mobility Model:** The model (Camp, Boleng & Davies, 2002), (Chen & Chen, 2003) is useful for scanning and search applications, e.g. a raw of soldiers marching together towards an enemy. The model represents a set of MNs moving around a given column (or line). First, an initial column (a reference grid) is defined. Each MN is then placed in relation to its reference point. MNs move around their reference points by employing a synthetic single node mobility model (mostly the RW model), see Figure 21. New reference points are then calculated by adding a predefined offset (termed advance vector) to current reference points as the figure illustrates.

Mobility patterns, where MNs follow one another are also possible by a slight modification of the column mobility model, e.g. a group of young students walking in a single-file line towards their classroom.

- **Nomadic Community Mobility Model:** The model (Chen & Chen, 2003) aims at modeling the movement patterns of ancient nomadic societies and communities. A community is in principle a group of MNs that collectively move from one location to another. Within each community, MNs maintain their personal movement space, where they move randomly. When implementing

Figure 21. An example movement pattern generated by the column mobility model

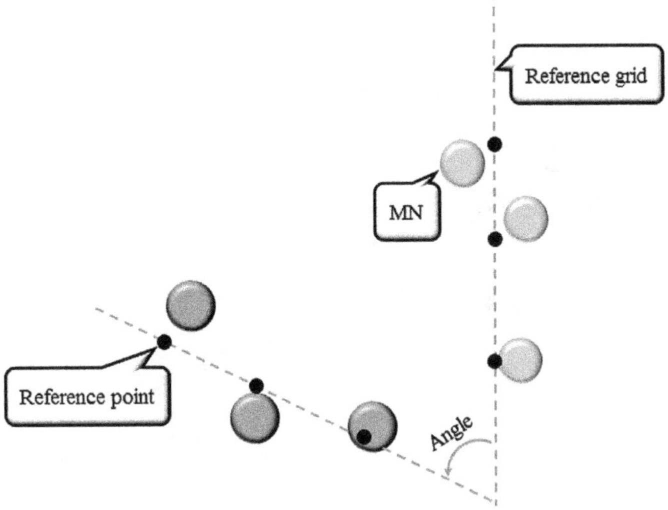

the nomadic community mobility model, MNs use a specific single node mobility model (mostly the RW model). However, they restrict their movements to the space around the reference point of the community they belong to. When the reference point moves, all group's members move to the new area defined by the reference point and start moving there. An example mobility pattern generated by the nomadic community model is illustrated in Figure 22. Note that the nomadic community mobility model is adequate for numeric application, e.g. a group of students touring a museum, a city, etc.

- **Pursue Mobility Model:** The model (Chen & Chen, 2003) aims at modeling the MNs that track a particular target, e.g. police officers attempting to catch a criminal. The pursued target is, in principle, a MN that moves based on a specific synthetic single node mobility model. Other MNs use also the same single node mobility model to determine their new locations. However, they consider the location of the pursued MN. An example movement pattern of a group of MNs while applying the pursue mobility model is illustrated in Figure 23.

Figure 22. An example movement pattern generated by the nomadic community mobility model

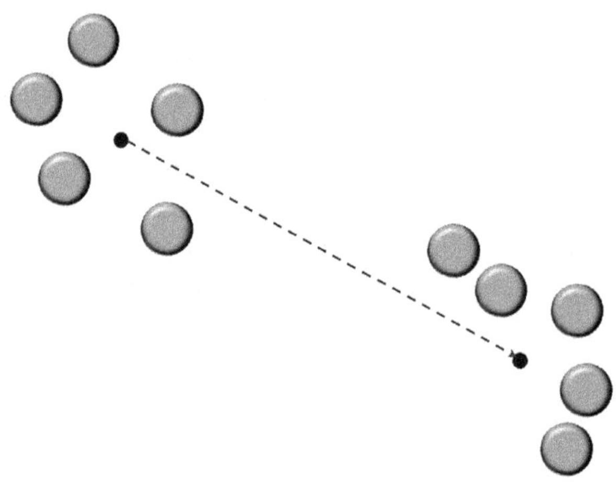

Figure 23. An example movement pattern generated by the pursue mobility model

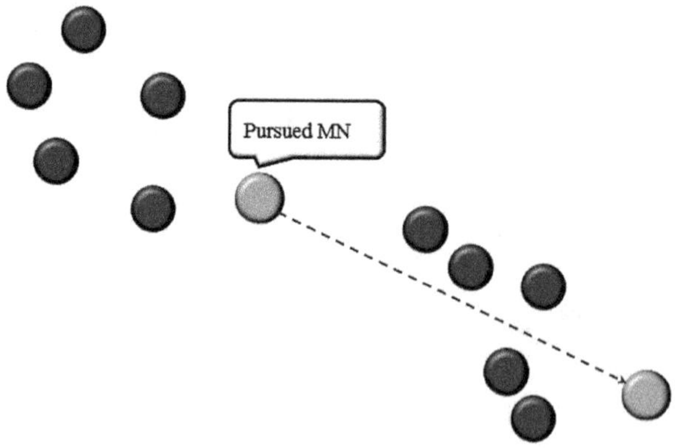

3.2. Reference Velocity Group (RVG) Mobility Model

The RVG mobility model (Wang & Li, 2002) characterizes groups mobility by extending the RPG mobility model investigated in the previous section. The RVG model is of a special interest for ad hoc networks, since it generates mobility patterns for simulation purposes where partitioning and merging of ad hoc networks are to be studied. Using the RVG model, partitions of ad hoc networks can be predicted and even mobility characteristics of each group can be estimated. The key idea of the RVG mobility model is the use of an omniscient observer to collect information about all mobiles and groups. The collected information is used then for prediction purposes.

Description: The RVG model extends the RPG model by additionally introducing a velocity representation of the groups, i.e. each group is characterized by its location and a mean group velocity $v = \left(v_x, v_y\right)$, where v_x and v_y are the velocity components in the *x*- and *y*-coordinate. The developers of the RVG model argue that such characterization of the velocity will characterize groups' movements much better. Note that the velocity assigned to each group is, in principle, the mean group velocity and it remains fixed during the simulation. Moreover, the coverage area of each group is set to a constant value regardless of its members. The members of the group are assigned velocities closed to the mean group velocity with a light deviation. This implies that the velocity of any member of any given group is characterized by two parameters, the velocity of the group and a local velocity deviation. Note that groups' velocities as well as local velocity deviations are modeled as random variables, whose distributions may be of any arbitrary type.

Formally, a member MN with an index j of a group i is characterized by following parameters:

- $W_i\left(t\right) \sim P_{ij}\left(w\right)$ the group velocity vector at time t. It is a random variable drawn from the distribution $P_{ij}\left(w\right)$ which can be of any arbitrary type, as mentioned above.
- $U_{ij}\left(t\right) \sim Q_{ij}\left(t\right)$: the local velocity deviation of the MN at time t. It is also a random variable and is drawn from the distribution $Q_{ij}\left(t\right)$. Again, the distribution $Q_{ij}\left(t\right)$ can be of any arbitrary type.
- $V_{ij}\left(t\right) = W_i\left(t\right) + U_{ij}\left(t\right)$: the velocity of the mentioned MN at time t.

The model assumes a clustering algorithm to cluster MNs based on their velocities, see (Wang & Li, 2002). All MNs of a specific group will have speeds closed to the mean group velocity. So, it is not necessary to join members to groups. The groups will be formed automatically by grouping the MNs based on their speeds. Note that groups' movements at different velocities towards their destinations cause often network partitioning and merging. This highly affects the performance of ad hoc networks protocols, applications, services, etc. Therefore, the model assumes that there is an omniscient observer that collects all information related to the mobility of MNs within the scenario. This enables to predict partitioning and merging before they occur (the prediction is done based on the distances between MNs, their velocities and also directions). Such prediction is helpful to optimize the performance of ad hoc networks. Sure, how such collection of information can be realized in real-life applications is another question. The model does not put restrictions on this issue. Again, a main feature of the RVG model is the transformation of MNs representation from physical (Figure 24) to velocity (Figure 25) space.

Discussion: As by the RPG model, the RVG model allows for a random motion of the groups and their members. However, the speed of both groups and their members are not selected randomly. In

Figure 24. MNs representation at physical space (Ranjan Roy, 2011)

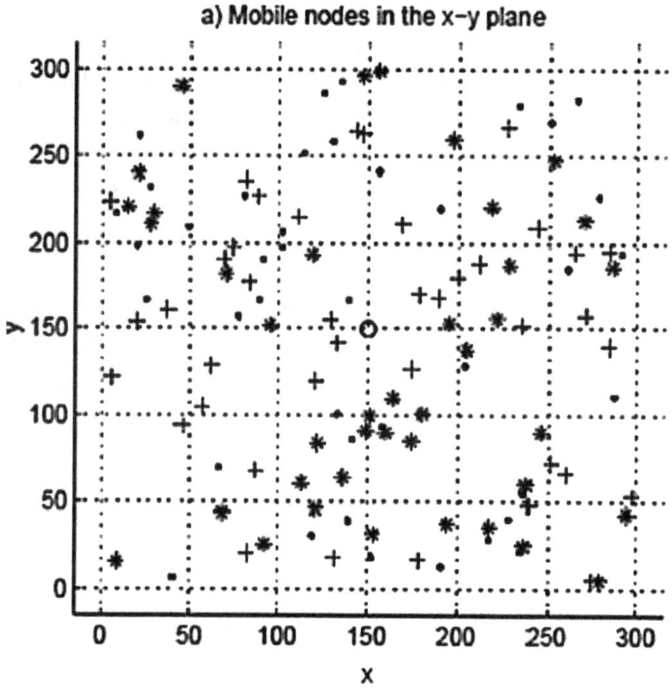

Figure 25. MNs representation at velocity space (Ranjan Roy, 2011)

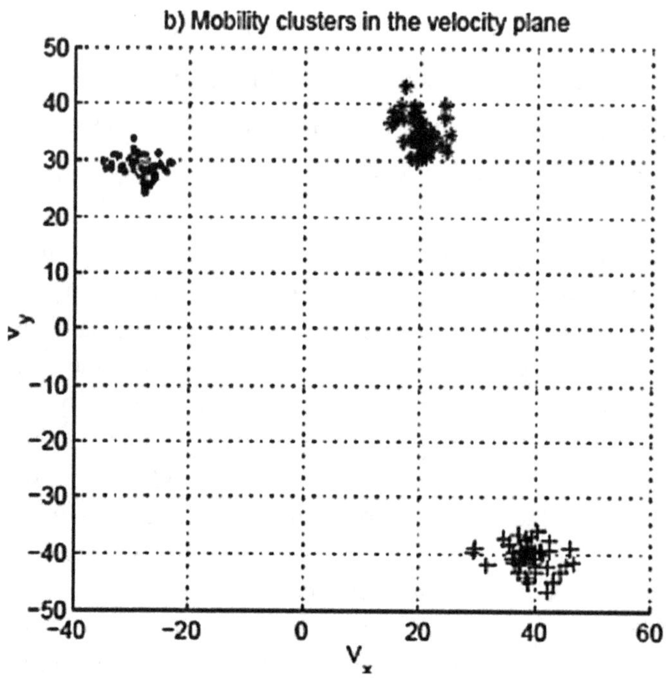

fact, a synthetic single node mobility model (with light modification sometimes) is used to model the movements of groups and their members. As we mentioned, the model is of a special interest for ad hoc networks where partitioning and merging of these networks are to be focused on. The capabilities of the prediction of partitioning and merging are a mean feature of the approach. However, as we highlighted, this depends on the use of an omniscient observer that has a global knowledge of the network. Sure, this is not practical.

The construction of the scenario as well as the grouping of MNs is simple, since one has to define the velocities of groups and MNs. The clustering algorithm of the RVG model will then take care of grouping MNs. A notable drawback of the RVG model is the assumption that the coverage area of each group is fixed regardless of its members, which is impractical. The assumption that groups' velocities do not change during the simulation is also impractical.

The transformation of MNs representation from the physical to the velocity space is seen as an advantage that simplifies the construction of scenarios and the implementation of the model. However, this is also a weakness of the model. Imagine that we have two groups with the same mean group velocity (or even two mean velocities closed to each other), however, with totally different mobility characteristics. The RVG model cannot distinguish between these two groups and will, due to the applied clustering algorithm, form either one group from the two groups or two groups different from the two groups we are aiming at simulating.

Limitations: The model inherits the strengths and weaknesses of the selected synthetic single node mobility model used to model the movements of groups and their members. Of course, strengths and weaknesses are not inherited one to one due to modifying the way applied to define the speed of all MNs in the scenario. This is also the reason why we may see different behavior than that seen by the selected synthetic single node mobility model applied.

As by the RPG model, there is no spatial and temporal dependency of the velocity and the direction between groups among each other and between members of each group (we assume the RW model). However, there is a velocity as well as location dependency between the members of a group and the group itself. Geographic restrictions of movements are possible neither among members nor between the members and groups. Concerning the pause times, groups may stop between movements for a specific time duration. In fact, this depends on the synthetic single node mobility model applied to model their movements. For groups' members, they normally do not have pause times. Again and similar to the RPG model, this behavior does not really reflect the real-life behavior.

3.3. Reference Velocity and Acceleration Group (RVAG) Mobility Model

The RVAG mobility model (Chen & Chen, 2003) extends the RVG model by additionally using the acceleration to characterize mobility of MNs[18]. Like the RVG model, the RVAG model assumes several groups moving independently within the simulation area, and focuses on ad hoc networks partitioning and merging issues. A key feature of the RVAG mobility model is the capability it has to predict ad hoc networks portioning by observing groups mobility patterns aggregated from MNs movements. A group-identifier-based clustering algorithm has also been proposed by the mobility model.

Description: Like the RVG model, a mean group velocity is assumed for each group. MNs are also grouped according to their logical relationships with the group's velocity. The velocities of group's members are chosen closed to the mean group velocity with a light deviation. Thus, the velocity of each member is characterized by two parameters, namely the mean group's velocity $W_i(t)$, where i is the group's index) and a local velocity deviation ($U_{ij}(t)$, where j is the index of the member within the group). Again, groups' velocities as well as local velocity deviations are modeled as random variables, whose distributions may be of any arbitrary type.

In addition to the mean group velocity, a group acceleration ($A_i(t)$) is defined as a velocity-dependent random variable with a distribution of any arbitrary type. Due to the definition of the acceleration, more accurate modeling of the relationships between groups' members is possible, which improves the prediction of networks partitioning.

Formally, the group velocity, group acceleration, local velocity deviation and the velocity of a member MN with an index j of a group i at a time t are defined as follows:

- $W_i(t) \sim P_{ij}(w)$: the group velocity vector of the group i at a time t. It is a random variable drawn from the distribution $P_{ij}(w)$, which can be of any arbitrary type.
- $A_i(t) \sim C_{ij}(a) = \frac{dW_i(t)}{dt}$: the group acceleration vector of the group i at a time t. It is also a random variable drawn from the distribution $C_{ij}(a)$, which can also be of any arbitrary type.
- $U_{ij}(t) \sim Q_{ij}(t)$: the local velocity deviation of the MN with the index j of the group i at a time t. It is a random variable, as well, and it is drawn from the distribution $Q_{ij}(t)$ that can be of any arbitrary type.
- $V_{ij}(t) = W_i(0) + U_{ij}(t) + \{A_i(t)\} * t$: the velocity or motion vector of the mentioned MN at a time t.

A group-identifier-based distributed clustering algorithm has been proposed to cluster MNs into groups, select a cluster head for each group and separate groups during the simulation. The clustering algorithm assumes that each group has a unique group identifier (group-ID) and one cluster head. Each MN will be assigned a group-ID that identifies the group the MN belongs to. Moreover, each MN is equipped with a Geo-Positioning Satellite (GPS) to accurately determine its location, which will be used later for partitioning prediction. The clustering algorithm is pretty simple and can be summarized as follows: MNs periodically exchange information with neighbors. Following that, each MN determines which neighbors are members of its group (based on the group-ID). The MN equipped with interfaces to other ad hoc networks (other groups, in principle) is selected to be a cluster head.

Another focus of the RVAG model is the prediction of ad hoc networks portioning. This is done based on the mobility parameters provided by the model, especially by the clustering algorithm, e.g. groups velocities, groups accelerations, the distance between two overlapping groups, etc. To accurately determine the partitioning time, each group (represents an ad hoc network) exchanges via the cluster head mobility parameters with its neighbor groups to calculate the overlapping area between them. The most important metric used is the acceleration, which represents the motion behavior of a group.

Discussion: Same as by the RPG and RVG models, the RVAG model also allows for a random motion of the groups and their members. The speed of both groups and their members are, however, not selected randomly. The mobility of groups can, in principle, be modeled using any synthetic single node mobility model (with light modification sometimes).

The RVAG mobility model uses the distance, velocity and acceleration to characterize mobility of groups and groups' members. The use of the acceleration enables more accurate modeling of the relationships between groups members and their clusters heads, which in turn improves the prediction of networks partitioning.

The proposed clustering algorithm is simple and distributed, which makes its implementation in real ad hoc networks possible. Moreover, the clustering of MNs and the selection of a cluster head for each group may help other protocols of ad hoc network, routing protocols for instance. The group-ID and location information of MNs help in separating groups during the simulation, and in determining the overlapping area between various ad hoc networks. A main feature of the clustering algorithm is that the information it provides significantly helps in an accurate network partitioning prediction. A disadvantage of the RVAG model is the need for GPS-equipped devices to deliver locations information and devices with multiple interfaces to operate as clusters heads.

Limitations: The model also inherits the limitations, advantages and disadvantages of the selected synthetic single node mobility model applied to model the movements of the groups and their members. Again, the inheritance is not one to one due to modifying the definition of the speed of all MNs in the scenario. As by the RPG and RVG models, there exist no spatial and temporal dependency of the velocity and the direction among groups and groups' members (we assume the RW model). However, there is a spatial dependency of the velocity and the direction between the members of a group and the group itself. There are also no geographic restrictions of movements. Concerning pause times, groups may stop between movements. This depends on the synthetic mobility model applied. However, groups' members do not stop, since the RW model is the model mostly applied to model their movements. Again and similar to the RPG and RVG models, this behavior does not reflect the real-life behavior.

3.4. Structured Group (SG) Mobility Model

The SG mobility model (Blakely & Lowekamp, 2004) aims at generating more realistic movement patterns for mobile communication networks, especially for ad hoc networks. The RPG, RVG and RVAG models assume that the members of any group are independent actors moving randomly without any relation between them, although they have relations to the group's leader. However, groups movements in real-life are oriented at common goals, and hence inherent groups structures. The parameterization of the structures of groups is the main contribution of the SG mobility model, which assigns MNs to groups based on a known organizational structure.

Description: The developers of the SG mobility model state that inherent structures of groups are always apparent. Groups normally achieve common goals, which necessitate that groups' members must be aware of the goal rather than move randomly around groups' leaders. This also implies that the structures of the groups are known in advance and relate to the goals to be achieved. Moreover, it must be possible to define a hierarchical structure of groups, where members of main groups are allowed to nstruct groups by themselves.

e SG model assigns no velocity vectors to individuals, whose location is determined by the distance r relationship to the reference points of their groups. The reference point of the group indexed e t ($R_i(t)$) may be the geographical center of the group, the location of the group's leader or the group's center of mass. $R_i(t)$ has a directional orientation of angle $\varphi_i(t)$ on a global coordination system regardless if the group moves or not. Thus, the members of the group and also its sub-ordinate groups determine their positions and orient themselves based on $\varphi_i(t)$.

The SG model derives the position of a MN by selecting a distance $d_{ij}(t)$ between the MN and $R_i(t)$, and an angle $\theta_{ij}(t)$ from $\varphi_i(t)$. The distance must be taken from a given distribution $D_{it}(d)$ and the angle $\theta_{ij}(t)$ have to be drawn from a given distribution $\varnothing_{it}(\theta)$, where j is the index of the MN that can be, as mentioned above, a group leader of a sub-ordinate group. Note that the relationship to the reference point is maintained by specifying the distributions $D_{it}(d)$ and $\varnothing_{it}(\theta)$. Sure, changing these distributions will change the resulting movement pattern. For instance, if a tight structure during movements is required, a normal distribution with a small variance may be used for both $D_{it}(d)$ and $\varnothing_{it}(\theta)$, e.g. to simulate a military unit moving for example. The distance $d_{ij}(t)$ and the angle $\theta_{ij}(t)$ are maintained during the simulation, so the structure of the group is also maintained.

Movements of $R_i(t)$ are modeled by a specific single node mobility model, see section 2.1. The movements of groups' members and sub-ordinate groups are then derived based on the relation to $R_i(t)$. Concrete, by using $d_{ij}(t)$ and $\theta_{ij}(t)$, which themselves are updated through a random selection from the distributions $D_{it}(d)$ and $\varnothing_{it}(\theta)$.

The SG model is of a special importance for military scenarios and rescue actions in case of disasters. For clarity, we will briefly give an insight into a firefighters operation inside a building. Figure 11 shows an example building and an example groups structure and movement pattern. Firefighters teams consist normally of small sub-teams that may have hierarchical structure as the figure shows. The sub-teams are equipped with modern electronic devices and form, in principle, an ad hoc network among each other. The structure of the groups is crucial and the communication between the members is of a major importance. The figure shows a small main team locating at the entrance of the building. The team contains a sub-ordinate team that has to enter the building and search for fire and victims. The task now is only to structure and locate the MNs in the scenario to be as depicted in the figure. Following that, the distributions $D_{it}(d)$ and $\varnothing_{it}(\theta)$ must be carefully determined for the main team and for the sub-ordinate team, so that the wanted movement pattern results.

Discussion: The SG mobility model parameterizes groups' structures dependencies aiming at generating traces closed to real-life observations as most as possible. Therefore, the SG model is marginally more complex than previous studied mobility models. Moreover, the derivation of movement traces when applying the SG model is also somewhat more complex than when using other model from the already studied ones. However, the simulation time the SG model consumes is insignificant despite the complexity mentioned, see (Ranjan Roy, 2011).

The mobility of groups is modeled using a specific synthetic single node mobility model, which enables a wide range of applications to be simulated. The movements of the members of the groups are derived, however, based on the relations to the reference points of the groups, which are maintained by specifying the distributions $D_{it}(d)$ and $\varnothing_{it}(\theta)$. This also enables controlling the movement patterns by controlling the distributions mentioned. However, this is not that simple and requires an accurate and proper definition of the distributions.

Figure 26. Firefighters operation inside a building

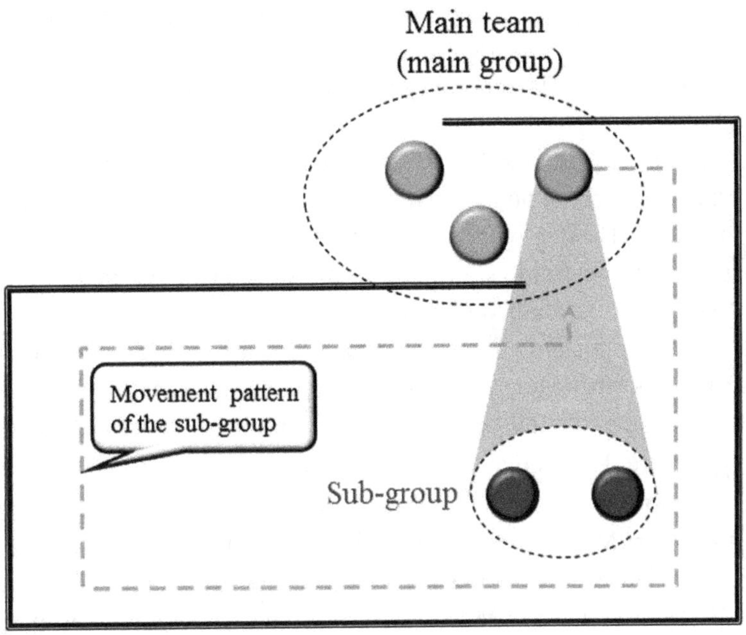

A mean feature of the SG model is its capability to represent hierarchical organizations by allowing groups to have sub-ordinate groups. As known, such hierarchical organizations are common in real-life.

Limitations: The SG model inherits the strengths and weaknesses of the synthetic single node mobility model used to model the movements of groups. There is a spatial dependency of the direction between groups among each other. Moreover, there is a spatial dependency of the direction between the members of each group and the group's leader. Concerning the spatial dependency of the velocity among groups, this depends on the synthetic single node mobility model applied. There is an implicit spatial dependency of the velocity between the groups' members and leaders, since the structures of the groups are maintained during the simulation. This means that the location of individuals are maintained in relation to the location of the groups' leader, which also implies that velocities relate to each other somehow. Note that there is no explicit definition of these dependencies in the SG model. For the temporal dependency, there is no temporal dependency of the direction and the velocity among groups' members. For the groups among each other, this depends of the mobility model applied to model their movements.

Geographic restrictions of movements are possible neither among members nor between the members and groups. For the pause times between movements, groups may stop between movements for a specific time duration when the synthetic single node mobility management scheme applied supports such pause times. Consequently, groups' members will pause due to maintaining the relation to the groups' reference points.

3.5. Virtual Track-Based Group (VTBG) Mobility Model

All group mobility models we have already investigated do not capture the dynamics one observes in real-life scenarios. For example, in large-scale military missions, mobility coherence among nodes is quite common, however, not a must. There may exist some nodes that move independently, while others form groups and move in a tight structure. Other nodes move also in groups, however, they dynamically reconfigure themselves, e.g. split into many groups, merge with other groups, etc. Furthermore, some nodes may remain static during the scenario. Such scenarios are termed heterogeneous, and it is, for sure, a challenge to model them. The VTBG mobility model (Zhou, Xu & Gerla, 2004) has been developed to face this challenge.

Description: The VTBG mobility model creates virtual tracks that may have certain constraints. The tracks can be derived from streets of an actual map of a specific region, city, etc. Some switch stations are deployed in the scenario. The deployment of the switch stations may be done randomly or based on some considerations related to certain geographic locations in an actual map. The switch stations are then interconnected via tracks with a given tracks width. MNs grouped together move on the defined tracks. At switch stations, groups are allowed to change dynamically, e.g. they may split into smaller groups or may be merged to form bigger groups. This dynamics may occur randomly or may be controlled by configured split and merge probabilities. The velocity of the groups are chosen randomly from the range $[V_{min} \rightarrow V_{max}]$, where V_{min} is the minimum and $V_{\max}$ is the maximum allowed speed.

The members of a group move along the same track and are allowed to have internal random mobility within the scope of the group. Multiple classes of MNs can be defined, e.g. pedestrians, cars, Unmanned Ground Vehicles (UGVs), etc. Sure, each class has its own mobility characteristics, e.g. velocity, direction, coverage, etc. Of course, only the MNs belonging to the same class can merge to form a single group. Individual moving MNs (not belonging to any group) and also individual stationary ones can be modeled as well. The movements of individual moving MNs are not restricted to tracks and switch stations, rather they move randomly in the whole area.

An example scenario is depicted in Figure 27. The scenario contains five switch stations interconnected via eight virtual tracks with equal track width. Note that groups move along the tracks, split and merge at switch stations. Individual moving MNs are free to move anywhere they want without caring of the tracks and switch stations.

Figure 27. An example scenario generated by the VTBG mobility model

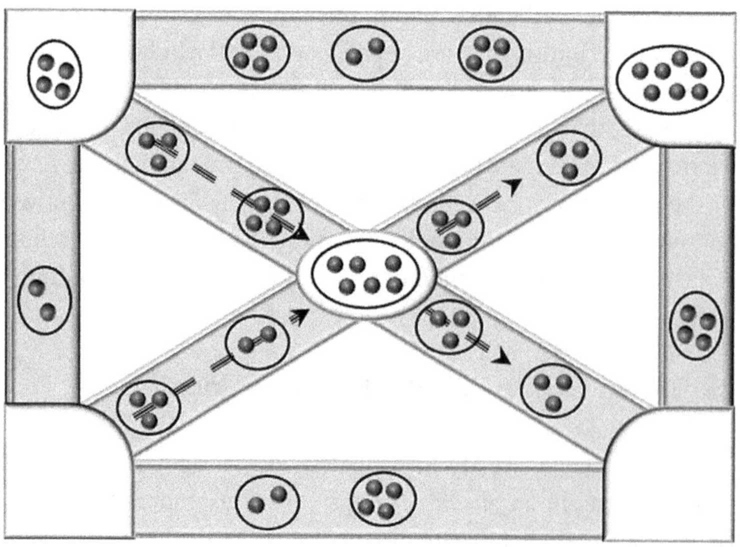

The VTBG model is especially adequate for military, urban environments, railway scenarios, etc. For urban scenarios for instance, virtual tracks represent the streets in a specific geographical area, they can be taken from a real map. Switch stations are then the intersections between streets. MNs stand for pedestrians and cars that are assigned different mobility characteristics to model real-life scenarios in urban environments. Note that cars split in intersections, new groups are also formed there.

Now, let us give more insight into the movements of groups along the tracks. First, all members of a group are placed on the same track. They also select the same switch station as their destination at one end of the track's ends. Following that, the group as a whole moves towards the destination. It is worth mentioning that the group does not move on a straight line towards its destination. Rather, the movement of the group on the track is modeled by means of the RWp model. Concrete, a waypoint is selected on the track towards the destination. Furthermore, each member of the group can also have a small internal mobility (normally random) under the constraints of the group and the track the group moves on. Once the group reaches a switch station, a new track is selected randomly from the tracks at that switch station.

The split and merge of groups occur at switch stations. Each group is parameterized with a stability threshold value. This value is used for split and merge purposes. Concrete, each group's member checks whether its stability value smaller than the stability value of the group. In this is the case, the member chooses a track different from the track selected by the group, i.e. a group split occurs. Note that when many groups select the same track for next movement, they will be merged with each other.

Discussion: The VTBG mobility model is a heterogeneous model and is capable of defining scenarios containing groups with a dynamic behavior and individual MNs, thus, generating more realistic scenarios than other studied group mobility models produce. The preliminary results provided in (Ranjan Roy, 2011) show that mixing groups and individuals has a significant impact on the performance of the network and makes the movement pattern resulting more realistic. The reason is simply because the connectivity between multiple groups is improved due to the individuals roaming inside the simulation area without caring of tracks and switch stations. Individuals operate in this case as relays between groups. The dynamic behavior (groups splitting and merging) the VTBG mobility model enables is a great advantage. However, to be able to merge at switch stations, two groups must arrive at a switch station at the same time. This constraint leads to a plethora of small groups. This also implies that large groups cannot be well represented, see (Ranjan Roy, 2011).

A drawback of the model is its complexity from the mathematical point of view. The implementation of the model is also not that simple. Moreover, the VTBG is not capable of expressing scenarios where groups move freely within a specific region, e.g. gardens, museums, etc. The movement of the groups are restricted to tracks and switch stations. Moreover, an essential drawback is the lack of pause times between movements of groups and individuals. This prohibits modeling scenarios where such pause times are present. Sure, this does not speak against the major developments achieved by the VTBG model against other models. It is expected that the model will play an essential role in simulating military and emergency recovery scenarios, where divers mobility behaviors typically coexist.

So as to enable the VTBG model to generate more realistic scenarios than currently obtained, it has to cover some open issues like the coexistence of multiple virtual tracks for different classes of MNs groups with different mobility characteristics, e.g. pedestrians, bicycles, cars, trains, etc., correlations between movement patterns in a complex scenario, etc.

Limitations: Concerning the individuals, the model inherits the limitations of the RW mobility model used to model their movements, see section 2.1.1. For the groups, they move on tracks employing the RWp mobility model, so they also inherit its limitations, however, not one to one. Concrete, there are no temporal dependency of the velocity and the direction. There are also no spatial dependency of the velocity and the direction among groups' members. However, such dependencies exist between the members of a group and the group itself, since their movements are inherent with the movement of the group. Geographic restrictions of movements exist in the VTBG model by concept. With respect to the pause times, neither groups nor individuals stop between movements.

3.6. Drift Group (DG) Mobility Model

The DG mobility model (Rossi, Badia, Giacon & Zorzi, 2005) attempts to present a general tunable mobility framework, within multiple different movement patterns can coexist and can be tuned to provide the desired behavior. The major feature of the DG model is the clear and compact representation of groups, since each group is described by means of an attractivity between members without fixed constrains, as by other studied group mobility models. The DG mobility model can be adapted to generate movement patterns for specific scenarios, e.g. users move beside obstacles, on streets, etc., by imposing superstructures on the top of the model, see (Rossi, Badia, Giacon & Zorzi, 2005).

Description: The DG mobility model characterizes each MN by a charge value termed "charisma". Leadership relations are also defined, so that each MN has a unique leader. The MN in this case is termed a follower. Movements of MNs are controlled by means of two components, namely a drift and a random movement component. The drift movement tends to follow the leader, while the random movement is regarded as a noise superimposed to the drift component. The random movement can theoretically be modeled using any one of the single nodes mobility models we mentioned in this chapter. This implies that a movement of any MN is, in principle, the sum of two components, a randomly generated movement and an attraction movement toward a leader.

The DG assumes two charismas, C_l for the leader and C_f for the follower. Thereafter, it defines an attraction force ($F_a(d)$) experienced along the axis that connects two MNs separated by a distance of *d* meters. The attraction force is derived from the following equation, where α and β are parameters used to tune the force. α is assumed constant, whereas β is drawn at each time step from a Gaussian distribution with a mean value μ_β and variance $\sigma_\beta{}^2$ that is applied to control the followers dispersion around the leader. u_a expresses the unit vector denoting the axis that connects the two MNs.

$$F_a(d)=\begin{cases}\beta\dfrac{C_l * C_f}{d^\alpha} * u_a & d \geq d_{min}\\ F_a(d_{min}) & d < d_{min}\end{cases} \tag{20}$$

By considering MNs as particles of unit mass, one can write the following equation, where Δv_a is the speed variation vector and is associated with a time interval Δt along the direction between two MNs (leader and follower).

$$\Delta v_a = \Delta t * \beta\frac{C_l * C_f}{d^\alpha} * u_a \tag{21}$$

Thus, the speed vector of a given follower at time $t + \Delta t$ is calculated from the following equations, where Δv_a is the speed variation resulting from the attraction present between the follower and the leader. The direction of the speed variation lies along the line connecting both the follower and the leader. In fact, Δv_a is used to model the follower-leader attractive behavior, which forces the follower to move in the same direction of the leader. Δv_i expresses the speed variation associated with an independent mobility pattern. The pattern is modeled by a single node mobility model and superimposed to have the attractive behavior. Δv_i is also used to implement the lead-independent mobility pattern.

$$v(t+\Delta t) = v(t) + \Delta v(\Delta t) \tag{22}$$

$$\Delta v(\Delta t) = \Delta v_a + \Delta v_i \tag{23}$$

So, the result of the described model is a pattern, within each follower follows its leader. So, movements of followers are modeled by the drift movement (with a random movement superimposed) that is controlled by the attraction force, which in turn heavily depends on the charismas C_l and C_f and also on the parameters α and β. The movements of leaders are modeled according to leader-independent factors be means of one of the single node mobility models. In brief, leaders move freely based on a specific single node mobility model and followers follow them by means of the attraction forces.

Lots of applications profit from the DG mobility model, e.g. a group of tourists moving in a historical place or museum, where the group leader moves and members follow him and also move around to take a look at the surroundings.

The speed vector of the leader at a given time $t + \Delta t$ is derived from the following equations.

$$v(t+\Delta t) = v(t) + \Delta v(\Delta t) \tag{24}$$

$$\Delta v(\Delta t) = \Delta v_i \tag{25}$$

Two example mobility patterns resulting from the DG mobility model are provided in Figure 28. The scenario contains a leader moving with a constant speed of 2 km/h and followed by three followers. The scenario contains no movements' constraints and models the movements of the leader by means of the BSA mobility model.

Note that the movements of the group occur as a compact unit without preserving a given structure, keeping the distance among MNs rigid or preventing MNs from having independent directions.

Discussion: The DG mobility model is a general model since there is no need to define reference points for the groups, whose locations are tracked by groups' members. Rather, a leader for the group is selected and all followers follow this leader. The attractivity between a leader and its followers is also done without fixed constraints. So, the simplicity and also generality of the DG mobility model make it very attractive to be used in simulations. Note that, different mobility patterns can be obtained by appropriately tuning the force field ($F_a(d)$), which is a great advantage. A main feature of the DG mobility model is the capability of imposing superstructures on the top of it to model streets, obstacles, etc.

Figure 28. Two example mobility patterns generated by the DG mobility model (Rossi, Badia, Bui & Zorzi, 2005) [19]

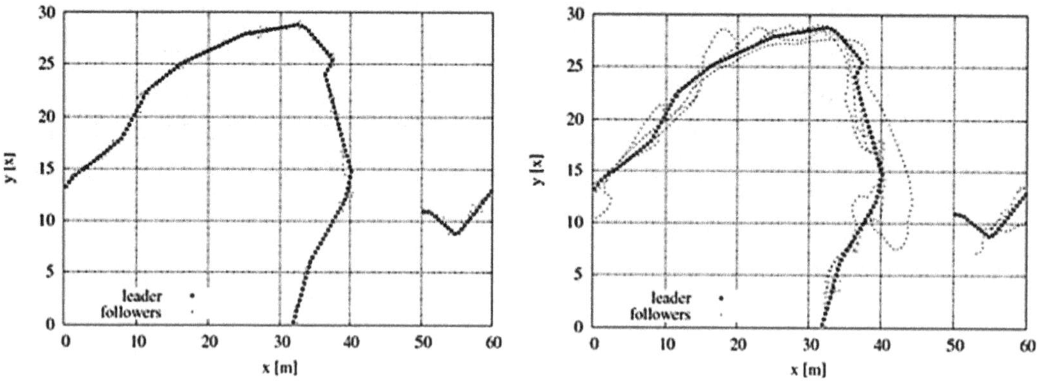

That the movements of followers within groups are achieved without fixed constraints can be seen as a drawback, as well, since this prohibits the usage of the model for specific applications, where a structure of the group must be maintained.

Limitations: As mentioned above, leaders move based on a specific single node mobility model. Thus, movements of the leaders inherit all limitations of the specific single node mobility model applied. Concerning the followers, each one follows its leader without caring of other followers in the scenario/neighborhood. So, there is no temporal as well as spatial dependency of the location among followers. There is also no spatial dependency of the velocity between followers since the velocity of a follower is not affected be the velocities of other followers. There is, however, a temporal dependency of the velocity when we speak about followers, since the model defines a speed variation vector associated with a time interval Δt. This means that the future speed of a follower depends on the current speed.

Concerning the relation between followers and leaders, one observes the existence of a spatial dependency of the velocity and the location, since leaders are followed by followers. There exist no geographical restrictions of movements in the classical model. However, they can be obtained by imposing superstructures on the top of the model, as we mentioned earlier. With respect to pause times, followers do not have pause times between movements. However, groups' leaders may have. In fact, this depends on the specific single node mobility model applied.

3.7. Gathering Group (GG) Mobility Model

The GG mobility model (Borrel, Dias De Amorim & Fdida, 2005) relies on a simple principle of dynamic networks, namely the preferential attachment (Barabási & Albert, 1999). It focuses on behavioral aspects of individual MNs and on interactions between them, i.e. MNs move independently, however, the resulting movement pattern exhibits a collective behavior. Such situation is obvious in ad hoc networks, where MNs move freely. However, due to the communication between MNs, clusters and groups appear.

A main property of the GG mobility model is its scale-free spatial distribution that makes it adequate for a population growth. This helps in solving many problems of large-scale mobile ad hoc networks that may be constructed in metropolitan areas for instance. As known, such networks face serious challenges concerning the MNs density and the complexity of communication environments.

Description: From the discussions, we induced in previous sections, one notes that the population spatial density dynamics has not been properly considered. In real-life scenarios people form groups randomly, mimic others, etc. So, population grows or diminishes. The dynamic itself arises from meeting of individuals in space with some level of interactions among them, which in turn results in a collective behavior. The GG mobility model emphasizes that a common particular property is pertinent to human organizations (e.g. social networks, information networks, etc.), namely they are scale-free. This implies that human organizations follow a power law distribution (Barabási & Albert, 1999), which says that the probability of having a node of k degree is given by the following equation, where η is a constant, and λ is the exponent.

$$P[k] = \eta * k^{-\lambda} \tag{26}$$

The GG mobility model is interested in the mobility of individuals, whose displacement has well-identified centers of interest, e.g. shopping centers, metro stations, restaurants, etc. In reality, humans can be aggregated within a crowd of people. Furthermore, aggregated individuals assume that the preferential attachment to centers of interest is a valuable assumption.

So, the scenario, from the GG mobility model point of view, consists of two components, namely individuals (MNs) and attractors. Attractors are the centers of interest, towards which individuals move. They appear for a specific time duration, do not move, and disappear following that. Concerning the individuals, they arrive, behave in cycles for a given time duration and then depart. A cycle consists of a displacement followed by a pause duration. Note that displacements are characterized by the origin and target attractors.

The GG model assumes two sets, namely individuals set ($Z(t) = \{z_1, z_1, \ldots, z_l\}$, where l is the number of individuals in the scenario) and attractors set ($A(t) = \{a_1, a_1, \ldots, a_m\}$, where m is the number of attractors) at time t. The initial set of individuals is Z_0, while the initial set of attractors is A_0. New attractors appear following a Poisson process with a rate τ_{α_a}. The life time of the attractors is chosen randomly from a range $[D_a^{min} \to D_a^{max}]$. In a similar way, the arrival of new individuals also follows a Poisson process with a rate τ_{α_z}. The displacement speed of each individual is randomly chosen within a fixed interval. For each new cycle, the corresponding pause duration is also selected randomly. Each individual leaves after the expiration of a specific sojourn duration, which is also randomly chosen from a range $[D_z^{min} \to D_z^{max}]$.

Let us assume that there is an individual z_i under the influence of an attractor a_j. Let us also assume that the individual decides to move towards another attractor a_k. The individual calculates, first, the attractiveness ($A_{(z_i,a_k)}$) the attractor a_k exerts on it using the following equation.

$$A_{(z_i,a_k)} = \frac{\left(1 + \sum_{z_n \in Z} B(z_n, a_k)\right)}{\sqrt{\left(X_{a_k} - X_{z_i}\right)^2 + \left(Y_{a_k} - Y_{z_i}\right)^2}} \tag{27}$$

$B(z_n, a_k)$ is a Bernoulli variable. This variable equals to 1 if the individual moves towards or is at the attractor a_k, and 0 otherwise. X and Y are the coordinates of the individual/attractor. Note that the previous equation includes the impact of other individuals and also attractors on the individual under study, which leads to the collective behavior we mentioned earlier. The GG model assumes that attractors are selected proportionally to their popularity and inversely to the distance between them and the individuals.

Attractors leave the scenario following a process with a rate $\tau_{\beta_a}(t)$. This rate is derived from the following equation.

$$\tau_{\beta_a}(t) = \frac{2 * A(t)}{D_a^{min} + D_a^{max}} \tag{28}$$

$[\![A(t)]\!]$ is the attractors population size at time t. It follows the following semantic:

$$A(t): \begin{cases} growing & if \quad \tau_{\beta_a}(t) < \tau_{\alpha_a} \\ steady & if \quad \tau_{\beta_a}(t) = \tau_{\alpha_a} \\ decreasing & if \quad \tau_{\beta_a}(t) > \tau_{\alpha_a} \end{cases} \tag{29}$$

In a similar way, individuals depart also following a process with a rate $\tau_{\beta_z}(t)$. This rate is derived from the following equation.

$$\tau_{\beta_z}(t) = \frac{2 * Z(t)}{D_z^{min} + D_z^{max}} \tag{30}$$

$[\![Z(t)]\!]$ is the individuals population size at time t. It follows the following semantic:

$$Z(t): \begin{cases} growing & if \quad \tau_{\beta_z}(t) < \tau_{\alpha_z} \\ steady & if \quad \tau_{\beta_z}(t) = \tau_{\alpha_z} \\ decreasing & if \quad \tau_{\beta_z}(t) > \tau_{\alpha_z} \end{cases} \tag{31}$$

One can deduce from the above listed equations that both attractors and individuals' populations tend to a finite size. So one can write:

$$\lim_{t \to \infty} A(t) = \tau_{\alpha_a} \left(\frac{D_a^{min} + D_a^{max}}{2} \right) \tag{32}$$

$$\lim_{t \to \infty} Z(t) = \tau_{\alpha_z} \left(\frac{D_z^{min} + D_z^{max}}{2} \right) \tag{33}$$

Note that the GG mobility model allows for spatial and temporal traces of the displacements of individuals. The simulation area can be either bounded or boundless. In case of a boundless simulation area, the area will be wrapped around its opposite sides, also in a toroidal-like manner, see (Ranjan Roy, 2011).

The experiments done in (Borrel, Dias De Amorim & Fdida, 2005) show that the GG mobility model produces bipartite graphs that have an equiprobable spatial distribution. Note that the GG model weights the preferential attachment by the inverse of the distance. Projections of the spatial distribution of individuals and attractors following the GG mobility model are depicted in Figure 29. The green points express individuals and the red circles stand for attractors. Note that attractors locate in the coordinates, where individuals converge. One clearly sees from the figure that there are differences in the spatial density and pathways are established between popular attractors.

Discussion: The GG mobility model is devised from observations in nature, which show that human organizations have scale-free spatial distributions. Such distributions are adequate for population growth. So, the GG model helps in solving many problems of large-scale mobile ad hoc networks. As known, such networks appear in metropolitan areas, cities centers, etc. and face serious challenges when MNs density clearly increases.

Although individuals do not belong explicitly to groups, they evolve independently and exhibit a collective behavior resulting implicitly from the impacts of individuals and attractors on other individuals. It is deduced from the model that individuals gather around the centers of interest, especially those with high popularity.

Beside the benefits we mentioned, there are also several weaknesses since there is no possibility to represent rigid dedicated movements of groups or individuals, such as movements on streets, within gardens, etc. Moreover, no structures of groups can be maintained, this prohibits the usage of the GG mobility model to model a wide range of applications, where the structure of groups is important.

Further research are necessary to accurately characterize the parameters space of the GG mobility model. Other types of population dynamics, e.g. steady populations, renewing populations, decreasing populations, etc., must also be investigated. It is important to investigate if the scale-free distribution of attractors remains maintained under these populations.

Limitations: As mentioned, the GG model has scale-free spatial distributions, which implies that there is a spatial dependency of the location of individuals among each other and with attractors. There is,

Figure 29. Two example spatial distributions of individuals and attractors generated by the GG mobility model in bounded (a) and boundless space (b) (Borrel, Dias De Amorim & Fdida, 2005)

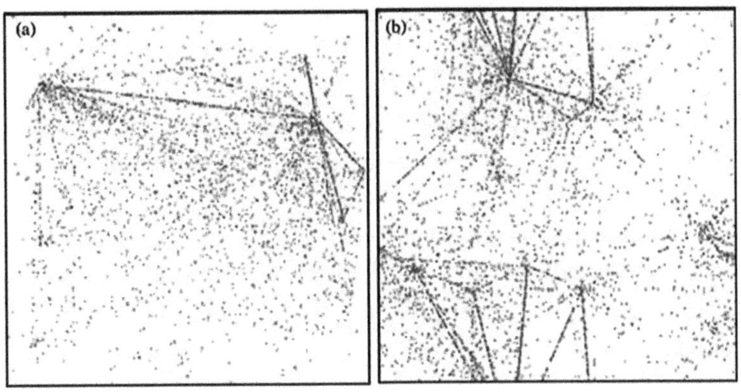

however, no temporal dependency of the location. Concerning the velocity, there is no temporal as well as spatial dependency of the velocity. There exist no geographic restrictions of movements. However, pause times between movements are included.

3.8. Summary

Table 2 summarizes the survey we introduced in this section and provides a qualitative comparison between the reviewed models with respect to the temporal and spatial dependency. When taking a look at the temporal dependency, one notes that no temporal dependency is supported when considering movements of groups' members. This is because all studied models assume the RW model as the standard model to be used to model the movements of the members. Of course, other models may be used as we mentioned. Concerning the temporal dependency of groups' movements, most studied mobility models support conditionally this dependency. In fact, this depends on the mobility model applied to model the movements of the groups. Exceptions are found in the column model, VTBG model and GG model. The column model supports such dependency, while other both mobility models do not.

Let us now consider the spatial dependency. When we consider the movements of groups' members, one notes that there is in general no spatial dependency of the velocity. However, there is a spatial dependency of the direction between the members and the groups' leaders/reference points. This dependency is expected since members have to move in groups. For the movement of groups themselves, the existence of a spatial dependency of the velocity as well as direction between groups depends on the mobility model applied to implement groups' movements. Exceptions exist, of course. The SG model supports such dependency, while the column model, VTBG model and GG model do not.

Table 3 provides a qualitative comparison between the reviewed models with respect to the support of geographical restrictions and pause times. Concerning the geographic restrictions, one notes that the majority of the models do not support the definition of streets with specific mobility characteristics. There is in general no support of obstacles as well. Concerning the pause times, the majority of the mobility models studied do not include pause times between movements of groups' members. However, pause times may exist between movements of groups. This depends in fact on the mobility model applied to model the movements of the groups.

4. OBSTACLES MODELING

Another key issue heavily related to the modeling of human patterns is the modeling of obstacles, e.g. buildings, walls, regions where moving inside is forbidden, etc. This problem is highly intertwined with the definition of realistic radio propagation models (Sarkar, Wicks, Salazar-Palma & Bonneau, 2004). Until now, there are very few mobility models that considered the modeling of obstacles by concept. The modeling of obstacles is, in fact, an open research issue that necessitates deeper researches and investigations.

When taking a look at the survey we introduced above, one notes that all mobility models that introduce geographic restrictions on movements include somehow a definition of obstacles either explicitly or implicitly. One also notes that two main principles followed to express the existence of obstacles inside simulated scenarios.

Table 2. A qualitative comparison between the synthetic group mobility models investigated

Model	Temporal Dependency				Spatial Dependency			
	Groups' Members[20]		Groups		Groups' Members		Groups	
	Velocity	Direction	Velocity	Direction	Velocity	Direction	Velocity	Direction
RPG model	N	N	Mobility model-dependent		N	Between group's members and the reference point	Mobility model-dependent	
In-place model	N	N	Mobility model-dependent		N	Between group's members and the reference point	Mobility model-dependent	
Overlay model	N	N	Mobility model-dependent		N	Between group's members and the reference point	Mobility model-dependent	
Convention model	N	N	N (for fixed groups) Mobility model-dependent (for moving groups)		N	Between group's members and the reference point	N (for fixed groups) Mobility model-dependent (for moving groups)	
Modified RPG model	N	N	Mobility model-dependent		N	Between group's members and the reference point	Mobility model-dependent	
Column model	N	N	Y	Y	N	Between group's members and the reference point	N	N
Nomadic community model	N	N	Mobility model-dependent		N	Between group's members and the reference point	Mobility model-dependent	
Pursue model	N	N	Mobility model-dependent		N	Between group's members and the reference point	Mobility model-dependent	
RVG model	N	N	Mobility model-dependent		Between group's members and the reference point		Mobility model-dependent	
RVAG model	N	N	Mobility model-dependent		Between group's members and the reference point		Mobility model-dependent	
SG model	N	N	Mobility model-dependent		Between group's members and the reference point		Y	Y
VTBG model	N	N	N	N	Between group's members and the reference point		N	N
DG model	Y	N	Mobility model-dependent		Between followers and leaders		Mobility model-dependent	
GG model	N	N	N	N	N	Yes (among individuals) Between individuals and attractors	N	N

1. **Implicit Definition of Obstacles:** here, obstacles are not clearly defined in the simulation area. Rather, the scenario introduces restrictions that affect the movements of MNs, so that an obstacle-like behavior results. This is done by utilizing one of the following methods.
    a. **Division of the Simulation Area into Adjacent/Overlapping Regions:** This is the weakest method applied to express obstacles, since it restricts somehow the movements of MNs. Examples of models following such method include the in-place and overlapping mobility models.
    b. **Definition of Tracks and Streets:** In this way, users are allowed to move only on streets or tracks. This implies that no movements are possible in areas where no streets/tracks are defined. Sure, these areas express then obstacles that restrict the movements of MNs. Example

Table 3. A qualitative comparison between the synthetic group mobility models investigated

Model	Geographical Restrictions				Pause Times	
	Groups' Members		Groups		Groups Members	Groups
	Paths, Streets, etc.	Obstacles	Paths, Streets, etc.	Obstacles		
RPG model	N	N	Mobility model-dependent		N	Mobility model-dependent
In-place model	Y (area is divided into adjacent regions)		Y (area is divided into adjacent regions)		N	Mobility model-dependent
Overlay model	Y (area is divided into overlapped regions)		Y (area is divided into overlapped regions)		N	Mobility model-dependent
Convention model	N	N	N	N	N	Y (moving groups stop at the location of fixed groups)
Modified RPG model	N	N	Mobility model-dependent		N	Mobility model-dependent
Column model	N	N	N	N	N	N
Nomadic community model	N	N	Mobility model-dependent		N	Mobility model-dependent
Pursue model	Mobility model-dependent		Mobility model-dependent		Mobility model-dependent	
RVG model	N	N	Mobility model-dependent		N	Mobility model-dependent
RVAG model	N	N	Mobility model-dependent		N	Mobility model-dependent
SG model	N	N	Mobility model-dependent		Y (only, if groups has pause times)	Mobility model-dependent
VTBG model	Y (for groups members) N (for individual moving MNs)	N	Y	N	N	N
DG model	Can be obtained by imposing superstructures on the top of the model		Can be obtained by imposing superstructures on the top of the model		N	Mobility model-dependent
GG model	N	N	N	N	Y	Y

mobility models include the VTBG model, CS model, Graph-based model, AGb model, CosMos model, Freeway model, etc.

c. **Definition of Rules to Control the Movement Behavior of MNs:** The mobility models that follow this method do not construct streets or regions with specific mobility characters inside the simulation area. Rather, they determine movements' characteristics of MNs via rules or motion profiles. So, restrictions may be included in the rules or inside the motion profiles. The KD mobility model is an example of such models.

2. **Explicit Definition of Obstacles**: There are very few models that define obstacles explicitly. The well-known mobility models that consider obstacles are the Voronoi-based and the obstacle mobility models. These models consider buildings and convert their entrances to intersections in the Voronoi-diagrams they construct. Another important model that significantly contributed to the issue of modeling obstacles is the behavioral model, which enables defining any type of obstacles inside the simulation scenario. The impact of this obstacle is reflected via a force controlled by a specific rule. Other mobility models that consider obstacles are the disaster-area and the UoMM mobility models.

REFERENCES

Ali, S., Mitschele-Thiel, A., Diab, A., & Rasheed, A. (2010). *A Survey of Services Placement Mechanisms for Future Mobile Communication Networks*. In FIT 10, 8th International Conference on Frontiers of Information Technology, Islamabad, Pakistan.

Alnasouri, E., Mitschele-Thiel, A. & Diab, A. (2009). A New Distributed Dynamic Algorithm for Mobility Patterns Prediction. In *Proceeding of PE-WASUN'09*.

Aschenbruck, N., Gerhards-Padilla, E., Gerharz, M., Frank, M., & Martini, P. (2007). Modelling Mobility in Disaster Area Scenarios. In *Proc. 10th ACM IEEE Int. Symp. Model. Anal. Simul. Wirel. Mob. Syst. MSWIM*. doi:10.1145/1298126.1298131

Aschenbruck, N., Gerhards-Padilla, E. & Martini, P. (2008). A Survey on Mobility Models for Performance Analysis in Tactical Mobile Networks. *Journal of Telecommunication and Information Technology*.

Bai, F., Sadagopan, N., & Helmy, A. (2003). IMPORTANT: A Framework to Systematically Analyze the Impact of Mobility on Performance of Routing Protocols for Ad-Hoc Networks. In *Proc. IEEE INFOCOM*. doi:10.1109/INFCOM.2003.1208920

Barabási, A. L., & Albert, R. (1999). Emergence of Scaling in Random Networks. *Science*, 286. PMID:10521342

Basu, P., Redi, J., & Shurbanov, V. (2004). Coordinated Flocking of UAV for Improved Connectivity of Mobile Ground Nodes. In *Proceeding of the Military Communications Conference (MILCOM)*. doi:10.1109/MILCOM.2004.1495182

Bellavista, P., Corradi, A., & Magistretti, E. (2005). REDMAN: An Optimistic Replication Middleware for Read-only Resources in Dense MANETs. *Journal on Pervasive and Mobile Computing*, *1*(3), 279–310. doi:10.1016/j.pmcj.2005.06.002

Bettstetter, C. (2001). Mobility Modeling in Wireless Networks: Categorization, Smooth Movement, and Border Effects. *ACM SIGMOBILE Mob. Comp. Commun. Rev.*, *5*(3), 55–66. doi:10.1145/584051.584056

Bettstetter, C. (2001a). Smooth is Better Than Sharp: A Random Mobility Model for Simulation of Wireless Networks. In *Proc. 4th Int. Symp. Model. Anal. Simul. Wirel. Mob. Syst. MSWIM*.

Bettstetter, C., Hartenstein, H., & Perez-Costa, X. (2003). *Stochastic Properties of the Random Waypoint Mobility Model*. Kluwer Academic Publishers.

Bettstetter, C., & Wagner, C. (2002). The Spatial Node Distribution of the Random Waypoint Mobility Model. In *Proceeding of 1st German Workshop on Mobile Ad-Hoc Networks. WMAN'02*.

Bittner, S., Raffel, W.-U., & Scholz, M. (2005). The Area Graph-based Mobility Model and its Impact on Data Dissemination. In *Proc. IEEE PerCom*. doi:10.1109/PERCOMW.2005.79

Blakely, K., & Lowekamp, B. (2004). *A Structured Group Mobility Model for the Simulation of Mobile Ad Hoc Networks*. MobiWac. doi:10.1145/1023783.1023805

Blazevic, L., Le Boudec, J.-Y., & Giordano, S. (2005). A Location-Based Routing Method for Mobile Ad Hoc Networks. *IEEE Transactions on Mobile Computing*, *3*(4).

Borrel, V., Dias De Amorim, M., & Fdida, S. (2005). A Preferential Attachment Gathering Mobility Model. *IEEE Communications Letters*, *9*(10), 900–902. doi:10.1109/LCOMM.2005.10023

Boudec, J. L., & Vojnovic, M. (2005). Perfect Simulation and Stationary of a Class of Mobility Models. In *Proc. IEEE INFOCOM*.

Camp, T., Boleng, J., & Davies, V. (2002). A Survey of Mobility Models for Ad Hoc Network Research. *Wireless Communication and Mobile Computing Special Issue on Mobile Ad Hoc Networking: Research. Trends and Applications*, *2*(5), 483–502.

Chandra, R., Ramasubramanian, V., & Birman, K. (2001). *Anonymous Gossip: Improving Multicast Reliability in Mobile Ad-Hoc Networks. 21st International Conference on Distributed Computing Systems*, Mesa, Arizona. doi:10.1109/ICDSC.2001.918957

Chen, W., & Chen, P. (2003). *Group Mobility Management in Wireless Ad Hoc Networks*. VTC.

Chiang, C. (1998). *Wireless Network Multicasting*. (PhD Thesis). Department of Computer Science, University of California, Los Angeles, CA.

Choffnes, D. R., & Bustamante, F. E. (2005). An Integrated Mobility and Traffic Model for Vehicular Wireless Networks. In *Int. Conf. Mob. Comp. Netw., Proc. 2nd ACM Int. Worksh. Veh. Ad Hoc Netw*. doi:10.1145/1080754.1080765

Coppens, J., Wauters, T., De Turck, F., Dhoedt, B., & Demeester, P. (2005). Evaluation of Replica Placement and Retrieval Algorithms in Self-Organizing CDNs. In *Proceeding of the IFIP/IEEE International Workshop on Self-Managed Systems & Services (SelfMan'05)*.

Davies, V. (2000). *Evaluating Mobility Models Within An Ad Hoc Network*. (Master's thesis). Colorado School of Mines.

Diab, A. (2010). *Mobility Management in IP-Based Networks: Analysis, Design, Programming and Computer-based Learning Modules*. (Dissertation thesis). Ilmenau University of Technology.

Diab, A., & Mitschele-Thiel, A. (2014). Self-Organization Activities in LTE-Advanced Networks. In Handbook of Research on Progressive Trends in Wireless Communications and Networking. IGI Global. Doi:10.4018/978-1-4666-5170-8

Dobson, S., et al. (2006). A Survey of Autonomic Communications. *ACM Transaction on Autonomous and Adaptive Systems, 1*(2).

ETSI TR 101 112. (1998). *Universal Mobile Telecommunicatios System (UMTS); Selection Procedures for the Choice of Radio Transmission Technologies of the UMTS (UMTS 30.03 version 3.2.0). ETSI TR 101 112 V3.2.0 (1998-04).*

Franois, J. M. (2007). *Performing and Making Use of Mobility Prediction.* (Ph.D. Dissertation). University of Lige.

Garza, C., Ashai, B. H., Monturus, E., & Syputa, R. (2010). *LTE Operator Commitments: Deployment Scenarios and Growth Opportunities.* Research report. MARAVEDIS Wireless Market Research & Analysis.

Giner, V. C. (2004). *Study of mobility behavior of new Internet mobile users. Delivery Version No. 1.* Euro-NGI.

Gonzalez, M. C., Hidalgo, C. A., & Barabasi, A.-L. (2008). *Understanding Individual Human Mobility Patterns. Nature Journal.*

Gramoli, V., Kermarrec, A. M., & Le Merrer, E. (2008). SONDe, a Self-Organizing Object Deployment Algorithm in Large-Scale Dynamic Systems. In *Proceeding of 7th conference on European Dependable Computing Conference.* doi:10.1109/EDCC-7.2008.17

Guenes, M., & Siekermann, J. (2005). CosMos – Communication Scenario and Mobility Scenario Generator for Mobile Ad-Hoc Networks. In Proc. 2nd Int. Worksh. MANETs Interoper. Iss. MANETII'05.

Haas, Z. (1997). A New Routing Protocol for Reconfigurable Wireless Networks. In *Proceedings of the IEEE International Conference on Universal Personal Communications (ICUPC).* doi:10.1109/ICUPC.1997.627227

Handorean, R., Gill, C., & Roman, G-C. (2004). Accommodating Transient Connectivity in Ad Hoc and Mobile Settings. *LNCS, 3001.*

Helbing, D. (2001). Traffic and Related Self-Driven Many-Particle Systems. *Journal of Review of Modern Physic. Reviews of Modern Physics, 73*(4), 1067–1141. doi:10.1103/RevModPhys.73.1067

Helbing, D., Farkas, I., & Vicsek, T. (2000). Simulating Dynamical Features of Escape Panic. *Nature,* 407. PMID:11028994

Holzer, R. (2009). *Modeling and Control of Complex and Self-Organizing Systems. Technical Report, Number MIP-0916.* Department of Informatics and Mathematics, University of Passau.

Hong, X., Gerla, M., Pei, G., & Chiang, C.-C. (1999). A Group Mobility Model for Ad-Hoc Wireless Networks. In *Proceedings of the 2nd ACM international workshop on Modeling, analysis and simulation of wireless and mobile systems (MSWiM'99).* doi:10.1145/313237.313248

Hsu, W., Merchant, K., Shu, H.-W., Hsu, C.-H., & Helmy, A. (2004). *Weighted Waypoint Mobility Model and its Impact on Ad Hoc Networks.* MobiCom.

Isaacman, S. (2011). Identifying Important Places in People's Lives from Cellular Network Data. In *Proceeding of the 9th International Conference on Pervasive Computing (Pervasive'11).*

Isaacman, S. (2011a). *Ranges of Human Mobility in Los Angeles and New York.* In 8th IEEE Workshop on Managing Ubiquitous Communications and Services (MUCS' 11), Seattle, WA. doi:10.1109/PERCOMW.2011.5766977

Isaacman, S. (2012). Human Mobility Modeling at Metropolitan Scales. In *Proceeding of the 10th International Conference on Mobile Systems (MobiSys'12).*

Isaacman, S., Becker, R., Cáceres, R., Kobourov, S., Rowland, J., & Varshavsky, A. (2010). *A Tale of Two Cities.* In *Workshop on Mobile Computing Systems and Applications (HotMobile'10).*

Jardosh, A., Belding-Royer, E. M., Almeroth, K. C., & Suri, S. (2003). Towards Realistic Mobility Models for Mobile ad hoc Networks. In *Proc. IEEE MobiCom.* doi:10.1145/938985.939008

Karp, B. (2000). *Geographic Routing for Wireless Network.* (PhD thesis). Harvard University.

Khan, S. N., Diab, A., Brosch, C., & Ahmad, M. (2012). Self-Organized Disjoint Service Placement in Future Mobile Communication Networks. *International Journal of Communication Networks and Information Security, 4*(2).

Kim, T., Chung, M.-Y., & Sung, D.-K. (1998). Mobility and traffic analyses in three-dimensional PCS environments. *IEEE Transactions on Vehicular Technology, 47*(2).

Kraaier, J., & Killat, U. (2005). The Random Waypoint City Model – User Distribution in a Street-based Mobility Model for Wireless Network Simulations. In *Proc. 3rd ACM Int. Worksh. Wirel. Mob. Appl. Serv. WLAN Hotsp.*

Krbalek, M., & Helbing, D. (2004). Determination of Interaction Potentials in Freeway Traffic from Steady-State Statistics. *Journal of Physica A-statistical Mechanics and Its Applications - Physica A, 333*, 370–378. doi:10.1016/j.physa.2003.10.059

Legendre, F., Borrel, V., Dias de Amorim, M., & Fdida, S. (2006). Reconsidering Microscopic Mobility Modeling for Self-Organizing Networks. *Network.*

Li, Q., & Rus, D. (2003). Communication in disconnected ad hoc networks using message relay. *Journal of Parallel and Distributed Computing, 63*(1), 75–86. doi:10.1016/S0743-7315(02)00033-3

Liang, B., & Haas, Z. (1999). Predictive Distance-Based Mobility Management for PCS Networks. In *Proceedings of the Joint Conference of the IEEE Computer and Communications Societies (INFOCOM).* doi:10.1109/INFCOM.1999.752157

Lim, S., Yu, C., & Da, C. R. (2006). Clustered Mobility Model for Scale-free Wireless Networks. In *Proc. IEEE Conf. Loc. Comput. Netw. LCN.*

Lopez-Nores, M., García-Duque, J., Pazos-Aria, J.-J., & Díaz Redondo, R.-P. (2008). Qualitative Assessment of Approaches to Coordinate Activities of Mobile Hosts in Ad Hoc Networks. *IEEE Communications Magazine, 46*(12), 108–111. doi:10.1109/MCOM.2008.4689216

Markoulidakis, J.-G., Lyberopoulos, G.-L., Tsirkas, D.-F., & Sykas, E.-D. (1997). *Mobility modeling in third-generation mobile telecommunications systems*. IEEE Personal Communications.

McNett, M., & Voelker, G. (2005). Access and Mobility of Wireless PDA Users. *Mobile Computing and Communications Review*, *9*(2), 40. doi:10.1145/1072989.1072995

Mishra, A., Shin, M., & Arbaugh, W.-A. (2004). *Pro-Active Key Distribution Using Neighbor Graphs*. IEEE Wireless Communications.

Musolesi, M., & Mascolo, C. (2009). *Mobility models for systems evaluation: A Survey*. In B. Garbinato, H. Miranda, & L. Rodrigues (Eds.), *Middleware for Network Eccentric and Mobile Applications* (pp. 43–62). Springer. doi:10.1007/978-3-540-89707-1_3

Navidi, W., & Camp, T. (2004). Stationary Distributions for the Random Waypoint Mobility Model. *IEEE Transactions on Mobile Computing*, *3*(1), 99–108. doi:10.1109/TMC.2004.1261820

Noulas, A., Scellato, S., Lambiotte, R., Pontil, M., & Mascolo, C. (2012). A Tale of Many Cities: Universal Patterns in Human Urban Mobility. *PLoS ONE* *7*(9). doi: .10.1371/annotation/ca85bf7a-7922-47d5-8bfb-bcdf25af8c72

Ranjan Roy, R. (2011). Handbook of Mobile Ad Hoc Networks for Mobility Models. Springer Science+Business Media.

Rossi, M., Badia, L., Bui, N., & Zorzi, M. (2005). *On Group Mobility Patterns and their Exploitation to Logically Aggregate Terminals in Wireless Networks*. VTC. doi:10.1109/VETECF.2005.1558429

Rossi, M., Badia, L., Giacon, P., & Zorzi, M. (2005). *On the Effectiveness of Logical Device Aggregation in Multi-Radio Multi-hop Networks*. MobiWac. doi:10.1109/WIRLES.2005.1549435

Royer, E., Melliar-Smith, P. M., & Moser, L. (2001). An Analysis of the Optimum Node Density for Ad Hoc Mobile Networks. In *Proceedings of the IEEE International Conference on Communications (ICC)*. doi:10.1109/ICC.2001.937360

Sanchez, M., & Manzoni, P. (2001). Anejos: A java based simulator for ad-hoc networks. *Future Generation Computer Systems*, *17*(5), 573–583. doi:10.1016/S0167-739X(00)00040-6

Sarkar, T. K., Wicks, M. C., Salazar-Palma, M., & Bonneau, R. J. (2004). *A Survey of Various Propagation Models for Mobile Communication*. Smart Antennas.

Sen, R., Handorean, R., Roman, G.-C., & Hackmann, G. (2004). *Knowledge-Driven Interactions with Service Across Mobile Ad Hoc Networks*. Technical Report WUCSE-2004-34. Washington University, Department of Computer Science.

Song, C., Qu, Z., Blumm, N., & Barabási, A.-L. (2010). Limits of Predictability in Human Mobility. *Science*, 327. PMID:20167789

Stepanov, I., Haehner, J., Becker, C., Tian, J., & Rothermel, K. (2003). A Meta-Model and Framework for User Mobility in Mobile Networks. In *Proc. 11th Int. Conf. Netw. ICON 2003*. doi:10.1109/ICON.2003.1266195

Stepanov, I., Marron, P. J., & Rothermel, K. (2005). Mobility Modeling of Outdoor Scenarios for MANETs. In *Proc. 38th Ann. Simul. Symp. ANSS'38*. doi:10.1109/ANSS.2005.29

Su, W., Lee, S. & Gerla, M. (2001). Mobility Prediction and Routing in Ad-Hoc Wireless Networks. *International Journal of Networking Management*.

Tian, J., Haehner, J., Becker, C., Stepanov, I., & Rothermel, K. (2002). Graph-based Mobility Model for Mobile Ad Hoc Network Simulation. In *Proc. 35th Ann. Simul. Symp.*

Tolety, V. (1999). *Load Reduction in Ad Hoc Networks using Mobile Servers*. (Master's thesis). Colorado School of Mines.

Turgut, D., Das, S.-K., & Chatterjee, M. (2001). *Longevity of Routes in Mobile Ad Hoc Networks*. VTC. doi:10.1109/VETECS.2001.944118

Wang, K. H., & Li, B. (2002). Group Mobility and Partition Prediction in Wireless ad hoc. *Networks*.

Weisstein, E. W. (1998). *The CRC Concise Encyclopedia of Mathematics*. CRC Press.

Yoon, J., Liu, M., & Noble, B. (2003). Random Waypoint Considered Harmful. In *Proceeding of IEEE INFOCOM*.

Zhou, B., Xu, K., & Gerla, M. (2004). *Group and Swarm Mobility Models for Ad Hoc Network Scenarios using Virtual Tracks*. MILCOM'04.

Zimmermann, H.-M., & Gruber, I. (2005). A Voronoi-based Mobility Model for Urban Environments. In *Eur. Wirel Conf.*, Zypern, Greece.

Zonoozi, M., & Dassanayake, P. (1997). User Mobility Modeling and Characterization of Mobility Pattern. *IEEE Journal on Selected Areas in Communications*, *15*(7), 1239–1252. doi:10.1109/49.622908

ENDNOTES

1 The MN locates initially in the center of the simulation area. At each movement, the MN selects a direction between 0 and 2π and a speed between 0 and 10 m/s. Δt equals to 60 ms.

2 The initial location of the MN is selected randomly. At each movement, the MN selects a direction between 0 and 2π and a speed between 0 and 10 m/s.

3 The MN locates initially in the center of the simulation area and moves following that.

4 The MN locates initially in the center of the simulation area and moves following that. A movement happens at each 10 s.

5 The MN locates initially in the center of the simulation area and moves for 1000 s. A movement happens at each 1 s, α is 0.75, $S_{x_{n-1}}$ and $d_{x_{n-1}}$ are random variables from Gaussian distribution (mean value = 0 and standard deviation = 1), $\bar{S}$ is 10 m/s and $\bar{d}$ is set initially to 90 degrees.

6 V_{max} is 10 m/s, A_{max} is 10 m/s^2, α is 90 degrees and Δt is 0.1 s. The MN locates initially in the center of the simulation area and moves for 500 s.

7 The center-most vertical and horizontal streets are mid-speed roads (dashed lines), while other streets are slow-speed roads (solid lines). The MN locates initially at (1,1), moves then to (5,4) and (1,4), respectively. The double lines represent the streets traveled by the MN.

8 The autonomy means self-governing, i.e. almost the system requires no external control, see (Holzer, 2009).

9 The new velocity is equal to the old one, since the velocity remains constant. So, there is no temporal dependency of the velocity.

10 Implicitly, the new location of a MN may be affected by the old location. This depends on the distance to the attraction points and on the scenario being running.

11 See the previous footnote.

12 The new velocity is equal to the old one, since the velocity remains constant. So, there is no temporal dependency of the velocity.

13 The behavior of streets can be defined via rules. However, not an explicit street with given mobility characteristics.

14 In fact, there is a temporal dependency of the location when considering the selection of the new cluster, since transitions between clusters are done based on assigned probabilities. For movements inside the clusters, there is, however, no such dependency.

15 This depends on the mobility model applied inside the non-overlapping zones.

16 Similar to the AGb model, there is a temporal dependency of the location when selecting the destination zone. Concerning the movements inside the zones, this depends of the mobility model applied.

17 This depends, in fact, on the selected parameters of the model.

18 The RVG model characterizes the mobility of MNs by the velocity and the distance.

19 $\alpha = 1$, $\mu_\beta = 0.15$, $\sigma_\beta = 0.25$ (top), $\alpha_\beta = 0.5$ (bottom), $C_l = C_f = 2$, $\Delta t = 1$ s.

20 We assume here the RW model for groups' members. This is the most used model.

Chapter 12
Trace- and Social-Based Modeling of Human Mobility Patterns

Ali Diab
Al-Baath University, Syria & Ilmenau University of Technology, Germany

Andreas Mitschele-Thiel
Ilmenau University of Technology, Germany

ABSTRACT

The 5th Generation (5G) of mobile communication networks is being developed to address the demands and business contexts of 2020 and beyond. Its vision is to enable a fully mobile and connected society and also to trigger socio-economic transformations in ways eventually unimagined today. This means that the physical world to be covered with planned 5G networks including communication networks, humans and objects is becoming a type of an information system. So as to improve the experience of individuals, communities, societies, etc. within such systems, a thorough comprehension of intelligence processes responsible of generating, handling and controlling those data is fundamental. One of the major aspects in this context and also the focus of this chapter is the development of novel methods to model human mobility patterns, which have crucial role in forthcoming communication technologies. Human mobility patterns models can be categorized into synthetic, trace-based and community-based models. Synthetic models are largely preferred and widely applied to simulate mobile communication networks. They try to capture the patterns of human movements by means of a set of equations. These models are traceable, however, not capable of generating realistic mobility models. The key idea of trace-based models is the exploitation of available measurements and traces achieved in deployed systems to reproduce synthetic traces characterized by the same statistical properties of real traces. A main drawback of trace-based modeling of human patterns is the tight coupling between the trace-based model and the traces collected, the network topology deployed and even the geographic location, where the traces were collected. This is why the results of various trace-based models deviate clearly from each other. Sure, this prohibits the generalization of trace-based models. When one also considers that the traces themselves are rarely available, one can understand why synthetic models are preferred over trace-based ones. Community-based modeling of human movements depends on the fact stating that

DOI: 10.4018/978-1-5225-0239-5.ch012

mobile devices are usually carried by humans, which implies that movement patterns of such devices are necessarily related to human decisions and socialization behaviors. So, human movement routines heavily affect the overall movement patterns resulting. One of the major contributions in this context is the application of social networks theory to generate more realistic human movement patterns. The chapter highlights the state of art and provides a comprehensive investigation of current research efforts in the field of trace- and social-based modeling of human mobility patterns. It reviews well-known approaches going through their pros and cons. In addition, the chapter studies an aspect that heavily relates to human mobility patterns, namely the prediction of future locations of users.

1. INTRODUCTION

The 5th Generation (5G) of mobile communication networks is being developed to address the demands and business contexts of 2020 and beyond. It`s vision is to enable a fully mobile and connected society and also to trigger socio-economic transformations in ways eventually unimagined today. This means that the physical world to be covered with planned 5G networks including communication networks, humans and objects is becoming a type of an information system. So as to improve the experience of individuals, communities, societies, etc. within such systems, a thorough comprehension of intelligence processes responsible of generating, handling and controlling those data is fundamental. One of the major aspects in this context and also the focus of this chapter is the development of novel methods to model human mobility patterns, which have crucial role in forthcoming communication technologies.

One of the major aspects in this context and also the focus of this chapter is the development of novel methods to model human mobility patterns, which have myriad uses in crucial fields such as mobile communication research, urban planning, ecology, epidemiology, etc. It is proven that human mobility patterns are far from being random, see (Gonzalez, Hidalgo, & Barabasi, 2008). A high degree of temporal and spatial regularity is observed in these patterns, which implies that they follow reproducible patterns. So, faithfully reproduction of the movements of real people significantly helps in answering crucial questions necessary to improve future technologies.

Models of human mobility patterns are grouped in three groups, namely synthetic, trace-based and community-based models, see (Musolesi, & Mascolo, 2009), (Camp, Boleng, & Davies, 2002) and (Ranjan Roy, 2011). Synthetic models are mathematical models aiming at capturing the patterns of human movements by means of a set of equations. Trace-based models exploit available measurements and traces achieved in deployed systems, e.g. connectivity logs, location information, etc., to reproduce synthetic traces characterized by the same statistical properties of real traces. Community-based modeling of human movements depends on the fact that mobile devices are normally carried by humans. This means that movement patterns of such devices relate to human decisions and socialization behaviors. A major contribution in this context is the application of social network theory to generate more realistic human movement patterns, see (Musolesi, & Mascolo, 2009).

The chapter highlights the state of art and provides a comprehensive investigation of current research efforts in the field of trace- and social-based modeling of human mobility patterns. It reviews well-known approaches going through their pros and cons. In addition, the chapter studies an aspect that heavily relates to human mobility patterns, namely the prediction of future locations of users.

2. TRACE-BASED MODELING OF HUMAN MOBILITY PATTERNS

Trace-based mobility models, also termed empirical models, characterize movement patterns based on connectivity information that represent the distribution of users' locations within the environment and the movements of humans carrying wireless devices. These models depend on traces gathered by already deployed cellular networks or GPS-enabled devices to obtain information about the locations of users. A well-known open-source project issued to hold such sets of traces is the Community Resource for Archiving Wireless Data At Dartmouth (CRAWDAD) (Kotz, Henderson & Abyzov, 2015). The project is one of the largest projects in this scope. It contains significant traces that are considered by many researchers. Another relevant set of traces were gathered by the Networking Research Lab of North Carolina State University (NCSU) from 100 volunteers in five different sites (two university campuses, New York City, Disney World, and North Carolina state fair) in the duration between 2006 and 2008 (Networking Research Lab, 2015). Out of these traces, two main properties were extracted, namely the flight length and pause time, see (Ribeiro, 2011).

The main goal of trace-based models is to refine existing mobility models based on available mobility traces to make them more realistic (Kotz & Henderson, 2005) and capable of generating synthetic traces characterized by the same statistical properties of real ones. The following provides an insight into the main results derived from traces and reviews well-known empirical mobility models.

2.1. Users Behavior in Mobility Traces

The authors in (Tang & Baker, 2000) analyzed a twelve-week trace of a building-wide Wireless Local Area Network (WLAN). The analysis focused on studying the overall user behavior and network traffic, in addition to, load characteristics. Questions of interest included, when and how intensively do users access to the network, how much do they move around, what is the observed throughput, what are the types of traffic users exchange, etc. The authors have noted that most users access the network to surf the web, do session-oriented activities and chat with other users. In fact, there was a high number of chats, which showed that synchronous communication with others is preferred by users. Experienced peak throughput is usually caused by a single user and application. During the peak throughput duration, the outgoing traffic dominates incoming traffic. During other times, outgoing traffic is significantly less that incoming one. The main result deduced from the analysis is that users are divided into distinct location-based sub-communities. Each sub-community has its own movements, activities, and usage characteristics.

So as to extend the understanding of users' behaviors and wireless networks performance, traces captured at a well-attended ACM conference (the ACM SIGCOMM'01 conference held at U.C. San Diego, August 2001) were investigated in (Balachandran, Voelker, Bahl & Rangan, 2002). The network monitored is an IEEE 802.11b WLAN (IEEE specification, 1999) with 4 APs installed in a large auditorium. The APs were operating at 11 Mbps data rate and at a power of 100 mW. The users monitored are 195 users, roughly 40% of the total attendees. The results of the analysis have shown that most sessions are relatively short (< 10 minutes in the analysis). Although there are some longer sessions, however, they tend to be idle for the majority of the time. These results imply that in networks operating DHCP to dynamically configure new attending devices, it is useful to provide short-term leases, after which IP addresses are reclaimed or renewed. The dominant applications are the web surfing and secure shell applications (64% of all bytes, 58% of all flows). The authors also noted that average as well as peak bandwidth requirements of individual users are relatively small. This means that few APs are, in principle, adequate to serve relatively large number of users in such situations. An important result concerning the load distribution, namely the bandwidth is highly uneven and not well correlated with the number of users those APs serve. This implies that load balancing algorithms should not consider only the number of users. Concerning users' behavior, the authors noted that user arrivals into the network can be correlated according to a two-state Markov-modulated Poisson process with a mean inter-arrival time of 38s during the ON state and a mean duration of 6 minutes for the OFF state. The distribution of the session time can be approximated to a general Pareto distribution with a shape parameter of 0.78 and a scale parameter of 30.76.

Another attempt to understand user movement patterns as well as wireless network usage characteristics is presented in (Balazinska & Castro, 2003). The work presented the results obtained from a four week trace data collected in a large corporate environment. More concrete, 1366 users distributed in three buildings are tracked. The pattern used reflects a usual office environment traced during normal office work hours. The authors found that average user transfer-rates follow a power law. Network load is unevenly distributed across APs. Furthermore, the load is more affected by which users are present, rather than how many. An important result of the study is the note that the variations in the number of users over time adherence the patterns of the underlying population, although most users do not continuously remain online, they access the network for a fraction of time. So, the number of network users might be adequately modeled by scaling down general population models.

The results showed also that there are large differences in user movement patterns and data transfer rates. Usually, users spend a large fraction of their time at a single location, referred to as a home location, and do not change their network usage when being a way of their home or when changing their location frequently. The authors, therefore, proposed to model users' mobility by means of a persistence and prevalence. The persistence determines how long users remain associated with the same AP, while the prevalence defines how often users visit various APs. The authors state that the probability distribution of both persistence and prevalence follows a power law distribution. The prevalence is further used to classify users into various mobility categories. The categories are derived from the traces, since the authors mentioned that 50 to 80% of users are occasionally mobile. That is, they spend most of their time at a single location. The results of the investigation achieved in (Balazinska & Castro, 2003) were compared with other studies from the literature. The authors found similarities, especially to those observed at university campuses, offices, libraries, academic buildings, etc., see (Kotz & Essien, 2002), (Kotz & Essien, 2002a) and (Tang & Baker, 2000). They found, however, differences compared to the results obtained from large-scaled scenarios such as dormitories (Kotz & Essien, 2002), (Kotz & Essien, 2002a) and public metropolitan networks (Tang & Baker, 2002). In these scenarios for instance, the number of users clearly increases during evening hours and much lower reductions in the number of users are observed on weekends. Small-scale networks, e.g. a single conference room (Balachandran, Voelker, Bahl & Rangan, 2002), show clear variability in terms of the number of users due to the impact of users scheduled activities on the network.

An analysis of the usage of the Dartmouth College campus WLAN is provided in (Henderson, Kotz & Abyzov, 2004). This work analyzed more than 550 APs and 7000 users over seventeen weeks. A special important contribution of this study was the comparison of traces results to those taken from the same network direct after the initial deployment, two years ago. The authors found that the applications used in the network changed drastically. Initially, the network was mainly used for web surfing. After that, other applications started to appear and be widely used in the network, e.g. Peer-to-Peer (P2P), streaming multimedia, Voice over IP (VoIP) traffic, etc. Off-campus traffic was initially dominating, however, exceeded by on-campus traffic following that. The VoIP traffic was mainly used on the wired network and the calls were mainly short, most calls took less than 1 minute. The analysis of the devices used by clients has shown a considerable heterogeneity in the types of clients used including embedded devices, laptops, PDAs, etc.

To trace mobility of users, the authors defined a mobile session to be the session, whose diameter is larger than a threshold D. The diameter of a session is the maximum horizontal distance between any two consecutive APs visited during the session. Note that the definition of a diameter in such a way is necessary, since it is not possible to measure mobility of users only from logs into APs. There will be ping-pong handoffs many times in succession. Such behavior results, for instance, when the user locates in an overlapping area between two APs. Sure, there are other definitions of a session, e.g. the session is that associated with APs in more than one building (Kotz & Essien, 2003). Such definition was not accurate in the trace analysis done in (Henderson, Kotz & Abyzov, 2004), since ping-pong handoffs were observed between APs locating in different buildings. In general, the surprising result is that most users were non-mobile. Even those users, who moved around, remained close to their home approximately 98% of the time. Furthermore, different devices travel also differently. VoIP devices visited the largest number of APs, since they are "always on" and ready to receive calls. Laptops associated with a small number of APs since when a user wants to use his laptop, he will normally stop and connect to the network. Another result said that different devices have different session characteristics. In general, all device types had short session durations, less than 10 minutes in the traces. The shortest sessions' durations were seen by PDAs, while the broadest were observed by VoIP devices. Another interesting result says that mobility characteristics are affected by the applications being running. Figure 1 illustrates the distance travelled during VoIP, P2P and streaming sessions, see (Henderson, Kotz & Abyzov, 2004).

Figure 1. The distance travelled during VoIP, P2P and streaming sessions (Henderson, Kotz & Abyzov, 2004)

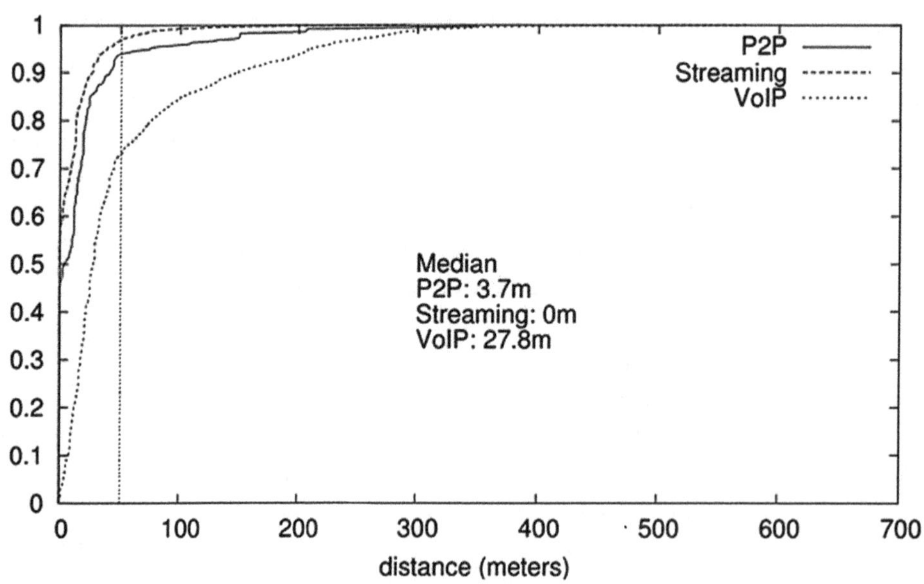

The figure shows that mobile devices running VoIP sessions travel in general further. The devices operating P2P sessions are less mobile than those running VoIP, however, more-mobile than those running streaming sessions. The reasons lay in the way humans interact with the applications being running. Video streaming may involve active user participations and normally requires that the user remains looking at the device. P2P applications necessitate less interactivity with users and can run in the background.

The authors in (Leguay, Friedman & Conan, 2005) focused on routing issues in ad hoc networks. They state that the routing of data can be considerably improved if knowledge concerning movement patterns could be utilized. The authors provided a generic routing algorithm based on a high-dimensional Euclidean space, termed MobySpace, and constructed based on movement patterns. Data about movement patterns were collected from traces derived from Dartmouth WLAN between January 26th, 2004 and March 11th, 2004. The distributions that illustrate users' behavior within the mentioned period are plotted in Figure 2.

Figure 2. Statistics illustrating the users' behavior in the applied mobility traces (Leguay, Friedman & Conan, 2005)

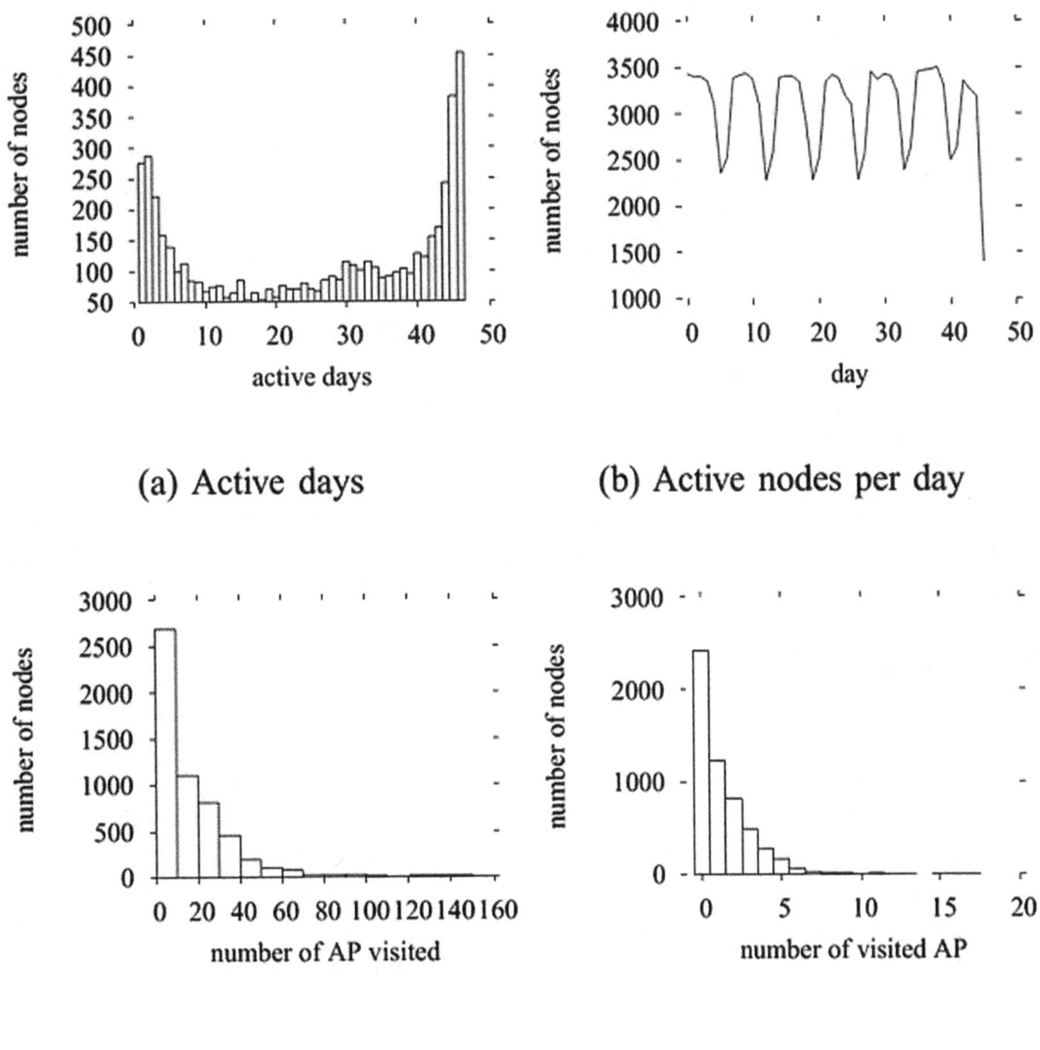

The total number of tracked mobile users was 5,545 users. They were allowed to visit 536 locations, where each location is served by an AP. The figure shows that the average number of locations visited is 16.66 in the whole period and 1.75 per day. The authors noted that the distributions of the number of locations visited follow heavy tailed distribution, which implies that the majority of users have a low level of mobility, while a small amount of users move actively. Note that users with low level of mobility are those not often involved in data exchange or those who prefer to remain or leave their wireless device in a specific place for a long time.

The mobility traces show that there is regularity in terms of the network usage, see the number of active users per day. Note that there are regular weekly cycles and approximately a constant number of active users, 2,901 users per day in the traces. This simply means that users move based on a number of approximately constant habits, which results in regular movement patterns. The analysis of the appearance of users in the network has shown that it is high probable that MNs will not disappear close to the beginning of the trace. Similarly, the probability that MNs will appear for the first time close to the end of the trace is also low. The appearance as well as disappearance of users is distributed over the whole time.

In brief, following results can be derived when abstracting the similarities between the traces we mentioned:

- The variations in the number of users over time are proper representations of the patterns of the underlying population.
- Users are divided into distinct location-based sub-communities, each has its own movements, activities, and usage characteristics.
- The majority of users have a low level of mobility, while a small amount of users move actively.
- Users move based on a number of approximately constant habits. This results in regular movement patterns that are, for sure, reproducible.
- Mobility characteristics are affected by the applications being running.
- There is regularity in the network usage, there are even regular weekly cycles.
- The load of the network is unevenly distributed across APs. Moreover, the load is more affected by which users are present, rather than how many.
- The dominant applications are the web surfing and synchronous communications, chat for instance. The sessions users initiate are, in general, short.

The following discusses well-known empirical models developed on the basis of traces collected from already deployed networks. Within each model, the results of traces analysis the developers noted will be mentioned.

2.2. WLAN Mobility Model

One of the first attempts to model users movements based on traces collected from already deployed networks is provided in (Tuduce & Gross, 2005). The authors measured mobility characteristics of users of a WLAN campus network of ETH Zurich. The WLAN monitored spread across 32 buildings with seminar and lecture rooms in addition to offices, however, without student dormitories. There were 166 APs (Cisco Aironet 340 and 350) installed, so that overlapping areas were minimized. Exceptions existed in seven heavily occupied lecture halls, where each hall was covered by more than one AP. Mobility data were gathered in two traces (each took nine weeks long) at different times (in 2003 and 2004). The first trace captured 3,073 users, who connected to the WLAN 97,575 times. The second trace captured 4,762 users, who connected to the WLAN 343,626 times.

The authors in (Tuduce & Gross, 2005) analyzed the number of users accessing the WLAN simultaneously. They noted that the number of mentioned users tends to follow the weekly sociological trends expected. Users heavily access the network during working hours. There is a significant decrease in the number of users accessing the network at the weekends. Concerning daily routines, the number of connected users starts to increase at 6 am and diminishes slowly after 6 pm. Note that there is a slight decrease over lunch time. The mentioned cyclic pattern adherences with other observations from studies of other networks, see (Balazinska & Castro, 2003), (Kotz & Essien, 2002) and (Tang & Baker, 2000).

The methodology the authors followed simply includes the analysis of mobility traces and the derivation of mobility model parameters. These parameters are applied, following that, to generate movement patterns for simulative studies. The mobility model from the authors point of view is seen as a collaboration between two processes, a spatial process (P_S) and a temporal process (P_T). The result of this collaboration will be a trajectory in the described environment. The environment itself is described by the boundaries of the physical space, inside which users are allowed to move, together with restrictions on user movements, e.g. obstacles, the paths users must move on, etc. The spatial process defines the subset of points the user has to visit during the simulation, i.e. the user's trajectory. The temporal process determines the time, at which the user must be at his destination. That is, the temporal process implies the speed of the user. Note that a spatial as well as a temporal process is determined for each movement between two points, a starting and a destination point.

Based on the analysis of mobility data traces, the topology of the network is derived, e.g. the number of cells in the scenario, cells coverage areas, the population size, etc. Based on the analysis of users movements in the traces, many transition probabilities and parameters are derived, e.g. the probability to move to a neighbor cell, the probability to remain in the same cell, the probability to move to a non-neighbor cell, the average speed, etc. So as to capture the spatial as well as the temporal process, the authors analyzed the mobility of users based on two metrics, namely the prevalence and the persistence. The prevalence captures the spatial process, while the persistence captures the temporal process. The analysis of mobility traces has shown that the prevalence and persistence follow power-law distributions with low exponents. The distribution of prevalence values shows that the time durations users spend in different cells is unbalanced. Users spend the most of their time in one cell and move shortly to other cells.

Based on the analysis of mobility data traces, the authors decided to assign each user a set of cells to be visited during the simulation. The destination cell is selected randomly from this list and, following that, the user moves towards this cell. During the simulation, users follow inactive-active cyclic behavior. During the active phase, the user associates with a cell and actively uses the network, while during the inactive phase, the user will not be a part of the network, although he keeps moving. The temporal process can, therefore, be seen as a process invoking an active and inactive timer processes. The inactive phase duration is modeled using a uniform random distribution. Note that the inactive phase duration is adjusted based on the travel distance, so that the velocity of users do not exceed a maximum speed (V_{max}), which is derived from the mobility traces, as we mentioned above. The active phase is generated using a general Pareto distribution. The parameters of this distribution are derived from the mobility traces as well. The active phase is, in principle, the phase, during which sessions are established. The authors noted that most of the sessions are short, and the probability that a session takes longer decreases rapidly.

To sum up, the proposed WLAN mobility model works briefly as follows:

1. Mobility data traces are analyzed to derive following parameters, see (Tuduce & Gross, 2005).
    a. The simulation time.
    b. The simulated population size.
    c. Cells coverage areas.
    d. Cells numbers and locations.
    e. The maximum number of cells the user is allowed to visit.
    f. The distribution of the number of visited cells.
    g. The probability to move to a neighbor cell.
    h. The probability to remain in the same cell.
    i. The probability to move to a non-neighbor cell.
    j. The maximum user's speed.
2. Following that, these parameters are used to generate various movement patterns with the same statistical characteristics of the analyzed data traces.
3. Resulting movement patterns are used then for simulation purposes.

2.3. The Model T and Its Evolution, the Model T++

This work depends on the data collected from the WLAN of Dartmouth College (Kotz & Essien, 2002). The network had a dense topology with 586 APs in 161 buildings spread over approximately 200 acres. Each AP in the network had an average number of 65 other APs in its neighborhood. The data collected represented a two-year long record of AP association, diassociation and re-association requests resulting from 6,202 users. Note that the collected data do not provide any exact geographical locations. Moreover, the associations with APs do not necessarily imply a physical movement of humans, due to the existence of ping-pong effects. Another limitation in the data collected is the inaccuracy of the residence time before the user transits himself into an OFF state. Concerning the diassociation, it is not necessary that users diassociate themselves explicitly. Therefore, users that remain inactive for more than 30 minutes are considered diassociated. For more details about the data collection and the network, readers are directed to (Kotz & Essien, 2002) and (Song, Kotz, Jain & He, 2004).

The methodology the authors followed can be described briefly as follows: the authors first examined the traced data to better understand user behaviors and to extract the dominant features. Following that, the authors divided the traces into two disjoint sets, namely a training and a test data set. The training data set contains the data collected at the initial phase of the network installation, while the test data set includes the data collected after the network gets stable from the operation point of view. Based on the results obtained, the authors developed a statistical mobility model with few equations and parameters that are capable of characterizing the training data set. Following that, the authors generated synthetic mobility traces based on the developed statistical mobility model. The generated mobility traces are compared to the test data set for validation purposes. The results of these studies are the development of the Model T (Jain, Lelescu & Balakrishnan, 2005) and the Model T++ (Lelescu, Kozat, Jain & Balakrishnan, 2006) later on. The following provides more insight into both models.

The Model T captures the spatial movements of users across the APs monitored. It shows that the observed patterns of user registrations exhibit a distinct two-level hierarchy. The APs of the WLAN of the Dartmouth College can be clustered based on a transition rate matrix. The distributions of clusters sizes are highly skewed. Moreover, intra-cluster transition probabilities and also trace lengths follow the heavy-tailed Weibull distribution. Inter-cluster transition probabilities have similar characteristics to intra-cluster ones. The popularities of APs are considered important factors. The main drawback of the Model T is the lack of a time model. The Model T states that any independent time model can simply be superimposed on the empirical mobility model developed, i.e. the Model T. In fact, the residence time model, derived from the same traces studied, is available in (Jain, Shivaprasad, Lelescu & He, 2004).

The authors compared the results of the Model T to those produced by the WLAN mobility model discussed earlier. The comparison results were, in fact, not surprising since different empirical models from different WLAN networks will significantly deviate from each other. Sure, this is not desirable and speaks against the generalization of empirical mobility models. The authors state that both models seem to well reflect the spatial mobility with a few parameters. However, they lack of a representation of the dependency between space and time aspects. Both models do not consider the changes in the velocity of users, which result from accelerations or decelerations due to present restrictions such as obstacles, characteristics of paths, etc. The conclusion of the comparison emphasized that an independent handling of space and time dimensions is not adequate, rather the relation between them must be reflected in the mobility model. The authors expected a strong correlation between both space and time domains.

So as to express the correlation between space and time aspects, the authors have extended the Model T and developed the Model T++. The authors defined the concept of popularity gradient between different APs. They state that APs popularities and well-defined popularity gradients offer an effective way to reflect the relations between space and time without going into the details of obstacle modeling or context information. The AP popularity is defined as the number of visits the AP receives during the time duration defined for the study. So, the Model T++ is described as a joint time-space model that can regenerate the inter-dependence of time and space aspects by means of few methodical equations. The Model T++ suggests that when there is a transition from a popular to an unpopular AP, the residence time before the transition takes place will be longer than by other transitions, e.g. transitions from an unpopular to a popular AP or transitions between APs with similar popularities. The authors compared the Model T++ to the Model T and summarized that there is a small increase in the complexity. However, there are significant improvements and, thus, better capabilities to capture real mobility scenarios.

In spite of the great features of the Model T++, one must be cautious when using this model to generate mobility scenarios for ad hoc networks. This model is, in fact, not adequate for such networks, especially when the users of these networks are heavily mobiles. Another drawback is observed when geographical locations of users are necessary, since the model cannot reproduce such information.

2.4. Statistical Mobility Model (SMM)

So as to provide more realistic empirical mobility model, the SMM (Yoon, Noble, Liu & Kim, 2006) combines between coarse-grained wireless traces and an actual map of the geographical space, where the traces were collected. Following a sequence of data processing steps including filtering of traces and mapping to real maps, the SMM constructs a probabilistic model that generates movement patterns to be used in simulations. The generated movement patterns have similar statistical properties to those seen in the real traces studied.

First, the traces used in the SMM are also collected from the WLAN of Dartmouth College (Kotz, Henderson & Abyzov, 2015), see the previous section. The traces were filtered and a chronological list of "trips" taken by each user is generated. A trip includes an origin, destination and possibly intermediate positions, i.e. the APs where users pass during the trip. The applied filtering process comprises various steps. First, discontinuous trips are separated. This simply means the abstraction of the movement trip of each user including the start, end and intermediate points in addition to the paths the user has moved on. Note that the term point in this context refers to an AP. Following that, the defined points in each trip will be checked to filter undesired ping-pong handoffs. In the last step, the trip is identified in a form of stationary and transient points.

Following the filtering process described above, the route the user takes along his trip is inferred, i.e. the trip is converted to a geographical representation that is further converted following that to a graph with vertices (APs, buildings, etc.) and links (routes between the vertices). Figure 3 provides an example scenario to illustrate the steps done to obtain the graph mentioned. As the figure shows, the buildings, where the APs are installed are enumerated on the map as vertices, while the major routes are enumerated as segments. Following that, the vertices are interconnected with the major routes. In this step, intersections are enumerated as vertices as well.

After the construction of the graph with vertices and segments, the route users take on the graph must be inferred. First, the route candidates between each two points (APs) must be determined. In general, the shortest route is preferred. However, this is not an absolute rule. The selection of the preferred route highly depends on the metrics people use. Moreover, the shortest distance route may be different from the shortest time route, see (MSN maps & directions, 2015). The SMM developers decided to use the distance as a metric to determine the preferred routes. Furthermore, they used an algorithm to find N shortest paths, see (Lawler, 1972) and (Lawler, 2001). After the generation of route candidates, the

Figure 3. An example illustrating how to convert a map to a graph (Yoon, Noble, Liu & Kim, 2006)

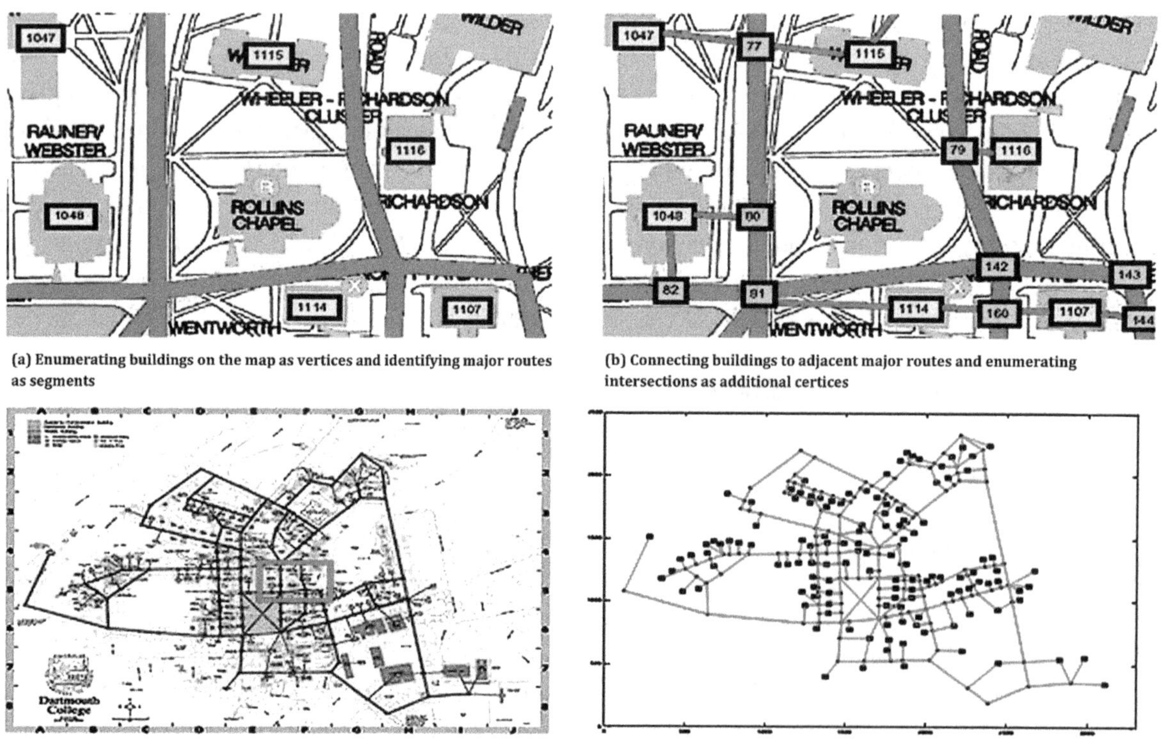

(a) Enumerating buildings on the map as vertices and identifying major routes as segments

(b) Connecting buildings to adjacent major routes and enumerating intersections as additional certices

(c) Final map (the rectangle on the map is the area shown in (a) and (b))

(d) The graph with the vertices and segments

likelihood of using each of these candidates is derived based on the traces. This step results in a route frequency statistics that are transferred to a set of transition probabilities associated with graph vertices. It is worth mentioning that this is done by modeling the mobility of users as a Markov chain. Note also that when the transition probabilities are available, no route frequency statistics are needed anymore to generate movement patterns, which will have similar statistics to those seen in the traces. Of course, accurate transitions probabilities are assumed. The last step one has to accomplish is the determination of the initial distribution of users, which in turn corresponds to the initial states distribution of the Markov model. This is done by determining the initial distribution of origins and destination. The initial distribution of origins can be simply derived by counting the frequency of visits the buildings receive in the data traces. The initial distribution of destinations for each origin can be calculated by counting the relative frequency of each building that appears as a destination from the given origin.

The authors evaluated the developed model (i.e. the SMM) by comparing the movement patterns statistics the model generates to those observed in the real trace data of the WLAN of Dartmouth College. The authors concluded that the SMM presents a realistic mobility model, since it is constructed within a geographic environment given by the real map of the network being monitored. Generated movement patterns statistics are similar to those seen in the real traces collected from the network being in reality monitored. The model provides various important statistics, e.g. user densities on roads as well as in buildings, transition probabilities at intersections, etc.

2.5. User Mobility Characteristics-Based Mobility Model

A method to estimate the physical location and to generate movement patterns based on mobile traces is illustrated in (Kim, Kotz & Kim, 2006). The traces are collected by tracking 198 distinct users of the WLAN of Dartmouth College. The traces also covered 13 months long (the beginning of June 2003 to the end of June 2004) and focused on tracking users, who used always-on devices (i.e. VoIP devices) during workdays. That is, measurements before 8 AM and after 6 PM are ignored.

It is worth mentioning that the Dartmouth trace delivers a map with APs coordinates. Note that if a map is not available, locations of APs can be derived using methods such as war-driving, see (LaMarca, et al., 2005) and (Cheng, Chawathe, LaMarca & Krumm, 2005), or AP positions estimation techniques provided in (Satoh, Ito, & Kawaguchi, 2005).

Tracking results are parsed to generate mobility traces, which list locations of APs the MN associated with, along with a timestamp for each association. The authors defined then popular regions, termed hotspots, where the transitions among these regions are characterized by means of a Markovian model. The key finding of the mobility traces was the detection that the speed and pause times follow a log-normal distribution, and that the direction of movements closely represents the direction of roads on the map.

The mobility traces derived are used to generate the mobility characteristics required for the mobility model developed. The characteristics include the ratio of mobile set size to stationary set size, where the mobile set is the set of users who move in the network. On the contrary, the stationary set includes users that do not move. For the mobile set, the following mobility characteristics are provided:

- The overall speed distribution,
- The overall start time distribution for workdays,
- The initial region distribution, and
- The pause time distribution per region.

Concerning the stationary set, the following mobility characteristics are generated:

- The initial location map,
- The start time distribution for workdays, and
- The duration of the stay connected with the network.

By means of the parameters listed above, synthetic movement traces are produced. Users are configured first either as stationary or mobile, based on the mobile set to stationary set ratio. Stationary users appear in the network at specific locations selected from the initial location map. Each stationary user enters the network at a time selected from the overall start time distribution for workdays, and remains at his location for a duration selected from the duration of the stay mentioned above. Each mobile user appears at a region selected from the initial region distribution. He accesses the network at a time selected from the defined start time distribution. The destination of the user is then chosen based on the probabilities present in the region transition matrix. Note that the speed of the user is selected from the speed distribution mentioned above. Once a user reaches his destination, he pauses for a pause time selected from the measured pause time distribution per region. When the pause time expires, the user begins a new movement and so on.

The model was evaluated compared to real traces measured from the WLAN of Dartmouth College. The main result says that the empirical model developed generates synthetic tracks that match real tracks with a median relative error of 17%. A drawback of the model is its lack of expressing temporal effects, since temporal variations, in the speed for instance, cannot be modeled.

2.6. Work and Home Extracted REgions (WHERE) Mobility Model

WHERE (Isaacman, et al., 2012) is novel model attempting to model how large populations move within various metropolitan areas. The model utilizes spatial and temporal probability distributions derived from large amount of empirical data representing real people living across wide geographic areas. The source of these distributions are the Call Detail Records (CDRs) taken from a cellular telephone network. The motivation behind the usage of CDRs to characterize aspects of human mobility are the results of many premium efforts, which depicted that joining the spatial and temporal probability distributions taken from CDRs with geographical locations of the towers of mobile communication networks results in sporadic samples of users approximate locations, see (Gonzalez, Hidalgo & Barabasi, 2008), (Isaacman, et al., 2011), (Song, Qu, Blumm & Barabási, 2010) and (Bayir, Demirbas & Eagle, 2009). The analysis of the data CDRs offer results in synthetic CDRs for a synthetic population.

Although the existence of CDRs helps significantly in modeling human movement patterns, it is tempting to think that modeling human movement patterns based on such data is an easy task. It is correct that CDRs give an insight into population densities. However, they do not convey whether users' locations correspond to home, offices, etc. Therefore, without such semantic data, it is difficult to generate mobility models applicable to scenarios different from those, for which the CDRs were collected. Keep also in mind that the spatial as well as temporal granularity of CDRs data is quite coarse. The spatial granularity of CDRs data is only accurate to the granularity of cells' towers spacing, while the temporal granularity is accurate only to users actively involved in communications.

The authors claim that their model makes a key contribution in overcoming the lack of semantic information and coarse granularity, thus, producing accurate models for arbitrary metropolitan areas. It is nowadays widely accepted that there is a tight coupling between human movement patterns and the geographic of the city people live in, see (Isaacman, et al., 2011a), (Isaacman, Becker, Cáceres, Kobourov, Rowland & Varshavsky, 2010) and (Noulas, Scellato, Lambiotte, Pontil & Mascolo, 2012). As mentioned above, the WHERE model captures spatial and temporal information present in the CDRs or any other source of data. The spatial data include the definition of important locations from users' point of view and also hourly population densities. The temporal data required for the model cover the calling patterns issued within the network.

Concerning the definition of important locations, the authors argue that a good accuracy is expected when only considering the probability distributions for the few locations, where users spend the majority of their time. This statement is motivated from prior investigations that demonstrated that the majority of user movements occur between a small number of locations. Thus, the authors decided to focus on the most important two locations, "work" and "home". So, to estimate spatial home-work densities, an approach with three stages is followed. First, home locations are randomly pulled from a probability distribution across latitude and longitude that express the likelihoods of where humans live. Second, a probability distribution of commute distances is calculated for each point resulting from the first stage. A commute distance (d) can be envisioned as a circle of radius d centered on the selected home location. Third, the work location is selected somewhere along this circle. Thus, a probability distribution of different work locations around the mentioned circle is derived, see Figure 4. Note that the mentioned densities are derived from the population densities corresponding to the city under study.

So as to improve the fidelity of the proposed model, the mentioned population densities are linked to temporal information. Note that in reality, heavily residential areas are likely to be more populated at night, while city centers and commercial locations are likely to be more populated during a day. Previous work (Isaacman, et al., 2011a) has shown that cellular calls are an accurate representation of users' locations. When linking the calls to the towers handling the calls, one will be able to represent the density of users in the network, i.e. in the city as well. A step more is necessary to consider the time, namely the density is represented hourly. Figure 5 presents instances of one-hour population density estimated in the Los Angeles and New York cities, see (Isaacman, et al., 2012) for details. The final step is to combine the spatial probabilities discussed above with information about calling patterns, so that accurate synthetic CDRs can be produced.

Figure 4. Selecting user's home and work locations in the WHERE empirical mobility model

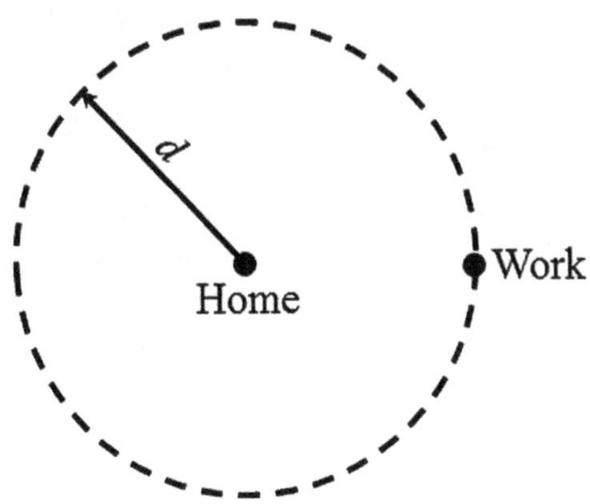

Figure 5. Instances of one-hour population density estimated in Los Angeles and New York (Isaacman, et al., 2012)

(a) Los Angeles (b) New York

The authors developed based on the investigation highlighted above a two-place model and called it WHERE2. The WHERE2 utilizes the fact saying: people spend the majority of their time either at home or at work. Thus, synthetic CDR traces for the model WHERE2 are generated in two stages, namely a user creation and movement. The user creation stage includes creating a synthetic user and assigning a home to him. Following that, a commute distance is chosen for the synthetic user. Note that the commute distance selected is conditioned on the user's particular home, which is selected based on the analysis of the real CDR traces in relation to the geographical map. Following that, possible work places are defined for the synthetic user based on the probability distribution of different work locations derived from the analysis of real CDRs. Finally, the user is assigned a calling pattern according to the distribution of calls present in the investigated data. In the second stage, the user moves between work and home according to the probability of being either at work or at home at that time of day (derived from real traces). Note that a movement means making a call from either the work or the home. So, the result will be synthetic traces with statistical properties similar to those seen in real CDRs. The authors further extended the WHERE2 to cover more than two places, see (Isaacman, et al., 2012) for details.

The authors have validated the WHERE model against billions of anonymous location samples (stand for hundreds of thousands of phones) in the New York and Los Angeles metropolitan areas. The validation results have shown that the WHERE model shows significantly higher fidelity than other modeling approaches. For instance, the statistics of the daily range of travel resulting from the model fall within one mile of their true values. The WHERE model has been also compared to the WWp model. The comparison has shown more than 14 times improvements over the WWp model.

Main advantages of the WHERE model is its application to metropolitan areas, since rarely we see models to represent movement patterns in such big areas. This helps in answering many questions related to human movement patterns in such places. Another main advantage of the model is its privacy concerning real CDRs, since it necessitates anonymous CDRs and generates then a wide range of synthetic CDRs for investigation purposes.

2.7. Summary

This section has focused on trace-based mobility modeling. It has reviewed the major projects issued to track real movement patterns. Well-known empirical models aiming at utilizing these traces to generate synthesis traces characterized by the same statistical characteristics the real traces have are also discussed in the section.

The results derived from analyzing real traces have shown that users are in general non mobile. They spend the majority of their time at a limited amount of locations that have sociological relation to them, e.g. work, home, etc. The variations in the number of users over time are proper representations of the patterns of the underlying population. There is regularity in movement patterns as well as network usage. This regularity comes from the cyclic habits users follow in their life.

The section has described well-known empirical models that tried to capture the observations we mentioned to provide more realistic movement patterns for simulation purposes. The empirical models reviewed have modeled mobility of users by reproducing the spatial relations between users and the environment, e.g. the WLAN mobility model and the model T, or by trying to capture the relation between time and space domains in the traces, e.g. the Model T++, WHERE, etc. Sure, the empirical models that extract the time-space relations are more accurate.

A main result from the investigation achieved and also a main drawback is the tight coupling between the empirical model developed and the traces collected, the network topology deployed and even the geographic location, where the data were collected. This implies that changing the traces, network topology or the geographical area will change the results of the empirical model. This is why the results of empirical models deviate clearly from each other. Sure, this prohibits the generalization of empirical models. When one also considers that the traces themselves are rarely available, one can understand why synthetic models are preferred over empirical ones.

3. SOCIAL-BASED COMMUNITY MODELING OF MOBILITY PATTERNS

Social-based community modeling of human patterns is currently a research trend gaining more and more interest. The aim is, as before, the generation of as realistic models as possible, which is crucial for divers fields, as we often mentioned. Researchers have noted that a realistic mobility behavior inferred from traces is seldom applicable to simulate large-scale mobile communication networks, as we mentioned in the previous section. This was the main reason behind the wide application of synthetic mobility models. It is, however, observed that synthetic mobility models generate movement patterns that are, in general, inhumane-like (Musolesi, Hailes & Mascolo, 2005), (Musolesi & Mascolo, 2007). The word 'human' is, in fact, the key. Mobile devices are usually carried by humans, so movement patterns of such devices are necessarily related to human decisions and socialization behavior. The studies achieved in (Gonzalez, Hidalgo & Barabasi, 2008) and (Rhee, Shin, Hong, Lee & Chong, 2008) show that human movements are far from being random. A high degree of temporal and spatial regularity is observed by these patterns. So, human cyclic routines heavily affect the movement pattern resulting, e.g. people move daily morning from their houses to their offices and come back at the afternoon for instance. Concerning mobile communication networks for instance, their resources will be best used if it is possible to capture realistic human movement patterns.

A social network is defined as a set of humans with some patterns of contacts or interactions among them (Scott, 2000). The first significant quantitative results were published in the 1950s and 1960s by Rapoport and his colleagues (Rapoport, 1957), who analyzed the statistics of epidemic diffusion in populations that are characterized by different social structures. In the same period, the interest in the graph theory was renewed and the so-called random graph was defined by Paul Erdoes and Alfred Rényi (Erdos & Renyi, 1959). These theoretical studies represented the beginning of the research in the complex network area and were applied to various disciplines, e.g. biology, ecology, linguistics, sociology, etc. In the last decade, various studies have been conducted to assess the validity of these theories, especially for the analysis of social networks. One of the most interesting approaches is the so-called small world model proposed in (Watts, 1999). Researchers believe nowadays that the results of social networks analysis can be used to generate more realistic mobility models.

In this section we will give an insight into social-based community mobility models, which are associated with communities of groups within societies. The section will also goes through the strengths and weaknesses of the presented models and summarizes with a qualitative comparison.

3.1. Time-Variant Community (TVC) Mobility Model

The TVC mobility model (Hsu, Spyropoulos, Psounis & Helmy, 2007) relays on the concept of Community of Interest (CoI)-based[1] location preferences and time-dependent periodical appearance of MNs. It is a realistic, flexible and mathematically tractable model.

Description: for clarity, Figure 6 illustrates an example scenario for the TVC mobility model. As the figure shows, there are Time Periods (TPs), three in the figure, each with a specific duration, e.g. TP1, TP2 and TP3 are of duration T^1, T^2 and T^3 time unit, respectively. During each time period, there are some communities that represent geographical areas heavily visited[2]. Sure, the number of communities differs from a TP to another.

Let us assume that $Comm_j^t$ is the community indexed with j at time t, and the total number of communities at time t is S^t. Each community is presented in the figure as a square of edge length $C_j^t\left(t\right)$. Note that the communities may overlap, be included in bigger communities, etc., see the figure. Note also that TPs are arranged in a recursive sequence since they present, in fact, a time schedule, e.g. daily, monthly, etc.

Based on the discussion induced above, movements of MNs are sequences of epochs that reflect their behaviors during these periods. The TVC model assumes that MNs residing within each epoch randomly select their directions (uniformly distributed from the range $\left[0 \rightarrow 2\pi\right]$), speeds (uniformly distributed from the range $\left[v_{min} \rightarrow v_{max}\right]$) and length of their movements (exponentially distributed). At the end of the epoch, MNs wait for a pause time duration selected also randomly and proceed, following that, to the next epoch.

In this context, one notes similar behavior to that observed by the studied RW (within the community) and RWp (between communities) mobility models, see section 2.1. The difference between the TVC model and others, mentioned above, is that the TVC mobility model additionally, and based on specified time-variant probabilities (δ_j^t), selects the community, inside which the next epoch will take place.

Figure 6. TVC mobility model illustration

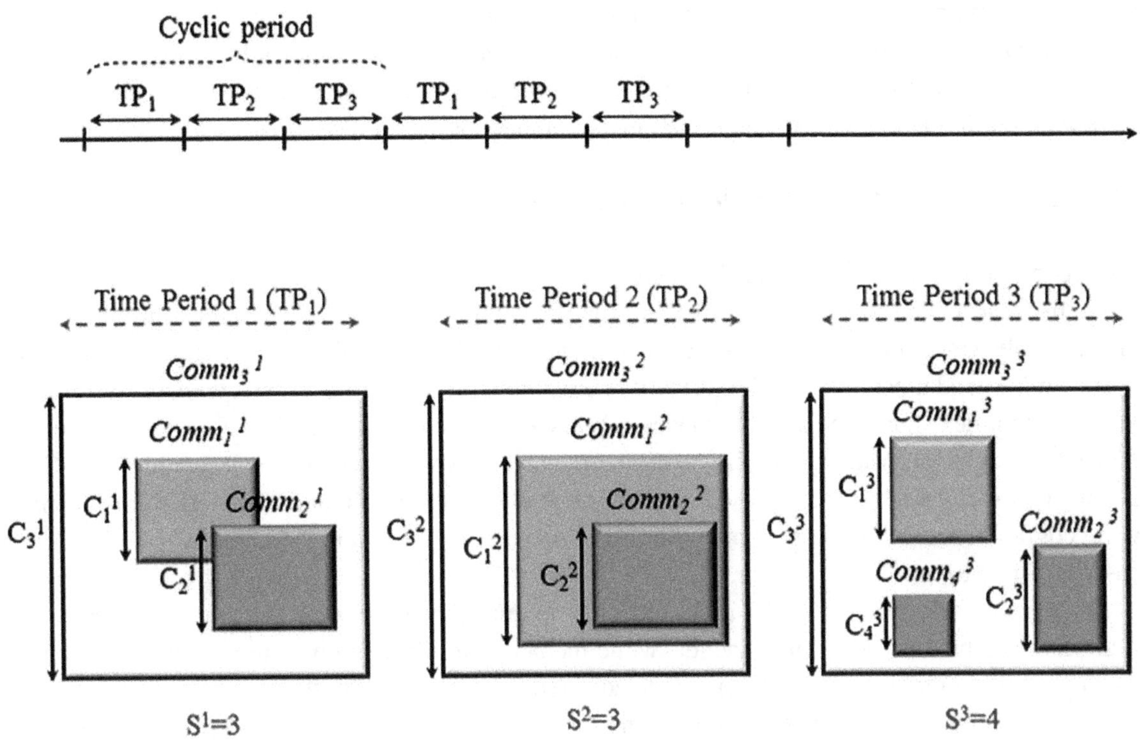

Note that the whole area can be determined as a community that contains all existing communities within a given epoch. This epoch is termed then a roaming epoch, see the figure.

In brief, the TVC model matches realistic scenarios when the following steps are accomplished:

- Determining the structure in space (communities) and time (epochs).
- Assigning communities' characteristics, e.g. probability, average pause time, average epoch length, etc.
- Determining the use on–off behavior.

The TVC model was analyzed mathematically, and the mathematical model was also validated through simulations. The mathematical model of the TVC enables calculating many metrics of interest, e.g. MNs' spatial distribution, MNs' degree, the hitting time[3], meeting time[4], etc. For more information about the mathematical analysis and the validation of the model, readers are referred to (Hsu, Spyropoulos, Psounis & Helmy, 2007a), (Zhang, Neglia, Kurose & Towsley, 2006) and (Ranjan Roy, 2011).

Discussion: the TVC mobility model is a time-variant model that expresses movements between communities during specific and recursive time durations. It is flexible and can be characterized to match realistic scenarios. The TVC model preserves common mobility characteristics known from other mobility models, e.g. the RW, RWp, etc.

A great feature of the TVC mobility model is the dynamic it allows for the structures of communities, since their size, number and location can be dynamically changed during time. Moreover, the TVC model allows individuals to have different set of parameters including own TPs' structures. This also enables reflecting the time-dependent behavior of humans more accurately and enables describing heterogeneous environments with users having diverse mobility characteristics.

Limitations: because the TVC mobility model behaves similar to the studied RW model within communities and similar to the RWp mobility model when considering the movements between communities, the TVC mobility model inherits the limitations of both the RW and RWp model.

3.2. Community-Based (CB) Mobility Model

The CB mobility model (Musolesi & Mascolo, 2006), (Boldrini, Conti & Passarella, 2007) aims to mimic the real scenarios we live in, since in everyone's life, daily movement patterns heavily depend on the people we interact with, thus, on the communities we belong to from time to time. For instance, most people spend a part of their day at work interacting with their colleges, and another part at home with their families or at specific social places, e.g. restaurants, cinemas, etc. Therefore, the CB mobility model depends in constructing mobility patterns on social networks theory (Musolesi & Mascolo, 2006) and exploits social relationships present in these networks. It assumes that communities are associated with geographical areas since humans with strong social links are often geographically co-located, at least from time to time.

Description: as mentioned, MNs in the CB mobility model belong to social communities. Members of the same community are called friends, while members of different communities are referred to as non-friends. Relations between MNs are modeled via social links with specific weights (Scott, 2000), see Figure 7 that shows an example graph presenting MNs of a community with the weights that express the social attractivity. Note that MNs express the vertices of the graph, while relations are represented by the edges. The weights themselves refer to the social attractivity between MNs, that is the greater the weight, the more strong the friendship. For simplicity, the weights in the figure are set bidirectional. That is, both friends have the same social attractivity to each other. Sure, the weights can also be unidirectional.

When starting up, all friends have social links to each other and form a graph similar to that shown in the figure. Links with non-friends are established according to a rewiring probability (P_r). This means that each link with a friend is moved to a non-friend with a rewiring probability. The mentioned non-

Figure 7. An example graph of a community generated by the CB mobility model

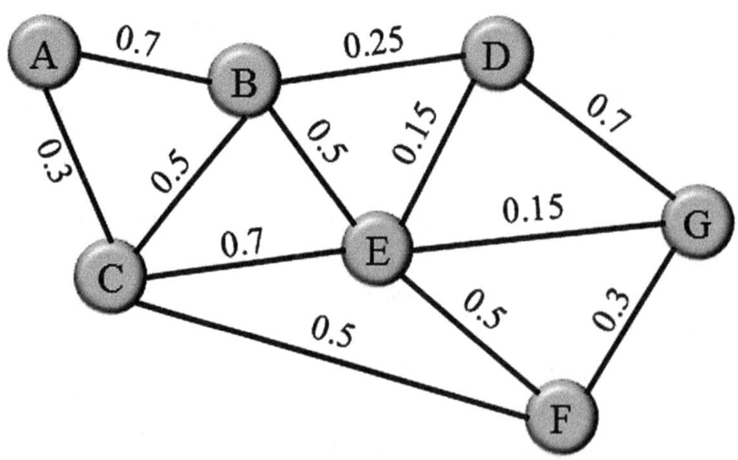

friend itself is selected randomly from a given community according to a uniform distribution, which creates time-dependent changing communities structure. In other words, social networks change from time to time. The interval time between such changes is an input parameter for the CB mobility model.

MNs are placed within a grid that consists of cells, each represents a community. One possible approach to initially setup communities is to randomly place MNs inside the cells of the grid. Another approach relies on a given clustering algorithm (Newman & Girvan, 2004) that depends on centralities values assigned to MNs. So, MNs characterized with high values of the centrality are progressively detected in subsequent iterations. At each iteration, one of the edges of the MN (the edge with the highest centrality value) is rewired. The result will be a network composed of groups of MNs (communities) as illustrated in Figure 8, which shows an example scenario with three groups.

Concerning nodes' movements, they can be considered as a sequence of steps, where each step means a movement toward a community (referred to as the target community) and an association with a MN (termed the target MN) in this community. The destination of the movement is determined by the social attractivity towards other communities. Note that MNs move toward their target cells gradually, where other MNs having strong interactions among them converge. This represents the behavior of most people, since they travel home every evening and join their families for instance.

Movements towards cells are done based on various rules. One says that the social attraction exerted by each cell on the MN must be, first, calculated. Following that, the MN moves towards the cell with the highest social attraction. The attraction exerted by a given cell on a MN, in this context, is the sum of the links' weights between the MN and all MNs currently moving in or toward the given cell, see (Ranjan Roy, 2011) for more details. Another rule considers the attraction as the sum of the weights of all links interacting the MNs inside a given community, being divided by the number of MNs in that community (Ribeiro, 2011). The problem of both discussed rules is that no new communities will be formed during the simulation. All MNs will end up in existing communities. The third rule is a probabilistic one. The target cell is selected proportional to the attractivity of each community. The application of this rule results in more realistic scenarios than when using the other two rules, since it enables changing com-

Figure 8. An example scenario with three groups applying the CB mobility model

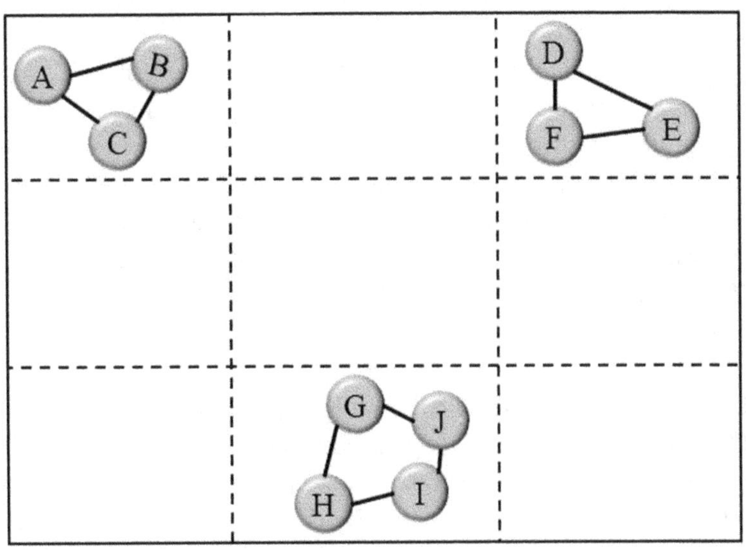

munities and creating new ones during the simulation. The target MN within the target cell is chosen randomly according to a uniform distribution, as we mentioned earlier. Note that the MN's velocity is also selected accordingly to a uniform distribution within a user-defined range.

An important feature of the CB mobility model is the avalanche effect resulting from movements of MNs. When a MN decides to move towards a target community, a probability termed remaining probability (P_{rem}) is calculated. This probability is defined as the probability that no other members of the community will follow the MN moving. When P_{rem} approaches 0, at least one MN will follow the MN moving towards its target cell. This behavior is termed avalanche effect and reveals a gregarious behavior.

Discussion: the CB mobility model presents the reference point for social-aware community models. It depends on social networks theory to mimic the real scenarios we live in. Movements from the CB point of view include changing the community and associating with a MN within the new community. That the communities are associated with geographical areas and movements between these communities are done gradually is an advantage that gives the CB mobility model flexibility and capability to match a wide range of real applications.

A great feature of the CB mobility model is its dynamic behavior concerning the structures of communities, since their size, number and location can be dynamically changed during the simulation. This helps the model in describing heterogeneous environments. However, the sequence of movements is not pre-defined as by the TVC mobility model. This can be considered as a strength since the scenario construction gets simpler and more patterns can be studied. However, this can also be considered as a weakness since it may prohibit the use of the CB model to represent pre-defined movements between specific communities. Furthermore, the investigation achieved in (Ranjan Roy, 2011) has shown that the CB model cannot capture locations' attractions on MNs.

The avalanche effect resulting from movements of MNs is a feature pertinent to the CB mobility model that must be carefully considered when designing scenarios.

Limitations: as mentioned above, the avalanche effect results in a gregarious behavior, which also means that there is a spatial and also a temporal dependency of MNs locations when we consider movements between communities. Concerning the velocity, there is neither a spatial nor a temporal dependency of the velocity since the velocity of MNs is selected randomly based on a uniform distribution from a user-defined range.

For the selection of communities, one expects that there will be a temporal and spatial dependency of the location of the target community due to the calculation of the attraction each cell has on the MN.

There exist no geographic restrictions of movements, which prohibits designing scenarios where streets, gardens, etc. exist. With respect to pause times, they are not included in the model. There is another representation of the time MNs spend within communities, namely the time interval to change the community. This interval is, however, an input parameter for the model and is also constant for all cells.

Variations of CB Mobility Model: there exist following extensions in the literature to further develop the CB mobility model:

- **Home-Cell Community-Based (HCCB) Mobility Model:** The HCCB model (Boldrini, Conti & Passarella, 2007) extends the CB model by introducing a home cell representing a home community that has stronger attractiveness than other communities. The home cell is the initial cell, the MN is placed in, when initiating the scenario. By the introduction of the home cell principle, not only the attraction towards other MNs with social links is considered, but also the attraction towards locations, namely the home cell. When the MN locates inside the home cell, the selection of the target cell depends on the CB model. Once the MN resides in a foreign cell, the probability of going back to the home cell equals to $1 - P_e$, where P_e is the probability or remaining in the foreign cell. So, the HCCB model is adequate for scenarios, where MNs gravitate across preferred locations.

- **Modified HCCB (MHCCB) Mobility Model:** The MHCCB model (Boldrini, Conti & Passarella, 2009) generates movement patterns closer to realistic human mobility patterns, particularly in terms of roaming. The developers state that in HCCB mobility model, there is no preferential selection of short paths (closer cells) over long paths (distant cells). The MHCCB resembles the HCCB model and differs only in the way applied to select the target cell, so that closer cells are preferred over distant cells.

3.3. Orbit-Based (OB) Mobility Model

The OB mobility model (Ghosh, 2006), (Ghosh, Yoon, Ngo & Qiao, 2005), (Ghosh, Sumesh & Qiao, 2007), (Ghosh, Ngo & Qiao, 2006) is a semi-deterministic model that depends on macro-level sociological orbits. The model considers hierarchical orbital levels, where each level considers a mobility model best suited to the own purposes of the level. In fact, each level is seen from the mobility model point of view as a black box that gives only inputs to the higher level.

Description: so as to illustrate the OB model, an example with three hierarchical orbital levels will be discussed, see Figure 9.

The first, second and third hierarchical orbital levels in the figure are termed Local Area Orbit (LAO), Medium Area Orbit (MAO) and Global Orbit (GO), respectively. In the LAO level, any mobility model, e.g. the RW model, RWp model, RD model, etc., can be applied. In the MAO level, movements within LAOs are not seen since only the movements between LAOs areas are of interest in this level. Again, any

Figure 9. An example scenario generated by the OB mobility model

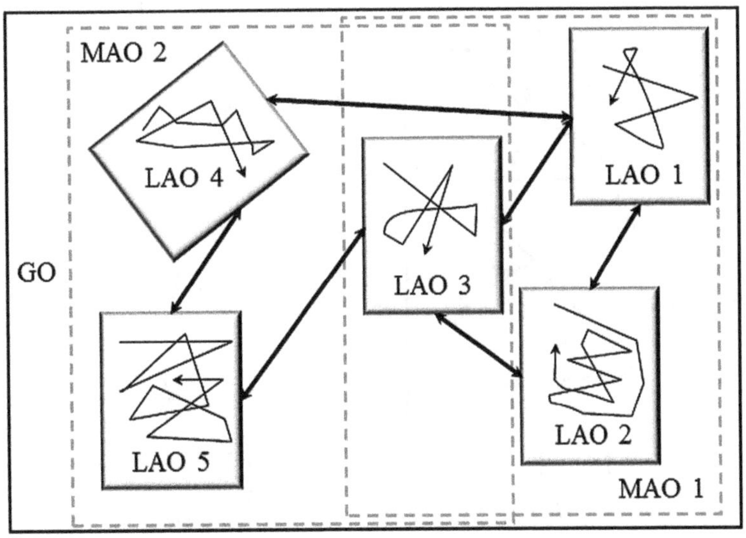

mobility model can be used to model movements between LAOs. Note that the mobility model used in the MAO level may differ from that used in the LAO level. In a similar way, the GO level only considers movements between MAOs that are also modeled by a specific mobility model.

Initially, a specific number of LAOs is assigned to each MN to move between in the second level, i.e. the MAO. These LAOs present, in principle, the communities the MN is allowed to visit. It is worth mentioning that an LAO is referred to as a hub as well. The MN locates initially inside a randomly selected LAO and starts moving based on the mobility model applied. Each LAO has its own speed range and timeout, which is determined as the time duration each visiting MN has to spend inside the LAO. When a given LAO timeout expires, the MN randomly selects another LAO from the LAOs it is allowed to move into, and starts moving towards the chosen LAO. This movement is controlled by the mobility model applied in the MAO level. Note that the MAO has also a timeout bounded, after which the MN has to change the MAO and to move in the GO level.

A main advantage of the OB model is its capability to emulate existing mobility models that are commonly used to simulate mobile communication networks. For instance, if there is a single LAO in the scenario, and it covers the whole simulation area. Movement patterns of the RWp model will result if the RWp model is applied inside the LAO and the LAO timeout is set to the simulation time. In case many LAOs exist, however, they are placed beside each other, so that movements from one LAO to another happen in one step. The RW model will be generated if the timeout of all LAOs is set to zero and the visiting sequence is configured to include adjacent LAOs only. The two examples presented above are shown in Figure 10.

Discussion: the OB mobility model is a semi-deterministic and practical model that considers hierarchical orbital levels, where each level considers its own mobility model. This enables the use of various mobility models within one simulation, thus, modeling heterogeneous environments. A great

Figure 10. Emulation of existing mobility models applying the OB mobility model (a: the RWp model, b: the RW model)

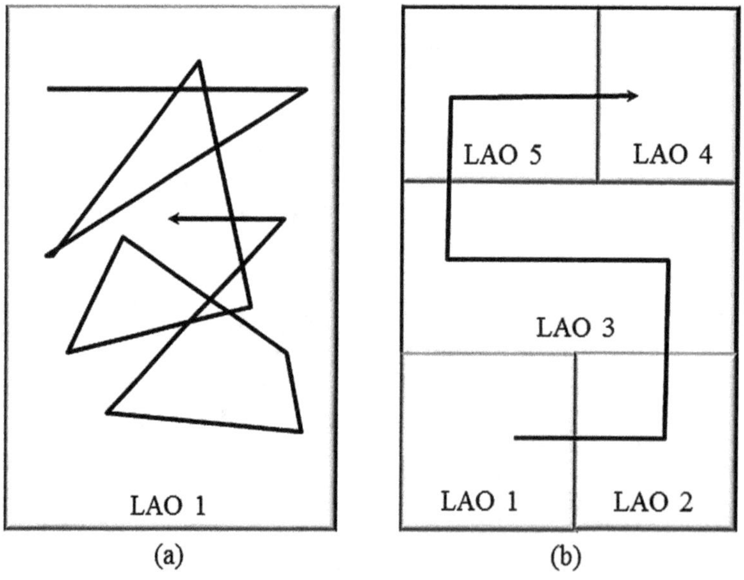

feature of the OB model is its capability to emulate existing mobility models, which enables modeling a wide range of applications and forming communities that interact with each others in specific patterns.

Limitations: because the OB mobility model utilizes existing mobility models to model movements inside LAOs and among them and also between MAOs, it inherits the limitations of the applied mobility models.

Variations of OB Mobility Model: so as to model specific scenarios known from real-life, following variations of the OB model have been proposed.

- **Random OB Mobility Model:** Similar to the classical OB model, the random OB mobility model assumes that LAOs are geographical areas with varying sizes. Within each LAO, MNs move using LAO-specific mobility parameters, e.g. a speed range, timeout, etc., and use the MAO-specific mobility parameters to move from one LAO to another under the same MAO. The main feature of the random OB model is that MNs visit LAOs of an MAO in a random sequence, see (Ghosh, 2006). This model is adequate to model regular city traffic, where each LAO expresses a geographical area and humans move in their sociological areas. Humans moving slowly remain inside their LAO, while those moving fast move around and cross LAOs.
- **Uniform OB Mobility Model:** This model resembles the features of the random OB model with one difference, namely the setup of the scenario. More concrete, the uniform OB model divides the simulation area into an asymmetric grid of LAOs with no overlapping. This model is adequate to model traffic in buildings. For instance, at a university building, where each LAO represents a classroom and an MAO represents a floor or maybe the whole building. So, students move inside a classroom and move from time to time to another classroom, see (Ghosh, 2006).
- **Restricted OB Mobility Model:** This model (Ghosh, 2006) is similar to the uniform OB mobility model. However, each MAO is set identical to an LAO. That is, each MAO consists of a single LAO. This model is adequate to model companies consisting of various departments. Each department is presented by an MAO. Employees restrict their movements to their communities within departments. Sometimes, their work requires them to shift to a different department, which is represented by changing their MAO.
- **Overlay OB Mobility Model:** This model (Ghosh, 2006) is also similar to the uniform OB mobility model. However, there is an LAO spanning over all LAOs on the top of the LAOs grid. Furthermore, movements are restricted to a single LAO. That is, MNs do not cross LAOs' borders. However, due to the overlay nature of the extra LAO placed on the top of the grid, MNs within the overlay LAO move, in principle, inside all other smaller LAOs. Such model is adequate to model exhibitions, where presenters install their small non-overlapping sections, while attendees move around from one section to another.
- **Probabilistic OB Mobility Model:** This model creates a certain number of hubs (i.e. LAOs) and places them randomly within the simulation area, so that they are not overlapped. Each MN forms a hub list to be visited. The hubs in the list are selected randomly and a sequence of movements among hubs is created by means of a specified transition probability matrix. The time each MN spends inside a hub is an exponential random variable with a mean value specified by the hub timeout. Movements of MNs inside hubs (referred to as intra-hub movements) are modeled by the RWp mobility model. The movements between hubs, termed inter-hub movements, are modeled [illegible] Point-to-Point (P2P) linear mobility model, which randomly selects a point in the destination [illegible]oses a transition time and moves linearly toward the destination point. Note that the tran-

sition time is exponentially distributed with a given mean inter-hub transition time, and the speed is obtained by dividing the distance to the destination point by the transition time. Interactions with other MNs are prohibited during inter-hub movements. The probabilistic OB model is adequate to model wireless devices carried by humans working in offices or attending conventions. The list of hubs recorded represents, in principle, the hubs visited by the wireless device while moving, i.e. the movement pattern. The transitions probabilities can be derived from the number of times the hubs are visited.

3.4. Entropy-Based Community (EBC) Mobility Model

The EBC mobility model (Ochiai & Esaki, 2008) utilizes the principle of entropy, which is a measure of the disorder of a system (Robert, 2009). Based on this, realistic mobility models can be expressed by the entropy of the movement patterns they generate (Ochiai & Esaki, 2008), (Gui, 2007), (Leguay, Friedman & Conan, 2005a), (Leguay, Friedman & Conan, 2005). More specifically, movement patterns with high entropies are random, which implies a high probability that MNs do not follow a reproducible pattern. On the contrary, sociologically organized movement patterns (Ghosh, 2006), (Ghosh, Yoon, Ngo & Qiao, 2005) produce smaller entropies, see (Ochiai & Esaki, 2008). In other words, the smaller the entropy, the more realistic the movement pattern, the better the mobility model applied.

Description: based on traces and observations of movements within ad hoc networks, researchers noted that movements of MNs are highly affected by specific Concentration Points (CPs) (Ghosh, 2006), (Ghosh, Yoon, Ngo & Qiao, 2005) that realize particular objectives to practical scenarios. In this sense, movements of MNs are not random, which implies that the entropies of these movements will not be high. Note that a CP corresponds to a community in the EBC mobility model. Note also that MNs move in small sets of communities when applying the ECB mobility model. So, the EBC mobility model measures the entropies of MNs movement patterns around the CPs, where MNs are likely to move around.

Let us assume that N is a set of MNs moving in the network, C is a set of communities and $C_n \in C$ is a subset of communities, to which a MN $n \in N$ belongs. The EBC defines two mobility states, namely a stay- and a transition-state. In the stay-state, the MN n remains at a community $c_i \in C_n$ (referred to also as the location (n)), while in the transition-state, the MN moves from the community c_i to a community c_j, where c_i and $c_j \in C_n$ and $i \neq j$.

The EBC mobility model makes the following assertion: MN n and MN k are in a direct transmission range $\Leftrightarrow$ MN n and MN k are in contact with each other $\Leftrightarrow \exists \mathrm{c} \in \mathrm{C}$, location ($n$), $\mathrm{c} \cap$ location (k)$= c$. Note that the location (n) is undefined when the MN n is in a transition-state.

The mobility in the EBC is defined as follows, where $[\![C_n]\!]$ is the number of elements in the subset C_n: a MN n residing in a community $c_i \in C_n$ selects a random value (r) uniformly distributed from the range $[0 \rightarrow 1]$ and compares r with the probability of transition from the stay- to the transition-state (p). In case the value of p is smaller than r, or $[\![C_n]\!]$ equals to 1, the MN n remains in the state $c_i \in C_n$. In case the value of p is greater than r, the MN n randomly selects a target community $c_t \in C_n$ and moves towards it. Once the MN n reaches the target community, it repeats the whole process again and so on.

The entropy of the EBC mobility model (S) is written formally as in the following equation.

$$S = \frac{1}{[\![N]\!]} \sum_{n \in N} \log_2 [\![C_n]\!] \tag{1}$$

Figure 11 provides various EBC mobility model instances with different entropies, namely $S = 0$ and 1.38. The figure shows an ad hoc network with 19 MNs and different community structures resulting from MNs' movements. Note that all MNs interconnecting with each other (via an edge in the figure) form a community. Note also that increasing the entropy means the increase in the randomness of the network, while stable and well-designed ad hoc network is given at very small entropy. Furthermore, community-to-community scenarios (Ghosh, 2006) are categorized by small entropy. In fact, one can generate many mobility models based on the value of the entropy, e.g. the RW, RWp, RD, etc.

The application of the EBC mobility model has huge advantages in various fields, especially the field of mobile communication networks since it enables the classification of mobiles within the network based on the entropy, thus, adapting protocols and services to the environment. In (Ochiai & Esaki, 2008) for instance, a routing scheme named Potential-based Entropy Adaptive Routing (PEAR) has been proposed to adapt the routing based on the entropy. The studies achieved in (Ranjan Roy, 2011) have proven that the Minimum Expected Delay (MED) and the Maximum Delivery Probability (MDP) routing schemes (Chen & Chen, 2007) work well only at small-entropies, where MNs randomly contact with each other. In contrary, the spray and wait (Spyropoulos, Psounis & Raghavendra, 2005) and epidemic routing (Vahdat & Becker, 2000) work well in networks with large entropies. The analysis concludes with the result that routing schemes must shift from a message transfer scheme to a mobility-based scheme. That is, utilizing mobility characteristics to improve routing events. Sure, not only routing, but also the performance of the network as a whole. The results gained from this analysis can be applied to other fields, e.g. movement of humans in metropolitan areas, movements inside cities, etc.

Discussion: the EBC mobility model is a simple model that depends on the entropy to measure the disorder of the resulting movement patterns and, is capable, therefore, of producing movement patterns

Figure 11. Communities organized due to movements of MNs at various mobility entropies, $S = 0$ and 1.38

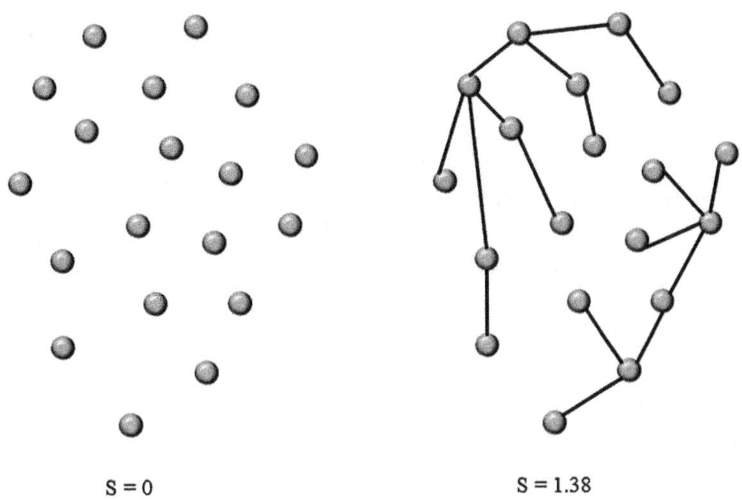

closed to those observed in realistic scenarios. Keep in mind that sociologically organized movement patterns produce smaller entropies, while random patterns generate large entropies.

The main feature of the EBC model is the capability it provides to classify MNs based on the entropies of movement patterns. This result can be utilized by systems to adapt to changes in the environment, since any change will be reflected by a change in the entropy. The main result of the research in this context states that researchers have to develop solutions that are mobility-based. That is, they depend on the mobility characteristics to optimize the performance of the system as a whole.

Weaknesses of the EBC model are seen in its incapability of defining obstacles, streets, pre-defined movement characteristics of specific groups or communities, etc. So, communities are established and destroyed randomly. Sure, this prohibits the use of this model to simulate a wide range of scenarios, where streets and highways are necessary for instance.

Limitations: as one notes from the description of the EBC model, there is no spatial as well as temporal dependency of MNs' locations and velocities, since MNs select their speeds and target locations without caring of previous values of the location and the speed and also without caring of what other MNs do. For the selection of communities, it is also done randomly without a spatial and temporal dependency. Concerning the geographic restrictions of movements, the EBC mobility model does not consider means to define them and does not consider pause times between movements.

3.5. Summary

Table 1 and Table 2 sum up the survey we provided in this section and provide a qualitative comparison between the reviewed models. Note that only the OB mobility model is capable of implementing a temporal dependency of the velocity, direction and community selection. This depends, as we mentioned, on the mobility models implemented. Note that the temporal dependency results in a smooth movement patterns. Concerning the spatial dependency and also geographical restrictions, the OB mobility model is also the model capable of representing all spatial dependencies studied and the geographical restrictions. The EBC model does not support any type of dependencies and does not enable representing geographic restrictions. This is due to the principles the EBC model follows, since it lets MNs move, measures the entropy and constructs the communities based on the measured entropies.

Table 1. A qualitative comparison between the social-based community mobility models investigated with respect to the temporal dependency, spatial dependency and geographical restrictions

Model	Temporal Dependency			Spatial Dependency			Geographical Restrictions	
	Velocity	Direction	Community	Velocity	Direction	Community	Paths, Streets, etc.	Obstacles
TVC	No	No	Yes	No	No	Yes	No	No
CB	No	Yes	Yes	No	Yes	Yes	No	No
HCCB	No	Yes	Yes	No	Yes	Yes	No	No
OB	Mobility model-dependent	Mobility model-dependent	Mobility model-dependent	Mobility model-dependent	Mobility model-dependent	Mobility model-dependent	Mobility model-dependent	Mobility model-dependent
EBC	No	No	No	No	No	No	No	No

Table 2. A qualitative comparison between the synthetic group mobility models investigated with respect to the individual MNs movements, group movements, direction, speed and pause times

Model	Individual MNs Movements	Groups Movements	Direction	Speed	Pause Times
TVC	Based on CoI-based location preferences and time-dependent periodical appearance of MNs	Community-based, product of the decision of individual MNs	Uniformly distributed from the range $[0 \rightarrow 2\pi]$	Uniformly distributed from a range $[v_{min} \rightarrow v_{max}]$	Partially (for movements between communities)
CB	Based on social attractiveness towards other MNs	Community-based (product of the decision of individual MNs)	Intrinsic (derived from the target attraction computation)	Uniformly distributed from a range $[v_{min} \rightarrow v_{max}]$	No
HCCB	Based on social attractiveness towards other MNs and locations	Community-based (product of the decision of individual MNs)	Intrinsic (derived from the target attraction computation)	Uniformly distributed from a range $[v_{min} \rightarrow v_{max}]$	No
OB	Based on the mobility model applied in each orbital level	Community-based	Mobility model-dependent	Mobility model-dependent	Mobility model-dependent
EBC	CP-based. The MN selects a community randomly and moves towards it	Community-based	Uniformly distributed from the range $[0 \rightarrow 2\pi]$	Uniformly distributed from a range $[v_{min} \rightarrow v_{max}]$	No

4. MOBILITY PATTERNS PREDICTION

One of the critical issues pertinent to human mobility patterns is the prediction of possible future locations of MNs. The performance of a wide range of protocols, services and applications in mobile communication networks will be optimized, and even better decisions for other concerns in other fields can be taken (e.g. economy, planning of metropolitan areas, etc.), if an accurate prediction of possible future MNs locations is realized. Note that from the infrastructure-based mobile communication networks point of view, new locations of MNs are predicted in terms of the new subnets the MN will move into their coverage areas in the future. The locations of these subnets are mapped, after that, to geographical information by means of maps for the studied area, so that possible movement patterns result. The paragraph focuses on the prediction of the mentioned future subnets and refers to each subnet as a Subnet Agent (SA) in the next following.

As we mentioned earlier, new SAs are determined either statically or dynamically based, for instance, on history caches and/or mechanisms like data mining to predict new mobility patterns based on old ones. The paragraph briefly highlights the state of art in this context along with the pros and cons of well-known prediction mechanisms.

4.1. State of Art

Neighbor Graph (NG) (Mishra, Shin & Arbaugh, 2004) is the widely used approach to predict the new SAs the MN may move to. NG of SA_i is a undirected graph $G = (V, E)$, where V is the set of all SAs including the current one. E is the set of edges between these SAs. It is defined that there is an edge between SA_i and SA_j if they have a relationship and, thus, it is possible for a MN to move into the coverage area served by SA_j. Let the association pattern $\varphi(c)$ for a MN c be defined as $\{(SA_i, t_1), (SA_j, t_2), \ldots\}$, which means that the MN c associates with SA_i at time t_1 and, after that, with SA_j at time t_2. The MN maintains continuous logical network connectivity from time t_1 to t_2. Obviously, the association pattern can to some extent be captured by the NG since the MN will move from SA_i to one of the neighbors in case the speed of the MN is not too high, so that the MN crosses a neighbor of SA_i without being registered with it. The association pattern discussed is mapped to a geographic area, which delivers predicted movement patterns, see Figure 12.

A dynamic selection algorithm to build groups of neighbors is proposed in (Alnasouri, Mitschele-Thiel, Boeringer & Diab, 2006). This algorithm is pretty simple and based on tracking movements of MNs inside the network. When a MN moves to a new SA, it includes the address of its old SA in the

Figure 12. Mobility pattern prediction applying the NG

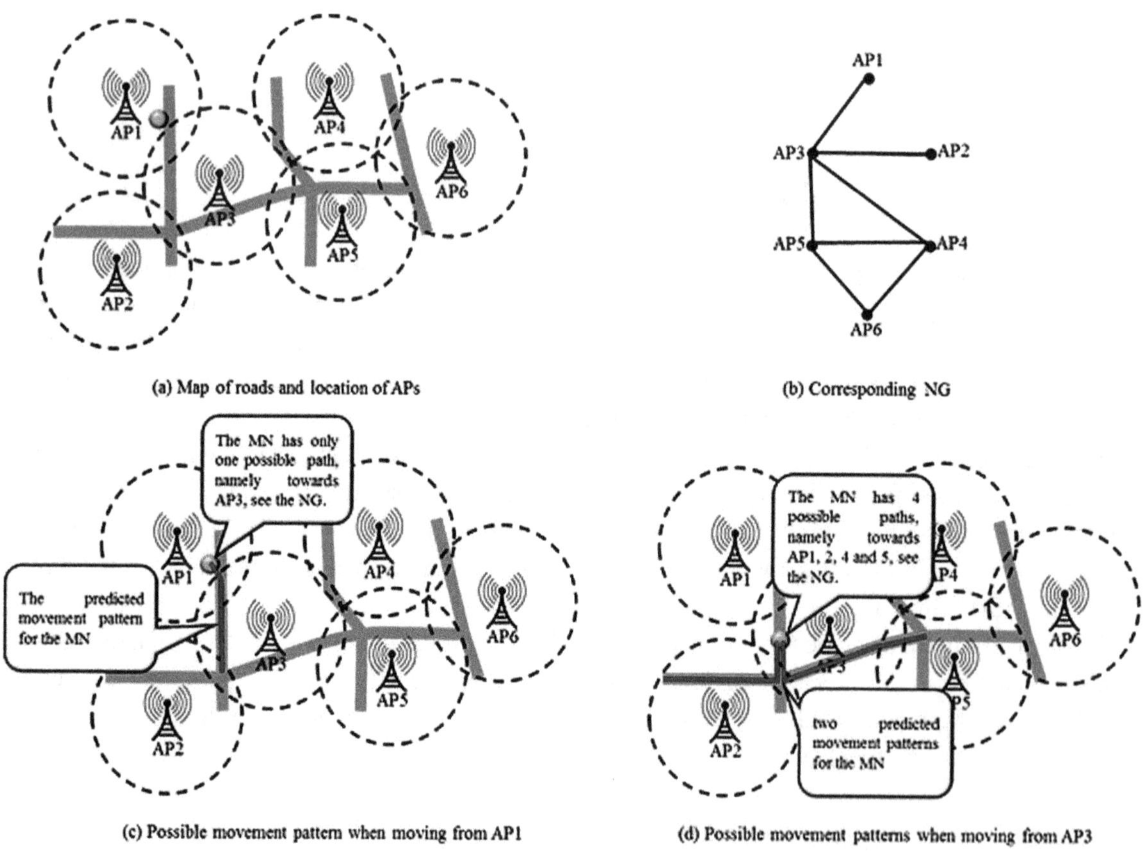

control messages the MN exchanges with the new SA during the re-association process. Consequently, if the new SA is not a member of the neighbor group of the old SA, it has to join itself to this neighbor group, see (Alnasouri, Mitschele-Thiel, Boeringer & Diab, 2006). An advantage of this method is that there is no need to take care of any changes in the network topology. The network dynamically adapts neighbor groups of SAs . Established neighbor groups are used for all MNs regardless of their movement patterns and traffic characteristics. So, NGs are constructed dynamically and adapted to changes in the environment. Corresponding movement patterns are predicted same as by the NG.

Many methods have tried to provide location-based prediction algorithms (Franois, 2007) that follow a different principle than the two mechanisms we reviewed. Concrete, they have addressed three tasks to be done. First, trajectories of MNs should be monitored. Second, monitored trajectories have to be extrapolated. Finally, extrapolated trajectories must be matched with next possible new SAs to be used further for optimization purposes. Linear approximation is used to extrapolate trajectories in (Su, Lee & Gerla, 2001). Obviously, this extrapolation form is not satisfactory since MNs trajectories are mostly not linear. To provide more accurate location-based prediction, a hybrid method based on location information and road topology has been proposed in (Soh & Kim, 2003) and (Soh & Kim, 2004). This method assumes that MNs periodically send their location information to the SAs . Each SA maintains a database containing the roads map in its coverage area, average time to transit each road and probability to transit from one road to another. Continuous tracking of MNs enables then determining the pieces of roads locating at cell boundaries. Based on these data, the SA predicts MNs paths.

Automatic learning and data mining are used sometimes to improve the performance of mobility patterns predictors. Basic idea is to extract possible MNs motion patterns from mobility traces. More concrete, a trajectory of a given MN is compared to those mobility patterns, which enables predicting new possible candidate SAs. These methods assume, however, two assumptions (Franois, 2007). First, MNs movements exhibit certain regularity, which is argued in many studies, see (Gonzalez, Hidalgo & Barabasi, 2008). Second, it should be possible to map MNs trajectories to SAs . Keep in mind that trajectories are defined mostly as physical locations of MNs. Under certain circumstances, the mapping between physical locations and SAs is not trivial, e.g. overlapping areas in cellular cells may be large, thus a certain physical location of a MN inside such overlapping area may be mapped to more than one SA. Well-known example of such methods is proposed in (Yavas, Katsaros, Ulusoy & Manolopoulos, 2005). This method collects motion traces, each is defined as a finite, ordered series of APs corresponding to a movement path of a MN. Motion traces are converted, after that, to mobility rules, e.g. select AP2 and AP3 if the MN is in the range of AP1, select AP5 if the MN moves from AP1 to AP2, etc. Thus, comparing the current MN movement path with recorded mobility traces and executing the defined rules enables predicting new candidate SAs .

4.2. Summary

The main obtained results of previous work review can be summarized as follows:

- Considering the methods predicting new SAs statically, e.g. NG, each SA uses the same neighbors group for all MNs locating in its range regardless of their mobility and traffic characteristics. Obviously, this implies wasting of network resources. Although, the dynamic selection algorithm proposed in (Alnasouri, Mitschele-Thiel, Boeringer & Diab, 2006) provides improvements over

NG, it suffers from similar problems since each SA uses the same neighbors group for all MNs it serves as well. In brief, predicted movement patterns are not personalized.

- The approaches that track MNs to determine possible new SAs, e.g. location-based prediction algorithms, require MNs to do lots of measurements, which consume MNs power. Moreover, they mostly require MNs to be equipped with devices capable of determining the location of the MN, e.g. GPS. A main drawback of these approaches is the discarding of traffic characteristics when predicting possible new SAs.
- The approaches attempting to utilize previous knowledge of mobility patterns of a MN to predict the new mobility pattern do not perform well if there is no or poor knowledge of the MN. Again, traffic characteristics of MNs are not considered in these approaches.

Based on the discussion above, the development of a mechanism capable of optimizing and dynamically self-organizing neighbor groups to accurately predict MNs mobility patterns while considering traffic characteristics without requiring the storage of mobility patterns of each single MN or forcing MNs to do lots of measurements is of a major interest. Obviously, such an algorithm will guarantee the best performance.

5. CONCLUSION

The chapter has focused on the development of novel methods to model human mobility patterns, which have myriad uses in crucial fields such as mobile communication research, urban planning, ecology, epidemiology, etc. The chapter has stated that human mobility patterns are far from being random. They show regularity and are reproducible.

The chapter has categorized mobility models into synthetic, trace-based and community-based models. Synthetic models are largely preferred and widely applied to simulate mobile communication networks. They are mathematical models that try to capture the patterns of human movements by means of a set of equations. These models are traceable, however, not capable of generating realistic mobility models.

The key idea of trace-based models is the exploitation of available measurements and traces achieved in deployed systems, e.g. connectivity logs, location information, etc., to reproduce synthetic traces characterized by the same statistical properties of real traces. The chapter has provided the main projects issued to track users movements and has highlighted the main results derived, which can be summarized as follows: users are in general non mobile. They spend the majority of their time at a limited amount of locations that have sociological relation to them. The variations in the number of users over time are proper representations of the patterns of the underlying population. Movement patterns as well as network usage show regularity. This regularity comes from the cyclic habits users follow in their life. The main drawback of trace-based models is the tight coupling between the empirical model developed and the traces collected, the network topology deployed and the geographic location. This implies that changing the traces, network topology or the geographical area will change the results of the empirical model. This is why the results of empirical models deviate clearly from each other. Therefore, a generalization of empirical models is not possible. When one also considers that the traces themselves are rarely available, one can understand why synthetic models are preferred over empirical ones.

Community-based modeling of human movements depends on the fact stating that mobile devices are usually carried by humans, which implies that movement patterns of such devices are necessarily related to human decisions and socialization behaviors. So, human movement routines heavily affect the overall movement patterns resulting. One of the major contributions in this context is the application of social network theory to generate more realistic human movement patterns. The chapter has discussed the models depending on social communities and finalizes with a qualitative comparison between them.

Another critical issue concerning human mobility patterns is the prediction of future locations of users. The chapter stated that better decisions for various fields, e.g. economy, planning of metropolitan areas, etc., can be taken if an accurate prediction of future users' locations is achievable. Currently, new users' locations are determined either statically or dynamically with the help of history caches and mechanisms like data mining to predict new mobility patterns based on old ones. The chapter has also provided a brief discussion of the mobility prediction issue.

REFERENCES

Alnasouri, E., Mitschele-Thiel, A., Boeringer, R., & Diab, A. (2006). QoMIFA: A QoS-enabled Mobility Management Framework in ALL-IP Network. In Proceeding of PIMRC'06.

Balachandran, A., Voelker, G. M., Bahl, P., & Rangan, P. V. (2002). Characterizing User Behavior and Network Performance in a Public Wireless LAN. In *Proceedings of SIG-METRICS'02*. ACM Press. doi:10.1145/511334.511359

Balazinska, M., & Castro, P. (2003). *Characterizing Mobility and Network Usage in a Corporate Wireless Local-Area Network*. In *1st International Conference on Mobile Systems, Applications, and Services (MobiSys'03)*, San Francisco, CA. doi:10.1145/1066116.1066127

Bayir, M. A., Demirbas, M., & Eagle, N. (2009). *Discovering Spatiotemporal Mobility Profiles of Cellphone Users. World of Wireless*. Mobile and Multimedia Networks and Workshops.

Boldrini, C., Conti, M., & Passarella, A. (2007). *Users Mobility Models for Opportunistic Networks: the Role of Physical Locations*. WRECOM.

Boldrini, C., Conti, M., & Passarella, A. (2009). The Sociable Traveller: Human Travelling Patterns in Social-Based Mobility. In ACM Mobiwac'09.

Camp, T., Boleng, J., & Davies, V. (2002). A Survey of Mobility Models for Ad Hoc Network Research. *Wireless Communication and Mobile Computing Special Issue on Mobile Ad Hoc Networking: Research, Trends and Applications*, *2*(5), 483–502.

Chen, C., & Chen, Z. (2007). Evaluating Contacts for Routing in Highly Partitioned Mobile Networks. *In Proceedings of the 1st international MobiSys workshop on Mobile opportunistic networking* (MobiOpp '07). doi:10.1145/1247694.1247698

Cheng, Y.-C., Chawathe, Y., LaMarca, A., & Krumm, J. (2005). *Accuracy Characterization for Metropolitan-Scale Wi-Fi Localization*. Intel Research Seattle, Tech. Rep. IRS-TR-05-003.

Erdos, P., & Renyi, A. (1959). On Random Graphs. *Publicationes Mathematicae, 6*, 290–297.

Franois, J. M. (2007). *Performing and Making Use of Mobility Prediction*. (Ph.D. Dissertation). University of Lige.

Ghosh, J. (2006). *Sociological Orbit Based Mobility Profiling and Routing for Wireless Networks*. (PhD Thesis). Department of Computer Science and Engineering, the State University of New York, Buffalo, NY.

Ghosh, J., Ngo, H.-Q., & Qiao, C. (2006). Mobility Profile-Based Routing within Intermittently Connected Mobile Ad hoc Networks (ICMAN). In Proceeding of WCMC'06.

Ghosh, J., Sumesh, J.-P., & Qiao, C. (2007). *Sociological Orbit Aware Location Approximation and Routing (SOLAR) in MANET. Ad Hoc Networks*. Elsevier.

Ghosh, J., Yoon, S., Ngo, H.-O., & Qiao, C. (2005). *Sociological Orbits for Efficient Routing in Intermittently Connected Mobile Ad Hoc Networks. UB CSE Technical Report, Department of Computer Science and Engineering University*. Buffalo, NY: The State University of New York.

Gonzalez, M. C., Hidalgo, C. A., & Barabasi, A.-L. (2008). *Understanding Individual Human Mobility Patterns*. Nature Journal.

Gui, C. (2007). An Entropy-based Stability QoS Routing with Priority Scheduler in MANET Using Fuzzy Controller. *Journal of Communication and Computer, 4*(3).

Henderson, T., Kotz, D., & Abyzov, I. (2004). The Changing Usage of A Mature Campus-Wide Wireless Network. In *Proceedings of MobiCom'04*. ACM Press. doi:10.1145/1023720.1023739

Hsu, W., Spyropoulos, T., Psounis, K., & Helmy, A. (2007). Modeling Time-variant User Mobility in Wireless Mobile Networks. In *Proceeding of the 26th IEEE International Conference on Computer Communications* (INFOCOM'07). doi:10.1109/INFCOM.2007.94

Hsu, W., Spyropoulos, T., Psounis, K., & Helmy, A. (2007a). *Modeling Spatial and Temporal Dependencies of Use Mobility in Wireless Mobile Networks. CENG-2008-1 Technical Report*. University of Southern California.

Isaacman, S. (2011). Identifying Important Places in People's Lives from Cellular Network Data. In *Proceeding of the 9th International Conference on Pervasive Computing (Pervasive'11)*.

Isaacman, S. (2011a). *Ranges of Human Mobility in Los Angeles and New York*. In 8th IEEE Workshop on Managing Ubiquitous Communications and Services (MUCS' 11), Seattle, WA. doi:10.1109/PERCOMW.2011.5766977

Isaacman, S. (2012). Human Mobility Modeling at Metropolitan Scales. In *Proceeding of the 10th International Conference on Mobile Systems (MobiSys'12)*.

Isaacman, S., Becker, R., Cáceres, R., Kobourov, S., Rowland, J., & Varshavsky, A. (2010). *A Tale of Two Cities*. In *Workshop on Mobile Computing Systems and Applications (HotMobile'10)*.

Jain, R., Lelescu, D., & Balakrishnan, M. (2005). Model T: An Empirical Model for User Registration Patterns in a Campus Wireless LAN. In *Proceedings of MobiCom'05*. doi:10.1145/1080829.1080848

Jain, R., Shivaprasad, A., Lelescu, D., & He, X. (2004). Towards A Model of User Mobility and Registration Patterns. *Mobile Computing and Communications Review, 8*(4).

Kim, M., Kotz, D., & Kim, S. (2006). Extracting A Mobility Model From Real User Traces. In Proceedings of IEEE INFOCOM'06. doi:10.1109/INFOCOM.2006.173

Kotz, D., & Essien, K. (2002). Analysis of A Campus-Wide Wireless Network. In *Proceeding of the 8th Annual International Conference on Mobile Computing and Networking (MobiCom).* ACM Press.

Kotz, D., & Essien, K. (2002a). *Characterizing Usage of a Campus-Wide Wireless Network.* Technical Report TR2002-423, Department of Computer Science, Dartmouth College.

Kotz, D., & Essien, K. (2003). *Analysis of A Campus-Wide Wireless Network. Mobile Networks and Applications.* Dartmouth College Technical Report TR2002-432.

Kotz, D., & Henderson, T. (2005). CRAWDAD: A Community Resource for Archiving Wireless Data at Dartmouth. *IEEE Pervasive Computing / IEEE Computer Society [and] IEEE Communications Society, 4*(4), 12–14. doi:10.1109/MPRV.2005.75

Kotz, D., Henderson, T., & Abyzov, I. (2015). *CRAWDAD Trace Dartmouth.* Retrieved from http://crawdad.cs.dartmouth.edu

LaMarca, A. (2005). Place Lab: Device Positioning Using Radio Beacons in the Wild. In Proceedings of Pervasive.

Lawler, E. L. (1972). A Procedure for Computing the K Best Solutions to Discrete Optimization Problems and Its Application to the Shortest Path Problem. *Management Science, 18*(7), 401–405. doi:10.1287/mnsc.18.7.401

Lawler, E. L. (2001). *Combinatorial Optimization: Networks and Matroids.* Dover Publications.

Leguay, J., Friedman, T., & Conan, V. (2005). *Evaluating Mobility Pattern Space Routing for DTNs.* Universite Pierre et Marie Curie, Laboratorie. Retrieved from http://arxiv.org/PS_cache/cs/pdf/0511/0511102v1.pdf

Leguay, J., Friedman, T., & Conan, V. (2005a). DTN Routing in a Mobility Pattern Space. In *Proceedings of the 2005 ACM SIGCOMM Workshop on Delay-Tolerant Networking* (WDTN '05). doi:10.1145/1080139.1080146

Lelescu, D., Kozat, U. C., Jain, R., & Balakrishnan, M. (2006). Model T++: An Empirical Joint Space-Time Registration Model. In *Proceedings of MobiHoc'06.* doi:10.1145/1132905.1132913

Mishra, A., Shin, M., & Arbaugh, W.-A. (2004). *Pro-Active Key Distribution Using Neighbor Graphs.* IEEE Wireless Communications.

MSN Maps and Directions. (2015). Retrieved from http://maps.msn.com/

Musolesi, M., Hailes, S., & Mascolo, C. (2005). Social Networks-Based Ad Hoc Mobility Models. In *3rd UK-UbiNet Workshop: Designing, evaluating and using ubiquitous computing systems.* University of Bath.

Musolesi, M., & Mascolo, C. (2006). *A Community Based Mobility Model for Ad Hoc Network Research.* RESLMAN. doi:10.1145/1132983.1132990

Musolesi, M., & Mascolo, C. (2007). Designing Mobility Models Based on Social Network Theory. *SIGMOBILE Mob. Comput. Commun. Rev.*

Musolesi, M., & Mascolo, C. (2009). *Mobility models for systems evaluation.* In B. Garbinato, H. Miranda, & L. Rodrigues (Eds.), *A Survey Book Chapter in n Middleware for Network Eccentric and Mobile Applications* (pp. 43–62). Springer. doi:10.1007/978-3-540-89707-1_3

Networking Research Lab. (2015). Retrieved from http://research.csc.ncsu.edu/netsrv/

Newman, M. E. J., & Girvan, M. (2004). Finding and Evaluating Community Structure in networks. *Physical Review E: Statistical, Nonlinear, and Soft Matter Physics, 69*(2), 026113. doi:10.1103/PhysRevE.69.026113 PMID:14995526

Noulas, A., Scellato, S., Lambiotte, R., Pontil, M., & Mascolo, C. (2012). A Tale of Many Cities: Universal Patterns in Human Urban Mobility. *PLoS ONE, 7*(9). doi: .10.1371/annotation/ca85bf7a-7922-47d5-8bfb-bcdf25af8c72

Ochiai, H., & Esaki, H. (2008). Mobility Entropy and Message Routing in Community-Structured Delay Tolerant Networks. In Proceeding of AINTEC'08. doi:10.1145/1503370.1503396

Rapoport, A. (1957). Contribution to the Theory of Random and Biased Networks. *The Bulletin of Mathematical Biophysics, 19*(4), 257–277. doi:10.1007/BF02478417

Rhee, I., Shin, M., Hong, S., Lee, K., & Chong, S. (2008). On the Levy-Walk Nature of Human Mobility. In *IEEE INFOCOM 2008 - IEEE Conference on Computer Communications.* IEEE. doi:10.1109/INFOCOM.2008.145

Ribeiro, A. G. (2011). Improving Mobile Networks Based on Social Mobility Modeling. In *Proceeding of 19th IEEE International Conference on Network Protocols* (ICNP). doi:10.1109/ICNP.2011.6089064

Robert, M. G. (2009). *Entropy and Information Theory.* Information Systems Laboratory, Electrical Engineering Department, Stanford University.

Roy, R. (2011). Handbook of Mobile Ad Hoc Networks for Mobility Models. Springer Science+Business Media.

Satoh, H., Ito, S., & Kawaguchi, N. (2005). Position Estimation of Wireless Access Point Using Directional Antennas. In *Proceedings of the International Workshop on Location- and Conext-Awareness* (LoCA). Oberpfaffenhofen, Germany: Springer-Verlag. doi:10.1007/11426646_14

Scott, J. (2000). *Social Networks Analysis: A Handbook* (2nd ed.). London: Sage Publications.

Soh, W., & Kim, H. S. (2003). QoS Provisioning in Cellular Networks Based on Mobility Prediction Techniques. *IEEE Communications Magazine.*

Soh, W., & Kim, H. S. (2004). Dynamic Bandwidth Reservation in Cellular Networks Using Road Topology-Based Mobility Predictions. In Proceeding of INFOCOM'04.

Song, C., Qu, Z., Blumm, N., & Barabási, A.-L. (2010). Limits of Predictability in Human Mobility. *Science*, 327. PMID:20167789

Song, L., Kotz, D., Jain, R., & He, X. (2004). Evaluating Next-Cell Predictors with Extensive Wi-Fi Mobility Data. *Proceedings - IEEE INFOCOM.*

Specification, I. E. E. E. (1999). *IEEE 802.11b/d3.0 Wireless LAN Medium Access Control (MAC) and Physical Layer (PHY).* Specification.

Spyropoulos, T., Psounis, K., & Raghavendra, C.-S. (2005). Spray and Wait: An Efficient Routing Scheme for Intermittently Connected Mobile Networks. In *Proceedings of the 2005 ACM SIGCOMM Workshop on Delay-Tolerant Networking* (WDTN '05).

Su, W., Lee, S., & Gerla, M. (2001). Mobility Prediction and Routing in Ad Hoc Wireless Networks. *International Journal of Networking Management.*

Tang, D., & Baker, M. (2000). Analysis of A Local-Area Wireless Network. In *Proceeding of the Sixth Annual International Conference on Mobile Computing and Networking (MobiCom).* ACM Press. doi:10.1145/345910.345912

Tang, D., & Baker, M. (2002). Analysis of A Metropolitan-Area Wireless Network. *Wireless Networks, 8*(2/3), 107–120. doi:10.1023/A:1013739407600

Tuduce, C., & Gross, T. (2005). A Mobility Model Based on WLAN Traces and its Validation. In *Proceedings of INFOCOM'05.* doi:10.1109/INFCOM.2005.1497932

Vahdat, A., & Becker, D. (2000). *Epidemic Routing for Partially Connected Ad Hoc Networks.* Technical Report CS-200006, Duke University.

Watts, D. J. (1999). *Small Worlds, the Dynamics of Networks between Order and Randomness. Princeton Studies on Complexity.* Princeton University Press.

Yavas, G., Katsaros, D., Ulusoy, O., & Manolopoulos, Y. (2005). A data mining approach for location prediction in mobile environments. *Data & Knowledge Engineering.*

Yoon, J., Noble, B. D., Liu, M., & Kim, M. (2006). Building Realistic Mobility Models From Coarse-Grained Traces. In *Proceedings of MobiSys'06.* doi:10.1145/1134680.1134699

Zhang, X., Neglia, G., Kurose, J., & Towsley, D. (2006). Performance Modeling of Epidemic Routing. IFIP Networking. *Computer Networks, 51*, 2867–2891.

ENDNOTES

1 This concept simply means that MNs attach to the communities they are interested in, although there may exist many other communities.

2 These geographical areas represent, in principle, areas where specific social communities exist. Therefore, many communities may coexist beside each other in the same area.

3 The hitting time is the time duration a MN consumes while moving within the transmission range of a fixed randomly chosen target coordinates. The MN starts from a stationary distribution.

4 The meeting time is the time duration until two MNs, starting from a stationary distribution, move to be in the transmission range of each other.

Section 4

Testbeds, Experiments, Applications, and Services

Chapter 13
Manifested Consumption:
Mobile Storefront

Wilson Ozuem
Regents University, London, UK

Bibi Nafiisah Mulloo
London Metropolitan University, UK

ABSTRACT

Recent years have witnessed a rapid growth in internet-enabled devices including smartphones which have enabled many tasks to be performed online. One of the major activities taking place on these online platforms is shopping. Indeed, shopping has changed the landscape of consumption from visiting conventional physical shops to consuming products and services in a digital arena. Despite a proliferation of both theoretical and empirical studies on smartphones marketplace, there is still a paucity of studies on how mobile phone devices enhance and facilitate consumption from physical marketplace to online market marketspace. This chapter aims to contribute to extant studies on how smartphones impact on consumption process.

INTRODUCTION

The mobile revolution in the last decade has recreated a new era of an always-connected society which has reshaped the retail industry including the clothing sector in a major way (Euromonitor, 2013). With mobile devices becoming more powerful with their own operating system, this has given rise to a new marketing strategy enabling companies to connect to their consumers on the go and on a 24 hours basis (Emarketer, 2013). This new phenomenon has impacted on the purchasing behaviour on consumers in the clothing sector as well who are increasingly adopting this new shopping medium. This emerging and ongoing trend has resulted in all major clothing retailers working to gain online presence in view to enforce a stronger customer relationship and to sustain in the clothing industry which is becoming highly competitive (Keynote, 2009). With the rise of more sophisticated mobile devices namely the smartphones, this new convergence has resulted in powerful consumers who now have their stores in their hands (Abrams, 2013).

DOI: 10.4018/978-1-5225-0239-5.ch013

The purchasing habits of consumers have undergone a drastic shift from the traditional physical shopping on high streets to home shopping (Keynote, 2013). The role of salesmen and shopping display is now eclipsed with modern ways of shopping using technology. Research by Euromonitor (2013) identified consumers who more increasingly favour seeking shopping informations on their smartphones rather than consulting salesperson or relying on adverts. The use of the smartphones is thus not limited to buying online but rather it is that device which accompanies the consumers in their shopping experience such as browsing for products information, product reviews, price comparison, consulting friends and families prior to shopping or simply buying online. "This unique ability has transformed the social and spatial aspects of shopping for many consumers" (Brown et al, 2003). In fact, the ongoing connectivity and exposure to shopping information are resulting in an "always shopping state of mind" (Babbitts, 2013). If the art of mobile marketing is effectively and efficiently mastered by retailers, this can lead to strong positioning in their respective market. Although, this new marketing trend has had a strong influence in different retail sectors, a survey carried out by Emarketer (2013) showed that consumers in the apparel sector still favour physical purchase or buying via the PC rather than smartphones when buying online. The main hindrance to buying via smartphones is usually the result of a non-user friendly mobile site (Emarketer, 2013).The answer to successful retailing via smartphones relies in a strong technique and applications of mobile commerce (Branki et al, 2008). Despite consumers in the clothing and apparel sector showing a preference to shop or buy in physical shops, consumers being "creatures of conveniences" are billions to reach first for their phones when they think about shopping (Boyle, 2013). Much is assumed about the supposed effects of smartphones on the luxury fashion sector, but much research in the field suffers from epistemological and ontological incompatibility and can be interpreted in various conflicting ways. The chapter focuses on the impact of smartphones on consumer buying behaviour, particularly on the UK luxury fashion sector.

Recent years have witnessed a rapid growth in internet-enabled devices including smartphones which have enabled many tasks to be performed online (Euromonitor, 2013). One of the major activities taking place on these online platforms is shopping. Indeed, shopping has changed the landscape of consumption from visiting conventional physical shops to consuming products and services in a digital arena (San-Martin & Lopez-Catalan, 2013). Boyle (2013) posited that, the number of shoppers using smartphones to shop is expanding and consumers are now more willing to reach for their smartphones first when the idea of shopping strikes. This global phenomenon is affecting different parts of the world. A survey conducted in June 2013 by Emarketer (2013), shows that China is the leading country in terms of internet users purchasing via mobile devices followed by India and the United Kingdom. UK was amongst the first nations to embrace digital shopping and buying and to date it continues to consume goods and services digitally at a pace greater than most other western European countries (Abrams, 2013). The UK's communications regulator Ofcom recently claimed that UK leads amongst all other European countries with the highest number of smartphones users (Euromonitor, 2013). UK retail ecommerce sales are estimated to have risen from £38.48 billion in 2012 to £45.40 billion in 2013 with a further increase of 15% forecasted for 2014 (Emarketer, 2013). This new trend of purchasing via mobile device has initiated a new type of e-commerce commonly referred to as retail mobile commerce (m-commerce) which is defined by eMarketer (2013) as "products and services purchased using a browser or app via any mobile device, [illegible]gardless of where fulfilment occurs." Whilst the growth of m-commerce is overwhelming in the UK, [illegible]ing via mobile devices is becoming the norm for internet shoppers in the UK with smartphones [illegible]le commerce sales forecasted to more than double by 2017 (McCarthy, 2013). With such a [illegible]f technologically advanced smartphones and the widespread availability of the internet in the UK, digital shoppers increasingly expect an enhanced mobile shopping experience (Boyle, 2013). This has highlighted the importance of mobile commerce for retailers as consumer buying behaviour has undergone a signifant transformation (Persaud & Azhar, 2012). To survive in this new era, several retailers in the UK are at the forefront of m-commerce with major clothing retailers such as Net-a-Porter and Asos are two examples of fashion organisations with the most innovative websites and apps which help them to maintain customer relationship (Abrams, 2013). However, m-commerce is still considered to be in its infancy with most internet users favouring the PC to buy online (Emarketer, 2013). This new shopping trend with its ongoing challenge makes it interesting to investigate the effects of smartphones usage on users purchase intentions.

THEORETICAL FRAMEWORK AND CONTEXT

Feature phones are now being gradually eclipsed by the proliferation of smartphones which are more sophisticated in terms of their functionality (Nokia, 2013; Li et al, 2012; Kumar & Mukherjee, 2013, Holmes et al, 2014). Technological advancement in mobile devices fuelled by the widespread availability of high-speed internet access has enhanced the various features and usage of mobile phones (Persaud & Azhar, 2012; Flint & Spaid, 2014; Schuster et al, 2013). In fact, the rapid growth of internet-enabled mobile devices have redefined the primary purpose of phones, from devices primarily designed for call-making to a product capable of performing different functions such as web browsing, emailing, shopping and other tasks (Hu et al, 2006; Ting et al, 2011; Swilley et al, 2012). A smartphone is a programmable phone which provides it's user with advanced capabilities and features to enhance their daily work and personal life (Yang & Kim, 2012).

The rolled out of 3G and EE 2012 launched of 4G in UK has boosted mobile internet speed which enable users to have a more enriching browsing experience (BBC, 2012; Kumar & Zhan, 2003; Yang, 2012). Although UK mobile operators have restrained their generous unlimited mobile data offerings, mobile internet subscriptions and smartphones usage are still witnessing constant growth (Guardian, 2012). This can be attributed to the fact that consumers are now less reliant on mobile networks in the UK as the number of Wi-Fi spots is increasingly available in cafes, transport hubs and other public places (Yang, 2012). The smartphone is now increasingly seen as the must-have device that permits users to perform a multitude of tasks with no time or location constraints (Wong, 2012). The growing dependency on smartphones has revolutionized the lifestyle and behaviour of consumers in different sectors creating a digital landscape where consumers are engulfed in their own worlds anywhere and anytime (Elkin, 2012). For many individuals the smartphone is not only a personal device but it is "an extension of their personality and individuality" (Persaud & Azhar, p.418). A mobile device, in this study, the smartphone can be accessed anytime, anywhere which enable the mobile web to be accessed in more diverse contexts (Kim et al, 2002). The capacity of seamlessly integrating Bluetooth, GPS and other technologies, the smartphone is considered as a useful tool that can provide a superior consumer experience (Swilley & Hofacker, 2006). Moreover, with smartphones at more affordable prices the rate of adoption is ever-increasing with no signs of abating which emphasises the central importance smartphones are gradually and surely taking in consumers' decision making process (Pasqua & Elkin, 2013). The rapid development and deployment in internet-enabled mobile devices led to the emergence of highly sophisticated applications commonly referred to as the mobile commerce (m-commerce) (Wang et al, 2007; Wheeler, 2013; San-Martin et al, 2012; Wong, 2012).

Mobile commerce is commonly referred to the ability to undertake commercial transactions through a wireless internet-enabled device (Swilley et al, 2012). Most definitions of m-commerce will be focused on an extension of e-commerce (Frolick & Chen, 2004; Siau et al, 2001; Dholakia & Dholakia, 2004). It is even considered as an unprecedented business opportunity for ecommerce (Wang et al, 2007; Yeh & Li, 2009; Wheeler, 2013). It is the medium through which retailers can develop their marketing strategies as it is the channel that has the potential to initiate purchase intentions in consumers. However, m-commerce is not regarded as a replacement to e-commerce but rather as a supplement (Clarke, 2008). Mobile commerce in the UK has been effectively adopted due to the widespread availability of high-speed internet access via mobile devices which render mobile shopping readily accessible to consumers (Mintel, 2013). Research conducted by Ipsos and RetailMeNot.com in August 2013 suggest that UK remains the highest users of m-commerce in the developed nations (eMarketer, 2013). In this respect, the importance of m-commerce cannot be ignored when conducting this research. Therefore, the ubiquity of smartphones and wireless technology are both important elements that indirectly affect consumers' purchase intentions.

Smartphones being the medium that has a growing influence on consumers purchase intention and buying habits, marketers has an increasing challenge to keep updated track on how to maximise consumers' retention by adopting appropriate mobile marketing strategies (Elkin, 2013). Ranging from online advertisement, QR codes, Apps development and Geo-fencing amongst others, marketers have the humongous responsibility of deterring consumers from being influenced by showrooming (BBC, 2013). Showrooming is defined as the activity of going online while in-store to determine which competitors are offering a better deal for the product a consumer is currently looking to buy (Abrams, 2013). From search engines to deal sites, it has become crucial for companies to capitalise fully on the digital landscape to enhance their online presence (Ahrens et al, 2013). This has become a necessity for the clothing sector as even though mobile shopping may not culminate in purchasing via mobile, consumers in this sector still rely heavily on their smartphones to browse for product information, deals and purchase-related queries (Cho & Fiorito, 2009). This ongoing thirst for information has emphasized to retailers the importance of enhancing their mobile commerce strategies to remain competitive in what is highly digitalised shopping era (Skeldon, 2012).

The main activities undertaken by consumers on their smartphones include contacting friends about shopping matters, carrying out price checks, looking at product availability, finding out product information while in-store, accessing social media, acquiring/ redeeming coupons, carrying out location-based search, consulting map, using apps, buying products using device and even buying products using device while in-store. These suggest the high influence of an effective mobile commerce on consumers' online shopping experience. For the purposes of this chapter, smartphones include all mobile phone devices which can access the internet via mobile networks or Wi-Fi spots. Mobile shopping and online consumer behaviour will be discussed further in this chapter in order to understand how smartphones can influence purchase intention.

ATTITUDES TOWARDS TECHNOLOGY

A number of studies have focused on research on key factors that influence the acceptance of new technologies (Ajzen, 1985; Davis et al; 1989; Lu et al, 2003). To understand how smartphones can affect consumers' purchase intentions and behaviour, it is primordial to understand the user perception towards the use of the technology (Kumar & Mukherjee, 2013). The Theory of Reasoned Action (TRA) suggests that the intention to perform a behaviour is influenced by a person's attitudes towards that behaviour and a subjective norm (Fishbein & Ajzen, 1975). Attitude in this context is further defined by Spaid and Flint (2014) as an individual positive or negative feeling towards a specific behaviour. Fishbein and Ajzen (1975, p.302) refers to a subjective norm as "perception that most people who are important to him think that he should not perform the behaviour in question." The TRA implies that any factors that influence a person's behaviour do so through a change of attitudes, the subjective norms or a combination of the two (Spaid & Flint, 2014). The theory of reasoned action (TRA) is considered fairly robust in predicting and explaining cognitively driven attitude (Spaid & flint, 2014). To address some of the limitations of the TRA, Ajzen (1991) attributes the element of control as another determinant to behavioural intention and actual behaviour. The theory of planned behaviour (TPB) was thus created by Ajzen (1991) to study a perceived lack of control when it comes to certain behaviours. Both theories assist Spaid and Flint (2014) in understanding shoppers' attitudes towards mobile devices.

The TRA was further modified by Davis et al (1989) where the element of subjective norm was dropped in the process. The Technology Acceptance Model (TAM) was thus devised where two main variables were included to the fundamental process of attitude, intention and behaviour. The TAM attempts to understand how users accept and use the technology (Lu et al, 2003). The two main variables in the original TAM are "perceived usefulness" and "perceived ease of use" (Davis, 1989 cited in Kumar & Mukherjee, 2013, p. 25). These two determinants are considered useful indicators of the attitudes of users towards a new technologies and their eventual adoption or rejection (Lu et al, 2003). Perceived usefulness is described by Davis (1989, p.320) as "the degree to which a person believes that using a particular system would enhance his or her job performance." A system high in perceived usefulness is considered to positively influence adoption of the specific system (Davis, 1989). On the other hand, perceived ease of use is referred to as "the degree to which a person believes that using a particular system would be free of effort" (Davis, 1989, p.320). A system scoring high in ease of use will encourage users to use the system (Davis, 1989).

The TAM model has been used in various recent research into smartphones and mobile consumer behaviour (Holmes et al, 2014). For instance, Lu and Su (2009) used the TAM model to explore factors affecting purchase intention in the context of mobile websites. Their research suggested that negative emotions such as anger have a negative impact on purchase intention whereas positive emotion such as enjoyment has a positive effect on purchase intention (Lamarre et al, 2012). Although TAM is a highly cited model by researchers, many have mixed opinions regarding its applications as being limited (Lu et al, 2003). Conversely to TRA, the TAM is criticised for lacking vigour and having limited variables (Davis et al, 1992; Venkatesh et al, 2013). Burton-Jones and Hubona (2006) stated that the original TAM can be modified to include external factors such as age, level of education and experience which according to their studies can affect the use of a technology. Mathieson (1991) on the other hand considers the TAM theoretical foundation to be poor and argues that there can be many uncertainties between the intention and adoption phase. However, the Technology Acceptance Model (TAM) remains a useful theoretical framework to assess users' acceptance or rejection of a particular technology (Davis, 1989 cited in Kumar & Mukherjee, 2013). Mobile shopping is an activity which involves interaction between the user and the technology which in this research is the smartphone hence the importance of the users' attitudes towards the said technology.

Researches in the past advocated positive attitudes and purchase intentions are inter-related (Shim et al, 2001; George, 2002; Goldsmiths & Bridges, 2005). Eagly and Chaiken (1993, p.1) define attitude as "a psychological tendency that is expressed by evaluating a particular entity with some degree of favour or disfavour." The many definitions of attitudes emphasise on the nature of the term and its connection with behaviour (Bohner & Bohner, 2001). For instance, Seok and Norton (2007) argue that consumers' attitudes towards a specific shopping channel will influence their likelihood to purchase through that channel. Therefore, consumers' perception of mobile internet to search for information about products and services will eventually influence their intentions to undertake those searches via their smartphones and likelihood to purchase online (Kim et al, 2012). It has been argued further that attitude towards a technology (i.e. mobile internet) will determine the probability of a consumer using that technology (i.e. smartphones) as a medium to make purchase. The consumers' intentions (Seock & Norton, 2007) and consumers' objectives (Nysveen & Pedersen, 2005) will contribute towards a high-degree goal-oriented search mode which will influence the purchase intentions in an online mobile environment. As pointed out by both George (2002) and Seock and Norton (2007), a positive attitudes towards online shopping will culminate into a positive purchase intention whereas a negative attitude towards online shopping will result in a negative purchase intentions.

SMARTPHONES AND MOBILE SHOPPING

Considerable evidence from industry sources reveals that mobile phones are playing a significant role in consumer's shopping process (Skeldon, 2012; Seock & Norton, 2007). With prediction of increase adoption of smartphones in future, it is inevitable to see a continuous growth in this new shopping trend. The use of smartphones for shopping is not restricted to purchase but is more thoroughly used for activities such as searching for products, gathering production information, checking prices or reading reviews (Holmes et al, 2014). The ability to conduct shopping activities on mobile devices has opened up opportunities for retailers to influence the purchase intentions of shoppers (Shankar et al, 2011).

To understand the term mobile shopping, it is important to understand consumer buying behaviour. Consumer buying behaviour involves the process and activities that consumers entail in when searching, evaluating, analysing, buying, reviewing, rejecting, accepting products or services in view to satisfy their personal needs. Solomon (2009, p.33) defines consumer behaviour as "the study of the processes involved when individuals or groups select, purchase, use, or dispose of products, services, ideas, or experiences to satisfy their needs and desires." According to Peter and Olson (1996), consumers buying decision are divided into interpretation, integration and behaviour. Interpretation deals with how consumers select information and derive meanings from that particular information (Peter & Olson, 1996). Then, the integration happens when the consumer evaluates the product using various sources of information among other alternative behaviours towards the product (Peter & Olson, 1996). This is an important stage in the decision-making process that will eventually result in either the customer buying or rejecting the product (Peter & Olson, 1996). Nysveen and Pedersen (2005) identify the consumers' objectives as the ultimate factor that will influence the consumers search for a product and hereafter the buying behaviour. In their opinion, consumers with high-degree goal-oriented search mode which is influenced by product knowledge, product risk and product involvement are more likely to complete a transaction than consumers with a low-degree search mode (Nysveen & Pedersen, 2005).

Mobile shopping is the process of consumers exhibiting particular buying behaviours in an online mobile environment (Skeldon, 2012). With easy access to internet on the go via smartphones, this has fuelled the ability of consumers to be more-goal oriented in their search for products (Nysveen & Pedersen, 2005). To assist them in their information search and in some cases buying process, the smartphones are now the convenient medium. With the increase adoption and functionality of the smartphones, consumers are making increase usage of their smartphones in the shopping process anywhere and anytime (Charlton, 2011). New technological developments in smartphones have indeed opened up opportunities for retailers to influence shopper attitudes and behaviours (Holmes et al, 2014).

ONLINE SHOPPING MOTIVATIONS

Salomon and Rabolt (2006) describe motivation as the processes that drive people to behave in a particular manner. In the shopping context, motivations will be the behavioural factor that will ultimately drive shopping behaviour (Jim & Kim, 2003). Babin et al (1994) identified two primary drivers of shopping motivations as being attributed to utilitarian and hedonic reasons. Adapting a traditional approach, shopping motivations was seen as highly task-related and rational from a utilitarian perspective (Kim, 2006). Babin et al (1994) however argued that shopping motivations are identified by the perception of shopping from the consumer point of view whether they see it as an activity of cognition or that of pleasure. Hence, the two perspectives of shopping motivations are identified as utilitarian and hedonic shopping motivations (Babin et al, 1994). The utilitarian aspects of shopping considered shopping as a task to be performed in a timely and efficient manner (Childers et al, 2001). These shoppers will use their mobile to search for product information and offers online at their convenience to attain their motives and this play an important role in shaping their shopping behaviour (Childers et al, 2001).

Similarly, Babin et al (1994) argue that shopping motivations need not be restricted to product or task-oriented but can also be hedonic. Consistent with this findings, Hoffman and Novak (1996) state that hedonic shopping motivation is associated with the entertainment and enjoyment derived from the shopping activity. Kim and Forsythe (2007) identify hedonic shoppers as those who seek pleasure and fun in the shopping process. Childers et al (2001) state that the social activity for hedonic shoppers is not limited to product acquisitions but involves the activity of browsing or seeking excitement in surfing. Hedonic shoppers are considered as more apt to engage in online shopping as their main motives are for non-economic reasons such as fun, social or emotional gratification (Park et al, 2006). Arnold and Reynolds (2003) conducted researches regarding the hedonic aspects of shopping which resulted in a range of motives from adventure, gratification, value, social, role and idea shopping motivations. Although these two aspects are considered to offer contrasting conceptual approaches to shopping, they are both considered important in shaping consumer buying behaviour (Babin et al, 1994). Scarpi (2006) suggests that consumers with a high level of hedonism tend to purchase on a frequent basis compared with consumers with a high level of utilitarianism. Childers et al (2001) confirmed that both usefulness and enjoyment associated with purchasing a product have a positive effect on consumers purchase intentions. Accordingly, these findings show that both hedonism and utilitarianism will shape consumers' perceptions and adoption of online shopping (Babin et al, 1994; Scarpi, 2006).

ESTABLISHING TRUST IN A MOBILE ENVIRONMENT

Trust is an important element that can promote or deter consumers from using smartphones for shopping processes as emphasised by (Yeh & Li, 2008). Trust is an important element which can ensure consumers' satisfaction and which can thus lead to consumers' retention for both online and offline retailers (Schiffman & Kanuk, 2010). Satisfaction and trust are closely related in online shopping experience and to benefit from this consumers place high importance on web site design, reliability, responsiveness, trust and personalisation when conducting online purchase (Yeh & Li, 2008). Satisfaction is a strong factor which will determine the level of trust established with consumers and this happens at the post-purchase stage (Yeh &Li, 2008). There should be a strong similarity in what consumers are expecting to be offered and what is being provided by the online service (Yeh & Li, 2008). Consequently, retailers need to focus on enhancing their mobile commerce with minimum hindrance in their mobile sites so as to encourage consumers to shop through their smartphones more frequently (Skeldon, 2011). Wang & Liao (2007) highlighted the fact that a successful online mobile experience will most likely encourage future online purchase. Other major concerns that influence consumers purchase intentions in a mobile environment are security and data protection concern (Yeh & li, 2009). This is being an on-going hindrance that is affecting purchase intentions on smartphones.

As mentioned earlier, consumers' attitudes towards their smartphones is an important consideration when studying the mobile shopping process. Consumers perceive their smartphones as personal and may consider it as an intrusion if they are bombarded with commercials or advertisings all the time (Holmes et al, 2014). The element of trust is thus considered crucial as consumers are becoming more conscious about source credibility (Spaid & flint, 2014). Consumers do intense research prior to purchase and thus online retailers need to provide genuine information to assist potential consumers (Solomon, 2009) Shopping experience via smartphones need to be perceived as relevant, useful and timely and possibly funny to establish a trust relationship (Hung et al, 2012). Another factor that affects the trust element in consumers' willingness to adopt online shopping can be attributed to the quality of service they are being offered online (Chemingui & Lallouna, 2013). In the absence of a salesperson to guide them in their shopping process, it is crucial for the quality of service online to be perceived as exceptionally good (Chemingui & Lallouna, 2013). The first online shopping experience contributes towards building an initial trust relationship between the retailers and the shoppers which will lead to future online shopping (Zhou, 2011). In the apparel sector, although customisation of clothes is now possible with body-scanning technology many consumers are reluctant to use this shopping format (Cho & Fiorito, 2009). Zhou (2011) further argues that it is even harder to establish a trust relationship in a virtual environment due to great uncertainty and trust. Therefore, it is important for retailers to concentrate on enhancing their mobile commerce strategies to maintain brand credibility and integrity (Chemingui & Lallouna, 2013). Considering that consumers still favour the PC and physical stores in the clothing sector, it is very interesting to see what affects their purchase intentions via smartphones and what changes they expect in the future by conducting this research. The element of trust is difficult to achieve in a mobile environment as the threats involved are considered very high and very difficult to eradicate.

SMARTPHONES AND PURCHASE INTENTIONS

The rapid adoption and use of smartphones has driven many cultural change including changes in consumers shopping patterns and shopping behaviour (Spaid & Flint, 2014). Most existing studies have been focused on identifying the main variables which are responsible for consumers purchasing behaviour in the mobile online environment (Kumar & Mukherjee, 2013; Spaid & Flint, 2014; Holmes et al, 2014; Ting et al, 2011). While some researchers attribute attitudes and consumers perception as the main determinants that influence the use of smartphones as a shopping tool (Holmes et al, 2014; Lu & Su, 2009; Lu et al, 2003), others consider motivation as a major factor contributing towards the use of smartphones as a shopping channel (Childers et al, 2001; Kim & Forsythe, 2007; Park et al, 2006). Whichever the variables that determine the use of smartphones to conduct shopping activities, it is evident that smartphones and purchase intentions are inter-related (Seok & Norton, 2007; Skeldon, 2012; Elkin, 2013).

The technology acceptance model (TAM) is indeed a useful tool in assessing users' willingness to adopt smartphone as the new purchase channel (Kumar & Mukherjee, 2013; Spaid & Flint, 2014). This model had been used and in certain research modified in view to ascertain a more defined understanding of how this new technology is continuously reshaping the shopping process. The common variables identified by various researches in this new field are attitudes, intention and behaviour (Spaid & flint, 2014, Lu et al, 2003; Seok & Norton, 2007; Holmes, 2014). Ting et al (2011) identified dependency and perception of the smart device as major attributes that influence user's attitude towards using a smartphone. Due to time scarcity, consumers find it convenient to use their smartphones anywhere and anytime to perform multiple tasks (Dibb & Carrigan, 2013). As such, they perceive their smartphones as the ultimate device that perform the task of both a mobile phone and a laptop (Lu & Su, 2009). Consequently, consumers having a positive perception of the multi-usage capabilities of a smartphones are identified as having a positive attitude towards adoption of smartphones as a multi-tasking device (Kumar & Mukherjee, 2013). Behavioural intention relates to the phase where an individual's attitude gradually controls or guides his or her behaviour (Ajzen, 1991). In the context of mobile shopping, attitude and motivation are considered as the determinants of behavioural intention (Kim, 2006 and Scarpi, 2006). Whether for utilitarianism or hedonism reasons, positive motivation and initial attitude are believed to trigger the use of smartphone as a shopping tool and henceforth purchase intention (Kim, 2006 and Scarpi, 2006). The actual behaviour is the last phase of the technology acceptance model where the users decision to use the device culminates (Ajzen, 1991). In the context of this research the actual behaviour happens when consumers actually use the device for shopping experience. Consumers' actual usage of this new shopping technology does not necessary guarantee ultimate adoption of smartphones as a shopping channel (Yeh & Li, 2008). The first mobile shopping experience is considered as a crucial touchpoint that will initiate future purchase (Skeldon, 2012). Hence, the growing importance of mobile commerce in this new digitalised marketplace (Zhou, 2011). The importance of a good mobile site, easy user interface, enhanced security and an overall sound mobile experience are the major contributors towards effective adoption of this new shopping tool (Yeh & Li, 2009). Considering the clothing sector, the challenge is even bigger as the sensory feelings of products constitute a major determinant of actual purchase (eMarketer, 2013). Moreover, as per research carried out by eMarketer (2013), although mobile commerce has shown considerable growth in the last few years, personal desktop remains the most favoured device for online purchase.

The theory of reasoned action (TRA) and the theory of planned behaviour (TPB) are both attitudinal theories that will assist the researcher in gaining in-depth knowledge about shoppers' attitude towards mobile technology. The technology acceptance model remains the ultimate tool that will be used in this research to understand the different phase that constitutes the shopping process in a mobile environment. The main gap in the literature lies between the behavioural intention phase and the actual behaviour. Though researches have focused on different variables that might be integrated in the model to have a clearer picture on how shopping behaviour might be influenced in a mobile environment, little relates to how smartphone can influence purchase intention. Hence, the focus of the research will be to establish a link between smartphones and purchase intention with particular emphasis on those variables that affect the acceptance of this new shopping technology.

CONCLUSION AND MANAGERIAL IMPLICATIONS

Understanding the motivational factors and features that contribute to an effective and efficient mobile retail channel is primordial for retailers, including the clothing sector, in view to sell their products successfully online and offline. For instance, consumers' pre-purchase activities through their smartphone are mainly directed towards deals and cheaper prices, both of which are driven by goal-oriented motivation. Therefore, marketers should revise their mobile commerce strategies to cater for and attract price sensitive consumers. The challenge is to be the price leader with active promotional offers, deals and online discounts to attract consumers to their websites and to deter them from moving to competitors' websites.

Motivation is considered strategic to entertain consumers in a mobile retail setting. Whether the initial motive for shopping is task-related or for enjoyment, the objective of retailers is to ensure consumers' satisfaction at all stages of the shopping process. Therefore, marketers should develop marketing strategies in view to encourage consumers to conduct shopping activities online. Such strategies might include promotional offers, click to buy option, click to collect option, free delivery service, next day delivery service or online coupons which are commonly favoured by online shoppers.

Although many consumers conduct direct purchase on their smartphones, the majority uses their devices for pre-purchases activities. These activities mostly include information queries regarding product specifications, product price, reviews and availability. In today's knowledge-based society, consumers are more willing to explore their personalized needs by using their smartphones prior to engaging in any form of transactions. Communications in the retail environment has evolved by involving consumers in the interactive platform of mobile commerce in view to provide information that meet consumers' expectation. Hence, marketers need to focus on allocation of accurate and updated database on their mobile site to assist mobile shoppers with information queries at all time.

Trust is an important element that contributes to the establishment of a good rapport between consumers and retailers. This has emerged as a key variable within this study and constitutes to be strategic to encourage adoption of smartphones both for pre-purchase activities and for immediate purchase. Online retailers should thus ensure that consumers enjoy the mobile retail channel by implementing enhanced mobile commerce. This involves investment of resources towards a mobile site with user-friendly interface, easy access to commonly researched information and safe click to buy or click to collect in-store options. Furthermore, considering consumers' trustworthiness in social interaction for authentic information, mobile site needs to integrate features of social media to enhance consumers' experience and improve transparency in the mobile retail environment. Overall, consumers need to perceive mobile shopping as a convenient and safe shopping medium.

Intense innovation in features of smartphones and wireless technologies has freed users from the tethers that initially bounded them to their desk for online shopping. Convenience and flexibility are key features that initiate the use of smartphones for shopping irrespective of time and location. This is a golden opportunity for retailers to establish a mobile presence by making efficient use of mobile technology. Smartphones are now equipped with wide and high resolution screen which emphasised the need for marketers to implement a mobile site interface that closely match the laptop or PC experience. Mobile sites need to improve navigation that will assist consumers whilst browsing for information they value at their convenience thus enriching their shopping experience. Marketers need to ensure that their mobile site is compatible and up-to-date with latest technologies such as fingerprint recognition, digital wallet, heartbeat rate, retinal recognition and other emerging technologies to attract and retain more consumers in the mobile environment. In successfully implementing such strategies, consumers will find it even more convenient to adopt mobile shopping and will also entertain a trust relationship with the mobile retail channel.

By understanding the link between smartphone and purchase intention, marketers can adapt their marketing strategies to positively influence more users to conduct at lease pre-purchase activities in a mobile environment. They need to acknowledge and analyse further variables that are valued by mobile consumers and evaluate ways to improve these variables in view to gain and retain mobile shoppers. Although the findings in this study are not exhaustive, it provides a snapshot into understanding the relationship between smartphone and mobile shoppers' behavioural intention.

REFERENCES

Abrams, K. V. (2012). *UK B2C Ecommerce: Mobile Takes Center Stage*. Available at: http://www.emarketer.com/Article/Mobile-Gains-Speed-UK-Ecommerce-Maintains-Momentum/1009093

Abrams, K. V. (2013a). *UK Digital Trends: Ecommerce, Video and Mobile Thrive, but Brands must work for loyalty and trust*. eMarketer.

Abrams, K. V. (2013b). *Western Europe B2C Ecommerce: Solid Growth Continues as Mobile Gains Ground*. eMarketer.

Ahrens, J., Coyle, J. R., & Strahilevitz, M. A. (2013). Electronic word of mouth: The effects of incentives on e-referrals by senders and receivers. *International Journal of Marketing*, *47*(7), 1034–1051.

Arnold, M. J., & Reynolds, K. E. (2003). Hedonic shopping motivations. *Journal of Retailing*, *79*(2), 77–95. doi:10.1016/S0022-4359(03)00007-1

Banerjee, S., & Dholakia, R. R. (2012). Location-based mobile advertisements and gender targeting. *Journal of Research in Interactive Marketing*, *6*(3), 198–214. doi:10.1108/17505931211274679

Barrutia, J. M., & Echebarria, C. (2013). Networks: A social marketing tool. *European Journal of Marketing*, *47*(1/2), 324–343. doi:10.1108/03090561311285574

BBC. (2012). *UK's first 4G mobile service launched in 11 cities by EE*. Available at: http://www.bbc.co.uk/news/technology-20121025

BBC. (2013). *The peril of "showrooming"*. Available at: http://www.bbc.co.uk/news/magazine-22098575

Bohner, M., & Bohner, G. (2001). *Intrapersonal processes*. Oxford, UK: Blackwell.

Boyatzis, R. E. (1998). *Transforming qualitative information: Thematic analysis and code development*. Sage Publications Inc.

Boyle, C. (2013). *Retail Mobile Commerce Forecast: Shoppers Turn to Smartphones First*. Available at: http://www.emarketer.com/Article/More-Shoppers-Reach-Mobile-Browse-Buy/1009615

Braun, V., & Clarke, V. (2006). Using thematic analysis in psychology. *Qualitative Research in Psychology, 3*(2), 77–101. doi:10.1191/1478088706qp063oa

Brown, M., Pope, N., & Vogues, K. (2003). Buying or browsing?: An exploration of shopping orientations and online purchase intention. *European Journal of Marketing, 37*(11/12), 1666–1684. doi:10.1108/03090560310495401

Bryman, A. (1988). *Contemporary Social Research Series.* London: Unwin Hyman Ltd.

Bryman, A. (1989). *Research Methods and Organization Studies.* London: Unwin Hyman Ltd. doi:10.4324/9780203359648

Burr, V. (2003). *Social Contructionism* (2nd ed.). London: Routledge.

Burton-Joners, A., & Hubona, G. S. (2006). The mediation of external variables in the technology acceptance model. *Information & Management, 43*(6), 706–717. doi:10.1016/j.im.2006.03.007

Butler, P., & Peppard, J. (1998). Consumer Purchasing on the Internet: Processes and Prospects. *European Journal of Marketing, 16*(5), 600–610.

Capgemini. (2013). *Online retail sales highest in two years as positive economic sign continues.* Available at: http://www.uk.capgemini.com/online-retail-sales-highest-in-two-years-as-positive-economic-signs-continue

Chan, F. T. S., & Chong, A. Y.-L. (2013). Analysis of the determinants of consumers' m-commerce usage activities. *Online Information Review, 37*(3), 443–461. doi:10.1108/OIR-01-2012-0012

Cheema, A., & Papatia, P. (2010). Relative importance of online versus offline information for Internet purchase: Product category and Internet experience effects. *Journal of Business Research, 63*(9), 979–985. doi:10.1016/j.jbusres.2009.01.021

Chemingui, H., & Lallouna, H. B. (2013). Resistence, motivation, trust and intention to use mobile finance services. *International Journal of Bank Marketing, 31*(7), 574–592. doi:10.1108/IJBM-12-2012-0124

Cheung, C. M. K., & Lee, M. K. O. (2012). What drives consumers to spread electronic word of mouth in online consumer-opinion platforms. *Decision Support Systems, 53*(1), 218–225. doi:10.1016/j.dss.2012.01.015

Childers, T. L., Carr, C. L., Peck, J., & Carson, S. (2001). Hedonic and utilitarian motivations for online retail shopping behaviour. *Journal of Retailing, 77*(4), 511–535. doi:10.1016/S0022-4359(01)00056-2

Cho, S., & Fiorito, S. S. (2009). Acceptance of online customization for apparel shopping. *International Journal of Retail & Distribution Management, 37*(5), 389–407. doi:10.1108/09590550910954892

Clarke, I. (2008). Emerging Value Propositions for M-commerce. *The Journal of Business Strategy, 25*(2), 41–57.

Collis, J., & Hussey, R. (2003). *Business Research: A practical guide for undergraduate and postgraduate students* (2nd ed.). New York: Palgrave Macmillan.

Collis, J., & Hussey, R. (2009). *Business Research: A Practical Guide for Undergraduate & Postgraduate Students* (3rd ed.). Hampshire, UK: Palgrave Macmillan.

Constantinides, E., Lorenzo-Romero, G. M., & Gómez, M. A. (2010). Effects of web expereience on consumer choice: A multicultural approach. *Internet Research*, *20*(2), 188–209. doi:10.1108/10662241011032245

Creswell, J. W. (1994). *Research design: Qualitative and Quantitative Approaches*. Thousand Oaks, CA: Sage.

Davis, F. D., Bagozzi, R. P., & Warshaw, P. R. (1989). User acceptance of computer technology: A comparison of two theoretical models. *Management Science*, *35*(8), 983–1003. doi:10.1287/mnsc.35.8.982

Demangeot, C., & Broderick, A. J. (2007). Conceptualising consumer behaviour in online shopping environments. *International Journal of Retail & Distribution Management*, *35*(11), 878–894. doi:10.1108/09590550710828218

Denzin, N. K. (1970). *The Research Act: A theoretical Introduction to Sociological Methods*. New York: McGraw-Hill.

Denzin, N. K., & Lincoln, Y. S. (2000). *The Handbook of Qualitative Research* (2nd ed.). Thousand Oaks, CA: Sage.

Dholakia, R. (1999). Going shopping: Key determinants of shopping behaviors and motivations. *International Journal of Retail & Distribution Management*, *27*(4), 140–165.

Dholakia, R. R., & Dholakia, N. (2004). Mobility and markets: Emerging outlines of m-commerce. *Journal of Business Research*, *57*(12), 1391–1396. doi:10.1016/S0148-2963(02)00427-7

Dibb, S., & Carrigan, M. (2013). Social marketing transformed. *European Journal of Marketing*, *47*(9), 1376–1398. doi:10.1108/EJM-05-2013-0248

Dittmar, H., & Drury, J. (2000). Self-image - is it in the bag? A qualitative comparison between 'ordinary' and 'excessive' consumers. *Journal of Economic Psychology*, *21*(2), 109–142. doi:10.1016/S0167-4870(99)00039-2

Eagly, A. H., & Chaiken, S. (1993). The psychology of Attitudes. *Public Opinion Quarterly*, *57*(3), 434. doi:10.1086/269388

Emarketer. (2013a). *Retail Drives UK B2C Ecommerce Sales Higher*. Available at: http://www.emarketer.com/Article/Retail-Drives-UK-B2C-Ecommerce-Sales-Higher/1010106

Emarketer. (2013b). *In-store Mobile Offers Still Finding Their Feet in the UK*. Available at: http://www.emarketer.com/Article/Retail-Drives-UK-B2C-Ecommerce-Sales-Higher/1010106

Emarketer. (2013c). *Mobile's Share of UK Digital Retail Sales Doubles*. Available at: http://www.emarketer.com/Article/Mobiles-Share-of-UK-Digital-Retail-Sales-Doubles/1010204

Emarketer. (2013d). *Three of Four UK Mobile Users Own Smartphones by 2016*. Available at: http://www.emarketer.com/Article/Three-of-Four-UK-Mobile-Users-Own-Smartphones-by-2016/1009614

Emarketer. (2013e). *UK Online Clothing Shoppers Still Favor the PC*. Available at: http://www.emarketer.com/Article/UK-Online-Clothing-Shoppers-Still-Favor-PC/1010274

Euromonitor. (2013). *Mobile Cocooning: How Growing Reliance on Smart Devices is Influencing Consumer Behaviour*. Available at: http://www.euromonitor.com/mobile-cocooning-how-growing-reliance-on-smart-devices-is-influencing-consumer-behaviour/report

Frolick, M. N., & Chen, L. (2004). Assessing M-Commerce Opportunities. *Information Systems Management*, *21*(2), 53–61. doi:10.1201/1078/44118.21.2.20040301/80422.8

Gamble, J., & Gilmore, A. (2013). A new era of consumer marketing? An application of co-creational marketing in the music industry. *European Journal of Marketing*, *47*(11/12), 1859–1888. doi:10.1108/EJM-10-2011-0584

George, J. F. (2002). Influence on the intent to make internet purchase. *Internet Research: Electronic Networking Applications and Policy*, *12*(2), 165-80.

Goldsmiths, R. E., & Bridges, E. (2000). E-tailing vs retailing: Using attitude to predict online buying behaviour. *Quarterly Journal of Electronic Commerce*, *1*(3), 245–253.

Gomm, R. (2004). *Social research methodology: a critical introduction.* Basingstoke, UK: Palgrave Macmillan.

Guest, G., MacQueen, K. M., & Namey, E. E. (2012). *Applied Thematic Analysis*. Los Angeles, CA: Sage publications Inc. doi:10.4135/9781483384436

Hartas, D. (2010). *Educational Research and Inquiry: Qualitative and Quantitative Approach*. London: Continuum International Publishing Group.

Hill, W. W., Beatty, S. E., & Walsh, G. (2013). A segmentation of adolescent users and shoppers. *Journal of Services Marketing*, *27*(5), 347–360. doi:10.1108/JSM-10-2011-0157

Hoffman, D. L., & Novak, T. P. (1996). Marketing in hype media computer-mediated environments: Conceptual foundations. *Journal of Marketing*, *60*(3), 50–68. doi:10.2307/1251841

Holmes, A., Byrne, A., & Rowley, J. (2014). Mobile shopping behaviour: Insights into attitudes, shopping process involvement and location. *International Journal of Retail & Distribution Management*, *42*(1), 25–39. doi:10.1108/IJRDM-10-2012-0096

Hsu, C. L., Lin, J. C. C., & Chiang, S. H. (2013). The effects of blogger recommendations on customers' online shopping intentions. *Internet Research*, *33*(1), 69–88. doi:10.1108/10662241311295782

Hu, W.-C., Yeh, J.-H., Fu, L., & Yang, H.-J. (2006). Handheld Computing and Programming for Mobile Commerce. *International Journal of Web Information Sytems*, *2*(3/4), 164–175. doi:10.1108/17440080780000297

Hung, M. C., Yang, S. T., & Hsieh, T. C. (2012). An examination of the determinants of mobile shopping continuance. *International Journal of Electronic Business Management*, *10*(1), 29–37.

Kelle, U. (2006). Combining qualitative and quantitative methods in research practice: Purpose and advantages. *Qualitative Research in Psychology*, *3*(4), 203–311.

Keynote. (2013). *Clothing Retailing Market Report.* Available at: http://www.keynote.co.uk/market-intelligence/view/product/10792/clothing-retailing?highlight=clothing&utm_source=kn.reports.search

Khalifa, M., & Shen, K. N. (2008). Explaining the adoption of transactional B2C mobile commerce. *Journal of Enterprise Information Management, 21*(2), 110–124. doi:10.1108/17410390810851372

Kim, H., Kim, J. L. Y., Chae, M., & Choi, Y. (2002). *An empirical study of the use context and usability problems in mobile internet.* IEEE Computer Society.

Kim, H.-W., Xu, Y., & Gupta, S. (2012). Which is more important in Internet Shopping, perceived price or trust? *Electronic Commerce Research and Applications, 11*(3), 241–252. doi:10.1016/j.elerap.2011.06.003

Kourouthanassis, P. E., & Giaglis, G. M. (2012). Introduction to the Special Issue Mobile Commerce: The Past, Present and Future of Mobile Commerce Research. *International Journal of Electronic Commerce, 16*(4), 5–17. doi:10.2753/JEC1086-4415160401

Kumar, A., & Mukherjee, A. (2013). Shop while you talk: Determinants of purchase intentions through a mobile device. Mobile Marketing Association, 8(1), 23-37.

Kumar, S., & Zhan, C. (2003). Mobile communications: Evolution and the impact on business operations. *Technovation, 23*(6), 515–520. doi:10.1016/S0166-4972(02)00120-7

Lamarre, A., Galarneau, S., & Boeck, H. (2012). Mobile Marketing and consumer Behaviour Current Research Trend. *International Journal of Latest Trends in Computing, 3*(1), 1–9.

Li, M., Dong, Z. Y., & Chen, X. (2012). Factors influencing consumptions experience of mobile commerce. *Internet Research, 22*(2), 120–141. doi:10.1108/10662241211214539

Lu, J., Yu, C.-S., Liu, C., & Yao, J. E. (2003). Technology acceptance model for wireless internet. *Internet Research: Electronic Networking Applications and Policy, 13*(3), 206–222. doi:10.1108/10662240310478222

Markovic, M. (2006). Analysing Qualitative Data: Health care xperiences of Women with Gynecological Cancer. *Field Methods, 18*(4), 413–429. doi:10.1177/1525822X06293126

Marks, D., & Yardley, L. (2004). *Research methods for clinical and health psychology.* London: Sage.

Mathieson, K. (1991). Predicting user intentions: Comparing the technology acceptance model with the theory of planned behaviour. *Information Systems Research, 2*(3), 173–191. doi:10.1287/isre.2.3.173

Miles, C. (2013). Persuasion, marketing, communication, and the metaphor of magic. *European Journal of Marketing, 47*(11/12), 2002–2019. doi:10.1108/EJM-11-2011-0632

Moth, D. (2013). *Mobile commerce now accounts for 23% of online salees: report.* Available at: https://econsultancy.com/blog/63338-mobile-commerce-now-accounts-for-23-of-online-sales-report

Nysveen, H., & Pedersen, P. E. (2005). Search mode and purchase intention in online shopping behaviour. *International Journal of Internet Marketing and Advertising, 2*(4), 288–306. doi:10.1504/IJIMA.2005.008103

Ozuem, W., Howell, K. E., & Lancaster, G. (2008). Communicating in the new interactive marketspace. *European Journal of Marketing, 42*(9/10), 1059–1083. doi:10.1108/03090560810891145

Parasuraman, A. (2000). Technology readiness index (TRI) a mutiple-item scale to measure readiness to embrace new technology. *Journal of Service Research*, *2*(4), 317–320. doi:10.1177/109467050024001

Park, E. J., Kim, E. Y., & Forney, J. C. (2006). A structural model of fashion-oriented impulse buying behaviour. *Journal of Fashion Marketing and Management*, *10*(4), 433–446. doi:10.1108/13612020610701965

Pasqua, R., & Elkin, N. (2013). *Mobile Marketing: An hour a day*. John Wiley & Sons Inc.

Patton, M. Q. (2002). *Qualitative Research and Evaluation Methods* (3rd ed.). London: Sage Publications Ltd.

Persaud, A., & Azhar, I. (2012). Innovative mobile marketing via smartphones: Are consumers ready? *Marketing Intelligence & Planning*, *30*(4), 418–443. doi:10.1108/02634501211231883

Peter, J. P., & Olson, J. (1996). *Consumer behaviour and marketing strategy*. Chicago: Irwin.

Qu, Z., Wang, Y., Wang, S., & Zhang, Y. (2013). Implications of online social activities for e-tailer's business performance. *European Journal of Marketing*, *47*(8), 1190–1212. doi:10.1108/03090561311324282

Robson, C. (2011). *Real World Research: A Resource for Users of Social Research Methods in Applied settings* (3rd ed.). Cornwall, UK: John Wiley & Sons Ltd.

Rogers, E. M. (1995). *Diffusion of Innovations*. New York: The Free Press.

San-Martin, S., & Lopez-Catalan, B. (2013). How can a mobile vendor get satisfied customers? *Industrial Management & Data Systems*, *113*(2), 156–170. doi:10.1108/02635571311303514

San-Martin, S., Lopez-Catalan, B., & Ramon-Jeronimo, M. A. (2012). Factors determining firms' perceived performance of mobile commerce. *Industrial Management & Data Systems*, *112*(6), 946–963. doi:10.1108/02635571211238536

Saunders, M., Lewis, P., & Thornhill, A. (2009). *Research methods for business students* (5th ed.). Harlow, UK: Pearson Education Ltd.

Scarpi, D. (2006). Fashion stores between fun and usefulness. *Journal of Fashion Marketing and Management*, *10*(1), 7–24. doi:10.1108/13612020610651097

Schiffman, L. G., & Kanuk, L. L. (2010). *Consumer Behaviour* (10th ed.). Pearson Education Inc.

Schuster, L., Drennan, J., & Lings, I. N. (2013). Consumer acceptance of m-wellbeing services: A social marketing perspective. *European Journal of Marketing*, *47*(9), 1439–1457. doi:10.1108/EJM-10-2011-0556

Seock, Y. K., & Norton, M. (2007). Attitude towards internet websites, online information search and channel choices for purchasing. *Journal of Fashion Marketing and Management*, *11*(4), 571–586. doi:10.1108/13612020710824616

Shim, S., Estlick, M. A., Lotz, S. L., & Warrrington, P. (2001). An online prepurchase intentions model: The role of intention to search. *Journal of Retailing*, *77*(2), 307–416.

Skeldon, P. (2011). *M-commerce: Boost your business with the power of mobile commerce*. Richmond, VA: Crimson Publishing.

Solomon, M. R. (2009). *Consumer Behaviour: Buying, Having and Being* (8th ed.). Pearson Education Inc.

Spaid, B. I., & Flint, D. J. (2014). The meaning of shopping experiences augmented by mobile internet devices. *Journal of Marketing Theory and Practice*, *22*(1), 73–89. doi:10.2753/MTP1069-6679220105

Swilley, E., & Hofacker, C. F. (2006). Defining mobile commerce in a marketing context. *International Journal of Mobile Marketing*, *1*(2), 18–23.

Swilley, E., Hofacker, C. F. & Lamont, B. T. (2012). The Evolution of E-Commerce to M-Commerce: Pressures, Firm Capabilities and Competitive Advantage in Strategic Decision Making. *International Journal of e-Business Research, 8*(1), 1-16.

Ting, D. H., Ling, S. F., Patanmacia, T. S., Low, C. G., & Ker, G. C. (2011). Dependency on smartphone and the impact on purchase behaviour. *Young Consumers*, *12*(3), 193–203. doi:10.1108/17473611111163250

Van Slyke, C., Belanger, F., & Comunale, C. L. (2004). Factors Influencing the adoption of web-based shopping: The impact of trust. *ACM SIGMIS Database*, *35*(2), 32–49. doi:10.1145/1007965.1007969

Venkatesh, V., Brown, S., & Bala, H. (2013). Bridging the Qualitative- Quantitative Divide: Guidelines for Conducting Mixed-Methodology Research in Information Systems. *Management Information Systems Quarterly*, *37*(1), 21–54.

Wang, M.-S., Chen, C. C., Chan, S.-C., & Yang, Y.-H. (2007). Effects of online shopping attitudes, subjective norms and control beliefs on online shopping intentions: A test of the theory of planned behaviour. *International Journal of Management*, *24*(2), 296–302.

Wang, Y., & Liao, Y. (2007). The conceptualization and measurement of m-commerce user satisfaction. *Computers in Human Behavior*, *23*(1), 381–398. doi:10.1016/j.chb.2004.10.017

Wheeler, E. (2013). *The emergence of M-Commerce and Geofencing*. Available at: http://www.bloomberg.com/video/from-e-commerce-to-m-commerce-bW0SIsCkRJaTiNTJ89zAnQ.html

Wong, S. H. R. (2012). Which platform do our users prefer: Website or mobile app? *RSR. Reference Services Review*, *40*(1), 113–115.

Yang, K., & Kim, H.-Y. (2012). Mobile shopping motivation: An application of multiple discriminant analysis. *International Journal of Retail & Distribution Management*, *40*(10), 778–789. doi:10.1108/09590551211263182

Yang, S. C. (2012). Mobile applications and 4G wireless networks: A framework for analysis. *Campus-Wide Information Systems*, *29*(5), 344–357. doi:10.1108/10650741211275107

Yeh, Y. S., & Li, Y.-M. (2009). Building trust in e-commerce: Contributions from quality and satisfaction. *Online Information Review*, *33*(6), 1066–1086. doi:10.1108/14684520911011016

Zhou, T. (2011). Examining the critical success factors of mobile site adoption. *Online Information Review*, *35*(4), 636–652. doi:10.1108/14684521111161972

Zhou, T. (2013). Understanding continuance usage of mobile site. *Industrial Management & Data Systems*, *113*(9), 1286–1299. doi:10.1108/IMDS-01-2013-0001

KEY TERMS AND DEFINITIONS

Mobile Commerce: Mobile commerce encompasses the buying and selling of goods and services online through wireless handheld devices usually smartphones or tablets.

Mobile Environment: The mobile environment is the move of the online environment towards the mobile platform to enable smartphones users to access and interact with online contents on the go.

Mobile Shopping: Mobile shopping refers to the action of using a smartphone to browse, compare or to shop for products and services online.

Mobile Storefront: Mobile storefront is when a business establishes an online presence by displaying their products and services on a mobile platform usually via a mobile website. It replaces the conventional storefront whereby consumers can browse for products' information and make purchases through their smartphones without having to physically go to a store.

Online Consumption: Online consumption refers to the action of using up resources, products and services in online environments or technology-enabled marketing platforms.

Showrooming: Showrooming is the action of using a smartphone to browse and compare goods and services online while being in a conventional store before considering a purchase.

Smartphones: A smartphone is an intelligent multi-functional handset capable of the traditional function of a mobile phone device (i.e making phone calls), along with highly advanced features such as internet access, high resolution cameras, GPS system and an operating system capable of running downloaded apps.

Compilation of References

16-2005, I. S. 802. (2005). *IEEE standard for local and metropolitan area networks part 16: air interface for Fixed broadband Wireless access system- physical and medium access control layer for combined mobile and Fixed Operation in licensed bands.*

3G Americas. (2011, July). *The benefit of SON in LTE: self-optimization and self-organizing networks.*

3GPP TR 36.902, "E-UTRA; Self-configuring and selfoptimizing network (SON) use cases and solutions (Release 9)," V9.3.1, March 2011.

3GPP TS 32.500, "Self-organizing networks (SON); Concepts and requirements (Release 9)," V9.0.0, December 2009.

3. GPP TS 36305: Evolved Universal Terrestrial Radio Access Network (E-UTRAN)

3. GPP. (2010). *Evolved Universal Terrestrial Radio Access (EUTRA); LTE Positioning Protocol (LPP).* 3GPP TS 36.355, Release 9 (V9.0.0).

3. GPP. (2010). *Evolved Universal Terrestrial Radio Access (EUTRA); LTE Positioning Protocol A (LPPa).* 3GPP TS 36.455, Release 9 (V9.0.0).

3. GPP. (2010). *LCS Application Protocol (LCS - AP) between the MME and E - SMLC; SLs interface.* 3GPP TS 29.171, Release 9 (V2.0.0).

3GPP. (n.d.). *Physical Layer Aspects for evolved Universal Terrestrial Radio Access (E-UTRAN), Technical Report TR 25.814.* Retrieved from http://www.3gpp.org

3. rd Generation Partnership Project. (2009a). *Primary component carrier selection, monitoring, and recovery.* Author.

3. rd Generation Partnership Project. (2009b). *Feasibility study for further advancements of E-UTRA (LTE-Advanced).* Author.

3. rd Generation Partnership Project. (2010). *Evolved Universal Terrestrial Radio Access Network (E-UTRAN); Self-Configuring and Self-Optimizing Network (SON) Use Cases and Solutions.* Author.

3. [rd] *Generation Project Partnership (3GPP) official website.* (2015). Retrieved May 07, 2015, from http://www.3gpp.org/

3. [rd] *Generation Project Partnership (3GPP).* (2008). *Automatic Physical Cell ID Assignment.* 3GPP TSG-SA5 (Telecom Management). S5-081185.

3. [rd] Generation Project Partnership (3GPP). (2009). *TS 36.211 v8.6.0: Technical Specification Group Radio Access Network; Evolved Universal Terrestrial Radio Access (E-UTRA); Physical Channels and Modulation (Release 8).* Technical Report.

3rd Generation Project Partnership (3GPP). (2009a). *TS 36.300v8.9.0: Evolved Universal Terrestrial Radio Access (E-UTRA) and Evolved Universal Terrestrial Radio Access Network (E-UTRAN); Overall Description (release 8).* Technical Specification.

3rd Generation Project Partnership (3GPP). (2010). *Tr 36.902v9.2.0: LTE; Evolved Universal Terrestrial Radio Access Network (E-UTRAN); Self-Configuring and Self-Optimizing Network (SON) Use Cases and Solutions (Release 9).* Technical Specification.

Aad, I., Hubaux, Y., & Knightlyz, E. (2004). Denial of Service Resilience in Ad Hoc Networks. ACM.

Abdullah, L. M., Baba, M. D., & Ali, S. G. A. (2014). Self-Configuration Concept to Solve Physical Cell ID Conflict for SON LTE-based Femtocell Environment. *Int. J. on Recent Trends in Engineering and Technology, 10*(2).

Abdul-Rahman, A., & Hailes, S. (1997). Using Recommendations for Managing Trust in Distributed Systems.*Proc. IEEE Malaysia Int'l Conf.on Communication.*

Abhayawardhana, V. S., Wassell, I. J., Crosby, D., & Sellars, M. P. (2005). Comparison of Empirical Propagation Path Loss Models for Fixed Wireless Access Systems. In *Vehicular Technology Conference, 2005. VTC 2005-Spring. 2005 IEEE.* doi:10.1109/VETECS.2005.1543252

Abou-Rjeily, C. (2011). Joint channel-diversity coding for the three-way free-space optical systems through atmospheric-turbulence channels. *International Journal on New Computer Architectures and Their Applications, 1*(1), 162–175.

Abrams, K. V. (2012). *UK B2C Ecommerce: Mobile Takes Center Stage.* Available at: http://www.emarketer.com/Article/Mobile-Gains-Speed-UK-Ecommerce-Maintains-Momentum/1009093

Abrams, K. V. (2013a). *UK Digital Trends: Ecommerce, Video and Mobile Thrive, but Brands must work for loyalty and trust.* eMarketer.

Abrams, K. V. (2013b). *Western Europe B2C Ecommerce: Solid Growth Continues as Mobile Gains Ground.* eMarketer.

Access, R., & Technologies, O. (2008). *Introduction to WiMAX Modeling for Network R & D and Planning.* Author.

Adams, J. C., Gregorwich, W., Capots, L., & Liccardo, D. (2001). Ultrawideband for navigation and communications. *Proceedings IEEE Aerospace Conference.*

Adams, W. J., Hadjichristofi, G. C., & Davis, N. J. (2005). Calculating a Node's Reputation in a Mobile Ad Hoc Network. *Proc. 24th IEEE Int'l Performance Computing and Communications Conference.*

Adnane, A., Bidan, C. & Junior. (2013). Trust based security for the OLSR Routing Protocol. Elsevier, 36, 1159-1171.

Aguado, Jacob, & Berbineau. (2010). QoS Challenges in Handover Process. In *WiMAX Security and Quality ofService: An End-to-End Perspective.* Academic Press.

Ahmadi, S. (2008). An Overview of Next Generation Mobile WiMAX Technology. *Communications Magazine, IEEE, 47*, 84–98.

Ahmed, F., & Tirkkonen, O. (2009). Local optimum based power allocation approach for spectrum sharing in unlicensed bands. In Spyropoulos, T., & Hummel, K. A. (Eds.), *Self-Organizing Systems:Proceedings of the 6th International Workshop (LNCS)* (*vol. 5918*, pp. 238-243). Berlin, Germany: Springer. doi:10.1007/978-3-642-10865-5_24

Ahmed, F., Dowhuszko, A. A., & Tirkkonen, O. (2012). Distributed algorithm for downlink resource allocation in multicarrier small cell networks. In *Proceedings of the International Conference on Communications.* Ottawa, Canada: IEEE. doi:10.1109/ICC.2012.6364716

Ahmed, F., Dowhuszko, A. A., Tirkkonen, O., & Berry, R. (2013). A distributed algorithm for network power minimization in multicarrier systems. In *Proceedings of the International Symposium on Personal Indoor and Mobile Radio Communications*. London, UK: IEEE. doi:10.1109/PIMRC.2013.6666456

Ahmed, F., & Tirkkonen, O. (2014). Topological aspects of greedy self-organization. In *Proceedings of the International Conference on Self-Adaptive and Self-Organizing Systems*. London, UK: IEEE.

Ahmed, F., & Tirkkonen, O. (2015). Self Organized Physical Cell ID Assignment in Multi-operator Heterogeneous Networks. In *Proceedings of the Vehicular Technology Conference (VTC-Spring)*. Glasgow, UK: IEEE. doi:10.1109/VTCSpring.2015.7146077

Ahmed, F., Tirkkonen, O., Dowhuszko, A. A., & Juntti, M. (2014). Distributed power allocation in cognitive radio networks under network power constraint. In *Proceedings of the International Conference on Cognitive Radio Oriented Wireless Networks and Communications*. Oulu, Finland: IEEE. doi:10.4108/icst.crowncom.2014.255738

Ahmed, F., Tirkkonen, O., Peltomäki, M., Koljonen, J. M., Yu, C. H., & Alava, M. (2010). Distributed graph coloring for self-organization in LTE networks. *Journal of Electrical and Computer Engineering*, 5.

Ahmed, F., Tirkkonen, O., Peltomäki, M., Koljonen, J.-M., Yu, C.-H., & Alava, M. (2010). Distributed Graph Coloring for Self-Organization in LTE Networks. *Journal of Electrical and Computer Engineering*. doi:10.1155/2010/402831

Ahrens, J., Coyle, J. R., & Strahilevitz, M. A. (2013). Electronic word of mouth: The effects of incentives on e-referrals by senders and receivers. *International Journal of Marketing*, *47*(7), 1034–1051.

Akyildiz Ian, F. (2006, September). Won-Yeol Lee, Mehmet C. Vuran, & Shantidev Mohanty. NeXt generation / dynamic spectrum access / cognitive radio wireless networks: A survey. *Computer Networks*, *50*(13), 2127–2159. doi:10.1016/j.comnet.2006.05.001

Akyildiz, I. F., Lee, W.-Y., & Chowdhury, K. R. (2009, July). CRAHNs: Cognitive radio ad hoc networks. *Ad Hoc Networks*, *7*(5), 810–836. doi:10.1016/j.adhoc.2009.01.001

Akyildiz, I. F., Lo, B. F., & Balakrishnan, R. (2011). Cooperative spectrum sensing in cognitive radio networks: A survey. *Physical Communication*, *4*(1), 40–62. doi:10.1016/j.phycom.2010.12.003

Akyildiz, I. F., Xie, J., & Mohanty, S. (2004). A survey of mobility management in next-generation all-IP-based wireless systems. *IEEE Wireless Communications*, *11*(4), 16–28. doi:10.1109/MWC.2004.1325888

Akyol, E., Rose, K., & Ramstad, T. (2010, January). Optimal mapping for joint source channel coding. In *Proceedings of 2010 IEEE Information Theory Workshop (ITW)* (pp. 1-5). doi:10.1109/ITWKSPS.2010.5503179

Alava, M., Ardelius, J., Aurell, E., Kaski, P., Krishnamurthy, S., Orponen, P., & Seitz, S. (2008). Circumspect descent prevails in solving random constraint satisfaction problems. *Proceedings of the National Academy of Sciences of the United States of America*, *105*(40), 15253–15257. doi:10.1073/pnas.0712263105 PMID:18832149

Alcatel-Lucent. (2004). *WiMAX, making ubiquitous high-speed data services a reality*. Retrieved from http://www1.alcatel-lucent.com/publications/abstract.jhtml?repositoryItem=tcm:172-44851635

Ali, S., Mitschele-Thiel, A., Diab, A., & Rasheed, A. (2010). *A Survey of Services Placement Mechanisms for Future Mobile Communication Networks*. In FIT 10, 8th International Conference on Frontiers of Information Technology, Islamabad, Pakistan.

Aliu, O. G., Imran, A., Imran, M. A., & Evans, B. (2013). A survey of self organisation in future cellular networks. *IEEE Communications Surveys and Tutorials*, *15*(1), 336–361. doi:10.1109/SURV.2012.021312.00116

Alnasouri, E., Mitschele-Thiel, A. & Diab, A. (2009). A New Distributed Dynamic Algorithm for Mobility Patterns Prediction. In *Proceeding of PE-WASUN'09.*

Alnasouri, E., Mitschele-Thiel, A., Boeringer, R., & Diab, A. (2006). QoMIFA: A QoS-enabled Mobility Management Framework in ALL-IP Network. In Proceeding of PIMRC'06.

Alvarion. (n.d.). *Wireless Connectivity Breaking the line-of-sight boundaries*. Author.

Amin, P., Ganesan, V. P. K., & Tirkkonen, O. (2012). Bridging Interference Barriers in Self-Organized Synchronization. In *Proceedings of the International Conference on Self-Adaptive and Self-Organizing Systems*. Lyon, France: IEEE.

Amirijoo, M., Frenger, P., Gunnarsson, F., Kallin, H., Moe, J., & Zetterberg, K. (2008). Neighbor Cell Relation List and Measured Cell Identity Management in LTEs. In *IEEE Symposium on Network Operations and Management* (NOMS`08). doi:10.1109/NOMS.2008.4575129

Amirijoo, M., Frenger, P., Gunnarsson, F., Kallin, H., Moe, J., & Zetterberg, K. (2008a). Neighbor Cell Relation List and Physical Cell Identity Self-Organization in LTE. In *Proceeding of IEEE International Conference on Communications Workshops* (ICC'08). doi:10.1109/ICCW.2008.12

Amundson, I., & Koutsoukos, X. D. (2009). A survey on localization for mobile wireless sensor networks. In *Proc. 2nd Int. Conf. Mobile Entity Localization Tracking GPS-Less Environ., MELT*. doi:10.1007/978-3-642-04385-7_16

Anand, P. M., & Bhaskar, V. (2013). A unified trust management strategy for content sharing in Peer-to-Peer networks. *Applied Mathematical Modelling*, *37*(4), 1992–2007. doi:10.1016/j.apm.2012.04.050

Anderson, J., & Rainie, L. (2014). *Digital Life in 2025: The Internet of Things Will Thrive By 2025*. Pew Research Center.

Andrews, & ArunabhaGhosh, R. (2007). Fundamentals of WiMAX: Understanding Broadband Wireless Networking. Prentice Hall.

Anita, X., Martin, J., Maickam, L., & Bhagyaveni, M, A. (2014). Two Way Acknowledgment based trust framework for wireless sensor networks. *International Journal of Distributed Sensor Network*. doi: 1155/2013/952905

Aphiraksatyakul, D., Seet, B. C., & Lau, C. T. (2008). Evaluation of terrain effects on Mobile WiMax in a Vehicular Environment. *Proceedings - 2008 8th International Conference on Intelligent Transport System Telecommunications, ITST 2008*. doi:10.1109/ITST.2008.4740290

Aqeel-ur-Rehman, Mehmood, K., & Baksh, A. (2013). Communication Technology That Suits IoT-A Critical Review. In Wireless Sensor Networks for Developing Countries (pp. 14-25). Springer.

Arnold, M. J., & Reynolds, K. E. (2003). Hedonic shopping motivations. *Journal of Retailing*, *79*(2), 77–95. doi:10.1016/S0022-4359(03)00007-1

Aschenbruck, N., Gerhards-Padilla, E. & Martini, P. (2008). A Survey on Mobility Models for Performance Analysis in Tactical Mobile Networks. *Journal of Telecommunication and Information Technology.*

Aschenbruck, N., Gerhards-Padilla, E., Gerharz, M., Frank, M., & Martini, P. (2007). Modelling Mobility in Disaster Area Scenarios. In *Proc. 10th ACM IEEE Int. Symp. Model. Anal. Simul. Wirel. Mob. Syst. MSWIM*. doi:10.1145/1298126.1298131

Ashby, W. R. (1947). Principles of the Self-Organizing Dynamic System. *The Journal of General Psychology*, *37*(2), 125–128. doi:10.1080/00221309.1947.9918144 PMID:20270223

Atayero, & Luka. (n.d.). Adaptive Neuro- Fuzzy Inference System for Dynamic Load Balancing in 3GPP LTE. International Journal of Advanced Research in Artificial Intelligence, 1(1), 11–16.

Azami, S., Duhamel, P., & Rioul, O. (1997). Combined source-channel coding for binary symmetric channels and uniform memoryless sources. In *16° Colloque sur le traitement du signal et des images, FRA, 1997*. GRETSI, Groupe d'Etudes du Traitement du Signal et des Images.

Babadi, B., & Tarokh, V. (2008). A Distributed Asynchronous Algorithm for Spectrum Sharing in Wireless Ad Hoc Networks. In *Proceedings of the 42nd Annual Conference on Information Sciences and Systems* (CISS '08). doi:10.1109/CISS.2008.4558635

Bai, F., Sadagopan, N., & Helmy, A. (2003). IMPORTANT: A Framework to Systematically Analyze the Impact of Mobility on Performance of Routing Protocols for Ad-Hoc Networks. In *Proc. IEEE INFOCOM*. doi:10.1109/INFCOM.2003.1208920

Balachandran, A., Voelker, G. M., Bahl, P., & Rangan, P. V. (2002). Characterizing User Behavior and Network Performance in a Public Wireless LAN. In *Proceedings of SIG-METRICS'02*. ACM Press. doi:10.1145/511334.511359

Balasundaram, B., & Butenko, S. (2006). Graph domination, coloring and cliques in telecommunications. In G. C. M. Resende & P. M. Pardalos (Eds.), *Handbook of Optimization in Telecommunications* (pp. 865–890). New York, NY: Springer. doi:10.1007/978-0-387-30165-5_30

Balazinska, M., & Castro, P. (2003). *Characterizing Mobility and Network Usage in a Corporate Wireless Local-Area Network*. In *1st International Conference on Mobile Systems, Applications, and Services (MobiSys'03)*, San Francisco, CA. doi:10.1145/1066116.1066127

Bamberger, W. (2010). Interpersonal Trust – Attempt of a Definition. *Scientific Report*.

Bandh, T., Carle, G., & Sanneck, H. (2009). Graph Coloring Based Physical-Cell-ID Assignment for LTE Networks. In *Proceeding of the ACM International Wireless Communications and Mobile Computing Conference* (IWCMC '09). doi:10.1145/1582379.1582406

Banerjee, S. (2008). Detection/Removal of Cooperative Black and Gray Hole Attack in Mobile Ad-Hoc Networks. *Proceedings of the World Congress on Engineering and Computer Science*.

Banerjee, S., & Dholakia, R. R. (2012). Location-based mobile advertisements and gender targeting. *Journal of Research in Interactive Marketing*, *6*(3), 198–214. doi:10.1108/17505931211274679

Bankovic, Z., Fraga, D., Moya, J., Vallejo, J., Malago, P., Araujo, A., & Nieto-Taladriz, O. et al. (2011). Improving security in WMNs with reputation systems and self-organizing maps. *Journal of Network and Computer Applications*, *34*(2), 455–463. doi:10.1016/j.jnca.2010.03.023

Bao, F., Chen, I., Chang, M., & Cho, J. (2012). Hierarchical Trust Management for Wireless Sensor Networks and its applications to Trust Based Routing and intrusion detection. *IEEE eTransactions on Network and Service Management*, *9*(2), 169–183. doi:10.1109/TCOMM.2012.031912.110179

Barabási, A. L., & Albert, R. (1999). Emergence of Scaling in Random Networks. *Science*, 286. PMID:10521342

Baran, P. (1964). On distributed communications networks. *IEEE Transactions on Communications Systems*, *12*(1), 1–9. doi:10.1109/TCOM.1964.1088883

Baras, J. S., & Jiang, T. (2005). Managing Trust in Self-Organized Mobile Ad Hoc Networks. In *Proc. 12th Annual Network and Distributed System Security Symposium Workshop*.

Barrenechea, G., Beferull-Lozano, B., Verma, A., Dragotti, P. L., & Vetterli, M. (2003, April). Multiple description source coding and diversity routing: A joint source channel coding approach to real-time service over dense networks. In *Proceedings of the Packet Video Workshop*.

Barrutia, J. M., & Echebarria, C. (2013). Networks: A social marketing tool. *European Journal of Marketing*, *47*(1/2), 324–343. doi:10.1108/03090561311285574

Barry, J. R., Lee, E. A., & Messerschmitt, D. G. (2004). *Digital communication*. Springer. doi:10.1007/978-1-4615-0227-2

Barua, S., Lam, S. C., Ghosa, P., Xing, S., & Sandrasegaran, K. (2015, June). A survey of Direction of Arrival estimation techniques and implementation of channel estimation based on SCME. In *Electrical Engineering/Electronics, Computer, Telecommunications and Information Technology (ECTI-CON), 2015 12th International Conference on* (pp. 1-5). IEEE. doi:10.1109/ECTICon.2015.7206986

Bassi, A., Bauer, M., Fiedler, M., Kramp, T., Kranenburg, R., Lange, S., & Meissner, S. (2013). *Enabling Things to Talk: Designing IoT Solutions with the IoT Architectural Reference Model*. Springer. doi:10.1007/978-3-642-40403-0

Basu, P., Redi, J., & Shurbanov, V. (2004). Coordinated Flocking of UAV for Improved Connectivity of Mobile Ground Nodes. In *Proceeding of the Military Communications Conference (MILCOM)*. doi:10.1109/MILCOM.2004.1495182

Bayir, M. A., Demirbas, M., & Eagle, N. (2009). *Discovering Spatiotemporal Mobility Profiles of Cellphone Users. World of Wireless*. Mobile and Multimedia Networks and Workshops.

BBC. (2012). *UK's first 4G mobile service launched in 11 cities by EE*. Available at: http://www.bbc.co.uk/news/technology-20121025

BBC. (2013). *The peril of "showrooming"*. Available at: http://www.bbc.co.uk/news/magazine-22098575

Bellavista, P., Corradi, A., & Magistretti, E. (2005). REDMAN: An Optimistic Replication Middleware for Read-only Resources in Dense MANETs. *Journal on Pervasive and Mobile Computing*, *1*(3), 279–310. doi:10.1016/j.pmcj.2005.06.002

Bellman, R. L. (n.d.). On a Routing Problem. *Quarterly of Applied Mathematics*.

Bettstetter, C. (2001a). Smooth is Better Than Sharp: A Random Mobility Model for Simulation of Wireless Networks. In *Proc. 4th Int. Symp. Model. Anal. Simul. Wirel. Mob. Syst. MSWIM*.

Bettstetter, C., Hartenstein, H., & Perez-Costa, X. (2003). *Stochastic Properties of the Random Waypoint Mobility Model*. Kluwer Academic Publishers.

Bettstetter, C. (2001). Mobility Modeling in Wireless Networks: Categorization, Smooth Movement, and Border Effects. *ACM SIGMOBILE Mob. Comp. Commun. Rev.*, *5*(3), 55–66. doi:10.1145/584051.584056

Bettstetter, C., & Wagner, C. (2002). The Spatial Node Distribution of the Random Waypoint Mobility Model. In *Proceeding of 1st German Workshop on Mobile Ad-Hoc Networks. WMAN'02*.

Bhalaji, N., & Shanmugam, A. (2012). Dynamic Trust Based Method to Mitigate Greyhole Attack in Mobile Ad Hoc Networks. *Procedia Engineering, Elsevier*, *30*, 881–888. doi:10.1016/j.proeng.2012.01.941

Bhunia, S., Misra, I. S., Sanyal, S. K., & Kundu, A. (2011). *Performance study of mobile WiMAX network with changing scenarios under different modulation and coding*. http://doi.org/<ALIGNMENT.qj></ALIGNMENT>10.1002/dac

Bienvenu, G., & Kopp, L. (1983). Optimality of high resolution array processing using the eigensystem approach. *Acoustics, Speech and Signal Processing. IEEE Transactions on*, *31*(5), 1235–1248.

Bittner, S., Raffel, W.-U., & Scholz, M. (2005). The Area Graph-based Mobility Model and its Impact on Data Dissemination. In *Proc. IEEE PerCom*. doi:10.1109/PERCOMW.2005.79

Blakely, K., & Lowekamp, B. (2004). *A Structured Group Mobility Model for the Simulation of Mobile Ad Hoc Networks*. MobiWac. doi:10.1145/1023783.1023805

Blaze, M., Feigenbaum, J., & Lacy, J. (1996). Decentralized trust Management. In *Proceedings of IEEE Symposium on security and Privacy.*

Blazevic, L., Le Boudec, J.-Y., & Giordano, S. (2005). A Location-Based Routing Method for Mobile Ad Hoc Networks. *IEEE Transactions on Mobile Computing*, *3*(4).

Bohner, M., & Bohner, G. (2001). *Intrapersonal processes*. Oxford, UK: Blackwell.

Boldrini, C., Conti, M., & Passarella, A. (2009). The Sociable Traveller: Human Travelling Patterns in Social-Based Mobility. In ACM Mobiwac'09.

Boldrini, C., Conti, M., & Passarella, A. (2007). *Users Mobility Models for Opportunistic Networks: the Role of Physical Locations*. WRECOM.

Borrel, V., Dias De Amorim, M., & Fdida, S. (2005). A Preferential Attachment Gathering Mobility Model. *IEEE Communications Letters*, *9*(10), 900–902. doi:10.1109/LCOMM.2005.10023

Boudec, J. L., & Vojnovic, M. (2005). Perfect Simulation and Stationary of a Class of Mobility Models. In *Proc.IEEE INFOCOM.*

Boyatzis, R. E. (1998). *Transforming qualitative information: Thematic analysis and code development*. Sage Publications Inc.

Boyle, C. (2013). *Retail Mobile Commerce Forecast: Shoppers Turn to Smartphones First.* Available at: http://www.emarketer.com/Article/More-Shoppers-Reach-Mobile-Browse-Buy/1009615

Braun, V., & Clarke, V. (2006). Using thematic analysis in psychology. *Qualitative Research in Psychology*, *3*(2), 77–101. doi:10.1191/1478088706qp063oa

Broadband, M. (2007). *Welcome to Your Internet Future*. Author.

Brown, M., Pope, N., & Vogues, K. (2003). Buying or browsing?: An exploration of shopping orientations and online purchase intention. *European Journal of Marketing*, *37*(11/12), 1666–1684. doi:10.1108/03090560310495401

Brunnenmeyer, D., Mills, S., Patel, S., Suarez, C., & Kung, L.-B. (2012). *Ka and Ku Operational Considerations for Military SATCOM Applications*. Paper presented at the The 2012 Military Communications Conference – Track 4 – System Perspectives, Milcom. doi:10.1109/MILCOM.2012.6415563

Bryman, A. (1988). *Contemporary Social Research Series*. London: Unwin Hyman Ltd.

Bryman, A. (1989). *Research Methods and Organization Studies*. London: Unwin Hyman Ltd. doi:10.4324/9780203359648

Burr, V. (2003). *Social Contructionism* (2nd ed.). London: Routledge.

Burton-Joners, A., & Hubona, G. S. (2006). The mediation of external variables in the technology acceptance model. *Information & Management*, *43*(6), 706–717. doi:10.1016/j.im.2006.03.007

Butler, P., & Peppard, J. (1998). Consumer Purchasing on the Internet: Processes and Prospects. *European Journal of Marketing*, *16*(5), 600–610.

Byers, J., & Nasser, G. (2000). Utility-based decision-making in wireless sensor networks. In *Proceedings of the Annual Workshop on Mobile and Ad Hoc Networking and Computing*. Boston, MA: IEEE. doi:10.1109/MOBHOC.2000.869228

Cacciapuoti, A., Calcagno, C., Caleffi, M., & Paura, L. (2010). CAODV: Routing in mobile ad-hoc cognitive radio networks. In *Wireless Days.*

Cacciapuoti, A. S., Caleffi, M., & Paura, L. (2012). Marcello Cale_, & Luigi Paura. Reactive routing for mobile cognitive radio ad hoc networks. *Ad Hoc Networks*, *10*(5), 803–815. doi:10.1016/j.adhoc.2011.04.004

Caffery, J. J. (2000). *Wireless location in CDMA Cellular Radio Systems*. Boston: Kluwer Academic.

Camazine, S., Deneubourg, J.-L., Franks, N. R., Sneyd, J., Theraulaz, G., & Bonabeau, E. (2003). *Self-Organization in Biological Systems*. Princeton, NJ: Princeton University Press.

Camiciotti, L., Lamy, C., Meilhac, L., Oliveieri, S., & Verdi, P. (2001, May). Joint source channel coding for 4G multimedia streaming. In *2nd WWRF meeting, WG3*.

Camp, T., Boleng, J., & Davies, V. (2002). A Survey of Mobility Models for Ad Hoc Network Research. *Wireless Communication and Mobile Computing Special Issue on Mobile Ad Hoc Networking: Research.Trends and Applications*, *2*(5), 483–502.

Capgemini. (2013). *Online retail sales highest in two years as positive economic sign continues*. Available at: http://www.uk.capgemini.com/online-retail-sales-highest-in-two-years-as-positive-economic-signs-continue

Capon, J. (1969). High-resolution frequency-wavenumber spectrum analysis. *Proceedings of the IEEE*, *57*(8), 1408–1418. doi:10.1109/PROC.1969.7278

Cavalcanti, D., Agrawal, D. P., Cordeiro, C., Xie, B., & Kumar, A. (2005). Issues in integrating cellular networks, WLANs, and MANETs: A futuristic heterogeneous wireless network. *IEEE Wireless Communications*, *12*(3), 30–41. doi:10.1109/MWC.2005.1452852

Cen, F. (2013). Distributed Joint Source and Channel Coding with Low-Density Parity-Check Codes. *IEEE Journal on Communications Letters*, *17*(12), 2336–2339. doi:10.1109/LCOMM.2013.101613.131616

Chandra, R., Ramasubramanian, V., & Birman, K. (2001). *Anonymous Gossip: Improving Multicast Reliability in Mobile Ad-Hoc Networks.21st International Conference on Distributed Computing Systems*, Mesa, Arizona. doi:10.1109/ICDSC.2001.918957

Chan, F. T. S., & Chong, A. Y.-L. (2013). Analysis of the determinants of consumers' m-commerce usage activities. *Online Information Review*, *37*(3), 443–461. doi:10.1108/OIR-01-2012-0012

Chase, J. (2013). *The Evolution of the Internet of Things*. Dallas, TX: Texas Instruments Incorporated.

Cheema, A., & Papatia, P. (2010). Relative importance of online versus offline information for Internet purchase: Product category and Internet experience effects. *Journal of Business Research*, *63*(9), 979–985. doi:10.1016/j.jbusres.2009.01.021

Chemingui, H., & Lallouna, H. B. (2013). Resistence, motivation, trust and intention to use mobile finance services. *International Journal of Bank Marketing*, *31*(7), 574–592. doi:10.1108/IJBM-12-2012-0124

Chen, C., & Chen, Z. (2007). Evaluating Contacts for Routing in Highly Partitioned Mobile Networks. *In Proceedings of the 1st international MobiSys workshop on Mobile opportunistic networking* (MobiOpp '07). doi:10.1145/1247694.1247698

Chen, C., Zhang, J., Cohen, R., & Ho, P. (2010). Secure and Efficient Trust Opinion Aggregation for Vehicular Ad-hoc Networks. *IEEE Digital Library*.

Chen, J.-C., Ting, P., Maa, C.-S., & Chen, J.-T. (2004). Wireless geolocation with TOA/AOA measurements using factor graph and sum-product algorithm. *IEEE 60th Vehicular Technology Conference*.

Cheng, Y.-C., Chawathe, Y., LaMarca, A., & Krumm, J. (2005). *Accuracy Characterization for Metropolitan-Scale Wi-Fi Localization*. Intel Research Seattle, Tech. Rep. IRS-TR-05-003.

Cheng, M. X., Li, Y., & Du, D.-Z. (2006). *Combinatorial Optimization in Communication Networks*. New York, NY: Springer. doi:10.1007/0-387-29026-5

Chen, J., Olafsson, S., Gu, X., & Yang, Y. (2008). A fast channel allocation scheme using simulated annealing in scalable WLANs. In *Proceedings of the International Conference on Broadband Communications, Networks, and Systems*. London, UK: IEEE. doi:10.1109/BROADNETS.2008.4769071

Chen, W., & Chen, P. (2003). *Group Mobility Management in Wireless Ad Hoc Networks*. VTC.

Chen, Z., Gokeda, G., & Yu, Y. (2010). *Introduction to Direction-of-arrival Estimation*. Artech House.

Cheung, C. M. K., & Lee, M. K. O. (2012). What drives consumers to spread electronic word of mouth in online consumer-opinion platforms. *Decision Support Systems*, *53*(1), 218–225. doi:10.1016/j.dss.2012.01.015

Cheung, G., & Zakhor, A. (2000). Bit allocation for joint source/channel coding of scalable video. *IEEE Transactions on Image Processing*, *9*(3), 340–356. doi:10.1109/83.826773 PMID:18255407

Chiang, C. (1998). *Wireless Network Multicasting*. (PhD Thesis). Department of Computer Science, University of California, Los Angeles, CA.

Chiang, M., Low, S. H., Calderbank, A. R., & Doyle, J. C. (2007). Layering as optimization decomposition: A mathematical theory of network architectures. *Proceedings of the IEEE*, *95*(1), 255–312. doi:10.1109/JPROC.2006.887322

Childers, T. L., Carr, C. L., Peck, J., & Carson, S. (2001). Hedonic and utilitarian motivations for online retail shopping behaviour. *Journal of Retailing*, *77*(4), 511–535. doi:10.1016/S0022-4359(01)00056-2

Choffnes, D. R., & Bustamante, F. E. (2005). An Integrated Mobility and Traffic Model for Vehicular Wireless Networks. In *Int. Conf. Mob. Comp. Netw., Proc. 2nd ACM Int. Worksh. Veh. Ad Hoc Netw.* doi:10.1145/1080754.1080765

Cho, S., & Fiorito, S. S. (2009). Acceptance of online customization for apparel shopping. *International Journal of Retail & Distribution Management*, *37*(5), 389–407. doi:10.1108/09590550910954892

Chouhan, N., & Yadav, S. (2011). Flooding Attacks Prevention in MANET. *International Journal of Computer Technology and Electronics Engineering*, *1*, 68–72.

Chowdhury, K., & Di Felice, M. (2009). SEARCH: A routing protocol for mobile cognitive radio ad-Hoc networks. In *Sarnoff Symposium, 2009. SARNOFF '09*. IEEE. doi:10.1109/SARNOF.2009.4850323

Chowdhury, K. R., & Akyildiz, I. F. (2011). Crp: A routing protocol for cognitive radio ad hoc networks. Selected Areas in Communications. *IEEE Journal on*, *29*(4), 794–804.

Chowdhury, K. R., & Felice, M. D. (2009). Search: A routing protocol for mobile cognitive radio ad-hoc networks. *Computer Communications*, *32*(18), 1983–1997. doi:10.1016/j.comcom.2009.06.011

Chowdhury, P., Misra, I. S., & Sanyal, S. K. (2012). *Cross Layer QoS Support Architecture with Integrated CAC and Scheduling Algorithms for WiMAX BWA Networks*. Academic Press.

Chuang, M., & Lee, J. (2013). TEAM: Trust-Extended Authentication Mechanism vehicular Ad-Hoc Networks. IEEE.

Clarke, I. (2008). Emerging Value Propositions for M-commerce. *The Journal of Business Strategy*, *25*(2), 41–57.

Clohessy, T., Acton, T., & Morgan, L. (2014). Smart City as a Service (SCaaS): A Future Roadmap for E-Government Smart City Cloud Computing Initiatives. In *Proceedings of the 2014 IEEE/ACM 7th International Conference on Utility and Cloud Computing*. doi:10.1109/UCC.2014.136

Collis, J., & Hussey, R. (2003). *Business Research: A practical guide for undergraduate and postgraduate students* (2nd ed.). New York: Palgrave Macmillan.

Collis, J., & Hussey, R. (2009). *Business Research: A Practical Guide for Undergraduate & Postgraduate Students* (3rd ed.). Hampshire, UK: Palgrave Macmillan.

Constantinides, E., Lorenzo-Romero, G. M., & Gómez, M. A. (2010). Effects of web expereience on consumer choice: A multicultural approach. *Internet Research, 20*(2), 188–209. doi:10.1108/10662241011032245

Coppens, J., Wauters, T., De Turck, F., Dhoedt, B., & Demeester, P. (2005). Evaluation of Replica Placement and Retrieval Algorithms in Self-Organizing CDNs. In *Proceeding of the IFIP/IEEE International Workshop on Self-Managed Systems & Services (SelfMan'05).*

Cordeiro & Challapali. (2007). C-mac: A cognitive mac protocol for multi-channel wireless networks. In *New Frontiers in Dynamic Spectrum Access Networks, 2007. DySPAN 2007. 2nd IEEE International Symposium on.*

Creswell, J. W. (1994). *Research design: Qualitative and Quantitative Approaches.* Thousand Oaks, CA: Sage.

Dahlman, E., Parkvall, S., & Skoeld, J. (2011). *4G - LTE/LTE-Advanced for Mobile Broadband.* Academic Press.

Davies, V. (2000). *Evaluating Mobility Models Within An Ad Hoc Network.* (Master's thesis). Colorado School of Mines.

Davis, F. D., Bagozzi, R. P., & Warshaw, P. R. (1989). User acceptance of computer technology: A comparison of two theoretical models. *Management Science, 35*(8), 983–1003. doi:10.1287/mnsc.35.8.982

Demangeot, C., & Broderick, A. J. (2007). Conceptualising consumer behaviour in online shopping environments. *International Journal of Retail & Distribution Management, 35*(11), 878–894. doi:10.1108/09590550710828218

Denzin, N. K. (1970). *The Research Act: A theoretical Introduction to Sociological Methods.* New York: McGraw-Hill.

Denzin, N. K., & Lincoln, Y. S. (2000). *The Handbook of Qualitative Research* (2nd ed.). Thousand Oaks, CA: Sage.

Dholakia, R. (1999). Going shopping: Key determinants of shopping behaviors and motivations. *International Journal of Retail & Distribution Management, 27*(4), 140–165.

Dholakia, R. R., & Dholakia, N. (2004). Mobility and markets: Emerging outlines of m-commerce. *Journal of Business Research, 57*(12), 1391–1396. doi:10.1016/S0148-2963(02)00427-7

Diab, A. (2010). *Mobility Management in IP-Based Networks: Analysis, Design, Programming and Computer-based Learning Modules.* (Dissertation thesis). Ilmenau University of Technology.

Diab, A., & Mitschele-Thiel, A. (2012). Comparative Evaluation of Distributed Physical Cell Identity Assignment Schemes for LTE-Advanced Systems. In *Proceeding of the Performance Monitoring, Measurement and Evaluation of Heterogeneous Wireless and Wired Networks Workshop.*

Diab, A., & Mitschele-Thiel, A. (2013). Development of Distributed and Self-Organized Physical Cell Identity Assignment Schemes for LTE-Advanced Systems. In *Proceeding of the 8th ACM workshop on Performance Monitoring and Measurement of Heterogeneous Wireless and Wired Networks* (PM2HW2N'13).

Diab, A., & Mitschele-Thiel, A. (2014). Self-Organization Activities in LTE-Advanced Networks. In Handbook of Research on Progressive Trends in Wireless Communications and Networking. IGI Global. Doi:10.4018/978-1-4666-5170-8

Diab, A., & Mitschele-Thiel, A. (2013). Development of distributed and self-organized physical cell identity assignment schemes for LTE-advanced systems. In *Proceedings of the workshop on Performance monitoring and measurement of heterogeneous wireless and wired networks.* ACM doi:10.1145/2512840.2512852

Díaz, J. M., Del Río, L. V., Cárdenas, O. Á., & Baldivia, M. G. (2014). Analysis of Propagation Models, Delay, and Throughput for WiMAX in Urban Environments. In *Broadband Wireless Access Networks for 4G: Theory, Application, and Experimentation* (pp. 1–13). Information Science Reference. http://doi.org/doi:10.4018/978-1-4666-4888-3

Dibb, S., & Carrigan, M. (2013). Social marketing transformed. *European Journal of Marketing*, *47*(9), 1376–1398. doi:10.1108/EJM-05-2013-0248

Digham, F. F., Alouini, M.-S., & Simon, M. K. (2007). On the energy detection of unknown signals over fading channels. *IEEE Transactions on Communications*, *55*(1), 21–24. doi:10.1109/TCOMM.2006.887483

Dijkstra, E. (1959). A note on two problems in connexion with graphs. *Numerische Mathematik, 1*(1), 269-271.

Dittmar, H., & Drury, J. (2000). Self-image - is it in the bag? A qualitative comparison between 'ordinary' and 'excessive' consumers. *Journal of Economic Psychology*, *21*(2), 109–142. doi:10.1016/S0167-4870(99)00039-2

Dobson, S., et al. (2006). A Survey of Autonomic Communications. *ACM Transaction on Autonomous and Adaptive Systems, 1*(2).

Dowhuszko, A. A., Ahmed, F., & Tirkkonen, O. (2013). Decentralized transmit beamforming scheme for interference coordination in small cell networks. In *Proceedings of the International Black Sea Conference on Communications and Networking*, Batumi, Georgia. IEEE. doi:10.1109/BlackSeaCom.2013.6623394

Dreher, A., & Mohl, D. (2015). *Precision Clock Synchronization*. The Standard IEEE 1588. White paper. Available at https://www.belden.com/docs/upload/Precision_Clock_Synchronization_WP.pdf

Dressler, F. (2008). A study of self-organization mechanisms in ad hoc and sensor networks. *Elsevier Computer Communications*, *31*(13), 3018–3029. doi:10.1016/j.comcom.2008.02.001

Duan, J., Yang, D., Zhu, H., Zhang, S., & Zhao, J. (2014). TSRF: A Trust aware secure routing framework in wireless sensor networks. *International Journal of Distributed Sensor Network*.

Du, L. (2003, November). Towards Intelligent Geographic Load Balancing for Mobile Cellular Networks. *IEEE Trans. on Systems, Man, and Cybernetics, Part C: Applications and Reviews*, *33*(4), 480–491. doi:10.1109/TSMCC.2003.818495

Eagly, A. H., & Chaiken, S. (1993). The psychology of Attitudes. *Public Opinion Quarterly*, *57*(3), 434. doi:10.1086/269388

Eisenblaetter, A. (2003). *Assigning Frequencies in GSM Networks*. Technical Report.

Eisenblätter, A., Grötschel, M., & Koster, A. M. C. A. (2002). Frequency assignment and ramifications of coloring. *Discussiones Mathematicae Graph Theory*, *22*, 51–88. doi:10.7151/dmgt.1158

Eklund, C., Marks, R. B., Ponnuswamy, S., & Stanwood, K. L. (2007). WirelessMAN: Inside the IEEE 802.16 Standard for Wireless Metropolitan Area Networks. IEEE.

Emarketer. (2013a). *Retail Drives UK B2C Ecommerce Sales Higher*. Available at: http://www.emarketer.com/Article/Retail-Drives-UK-B2C-Ecommerce-Sales-Higher/1010106

Emarketer. (2013b). *In-store Mobile Offers Still Finding Their Feet in the UK*. Available at: http://www.emarketer.com/Article/Retail-Drives-UK-B2C-Ecommerce-Sales-Higher/1010106

Emarketer. (2013c). *Mobile's Share of UK Digital Retail Sales Doubles*. Available at: http://www.emarketer.com/Article/Mobiles-Share-of-UK-Digital-Retail-Sales-Doubles/1010204

Emarketer. (2013d). *Three of Four UK Mobile Users Own Smartphones by 2016*. Available at: http://www.emarketer.com/Article/Three-of-Four-UK-Mobile-Users-Own-Smartphones-by-2016/1009614

Emarketer. (2013e). *UK Online Clothing Shoppers Still Favor the PC.* Available at: http://www.emarketer.com/Article/UK-Online-Clothing-Shoppers-Still-Favor-PC/1010274

Enserink, S., & Cochran, D. (1994). A cyclostationary feature detector. In *Signals, Systems and Computers, 1994.1994 Conference Record of the Twenty-Eighth Asilomar Conference on.* IEEE.

Erceg, V., Member, S., Greenstein, L. J., Tjandra, S. Y., Parkoff, S. R., Gupta, A., & Bianchi, R. et al. (1999). *An Empirically Based Path Loss Model for Wireless Channels in Suburban Environments.* Academic Press.

Erdos, P., & Renyi, A. (1959). On Random Graphs. *Publicationes Mathematicae, 6*, 290–297.

Eschenauer, L., Gligor, V. D., & Baras, J. (2002). On Trust Establishment in Mobile Ad Hoc Networks.*10th Int'l Security Protocols Workshop.*

Etemad, K., & Corporation, I. (2008). *Overview of Mobile WiMAX.* Author.

ETSI TR 101 112. (1998). *Universal Mobile Telecommunicatios System (UMTS); Selection Procedures for the Choice of Radio Transmission Technologies of the UMTS (UMTS 30.03 version 3.2.0). ETSI TR 101 112 V3.2.0 (1998-04).*

Euromonitor. (2013). *Mobile Cocooning: How Growing Reliance on Smart Devices is Influencing Consumer Behaviour.* Available at: http://www.euromonitor.com/mobile-cocooning-how-growing-reliance-on-smart-devices-is-influencing-consumer-behaviour/report

Felice, M. D., Chowdhury, K. R., & Akyildiz, I. F. (April2009). Tp-crahn: a transport protocol for cognitive radio ad-hoc networks. In *INFOCOM 2009* (pp. 2482–2490). IEEE.

Femg,H., & Liu,C. (2006). Design of a Joint Defense System for Mobile Ad Hoc Networks. *IEEE Digital Library*, 742-746.

FemtoForum. (2011). *HeNB (LTE Femto) Network Architecture.* White paper. Retrieved from http://www.smallcellforum.org/

Feng, R., Che, S., Wang, X., & Yu, N. (2013). *Trust management scheme based on D-S Evidence Theory for Wireless Sensor Networks. International Journal of Distributed Sensor Network.* doi:10.1155/2013/948641

Fitzek, F. H., Can, B., Prasad, R., Park, D. S., & Cho, Y. (2004). Application of Multiple Description Coding in 4G Wireless Communication Systems. *WWRF 8 bis.*

Foschini, G. J. (2002). Layered space-time architecture for wireless communication in a fading environment when using multi-element antennas. *Bell Labs Technical Journal, 1*(2), 41–59. doi:10.1002/bltj.2015

Franois, J. M. (2007). *Performing and Making Use of Mobility Prediction.* (Ph.D. Dissertation). University of Lige.

Fresnedo, O. (2015). Design of analog joint source-channel coding systems for multiple access channels. *IEEE Transactions on Communications.*

Frolick, M. N., & Chen, L. (2004). Assessing M-Commerce Opportunities. *Information Systems Management, 21*(2), 53–61. doi:10.1201/1078/44118.21.2.20040301/80422.8

Galinier, P., & Hertz, A. (2006). A survey of local search methods for graph coloring. *Computers & Operations Research, 33*(9), 2547–2562. doi:10.1016/j.cor.2005.07.028

Galshetwar, G. M., Jayakumar, A., & Mittal, Y. (2012). Comparative Study Of Different Scheduling Algorithms For Wimax MAC Scheduler Design. *International Journal of Engineering, 2*(2), 1031–1037.

Gamble, J., & Gilmore, A. (2013). A new era of consumer marketing? An application of co-creational marketing in the music industry. *European Journal of Marketing, 47*(11/12), 1859–1888. doi:10.1108/EJM-10-2011-0584

Ganesan, G., & Li, Y. (2005). Cooperative spectrum sensing in cognitive radio networks. *New Frontiers in Dynamic Spectrum Access Networks, 2005. DySPAN 2005. 2005 First IEEE International Symposium on*. IEEE. doi:10.1109/DYSPAN.2005.1542628

Gardner, W. A. (1991). Exploitation of spectral redundancy in cyclostationary signals. *Signal Processing Magazine, IEEE, 8*(2), 14–36. doi:10.1109/79.81007

Garrett, M. W., & Vetterli, M. (1993). Joint source/channel coding of statistically multiplexed real-time services on packet networks. *IEEE/ACM Transactions on Networking, 1*(1), 71–80. doi:10.1109/90.222908

Garza, C., Ashai, B. H., Monturus, E., & Syputa, R. (2010). *LTE Operator Commitments: Deployment Scenarios and Growth Opportunities*. Research report. MARAVEDIS Wireless Market Research & Analysis.

Gazi, V., & Passino, K. M. (2003). Stability analysis of swarms. *IEEE Transactions on Automatic Control, 48*(4), 692–697. doi:10.1109/TAC.2003.809765

George, J. F. (2002). Influence on the intent to make internet purchase. *Internet Research: Electronic Networking Applications and Policy, 12*(2), 165-80.

George, A., Kumar, A., Cavalcanti, D., & Agrawal, D. P. (2008). Protocols for mobility management in heterogeneous multi-hop wireless networks. *Pervasive and Mobile Computing, 4*(1), 92–116. doi:10.1016/j.pmcj.2007.03.002

Gerla, M., Lee, E.-K., Pau, G., & Lee, U. (2014). *Internet of vehicles: From intelligent grid to autonomous cars and vehicular clouds*. Paper presented at the IEEE World Forum on Internet of Things (WF-IoT). Available at http://www.idirect.net/Company/Resource-Center/Satellite-Basics.aspx

Ghosh, J. (2006). *Sociological Orbit Based Mobility Profiling and Routing for Wireless Networks*. (PhD Thesis). Department of Computer Science and Engineering, the State University of New York, Buffalo, NY.

Ghosh, J., Ngo, H.-Q., & Qiao, C. (2006). Mobility Profile-Based Routing within Intermittently Connected Mobile Ad hoc Networks (ICMAN). In Proceeding of WCMC'06.

Ghosh, J., Sumesh, J.-P., & Qiao, C. (2007). *Sociological Orbit Aware Location Approximation and Routing (SOLAR) in MANET. Ad Hoc Networks*. Elsevier.

Ghosh, J., Yoon, S., Ngo, H.-O., & Qiao, C. (2005). *Sociological Orbits for Efficient Routing in Intermittently Connected Mobile Ad Hoc Networks. UB CSE Technical Report, Department of Computer Science and Engineering University*. Buffalo, NY: The State University of New York.

Giner, V. C. (2004). *Study of mobility behavior of new Internet mobile users. Delivery Version No. 1*. Euro-NGI.

Goertz, N. (2007). *Joint source-channel coding of discrete-time signals with continuous amplitudes*. London: Imperial College Press.

Golaup, A., Mustapha, M., & Patanapongpibul, L. (2009). Femtocell Access Control Strategy in UMTS and LTE. IEEE Communications Magazine, 47(9). doi:10.1109/MCOM.2009.5277464

Goldsmiths, R. E., & Bridges, E. (2000). E-tailing vs retailing: Using attitude to predict online buying behaviour. *Quarterly Journal of Electronic Commerce, 1*(3), 245–253.

Gómez Mármol, F., & Martínez Pérez, G. (2012). TRIP, a trust based reputation infrastructure–based proposal for vehicular ad hoc networks. *Journal of Network and Computer Applications, 35*(3), 934–941. doi:10.1016/j.jnca.2011.03.028

Gomm, R. (2004). *Social research methodology: a critical introduction*. Basingstoke, UK: Palgrave Macmillan.

Gong, W., You, Z., Chen, D., Zhao, X., Gu, M., & Lam, K. (2009). Trust based malicious nodes Detection in MANET. *IEEE Digital Library.*10.1109/EBISS.2009.5138139

Gonzalez, M. C., Hidalgo, C. A., & Barabasi, A.-L. (2008). *Understanding Individual Human Mobility Patterns.Nature Journal.*

Gramoli, V., Kermarrec, A. M., & Le Merrer, E. (2008). SONDe, a Self-Organizing Object Deployment Algorithm in Large-Scale Dynamic Systems. In *Proceeding of 7th conference on European Dependable Computing Conference.* doi:10.1109/EDCC-7.2008.17

Guenes, M., & Siekermann, J. (2005). CosMos – Communication Scenario and Mobility Scenario Generator for Mobile Ad-Hoc Networks. In Proc. 2nd Int. Worksh. MANETs Interoper. Iss. MANETII'05.

Guest, G., MacQueen, K. M., & Namey, E. E. (2012). *Applied Thematic Analysis.* Los Angeles, CA: Sage publications Inc. doi:10.4135/9781483384436

Gui, C. (2007). An Entropy-based Stability QoS Routing with Priority Scheduler in MANET Using Fuzzy Controller. *Journal of Communication and Computer, 4*(3).

Guillemot, C., & Christ, P. (2003). Joint source channel coding as an element of a QoS framework for '4G' wireless multimedia. *ELSEVIER Journal on Computer Communication, 10*(5), 54–61.

Haardt, M., & Nossek, J. A. (1995). Unitary ESPRIT: How to obtain increased estimation accuracy with a reduced computational burden. *Signal Processing, IEEE Transactions on, 43*(5), 1232–1242. doi:10.1109/78.382406

Haas, Z. (1997). A New Routing Protocol for Reconfigurable Wireless Networks. In *Proceedings of the IEEE International Conference on Universal Personal Communications (ICUPC).* doi:10.1109/ICUPC.1997.627227

Haken, H. (2006). *Information and Self-Organization: A Macroscopic Approach to Complex Systems (Springer Series in Synergetics).* Berlin, Germany: Springer.

Hales, D., & Arteconi, S. (2006). SLACER: A Self-Organizing Protocol for Coordination in Peer-to-Peer Networks. *IEEE Intelligent Systems, 21*(2), 29–35. doi:10.1109/MIS.2006.35

Hämäläinen, S., Sanneck, H., & Sartori, C. (2011). *LTE Self-Organising Networks (SON): Network Management Automation for Operational Efficiency.* West Sussex, UK: Wiley & Sons. doi:10.1002/9781119961789

Hamza, A. S., Khalifa, S. S., Hamza, H. S., & Elsayed, K. (2013). A survey on inter-cell interference coordination techniques in OFDMA-based cellular networks. *IEEE Communications Surveys and Tutorials, 15*(4), 1642–1670. doi:10.1109/SURV.2013.013013.00028

Handorean, R., Gill, C., & Roman, G-C. (2004). Accommodating Transient Connectivity in Ad Hoc and Mobile Settings. *LNCS, 3001.*

Hartas, D. (2010). *Educational Research and Inquiry: Qualitative and Quantitative Approach.* London: Continuum International Publishing Group.

Hata, M. (n.d.). Empirical formula for propagation loss in land mobile radio services. *IEEE Trans. Veh. Tech., 29*, 317–325.

Helbing, D. (2001). Traffic and Related Self-Driven Many-Particle Systems. *Journal of Review of Modern Physic.Reviews of Modern Physics, 73*(4), 1067–1141. doi:10.1103/RevModPhys.73.1067

Helbing, D., Farkas, I., & Vicsek, T. (2000). Simulating Dynamical Features of Escape Panic. *Nature*, 407. PMID:11028994

Henderson, T., Kotz, D., & Abyzov, I. (2004). The Changing Usage of A Mature Campus-Wide Wireless Network. In *Proceedings of MobiCom'04*. ACM Press. doi:10.1145/1023720.1023739

He, T., Huang, C., Blum, B., Stankovic, J., & Abdelzaher, T. (2003). Range-Free Localization Schemes for Large Scale Sensor Networks.*International Conference on Mobile Computing and Networking: Proceedings of the 9th annual international conference on Mobile computing and networking*. doi:10.1145/938985.938995

He, Z., Cai, J., & Chen, C. W. (2002). Joint source channel rate-distortion analysis for adaptive mode selection and rate control in wireless video coding. *IEEE Transactions on Circuits and Systems for Video Technology*, *12*(6), 511–523. doi:10.1109/TCSVT.2002.800313

Hill, W. W., Beatty, S. E., & Walsh, G. (2013). A segmentation of adolescent users and shoppers. *Journal of Services Marketing*, *27*(5), 347–360. doi:10.1108/JSM-10-2011-0157

Hochwald, B. M., & Sweldens, W. (2000). Differential unitary space-time modulation. *IEEE Transactions on Communications*, *48*(12), 2041–2052. doi:10.1109/26.891215

Hoffman, D. L., & Novak, T. P. (1996). Marketing in hype media computer-mediated environments: Conceptual foundations. *Journal of Marketing*, *60*(3), 50–68. doi:10.2307/1251841

Hofmann-Wellenhof, B., Lichtenegger, H., & Collins, J. (2001). *Global Positioning System Theory and Practice* (5th ed.). Wien, Austria: Springer-Verlag. doi:10.1007/978-3-7091-6199-9

Holiday, T., Goldsmith, A. J., & Poor, H. V. (2008). Joint source and channel coding for MIMO systems: Is it Better to be robust or quick? *IEEE Transactions on Information Theory*, *54*(4), 1393–1405. doi:10.1109/TIT.2008.917725

Holmes, A., Byrne, A., & Rowley, J. (2014). Mobile shopping behaviour: Insights into attitudes, shopping process involvement and location. *International Journal of Retail & Distribution Management*, *42*(1), 25–39. doi:10.1108/IJRDM-10-2012-0096

Holzer, R. (2009). *Modeling and Control of Complex and Self-Organizing Systems. Technical Report, Number MIP-0916*. Department of Informatics and Mathematics, University of Passau.

Hong, X., Gerla, M., Pei, G., & Chiang, C.-C. (1999). A Group Mobility Model for Ad-Hoc Wireless Networks. In *Proceedings of the 2nd ACM international workshop on Modeling, analysis and simulation of wireless and mobile systems (MSWiM'99)*. doi:10.1145/313237.313248

Hossain, E., Rasti, M., Tabassum, H., & Abdelnasser, A. (2014). Evolution toward 5G multi-tier cellular wireless networks: An interference management perspective. *Wireless Communications, IEEE*, *21*(3), 118–127. doi:10.1109/MWC.2014.6845056

Hsu, W., Spyropoulos, T., Psounis, K., & Helmy, A. (2007). Modeling Time-variant User Mobility in Wireless Mobile Networks. In *Proceeding of the 26th IEEE International Conference on Computer Communications* (INFOCOM'07). doi:10.1109/INFCOM.2007.94

Hsu, C. L., Lin, J. C. C., & Chiang, S. H. (2013). The effects of blogger recommendations on customers' online shopping intentions. *Internet Research*, *33*(1), 69–88. doi:10.1108/10662241311295782

Hsu, W., Merchant, K., Shu, H.-W., Hsu, C.-H., & Helmy, A. (2004). *Weighted Waypoint Mobility Model and its Impact on Ad Hoc Networks*. MobiCom.

Hsu, W., Spyropoulos, T., Psounis, K., & Helmy, A. (2007a). *Modeling Spatial and Temporal Dependencies of Use Mobility in Wireless Mobile Networks. CENG-2008-1 Technical Report*. University of Southern California.

Huang, Y. C., & Narayanan, K. R. (2011, August). *Joint source channel coding with correlated interference*. Paper presented at the Conference of IEEE International Symposium on Information Theory (ISIT). St. Petersburg, Russia. doi:10.1109/ISIT.2011.6033709

Huang, D., Hong, X., & Gerla, M. (2012). Cognitive radio network security: A survey. *Journal of Network and Computer Applications*, *35*(6), 1691–1708. doi:10.1016/j.jnca.2012.06.006

Huang, J., Berry, R. A., & Honig, M. L. (2006). Distributed interference compensation for wireless networks. *IEEE Journal on Selected Areas in Communications*, *24*(5), 1074–1084. doi:10.1109/JSAC.2006.872889

Huang, J., Subramanian, V., Agrawal, R., & Berry, R. (2009). Joint scheduling and resource allocation in uplink OFDM systems for broadband wireless access networks. *Selected Areas in*, *27*(2), 226–234. doi:10.1109/JSAC.2009.090213

Hubaux, J.-P., Gross, T., Le Boudec, J.-Y., & Vetterli, M. (2001). Toward self-organized mobile ad hoc networks: The terminodes project. *IEEE Communications Magazine*, *39*(1), 118–124. doi:10.1109/35.894385

Hung, M. C., Yang, S. T., & Hsieh, T. C. (2012). An examination of the determinants of mobile shopping continuance. *International Journal of Electronic Business Management*, *10*(1), 29–37.

Hu, W.-C., Yeh, J.-H., Fu, L., & Yang, H.-J. (2006). Handheld Computing and Programming for Mobile Commerce. *International Journal of Web Information Sytems*, *2*(3/4), 164–175. doi:10.1108/17440080780000297

I. S. 802. 1.-2009 (Revision of I. S80216-2004). (n.d.). *IEEE Standard for Local and metropolitan area networks Part 16: Air Interface for Broadband Wireless Access Systems*. http://doi.org/<ALIGNMENT.qj></ALIGNMENT>10.1109/IEEESTD.2009.5062485

IEEE Standard. (2008). *IEEE Standard for a Precision Clock Synchronization Protocol for Networked Measurement and Control Systems*. DOI: .10.1109/IEEESTD.2008.4579760

IEEE Std 802.21-2008. (2009). *IEEE Standard for Local and metropolitan area networks - Media Independent Handover Services*. IEEE.

Irnich, T., Kronander, J., Selén, Y., & Li, G. (2013). Spectrum sharing scenarios and resulting technical requirements for 5G systems. In *Proceedings of the International Symposium on Personal, Indoor and Mobile Radio Communications (PIMRC Workshops)*.London, UK: IEEE. doi:10.1109/PIMRCW.2013.6707850

Isaacman, S. (2011a). *Ranges of Human Mobility in Los Angeles and New York*. In 8th IEEE Workshop on Managing Ubiquitous Communications and Services (MUCS' 11), Seattle, WA. doi:10.1109/PERCOMW.2011.5766977

Isaacman, S. (2011). Identifying Important Places in People's Lives from Cellular Network Data. In *Proceeding of the 9th International Conference on Pervasive Computing (Pervasive'11)*.

Isaacman, S. (2012). Human Mobility Modeling at Metropolitan Scales. In *Proceeding of the 10th International Conference on Mobile Systems (MobiSys'12)*.

Isaacman, S., Becker, R., Cáceres, R., Kobourov, S., Rowland, J., & Varshavsky, A. (2010). *A Tale of Two Cities*. In *Workshop on Mobile Computing Systems and Applications (HotMobile'10)*.

Islam, S., Rashid, M., & Tarique, M. (2011). *Performance Analysis of WiMax/WiFi System under Different Codecs*. Academic Press.

ITU-T Recommendation. (2003). One-way transmission time. *ITU-T Recommendation*, G.144.

Jain, A., & Verma, A. K. (n.d.). *Comparative Study of Scheduling Algorithms for WiMAX*. Academic Press.

Jain, R., Shivaprasad, A., Lelescu, D., & He, X. (2004). Towards A Model of User Mobility and Registration Patterns. *Mobile Computing and Communications Review, 8*(4).

Jain, R., Lelescu, D., & Balakrishnan, M. (2005). Model T: An Empirical Model for User Registration Patterns in a Campus Wireless LAN. In *Proceedings of MobiCom'05*. doi:10.1145/1080829.1080848

Jardosh, A., Belding-Royer, E. M., Almeroth, K. C., & Suri, S. (2003). Towards Realistic Mobility Models for Mobile ad hoc Networks. In *Proc. IEEE MobiCom*. doi:10.1145/938985.939008

Jee, & Madanapalli. (2008). *IP over IEEE 802.16 Problem Statement and Goals*. ETF RFC 5154.

Jennequin, C. (2004). *JOCO (Joint source-channel Coding-driven digital baseband design for 4G multimedia streaming)*. Paper presented at the Conference of JOCO Project. France.

Jia, J., Zhang, Q., & Shen, X. (2008, January). Hc-mac: A hardware-constrained cognitive mac for efficient spectrum management. *Selected Areas in Communications. IEEE Journal on, 26*(1), 106–117.

Jiang, M., & Hanzo, L. (2007). Multiuser MIMO-OFDM for next-generation wireless systems. *Proceedings of the IEEE, 95*(7), 1430–1469. doi:10.1109/JPROC.2007.898869

Johnson, D. S., Aragon, C. R., McGeoch, L. A., & Schevon, C. (1991). Optimization by simulated annealing: An experimental evaluation; part II, graph coloring and number partitioning. *Operations Research, 39*(3), 378–406. doi:10.1287/opre.39.3.378

Kang, X., & Wu, Y. (2014). A trust-based pollution attack prevention scheme in peer-to-peer streaming networks. *Computer Networks, 72*, 62–73. doi:10.1016/j.comnet.2014.07.012

Kannhavong, B., Nakayama, H., Nemoto, Y., & Kato, N. (2008). *SA-OLSR: Security Aware Optimized Link State Routing for Mobile Ad Hoc Networks*. IEEE. Doi.10.1109/ICC.2008.283

Kardar, M. (2007). *Statistical physics of fields*. Cambridge, UK: Cambridge University Press. doi:10.1017/CBO9780511815881

Karp, B. (2000). *Geographic Routing for Wireless Network*. (PhD thesis). Harvard University.

Karthika, A. L., Sumithram, G., & Shanmugam, A. (2013). *Performance of voice in integrated WiMAX- WLAN and UMTS-WLAN*. Academic Press.

Kasparick, M., & Wunder, G. (2014). Autonomous algorithms for centralized and distributed interference coordination: A virtual layer-based approach. *EURASIP Journal on Wireless Communications and Networking*, 1–20.

Kelle, U. (2006). Combining qualitative and quantitative methods in research practice: Purpose and advantages. *Qualitative Research in Psychology, 3*(4), 203–311.

Keynote. (2013). *Clothing Retailing Market Report*. Available at: http://www.keynote.co.uk/market-intelligence/view/product/10792/clothing-retailing?highlight=clothing&utm_source=kn.reports.search

Khalifa, M., & Shen, K. N. (2008). Explaining the adoption of transactional B2C mobile commerce. *Journal of Enterprise Information Management, 21*(2), 110–124. doi:10.1108/17410390810851372

Khalife, H., Malouch, N., & Fdida, S. (2009). Multihop cognitive radio networks: to route or not to route. *Network, IEEE, 23*(4).

Khalil, A., & Ksentini, A. (2007). *Classification of the Uplink Scheduling Algorithms*. Academic Press.

Khallaayoun, A. (2010). *High resolution direction of arrival estimation analysis and implementation in a smart antenna system.* (Doctoral dissertation). Montana State University-Bozeman, College of Engineering. Available at http://scholarworks.montana.edu/xmlui/handle/1/1632

Khan, S. N., Diab, A., Brosch, C., & Ahmad, M. (2012). Self-Organized Disjoint Service Placement in Future Mobile Communication Networks. *International Journal of Communication Networks and Information Security, 4*(2).

Kim, M., Kotz, D., & Kim, S. (2006). Extracting A Mobility Model From Real User Traces. In Proceedings of IEEE INFOCOM'06. doi:10.1109/INFOCOM.2006.173

Kim, H., Kim, J. L. Y., Chae, M., & Choi, Y. (2002). *An empirical study of the use context and usability problems in mobile internet.* IEEE Computer Society.

Kim, H.-W., Xu, Y., & Gupta, S. (2012). Which is more important in Internet Shopping, perceived price or trust? *Electronic Commerce Research and Applications, 11*(3), 241–252. doi:10.1016/j.elerap.2011.06.003

Kim, T., Chung, M.-Y., & Sung, D.-K. (1998). Mobility and traffic analyses in three-dimensional PCS environments. *IEEE Transactions on Vehicular Technology, 47*(2).

Kliewer, J., & Thobaben, R. (2005). Iterative joint source-channel decoding of variable-length codes using residual source redundancy. *IEEE Transactions on Wireless Communications, 4*(3), 919–929. doi:10.1109/TWC.2005.847032

Kotz, D., & Essien, K. (2002a). *Characterizing Usage of a Campus-Wide Wireless Network.* Technical Report TR2002-423, Department of Computer Science, Dartmouth College.

Kotz, D., & Essien, K. (2003). *Analysis of A Campus-Wide Wireless Network. Mobile Networks and Applications.* Dartmouth College Technical Report TR2002-432.

Kotz, D., Henderson, T., & Abyzov, I. (2015). *CRAWDAD Trace Dartmouth.* Retrieved from http://crawdad.cs.dartmouth.edu

Kotz, D., & Essien, K. (2002). Analysis of A Campus-Wide Wireless Network. In *Proceeding of the 8th Annual International Conference on Mobile Computing and Networking (MobiCom).* ACM Press.

Kotz, D., & Henderson, T. (2005). CRAWDAD: A Community Resource for Archiving Wireless Data at Dartmouth. *IEEE Pervasive Computing / IEEE Computer Society [and] IEEE Communications Society, 4*(4), 12–14. doi:10.1109/MPRV.2005.75

Kourouthanassis, P. E., & Giaglis, G. M. (2012). Introduction to the Special Issue Mobile Commerce: The Past, Present and Future of Mobile Commerce Research. *International Journal of Electronic Commerce, 16*(4), 5–17. doi:10.2753/JEC1086-4415160401

Kraaier, J., & Killat, U. (2005). The Random Waypoint City Model – User Distribution in a Street-based Mobility Model for Wireless Network Simulations. In *Proc. 3rd ACM Int. Worksh. Wirel. Mob. Appl. Serv. WLAN Hotsp.*

Krbalek, M., & Helbing, D. (2004). Determination of Interaction Potentials in Freeway Traffic from Steady-State Statistics. *Journal of Physica A-statistical Mechanics and Its Applications - Physica A, 333*, 370–378. doi:10.1016/j.physa.2003.10.059

Krim, H., & Viberg, M. (1996). Two decades of array signal processing research: The parametric approach. *Signal Processing Magazine, IEEE, 13*(4), 67–94. doi:10.1109/79.526899

Kumar, A., & Mukherjee, A. (2013). Shop while you talk: Determinants of purchase intentions through a mobile device. Mobile Marketing Association, 8(1), 23-37.

Kumari, M., Yadav, T., Yadav, P., & Sharma, D. (2011). *Comparative Study of Path Loss Models in Different Environments*. Academic Press.

Kumar, S., & Zhan, C. (2003). Mobile communications: Evolution and the impact on business operations. *Technovation*, *23*(6), 515–520. doi:10.1016/S0166-4972(02)00120-7

Kumar, V. (1992). Algorithms for constraint-satisfaction problems: A survey. *AI Magazine*, *13*(1), 32.

Kwak, H., Lee, P., Kim, Y., Saxena, N., & Shin, J. (2008). Mobility Management Survey for Home-eNB Based 3GPP LTE Systems. Journal of Information Processing Systems, 4(4). doi:10.3745/JIPS.2008.4.4.145

Kwon, S., & Lee, N. H. (2011). Virtual Extension of Cell IDs in a Femtocell Environment. In Proceeding of IEEE Wireless Communications and Networking Conference (WCNC`11). doi:10.1109/WCNC.2011.5779171

Kyösti, P., Meinilä, J., Hentilä, L., Zhao, X., Jämsä, T., Schneider, C., Narandzić, M., ... Rautiainen, T. (2008). *WINNER II channel models*. Academic Press.

Lacharite, Y., QuanNguyen, D., Wang, M., & Lamont, L. (2008). A Trust-based Security Architecture for Tactical MANET. *IEEE Digital Explorer*. 10.1109/MILCOM.2008.4753215

Ladd, A. M., Bekris, K., Rudys, A., Marceau, G., Kavraki, L. E., & Wallach, D. S. (2002). *Robotics-based location sensing using wireless Ethernet*. ACM MOBICOM.

Lam, S. C., Subramanian, R., Sandrasegaran, K., Ghosal, P., & Barua, S. (n.d.). *Performance of well-known frequency reuse algorithms in LTE downlink 3GPP LTE systems*. Academic Press.

LaMarca, A. (2005). Place Lab: Device Positioning Using Radio Beacons in the Wild. In Proceedings of Pervasive.

Lamarre, A., Galarneau, S., & Boeck, H. (2012). Mobile Marketing and consumer Behaviour Current Research Trend. *International Journal of Latest Trends in Computing*, *3*(1), 1–9.

Lapidoth, A., & Tinguely, S. (2010). Sending a bivariate Gaussian over a Gaussian MAC. *IEEE Transactions on Information Theory*, *56*(6), 2714–2752. doi:10.1109/TIT.2010.2044058

Lawler, E. L. (1972). A Procedure for Computing the K Best Solutions to Discrete Optimization Problems and Its Application to the Shortest Path Problem. *Management Science*, *18*(7), 401–405. doi:10.1287/mnsc.18.7.401

Lawler, E. L. (2001). *Combinatorial Optimization: Networks and Matroids*. Dover Publications.

Lee, P., Jeong, J., Saxena, N., & Shin, J. (2009). Dynamic Reservation Scheme of Physical Cell Identity for 3GPP LTE Femtocell Systems. Journal of Information Processing Systems, 5(4).

Lee, J.-Y., & Scholtz, R. A. (2004). Ranging in a dense multipath environment using an UWB radio link. *IEEE Transactions on Selected Areas in Communications*, *20*(9), 1677–1683.

Lee, W.-Y., & Akyildiz, I. F. (2012). Spectrum-aware mobility management in cognitive radio cellular networks. *Mobile Computing, IEEE Transactions on*, *11*(4), 529–542. doi:10.1109/TMC.2011.69

Legendre, F., Borrel, V., Dias de Amorim, M., & Fdida, S. (2006). Reconsidering Microscopic Mobility Modeling for Self-Organizing Networks. *Network*.

Leguay, J., Friedman, T., & Conan, V. (2005). *Evaluating Mobility Pattern Space Routing for DTNs*. Universite Pierre et Marie Curie, Laboratorie. Retrieved from http://arxiv.org/PS_cache/cs/pdf/0511/0511102v1.pdf

Leguay, J., Friedman, T., & Conan, V. (2005a). DTN Routing in a Mobility Pattern Space. In *Proceedings of the 2005 ACM SIGCOMM Workshop on Delay-Tolerant Networking* (WDTN '05). doi:10.1145/1080139.1080146

Lehser, F., Lei, L., Zecai, S., Zenshan, S., Djurica, M., Jorguseski, L., … Verschoor, A. (2008). *Next generation mobile networks recommendation on SON and O&M requirements*. NGMN Alliance.

Lehser, F. (Ed.). (2008). *Next generation mobile networks - recommendation on SON and O&M requirements*. Technical Report. NGMN Alliance.

Lelescu, D., Kozat, U. C., Jain, R., & Balakrishnan, M. (2006). Model T++: An Empirical Joint Space-Time Registration Model. In *Proceedings of MobiHoc'06*. doi:10.1145/1132905.1132913

Li, B., Li, D., Wu, & Li. (2009). Asar: Ant-based spectrum aware routing for cognitive radio networks. In *Wireless Communications Signal Processing, 2009. WCSP 2009. International Conference on.*

Li, Q., Hu, R. Q., Qian, Y., & Wu, G. (2012). Cooperative communications for wireless networks: techniques and applications in LTE-advanced systems. Wireless Communications, IEEE, 19(2).

Li, X., Zhou, F., & Du, J. (2013). LDTS: A Lightweight and Dependable Trust System for Clustered Wireeless Sensor Networks. *IEEE Transactions on Information Forensics and Security*, 8.

Liang, Y. W., & Columbia, B. (2010). *Ergodic and Outage Capacity of Narrowband MIMO Gaussian Channels*. Academic Press.

Liang, B., & Haas, Z. (1999). Predictive Distance-Based Mobility Management for PCS Networks. In *Proceedings of the Joint Conference of the IEEE Computer and Communications Societies (INFOCOM)*. doi:10.1109/INFCOM.1999.752157

Liberti, J. C., & Rappaport, T. S. (1999). *Smart antennas for wireless communications: IS-95 and third generation CDMA applications*. Prentice Hall PTR.

Lien, S. Y., Chen, K. C., Liang, Y. C., & Lin, Y. (2014). Cognitive radio resource management for future cellular networks. *IEEE Wireless Communications*, *21*(1), 70–79. doi:10.1109/MWC.2014.6757899

Lim, S., Yu, C., & Da, C. R. (2006). Clustered Mobility Model for Scale-free Wireless Networks. In *Proc. IEEE Conf. Loc. Comput. Netw. LCN.*

Li, M., Dong, Z. Y., & Chen, X. (2012). Factors influencing consumptions experience of mobile commerce. *Internet Research*, *22*(2), 120–141. doi:10.1108/10662241211214539

Lim, S. H., Minero, P., & Kim, Y. H. (2010, September). Lossy communication of correlated sources over multiple access channels. In *Proceedings of the 48th Annual Allerton Conference on Communications, Control, and Computing*. doi:10.1109/ALLERTON.2010.5706997

Li, Q., & Rus, D. (2003). Communication in disconnected ad hoc networks using message relay. *Journal of Parallel and Distributed Computing*, *63*(1), 75–86. doi:10.1016/S0743-7315(02)00033-3

Liu, Q., Member, S., Wang, X., & Giannakis, G. B. (2006). *A Cross-Layer Scheduling Algorithm With QoS Support in Wireless Networks*. Academic Press.

Liu, Y., Li, W., Zhang, H., & Yu, L. (2010). Distributed PCI Assignment in LTE Based on Consultation Mechanism. In *Proceeding of 6th International Conference on Wireless Communications Networking and Mobile Computing* (WiCOM`10). doi:10.1109/WICOM.2010.5601210

Liu. (2009). *Security and Trust Management in Self-organizing Wireless Adhoc Networks*. Retrieved from http://citeseerx.ist.psu.edu

Liu, Y., Li, W., Zhang, H., & Lu, W. (2010). *Graph-Based Automatic Centralized PCI Assignment in LTE.IEEE Symposium on Computers and Communications (ISCC'10)*, Riccione, Italy.

Li, W., & Liu, P. (2005). 3D AOA/TDOA emitter location by integrated passive radar/GPS/INS systems.*Proceedings of 2005 IEEE International Workshop on VLSI Design and Video Technology.*

Li, W., Parker, J., & Joshi, A. (2012). Security through Collaboration and Trust in MANETs. *Mobile Networks and Applications*, *17*(3), 342–352. doi:10.1007/s11036-010-0243-9

Lobinger, Stefanski, Jansen, & Balan. (2010). Load balancing in downlink LTE self-optimizing networks. In *Proc. of 2010 IEEE 71st Vehicular Technology Conference (VTC 2010-Spring)*. doi:10.1109/VETECS.2010.5493656

Lopez, J., Roman, R., Agudo, I., & Fernandez-Gago, C. (2010). Trust management systems for wireless sensor networks: Best practices. *Computer Communications*, *33*(9), 1086–1093. doi:10.1016/j.comcom.2010.02.006

Lopez-Nores, M., García-Duque, J., Pazos-Aria, J.-J., & Díaz Redondo, R.-P. (2008). Qualitative Assessment of Approaches to Coordinate Activities of Mobile Hosts in Ad Hoc Networks. *IEEE Communications Magazine*, *46*(12), 108–111. doi:10.1109/MCOM.2008.4689216

Luigi, G. (2011). Vehicle-to-Vehicle/Vehicle-to-Infrastructure Control. In T. Samad & A. M. Annaswamy (Eds.), The Impact of Control Technology (p. 211). IEEE Control Systems Society. Available at http://ieeecss.org/sites/ieeecss.org/files/documents/IoCT-Part4-13VehicleToVehicle-LR.pdf

Lu, J., Yu, C.-S., Liu, C., & Yao, J. E. (2003). Technology acceptance model for wireless internet. *Internet Research: Electronic Networking Applications and Policy*, *13*(3), 206–222. doi:10.1108/10662240310478222

M.1225, R. I.-R. (1997). *Guidelines for evaluation of radio transmission technologies for imt-2000.*

Ma, L., Han, X., & Shen, C.-C. (2005). Dynamic open spectrum sharing mac protocol for wireless ad hoc networks. In *New Frontiers in Dynamic Spectrum Access Networks, 2005. DySPAN 2005. 2005 First IEEE International Symposium on.*

Marabissi, D., Tarchi, D., Fantacci, R., & Balleri, F. (2008). Efficient Adaptive Modulation and Coding Techniques for WiMAX Systems.*2008 IEEE International Conference on Communications*. doi:10.1109/ICC.2008.636

Marcello Calefi, I. F. (2012). Akyildiz, & L. Paura. Opera: Optimal routing metric for cognitive radio ad hoc networks. *Wireless Communications. IEEE Transactions on*, *11*(8), 2884–2894.

Markoulidakis, J.-G., Lyberopoulos, G.-L., Tsirkas, D.-F., & Sykas, E.-D. (1997). *Mobility modeling in third-generation mobile telecommunications systems*. IEEE Personal Communications.

Markovic, M. (2006). Analysing Qualitative Data: Health care xperiences of Women with Gynecological Cancer. *Field Methods*, *18*(4), 413–429. doi:10.1177/1525822X06293126

Marks, D., & Yardley, L. (2004). *Research methods for clinical and health psychology*. London: Sage.

Mathieson, K. (1991). Predicting user intentions: Comparing the technology acceptance model with the theory of planned behaviour. *Information Systems Research*, *2*(3), 173–191. doi:10.1287/isre.2.3.173

Matinmikko, M., Palola, M., Saarnisaari, H., Heikkila, M., Prokkola, J., Kippola, T., … Yrjola, S. (2013). Cognitive Radio Trial Environment: First Live Authorized Shared Access-Based Spectrum-Sharing Demonstration. *Vehicular Technology Magazine, IEEE*, *8*(3).

Mayer, R. C., Davis, J. H., & Schoorman, F. D. (1995). An integrative model of organizational trust. *Academy of Management Review*, *20*, 709–734.

McAllester, D., Selman, B., & Kautz, H. (1997). Evidence for invariants in local search. In *Proceedings of the national conference on artificial intelligence and conference on Innovative applications of artificial intelligence of (AAAI/IAAI)*. Menlo Park, CA: AAAI press.

McNett, M., & Voelker, G. (2005). Access and Mobility of Wireless PDA Users. *Mobile Computing and Communications Review*, *9*(2), 40. doi:10.1145/1072989.1072995

Mejri, M., Ben-Othman, J., & Hamdi, M. (2014). Survey on VANET security challenges and possible cryptographic solutions. *Vehicular Communications*, *1*(2), 53–66. doi:10.1016/j.vehcom.2014.05.001

Meng, X., Ding, Y., & Gong, Y. (2012). Trust: A trust model based on feedback-arbitration in structured P2P network. *Computer Communications*, *35*(16), 2044–2053. doi:10.1016/j.comcom.2012.06.007

Milanovic, J. (2007). *Comparison of Propagation Models Accuracy*. Academic Press.

Miles, C. (2013). Persuasion, marketing, communication, and the metaphor of magic. *European Journal of Marketing*, *47*(11/12), 2002–2019. doi:10.1108/EJM-11-2011-0632

Mills, D. L. (1992). *Network Time Protocol (Version 3) Specification, Implementation and Analysis*. IETF RFC 1305.

Minero, P., Lim, S., & Kim, Y. H. (2011, July). Joint source-channel coding via hybrid coding. In *Information Theory Proceedings (ISIT), 2011 IEEE International Symposium on* (pp. 781-785). IEEE. doi:10.1109/ISIT.2011.6034241

Miorandi, D., Sicarib, S., & De Pellegrini, F. (2012). Internet of things: Vision, applications and research challenges. *Ad Hoc Networks*, *10*(7), 1497–1516. doi:10.1016/j.adhoc.2012.02.016

Mishra, A., Shin, M., & Arbaugh, W.-A. (2004). *Pro-Active Key Distribution Using Neighbor Graphs*. IEEE Wireless Communications.

Misra. (2009). *Wireless Communications and Networks 3G and beyond*. Tata McGraw Hill Education Private Limited.

Misra, P., & Enge, P. (2001). *Global Positioning System: Signals, Measurements, and Performance*. Ganga-Jamuna Press.

Mitton, N., Fleury, E., Lassous, I. G., & Tixeuil, S. (2005). Self-stabilization in self-organized multihop wireless networks, In *Proceedings of the International Conference on Distributed Computing Systems Workshops*. Columbus, OH: IEEE. doi:10.1109/ICDCSW.2005.122

Mogensen, P., Pajukoski, K., Tiirola, E., Lähetkangas, E., Vihriälä, J., Vesterinen, S., & Cattoni, A. F. et al. (2013). 5G small cell optimized radio design. In *Proceedings of the Global Communications Workshop on Emerging Technologies for LTE-Advanced and Beyond 4G*. Atlanta, GA: IEEE.

Moth, D. (2013). *Mobile commerce now accounts for 23% of online salees: report*. Available at: https://econsultancy.com/blog/63338-mobile-commerce-now-accounts-for-23-of-online-sales-report

Motorola. (2009). *LTE Operations and Maintenance Strategy - Using Self-Organizing Networks to Reduce OPEX*. White paper.

MSN Maps and Directions. (2015). Retrieved from http://maps.msn.com/

Musolesi, M., & Mascolo, C. (2007). Designing Mobility Models Based on Social Network Theory. *SIGMOBILE Mob. Comput. Commun. Rev.*

Musolesi, M., Hailes, S., & Mascolo, C. (2005). Social Networks-Based Ad Hoc Mobility Models. In *3rd UK-UbiNet Workshop: Designing, evaluating and using ubiquitous computing systems*. University of Bath.

Musolesi, M., & Mascolo, C. (2006). *A Community Based Mobility Model for Ad Hoc Network Research*. RESLMAN. doi:10.1145/1132983.1132990

Musolesi, M., & Mascolo, C. (2009). *Mobility models for systems evaluation: A Survey*. In B. Garbinato, H. Miranda, & L. Rodrigues (Eds.), *Middleware for Network Eccentric and Mobile Applications* (pp. 43–62). Springer. doi:10.1007/978-3-540-89707-1_3

Nakamura, V. M. (2008). *Standardization Activities for Mobile WiMAX*. Academic Press.

Narasimha, S., & Sivalingam, K. M. (2009). *Improved opportunistic scheduling algorithms for WiMAX Mobile Multihop Relay networks*. IEEE. doi:10.1109/HIPC.2009.5433228

Navidi, W., & Camp, T. (2004). Stationary Distributions for the Random Waypoint Mobility Model. *IEEE Transactions on Mobile Computing*, *3*(1), 99–108. doi:10.1109/TMC.2004.1261820

Neel, J. O., & Reed, J. H. (2006). Performance of Distributed Dynamic Frequency Selection Schemes for Interference Reducing Networks. In *Proceedings of the IEEE Military Communications Conference* (MILCOM '06). doi:10.1109/MILCOM.2006.302016

Networking Research Lab. (2015). Retrieved from http://research.csc.ncsu.edu/netsrv/

Newman, M. E. J., & Girvan, M. (2004). Finding and Evaluating Community Structure in networks. *Physical Review E: Statistical, Nonlinear, and Soft Matter Physics*, *69*(2), 026113. doi:10.1103/PhysRevE.69.026113 PMID:14995526

Next Generation Mobile Networks (NGMN) Official Website. (2015). Retrieved May 07, 2015, from https://www.ngmn.org/

Nokia Siemens Networks and Nokia. (2008). *SON Use Case: Cell Phy ID Automated Configuration*. R3-080376, Technical Report.

Noulas, A., Scellato, S., Lambiotte, R., Pontil, M., & Mascolo, C. (2012). A Tale of Many Cities: Universal Patterns in Human Urban Mobility. *PLoS ONE 7*(9). doi: .10.1371/annotation/ca85bf7a-7922-47d5-8bfb-bcdf25af8c72

Nuaymi, L. (2007). *WiMAX: Technology for Broadband Wireless Access*. John Wiley and Sons. doi:10.1002/9780470319055

Nysveen, H., & Pedersen, P. E. (2005). Search mode and purchase intention in online shopping behaviour. *International Journal of Internet Marketing and Advertising*, *2*(4), 288–306. doi:10.1504/IJIMA.2005.008103

Ochiai, H., & Esaki, H. (2008). Mobility Entropy and Message Routing in Community-Structured Delay Tolerant Networks. In Proceeding of AINTEC'08. doi:10.1145/1503370.1503396

Olfati-Saber, R., Fax, J. A., & Murray, R. M. (2007). Consensus and Cooperation in Networked Multi-Agent Systems. *Proceedings of the IEEE*, *95*(1), 215–233. doi:10.1109/JPROC.2006.887293

Open Mobile Alliance. (2009a). *Secure user plane location architecture*. OMA - AD - SUPL - V2_0, Candidate Version 2.0.

Open Mobile Alliance. (2009b). *Secure user plane location protocol*. OMA - TS - ULP - V2_0, Candidate Version 2.0.

Osa, V., Herranz, C., Monserrat, J. F., & Gelabert, X. (2012). Implementing opportunistic spectrum access in LTE-advanced. *EURASIP Journal on Wireless Communications and Networking*, *2012*(1), 1–17. doi:10.1186/1687-1499-2012-99

Østerbø, O., & Grøndalen, O. (2012). Benefits of Self-Organizing Networks (SON) for Mobile Operators. *Computer Networks and Communications*, *2012*. doi:10.1155/2012/862527

Oyman, O., & Foerster, J. (2010, September). Distortion-aware MIMO link adaptation for enhanced multimedia communications. In *Proceedings of 2010 IEEE 21st International Symposium on Personal, Indoor and Mobile Radio Communications Workshops (PIMRC Workshops)* (pp. 387-392). doi:10.1109/PIMRCW.2010.5670400

Ozuem, W., Howell, K. E., & Lancaster, G. (2008). Communicating in the new interactive marketspace. *European Journal of Marketing*, *42*(9/10), 1059–1083. doi:10.1108/03090560810891145

Palomar, D. P., & Chiang, M. (2006). A tutorial on decomposition methods for network utility maximization. *IEEE Journal on Selected Areas in Communications*, *24*(8), 1439–1451. doi:10.1109/JSAC.2006.879350

Parasuraman, A. (2000). Technology readiness index (TRI) a mutiple-item scale to measure readiness to embrace new technology. *Journal of Service Research*, *2*(4), 317–320. doi:10.1177/109467050024001

Pareit, D., Lannoo, B., Moerman, I., & Demeester, P. (2012). *The History of WiMAX : A Complete Survey of the Evolution in Certification*. Academic Press.

Park, E. J., Kim, E. Y., & Forney, J. C. (2006). A structural model of fashion-oriented impulse buying behaviour. *Journal of Fashion Marketing and Management*, *10*(4), 433–446. doi:10.1108/13612020610701965

Parker, R., & Valaee, S. (2007). Vehicular node localization using receivedsignal-strength indicator. *IEEE Transactions on Vehicular Technology*, *56*(6), 3371–3380. doi:10.1109/TVT.2007.907687

Parvazi, P. (2012). *Sensor Array Processing In Difficult And Non-Idealistic Conditions*. (PhD Thesis). TU Darmstadt. Available at http://tuprints.ulb.tu-darmstadt.de/view/person/Parvazi=3APouyan=3A=3A.html

Parvin, S., Hussain, F., Hussaina, O. K., & Faruquea, A. A. (2012). Trust-based Throughput in Cognitive Radio Networks. *Procedia Computer Science, Elsevier*, *10*, 713–720. doi:10.1016/j.procs.2012.06.091

Parvin, S., Hussain, F., & Hussain, O. (2013). Conjoint trust assessment for secure communication in cognitive radio networks. *Mathematical and Computer Modelling*, *58*(5-6), 1340–1350. doi:10.1016/j.mcm.2013.01.001

Pasqua, R., & Elkin, N. (2013). *Mobile Marketing: An hour a day*. John Wiley & Sons Inc.

Patil, S., Patil, V., & Bhat, P. (2012). A review on 5G technology. *International Journal of Engineering and Innovative Technology*, *1*(1), 26–30.

Patton, M. Q. (2002). *Qualitative Research and Evaluation Methods* (3rd ed.). London: Sage Publications Ltd.

Paulraj, A., Roy, R., & Kailath, T. (1986). A subspace rotation approach to signal parameter estimation. *Proceedings of the IEEE*, *74*(7), 1044–1046. doi:10.1109/PROC.1986.13583

Pei, Q., Yuan, B., Li, L., & Li, H. (2013). A sensing and etiquette reputation-based trust management for centralized cognitive radio networks. *Neurocomputing*, *101*, 129–138. doi:10.1016/j.neucom.2012.08.005

Peltomäki, M., Koljonen, J., Tirkkonen, O., & Alava, M. (2012). Algorithms for self-organized resource allocation in wireless networks. *IEEE Transactions on Vehicular Technology*, *61*(1), 346–359. doi:10.1109/TVT.2011.2151891

Perera, C., Zaslavsky, A., Christen, P., & Georgakopoulos, D. (2014). Context aware computing for the internet of things: A survey. *IEEE Communications Surveys and Tutorials*, *16*(1), 414–454. doi:10.1109/SURV.2013.042313.00197

Persaud, A., & Azhar, I. (2012). Innovative mobile marketing via smartphones: Are consumers ready? *Marketing Intelligence & Planning*, *30*(4), 418–443. doi:10.1108/02634501211231883

Pesavento, M., Mecklenbräuker, C. F., & Böhme, J. F. (2004). Multidimensional rank reduction estimator for parametric MIMO channel models. *EURASIP Journal on Applied Signal Processing*, *2004*(9), 1354–1363. doi:10.1155/S1110865704401036

Peter, J. P., & Olson, J. (1996). *Consumer behaviour and marketing strategy*. Chicago: Irwin.

Pinyi, R., Wang, Y., & Du, Q. (2014, February). Cad-mac: A channel-aggregation diversity based mac protocol for spectrum and energy efficient cognitive ad hoc networks. Selected Areas in Communications. *IEEE Journal on*, *32*(2), 237–250.

Poole, I. (2013). *4G LTE CoMP Coordinated Multiple Tutorials*. Retrieved November, 2013, from http://www.radio-electronics.com/info/cellulartelecomms/lte-long-term-evolution/4g-lte-advanced-comp-coordinated-multipoint.php

Pradhan, C., Sankhe, K., Kumar, S., & Murthy, G. R. (2014). Cognitive base station design for efficient spectrum utilization in cellular network. In *Wireless and Optical Communications Networks (WOCN), 2014Eleventh International Conference on*. IEEE. doi:10.1109/WOCN.2014.6923095

Prehofer, C., & Bettstetter, C. (2005). Self-organization in communication networks: Principles and design paradigms. *IEEE Communications Magazine*, *43*(7), 78–85. doi:10.1109/MCOM.2005.1470824

Prokopenko. (2009). Guided self-organization. *HFSP Journal*, *3*, 287-289.

QUALCOMM. (2001). *Synchronous vs. Asynchronous Deployment of WCDMA – Effect on Communication System Performance*. Author.

Qureshi, B., Min, G., & Kouvatsos, D. (2012). A distributed reputation and trust management scheme for mobile peer-to-peer networks. *Computer Communications*, *35*(5), 608–618. doi:10.1016/j.comcom.2011.07.008

Qu, Z., Wang, Y., Wang, S., & Zhang, Y. (2013). Implications of online social activities for e-tailer's business performance. *European Journal of Marketing*, *47*(8), 1190–1212. doi:10.1108/03090561311324282

Rahman, M., Caleffi, M., & Paura, L. (2012). Joint path and spectrum diversity in cognitive radio ad-hoc networks. *EURASIP Journal on Wireless Communications and Networking*, *2012*(1), 1–9. doi:10.1186/1687-1499-2012-235

Rahman, M., & Yanikomeroglu, H. (2010). Enhancing cell-edge performance: A downlink dynamic interference avoidance scheme with inter-cell coordination. *IEEE Transactions on Wireless Communications*, *9*(4), 1414–1425. doi:10.1109/TWC.2010.04.090256

Rai, A., Tewari, R., & Upadhyay, S. (2010). Different Types of Attacks on Integrated MANET-Internet Communication. *International Journal of Computer Science and Security*, *4*, 265–274.

Ramchandran, K., Ortega, A., Uz, K. M., & Vetterli, M. (1993). Multiresolution broadcast for digital HDTV using joint source/channel coding. *IEEE Transactions on Selected Areas in Communications*, *11*(1), 6–23. doi:10.1109/49.210540

Ramiro, J., & Hamied, K. (2012). *Self-Organizing Networks (SON) Self-Planning - Self-Optimization and Self-Healing for GSM, UMTS and LTE*. John Wiley & Sons.

Ranjan Roy, R. (2011). Handbook of Mobile Ad Hoc Networks for Mobility Models. Springer Science+Business Media.

Rapoport, A. (1957). Contribution to the Theory of Random and Biased Networks. *The Bulletin of Mathematical Biophysics*, *19*(4), 257–277. doi:10.1007/BF02478417

Rasheed, F. (n.d.). *Beamforming doa estimation for a uniform linear array*. Tech. Rep. Electrical and Computer Engineering Department, Drexel University.

Rashwan, A. H., Elbadawy, H. M., & Ali, H. H. (2009). *Comparative Assessments for Different WiMAX Scheduling Algorithms*. Academic Press.

Reddy, Y. B. (2012). Trust Based Approach in Wireless Sensor Networks using an Agent to Each Cluster. *International Journal of Security. Privacy and Trust Management*, *1*, 1.

Redwan, H., & Kim, K. (2008). Survey of Security Requirements, Attacks and Network Integration in Wireless Mesh Networks. *IEEE Digital Explorer*. .10.1109/FCST.2008.15,3-9

Ren, S., Ma, X., Yan, S., & Hao, C. (2013). 2-D unitary ESPRIT-like direction-of-arrival (DOA) estimation for coherent signals with a uniform rectangular array. *Sensors (Basel, Switzerland)*, *13*(4), 4272–4288. doi:10.3390/s130404272 PMID:23539031

Rhee, I., Shin, M., Hong, S., Lee, K., & Chong, S. (2008). On the Levy-Walk Nature of Human Mobility. In *IEEE INFOCOM 2008 - IEEE Conference on Computer Communications*. IEEE. doi:10.1109/INFOCOM.2008.145

Ribeiro, A. G. (2011). Improving Mobile Networks Based on Social Mobility Modeling. In *Proceeding of 19th IEEE International Conference on Network Protocols* (ICNP). doi:10.1109/ICNP.2011.6089064

Robert, M. G. (2009). *Entropy and Information Theory*. Information Systems Laboratory, Electrical Engineering Department, Stanford University.

Robert, W. (2004). Self-organizing innovation networks: Implications for globalization. *Technovation*, *24*(3), 187–197. doi:10.1016/S0166-4972(03)00092-0

Robson, C. (2011). *Real World Research: A Resource for Users of Social Research Methods in Applied settings* (3rd ed.). Cornwall, UK: John Wiley & Sons Ltd.

Rogers, E. M. (1995). *Diffusion of Innovations*. New York: The Free Press.

Rossi, M., Badia, L., Bui, N., & Zorzi, M. (2005). *On Group Mobility Patterns and their Exploitation to Logically Aggregate Terminals in Wireless Networks*. VTC. doi:10.1109/VETECF.2005.1558429

Rossi, M., Badia, L., Giacon, P., & Zorzi, M. (2005). *On the Effectiveness of Logical Device Aggregation in Multi-Radio Multi-hop Networks*. MobiWac. doi:10.1109/WIRLES.2005.1549435

Roy, R. (2011). Handbook of Mobile Ad Hoc Networks for Mobility Models. Springer Science+Business Media.

Royer, E., Melliar-Smith, P. M., & Moser, L. (2001). An Analysis of the Optimum Node Density for Ad Hoc Mobile Networks. In *Proceedings of the IEEE International Conference on Communications (ICC)*. doi:10.1109/ICC.2001.937360

Rutvij, H., Ashish, Patel, Parmar, & Shah. (2010). MANET Routing Protocols and Wormhole Attack against AODV. *International Journal of Computer Science and Network Security*, *10*, 12–18.

Sachs, J., Muñoz, L., Aguero, R., Choque, J., Koudouridis, G., Karimi, R., . . . Gebert, J. F. M. (2004). Future wireless communication based on multi-radio access. In *Proceedings Wireless World Research Forum Meeting*.

Saha, S. (2013). *Performance Analysis of Service Classes*. IEEE.

Saha. (2013). QoS and Mobility Management Issues on Next Generation Mobile WiMAX Networks. In *Broadband Wireless Access Networks for 4G: Theory, Application, and Experimentation* (pp. 298—323). Information Science Reference, IGI Global. http://doi.org/doi:10.4018/978-1-4666-4888-3.ch16

Saha, H., Bhattacharyya, D., & Banerjee, P. K. (2010). Semi-Centralized Multi-Authenticated RSSI Based Solution to Sybil Attack. *International Journal of Computer Science & Emerging Technologies*, *1*, 338–341.

SAI Global. (2003). *Standard Specification for Telecommunications and Information Exchange Between Roadside and Vehicle Systems—5.9 GHz Band Dedicated Short Range Communications (DSRC) Medium Access Control (MAC) and Physical Layer (PHY) Specifications*. ASTM E2213 -03(2010).

Saied, Y., Olivereau, A., Zeghlache, D., & Laurent, M. (2013). Trust management system design for the Internet of Things: A context-aware and multiservice approach. *Computers & Security*, *39*, 351–365. doi:10.1016/j.cose.2013.09.001

Saini, A., & Kumar, H. (2010). Comparison between Various Black Hole Detection Techniques in MANET.*Proceedings of National Conference on Computational Instrumentation.*

Saint-Exupery, D., & Antoine. (2009). Internet of things, strategic research roadmap. *Surrey: Internet of Things Initiative.*

Samaan, N., & Karmouch, A. (2009). Towards autonomic network management: An analysis of current and future research directions. *IEEE Communications Surveys and Tutorials*, *11*(3), 22–36. doi:10.1109/SURV.2009.090303

Sanchez, M., & Manzoni, P. (2001). Anejos: A java based simulator for ad-hoc networks. *Future Generation Computer Systems*, *17*(5), 573–583. doi:10.1016/S0167-739X(00)00040-6

San-Martin, S., & Lopez-Catalan, B. (2013). How can a mobile vendor get satisfied customers? *Industrial Management & Data Systems*, *113*(2), 156–170. doi:10.1108/02635571311303514

San-Martin, S., Lopez-Catalan, B., & Ramon-Jeronimo, M. A. (2012). Factors determining firms' perceived performance of mobile commerce. *Industrial Management & Data Systems*, *112*(6), 946–963. doi:10.1108/02635571211238536

Sari, A., & Necat, B. (2012). Securing Mobile Ad-Hoc Networks against Jamming Attacks through Unified Security Mechanism. *International Journal of Ad Hoc, Sensor & Ubiquitous Computing*, 3.

Sarkar, T. K., Wicks, M. C., Salazar-Palma, M., & Bonneau, R. J. (2004). *A Survey of Various Propagation Models for Mobile Communication.* Smart Antennas.

Satoh, H., Ito, S., & Kawaguchi, N. (2005). Position Estimation of Wireless Access Point Using Directional Antennas. In *Proceedings of the International Workshop on Location- and Conext-Awareness* (LoCA). Oberpfaffenhofen, Germany: Springer-Verlag. doi:10.1007/11426646_14

Saunders, M., Lewis, P., & Thornhill, A. (2009). *Research methods for business students* (5th ed.). Harlow, UK: Pearson Education Ltd.

Saxena, N., Denko, M., & Banerji, D. (2011). A hierarchical architecture for detecting selfish behavior in community wireless mesh networks. *Computer Communications*, *34*(4), 548–555. doi:10.1016/j.comcom.2010.04.040

Sayood, K., & Borkenhagen, J. C. (1991). Use of residual redundancy in the design of joint source/channel coders. *IEEE Transactions on Communications*, *39*(6), 838–846. doi:10.1109/26.87173

Scarpi, D. (2006). Fashion stores between fun and usefulness. *Journal of Fashion Marketing and Management*, *10*(1), 7–24. doi:10.1108/13612020610651097

Schiffman, L. G., & Kanuk, L. L. (2010). *Consumer Behaviour* (10th ed.). Pearson Education Inc.

Schmidt, R. O. (1986). Multiple emitter location and signal parameter estimation. *Antennas and Propagation. IEEE Transactions on*, *34*(3), 276–280. doi:10.1109/TAP.1986.1143830

Schuster, L., Drennan, J., & Lings, I. N. (2013). Consumer acceptance of m-wellbeing services: A social marketing perspective. *European Journal of Marketing*, *47*(9), 1439–1457. doi:10.1108/EJM-10-2011-0556

Scott, J. (2000). *Social Networks Analysis: A Handbook* (2nd ed.). London: Sage Publications.

Scully, N., Thiel, S., Litjens, R., Jorguseski, L., Nascimento, R., Linnell, O., … Schmelz, L. C. (2008). *Deliverable D2.1: Use Cases for Self-Organizing Networks.* SOCRATES.

Sen, R., Handorean, R., Roman, G.-C., & Hackmann, G. (2004). *Knowledge-Driven Interactions with Service Across Mobile Ad Hoc Networks.* Technical Report WUCSE-2004-34. Washington University, Department of Computer Science.

Seock, Y. K., & Norton, M. (2007). Attitude towards internet websites, online information search and channel choices for purchasing. *Journal of Fashion Marketing and Management, 11*(4), 571–586. doi:10.1108/13612020710824616

Sesia, S., Toufik, I., & Barker, M. (2011). *LTE, The UMTS Long Term Evolution: From Theory to Practice* (2nd ed.). Wiley. doi:10.1002/9780470978504

Shahab, M. B., & Bhatti, A. A. (2013). Neural Networks Based Physical Cell Identity Assignment for Self Organized 3GPP Long Term Evolution. *International Journal of Advances in Telecommunications, Electrotechnics, Signals and Systems, 2*(3).

Shannon, C. E. (2001). A mathematical theory of communication. *Mobile Computing and Communications Review, 5*(1), 3–55. doi:10.1145/584091.584093

Shan, T. J., Wax, M., & Kailath, T. (1985). On spatial smoothing for direction-of-arrival estimation of coherent signals. *IEEE Transactions on Acoustics, Speech, and Signal Processing, 33*(4), 806–811. doi:10.1109/TASSP.1985.1164649

Sheng, Yang, Zhang, & Li. (2014). Zone-Based Load Balancing in LTE Self-Optimizing Networks: A Game-Theoretic Approach. *Vehicular Technology, IEEE Transactions on, 63*(6).

Shengguang, L., Lin, T., Yuanshuo, Z., & Rucai, Z. (2013). *Internet of Things for special materials transportation vehicles*. Paper presented at the IEEE International Conference on Green Computing and Communications (GreenCom), and IEEE Internet of Things (iThings/CPSCom) and IEEE Cyber, Physical and Social Computing. doi:10.1109/GreenCom-iThings-CPSCom.2013.351

Shim, S., Estlick, M. A., Lotz, S. L., & Warrrington, P. (2001). An online prepurchase intentions model: The role of intention to search. *Journal of Retailing, 77*(2), 307–416.

Shladover, S. E., & Tan, S. K. (2006). Analysis of vehicle positioning accuracy requirements for communication-based cooperative collision warning. *J. Intell. Transp. Syst.: Technol., Plan. Oper., 10*(3), 131–140.

Singh, R. K. (2012). *WiMAX Networks: Services and their Quality Matrics*. Academic Press.

Singhal, D., & Garimella, R. M. (2014). Cognitive cross-layer multipath probabilistic routing for cognitive networks. *Springer's Wireless Networks*. DOI: 10.1007/s11276-014-0847-2

Siomina, P., Varbrand, P., & Yuan, D. (2006). Automated optimization of service coverage and base station antenna configuration in UMTS networks. *IEEE Wireless Communications, 13*(6), 16–25. doi:10.1109/MWC.2006.275194

Sivagurunthan, S., & Prathapchandran, K. (2014). *Trust based Security schemes in Mobile Ad Hoc Networks – A Review*. IEEE Digital Explorer; doi:10.1109/ICICA.2014.67

Skeldon, P. (2011). *M-commerce: Boost your business with the power of mobile commerce*. Richmond, VA: Crimson Publishing.

Skouby, K. E., & Lynggaard, P. (2014). *Smart home and smart city solutions enabled by 5G, IoT, AAI and CoT services*. Paper presented at the International Conference on Contemporary Computing and Informatics (IC3I).

SOCRATES. (n.d.). *Self-optimisation and self-configuration in wireless networks*. European Research Project. Retrieved from http://www.fp7-socrates.eu

Soh, W., & Kim, H. S. (2004). Dynamic Bandwidth Reservation in Cellular Networks Using Road Topology-Based Mobility Predictions. In Proceeding of INFOCOM'04.

Sohrabi, K., Gao, J., Ailawadhi, V., & Pottie, G. J. (2000). Protocols for self-organization of a wireless sensor network. *IEEE Personal Communications, 7*(5), 16-27.

Soh, W., & Kim, H. S. (2003). QoS Provisioning in Cellular Networks Based on Mobility Prediction Techniques. *IEEE Communications Magazine*.

So-in, C., Member, S., Jain, R., & Tamimi, A. (2009). Scheduling in IEEE 802. 16e Mobile WiMAX Networks. *Key Issues and a Survey*, *27*(2), 156–171.

Solomon, M. R. (2009). *Consumer Behaviour: Buying, Having and Being* (8th ed.). Pearson Education Inc.

Song, C., Qu, Z., Blumm, N., & Barabási, A.-L. (2010). Limits of Predictability in Human Mobility. *Science*, 327. PMID:20167789

Song, G., & Li, Y. (2005). Utility-based resource allocation and scheduling in OFDM-based wireless broadband networks. *IEEE Communications Magazine*, *43*(12), 127–134. doi:10.1109/MCOM.2005.1561930

Song, L., Kotz, D., Jain, R., & He, X. (2004). Evaluating Next-Cell Predictors with Extensive Wi-Fi Mobility Data. *Proceedings - IEEE INFOCOM*.

Son, K., Chong, S., & Veciana, G. (2009). Dynamic association for load balancing and interference avoidance in multi-cell networks. *IEEE Transactions on Wireless Communications*, *8*(7), 3566–3576. doi:10.1109/TWC.2009.071140

Spaid, B. I., & Flint, D. J. (2014). The meaning of shopping experiences augmented by mobile internet devices. *Journal of Marketing Theory and Practice*, *22*(1), 73–89. doi:10.2753/MTP1069-6679220105

Specification, I. E. E. E. (1999). *IEEE 802.11b/d3.0 Wireless LAN Medium Access Control (MAC) and Physical Layer (PHY)*. Specification.

Spyropoulos, T., Psounis, K., & Raghavendra, C.-S. (2005). Spray and Wait: An Efficient Routing Scheme for Intermittently Connected Mobile Networks. In *Proceedings of the 2005 ACM SIGCOMM Workshop on Delay-Tolerant Networking* (WDTN '05).

Srivastava, V., & Motani. (2005). Cross-layer design: a survey and the road ahead. *Communications Magazine, IEEE, 43*(12).

Staab, S. (2004). The Pudding of Trust. *IEEE Intelligent Systems*, *19*, 74–88.

Stallings, W. (2012). *Network Security Essentials*. Pearson.

Stepanov, I., Haehner, J., Becker, C., Tian, J., & Rothermel, K. (2003). A Meta-Model and Framework for User Mobility in Mobile Networks. In *Proc. 11th Int. Conf. Netw. ICON 2003*. doi:10.1109/ICON.2003.1266195

Stepanov, I., Marron, P. J., & Rothermel, K. (2005). Mobility Modeling of Outdoor Scenarios for MANETs. In *Proc. 38th Ann. Simul. Symp. ANSS'38*. doi:10.1109/ANSS.2005.29

Stolyar, A. L., & Viswanathan, H. (2009). Self-organizing dynamic fractional frequency reuse for best-effort traffic through distributed inter-cell coordination. In *Proceedings of the Conference on Computer Communications*. Rio de Janerio, Brazil: IEEE. doi:10.1109/INFCOM.2009.5062043

Strogatz, S. H. (2001). Exploring complex networks. *Nature*, *410*(6825), 268–276. doi:10.1038/35065725 PMID:11258382

Su, K., Li, J., & Fu, H. (2011). *Smart city and the applications*. Paper presented at the International Conference on Electronics, Communications and Control (ICECC). doi:10.1109/ICECC.2011.6066743

Su, W., Lee, S. & Gerla, M. (2001). Mobility Prediction and Routing in Ad-Hoc Wireless Networks. *International Journal of Networking Management*.

Su, W., Lee, S., & Gerla, M. (2001). Mobility Prediction and Routing in Ad Hoc Wireless Networks. *International Journal of Networking Management.*

Sun, J. (2001). Mobility management techniques for the next-generation wireless networks. *Proceedings of the Society for Photo-Instrumentation Engineers*, 155–166. doi:10.1117/12.445223

Sun, Y. L., Yu, W., Han, Z., & Liu, K. Jr. (2006). Information Theoretic Framework of Trust Modeling and Evaluation for Ad Hoc Networks. *IEEE Journal on Selected Areas in Communications*, *24*(2), 305–317. doi:10.1109/JSAC.2005.861389

Swilley, E., Hofacker, C. F. & Lamont, B. T. (2012). The Evolution of E-Commerce to M-Commerce: Pressures, Firm Capabilities and Competitive Advantage in Strategic Decision Making. *International Journal of e-Business Research, 8*(1), 1-16.

Swilley, E., & Hofacker, C. F. (2006). Defining mobile commerce in a marketing context. *International Journal of Mobile Marketing*, *1*(2), 18–23.

Szilágyi, P., Bandh, T., & Sanneck, H. (2013). Physical Cell ID Allocation in Multi-layer, Multi-vendor LTE Networks. In A. Timm-Giel, J. Strassner, R. Agüero, S. Sargento, & K. Pentikousis (Eds.), *Mobile Networks and Management* (pp. 156–168). Berlin, Germany: Springer. doi:10.1007/978-3-642-37935-2_12

Tang, D., & Baker, M. (2000). Analysis of A Local-Area Wireless Network. In *Proceeding of the Sixth Annual International Conference on Mobile Computing and Networking (MobiCom)*. ACM Press. doi:10.1145/345910.345912

Tang, D., & Baker, M. (2002). Analysis of A Metropolitan-Area Wireless Network. *Wireless Networks*, *8*(2/3), 107–120. doi:10.1023/A:1013739407600

Tayem, N. A. M. (2005). *Direction of arrival angel estimation schemes for wireless communication system.* (Master's thesis). Wichita State University. Available at http://soar-dev.wichita.edu/bitstream/handle/10057/688/d05020.pdf?sequence=1

Thomas, R.W., Friend, D.H., DaSilva, L.A., & MacKenzie, A.B. (2006). Cognitive networks: adaptation and learning to achieve end-to-end performance objectives. *Communications Magazine, IEEE, 44*(12).

Thomas, R. W., DaSilva, L. A., & Mackenzie, A. B. (2005). Cognitive networks. In *Proc. of IEEE DySPAN 2005.*

Tian, C., Yang, B., Zhongand, J., & Liu, X. (2014). Trust-based incentive mechanism to motivate cooperation in hybrid P2P networks. *Computer Networks*, *73*, 244–255.

Tian, J., Haehner, J., Becker, C., Stepanov, I., & Rothermel, K. (2002). Graph-based Mobility Model for Mobile Ad Hoc Network Simulation. In *Proc. 35th Ann. Simul. Symp.*

Timmers, M., Pollin, S., Dejonghe, A., Van der Perre, L., & Catthoor, F. (2010). A Distributed Multichannel MAC Protocol for Multihop Cognitive Radio Networks. *Vehicular Technology, IEEE Transactions on, 59*(1).

Ting, D. H., Ling, S. F., Patanmacia, T. S., Low, C. G., & Ker, G. C. (2011). Dependency on smartphone and the impact on purchase behaviour. *Young Consumers*, *12*(3), 193–203. doi:10.1108/17473611111163250

Tolety, V. (1999). *Load Reduction in Ad Hoc Networks using Mobile Servers*. (Master's thesis). Colorado School of Mines.

TruePosition. (2004). *An Examination of U-TDOA and Other Wireless Location Technologies: Their Evolution and Their Impact on Today's Wireless Market*. Author.

Tuduce, C., & Gross, T. (2005). A Mobility Model Based on WLAN Traces and its Validation. In *Proceedings of INFOCOM'05*. doi:10.1109/INFCOM.2005.1497932

Turgut, D., Das, S.-K., & Chatterjee, M. (2001). *Longevity of Routes in Mobile Ad Hoc Networks*. VTC. doi:10.1109/VETECS.2001.944118

TWT Whitepaper. (2010). *WiMAX QoS Classes*. TWT.

Ullah, H. (2013). *Multi-antenna aided positioning in LTE networks*. (Master's thesis). Technische Universitaet Darmstadt. Available at http://www.academia.edu/4256293/Theis_report

Urkowitz, H. (1967). Energy detection of unknown deterministic signals. *Proceedings of the IEEE*, *55*(4), 523–531. doi:10.1109/PROC.1967.5573

Vaghefi, R. M. (2014). Improving positioning in LTE through collaboration.*11th Workshop on Positioning, Navigation and Communication (WPNC)*.

Vahdat, A., & Becker, D. (2000). *Epidemic Routing for Partially Connected Ad Hoc Networks*. Technical Report CS-200006, Duke University.

Van den Bergh, J., & Viaene, S. (2015). *Key Challenges for the Smart City: Turning Ambition into Reality*. Paper presented at the 48th Hawaii International Conference on System Sciences (HICSS).

Van Slyke, C., Belanger, F., & Comunale, C. L. (2004). Factors Influencing the adoption of web-based shopping: The impact of trust. *ACM SIGMIS Database*, *35*(2), 32–49. doi:10.1145/1007965.1007969

Van Veen, B. D., & Buckley, K. M. (1988). Beamforming: A versatile approach to spatial filtering. *IEEE ASSP Magazine*, *5*(2), 4–24. doi:10.1109/53.665

Venkatesh, V., Brown, S., & Bala, H. (2013). Bridging the Qualitative- Quantitative Divide: Guidelines for Conducting Mixed-Methodology Research in Information Systems. *Management Information Systems Quarterly*, *37*(1), 21–54.

Viering, I., D¨ottling, M., & Lobinger, A. (2009). *A mathematical perspective of self-optimizing wireless networks*. IEEE International Conference on Communications 2009 (ICC), Dresden, Germany.

VIPIN. (2014). *5G Mobile Technology wiki - wikisedia*. Retrieved May 2015, from http://wikisedia.com/5g-mobile-technology-wiki/

Wanga, F., Wanga, F., Huang, B., & Yang, T. (2012). SONR: A reliable reputation system of self-organized network. *Journal of Network and Computer Applications*, *35*(3), 914–926. doi:10.1016/j.jnca.2011.03.008

Wang, B., Chen, X., & Chang, W. (2014). A Light Weight trust based QOS routing algorithm for Ad Hoc Networks. *Journal of Pervasive and Mobile Computing*, *13*, 64–180. doi:10.1016/j.pmcj.2013.06.004

Wang, K. H., & Li, B. (2002). Group Mobility and Partition Prediction in Wireless ad hoc. *Networks*.

Wang, M.-S., Chen, C. C., Chan, S.-C., & Yang, Y.-H. (2007). Effects of online shopping attitudes, subjective norms and control beliefs on online shopping intentions: A test of the theory of planned behaviour. *International Journal of Management*, *24*(2), 296–302.

Wang, X., Lau, V. K. N., & Ng, T. S. (2000). Performance analysis of multiresolution joint source-channel coding for mobile radio channels with feedback.*IEEE International Conference on Personal Wireless Communications* (pp. 157-161). doi:10.1109/ICPWC.2000.905793

Wang, Y., & Liao, Y. (2007). The conceptualization and measurement of m-commerce user satisfaction. *Computers in Human Behavior*, *23*(1), 381–398. doi:10.1016/j.chb.2004.10.017

Wannstrom, J. (2013). *LTE-Advanced for 3GPP*. Retrieved June, 2013, from http://www.3gpp.org/technologies/keywords-acronyms/97-lte-advanced

Warabino, T., Kaneko, S., Nanba, S., & Kishi, Y. (2012). Advanced load balancing in LTE/LTE-A cellular network. In *Personal Indoor and Mobile Radio Communications (PIMRC), 2012 IEEE 23rd International Symposium on*. doi:10.1109/PIMRC.2012.6362842

Watts, D. J. (1999). *Small Worlds, the Dynamics of Networks between Order and Randomness. Princeton Studies on Complexity*. Princeton University Press.

Weisstein, E. W. (1998). *The CRC Concise Encyclopedia of Mathematics*. CRC Press.

Wheeler, E. (2013). *The emergence of M-Commerce and Geofencing*. Available at: http://www.bloomberg.com/video/from-e-commerce-to-m-commerce-bW0SIsCkRJaTiNTJ89zAnQ.html

Wong, S. H. R. (2012). Which platform do our users prefer: Website or mobile app? *RSR. Reference Services Review*, *40*(1), 113–115.

Wu, T., Rui, L., Xiong, A., & Guo, S. (2010). An Automation PCI Allocation Method for eNodeB and Home eNodeB Cell. In *Proceeding of the 6th International Conference on Wireless Communications Networking and Mobile Computing* (WiCOM'10). doi:10.1109/WICOM.2010.5600764

Wu, Y., Jiang, H., Wu, Y., & Zhang, D. (2010). Physical Cell Identity Self-Organization for Home eNodeB Deployment in LTE. In *Proceeding of the 6th International Conference on Wireless Communications Networking and Mobile Computing* (WiCOM). doi:10.1109/WICOM.2010.5600778

Xi & Su. (2011, February). Cream-mac: Cognitive radio-enabled multi-channel mac protocol over dynamic spectrum access networks. *Selected Topics in Signal Processing. IEEE Journal of*, *5*(1), 110–123.

Xia, H., Jia, Z., Li, X., Ju, L., & Edwin, H. M. (2012). Trust Prediction and Trust based Source Routing in Mobile Ad Hoc Networks. *Ad Hoc Networks*, *11*(7), 2096–2114. doi:10.1016/j.adhoc.2012.02.009

Xiao, Hu, Qian, Gong, & Wang. (2013). Expanding LTE network spectrum with cognitive radios: From concept to implementation. *Wireless Communications, IEEE*, *20*(2). doi:10.1109/MWC.2013.6507389

Xiao, P., He, J., & Fu, Y. (2014). An access authentication protocol for trusted handoff in wireless mesh networks. *Computer Standards & Interfaces*, *36*(3), 480–488. doi:10.1016/j.csi.2013.08.016

Yang, K., & Kim, H.-Y. (2012). Mobile shopping motivation: An application of multiple discriminant analysis. *International Journal of Retail & Distribution Management*, *40*(10), 778–789. doi:10.1108/09590551211263182

Yang, S. C. (2012). Mobile applications and 4G wireless networks: A framework for analysis. *Campus-Wide Information Systems*, *29*(5), 344–357. doi:10.1108/10650741211275107

Yan, Z., Zhang, P., & Vasilakos, A. V. (2014). A survey on trust management for Internet of Things. *Journal of Network and Computer Applications*, *42*, 120–134. doi:10.1016/j.jnca.2014.01.014

Yarali, A., & Rahman, S. (2008). WiMAX broadband wireless access technology: Services, architecture and deployment models.*2008 Canadian Conference on Electrical and Computer Engineering*. doi:10.1109/CCECE.2008.4564499

Yavas, G., Katsaros, D., Ulusoy, O., & Manolopoulos, Y. (2005). A data mining approach for location prediction in mobile environments. *Data & Knowledge Engineering*.

Yeh, Y. S., & Li, Y.-M. (2009). Building trust in e-commerce: Contributions from quality and satisfaction. *Online Information Review*, *33*(6), 1066–1086. doi:10.1108/14684520911011016

Yokoo, M. (2012). *Distributed constraint satisfaction: foundations of cooperation in multi-agent systems*. Berlin, Germany: Springer.

Yokoo, M., & Hirayama, K. (2000). Algorithms for distributed constraint satisfaction: A review. *Autonomous Agents and Multi-Agent Systems*, *3*(2), 185–207. doi:10.1023/A:1010078712316

Yoon, J., Liu, M., & Noble, B. (2003). Random Waypoint Considered Harmful. In *Proceeding of IEEE INFOCOM*.

Yoon, J., Noble, B. D., Liu, M., & Kim, M. (2006). Building Realistic Mobility Models From Coarse-Grained Traces. In *Proceedings of MobiSys'06*. doi:10.1145/1134680.1134699

Youssef, M., Ibrahim, M., Abdelatif, M., Chen, & Vasilakos, A.V. (2014). Routing Metrics of Cognitive Radio Networks: A Survey. *Communications Surveys & Tutorials, IEEE, 16*(1).

Yucek, T., & Arslan, H. (2009). A survey of spectrum sensing algorithms for cognitive radio applications. *IEEE Communications Surveys and Tutorials*, *11*(1), 116–130. doi:10.1109/SURV.2009.090109

Yun, K., & Kim, D. (2007). Robust location tracking using a dual layer particle filter. *Pervasive and Mobile Computing*, *3*(3), 209–232. doi:10.1016/j.pmcj.2006.10.001

Zakrzewska, A., Berger, M. S., & Ruepp, S. (2008). *Modeling Multistandard Wireless Networks in OPNET*. Academic Press.

Zhang, H., & Zhu, L. (2011). *Internet of Things: Key technology, architecture and challenging problems*. Paper presented at the IEEE International Conference on Computer Science and Automation Engineering (CSAE).

Zhang, J., Shankaran, R., Orgun, M. A., Varadharajan, V., & Sattar, A. (2010). A trust management architecture for Hierarchical Wireless Sensor Network. *IEEE Digital Library*.

Zhang, W., Wang, G., Xing, Z., & Wittenburg, L. (2005). Distributed Stochastic Search and Distributed Breakout: Properties, Comparison and Applications to Constraint Optimization Problems in Sensor Networks. Journal of Artificial Intelligence, 161(1-2).

Zhang, X., Neglia, G., Kurose, J., & Towsley, D. (2006). Performance Modeling of Epidemic Routing. IFIP Networking. *Computer Networks*, *51*, 2867–2891.

Zhang, X., Zhou, D., Xiao, Z., Liu, E., Zhang, J., & Glasunov, A. A. (2011). Dynamic Group PCI Assignment Scheme. In *Proceeding of the 7th International Conference on Wireless and Mobile Communications*.

Zhang, J., Chen, C., & Cohen, R. (2010). A Scalable and Effective Trust-Based Framework for Vechicluar Ad Hoc Networks. *Journal of Wireless Mobile Networks Ubiquitous Computing and Dependable Applications*, *1*, 3–15.

Zhang, L., Qie, L., & Xu, J. (2013). Progressive linear precoder optimization for ARQ packet retransmissions in nonregenerative MIMO relay systems. *IEEE Transactions on Signal Processing*, *61*(1), 68–81. doi:10.1109/TSP.2012.2222385

Zhang, W., Wang, G., Xing, Z., & Wittenburg, L. (2005). Distributed stochastic search and distributed breakout: Properties, comparison and applications to constraint optimization problems in sensor networks. *Artificial Intelligence*, *161*(1), 55–87. doi:10.1016/j.artint.2004.10.004

Zhao, L., Zhang, W., Song, W., & Zhang, H. (2011). Cognitive radio csma/ca protocol for femto-wlans. In *Communication Software and Networks (ICCSN),2011IEEE 3rd International Conference on.*

Zhou, B., Xu, K., & Gerla, M. (2004). *Group and Swarm Mobility Models for Ad Hoc Network Scenarios using Virtual Tracks*. MILCOM'04.

Zhou, L., & Haas, Z. J. (1999). Securing Ad Hoc Networks. *IEEE Digital Library*.

Zhou, T. (2011). Examining the critical success factors of mobile site adoption. *Online Information Review*, *35*(4), 636–652. doi:10.1108/14684521111161972

Zhou, T. (2013). Understanding continuance usage of mobile site. *Industrial Management & Data Systems*, *113*(9), 1286–1299. doi:10.1108/IMDS-01-2013-0001

Zhu, H., & Ray Liu, K. J. (2008). *Resource Allocation for Wireless Networks: Basics, Techniques, and Applications*. New York, NY: Cambridge University Press.

Zia, N., & Mitschele-Thiel, A. (2012). Self-organized neighborhood mobility load balancing for LTE networks. Wireless Days (WD), 2013 IFIP.

Zimmermann, H.-M., & Gruber, I. (2005). A Voronoi-based Mobility Model for Urban Environments. In *Eur. Wirel Conf.*, Zypern, Greece.

Zoltowski, M. D., Haardt, M., & Mathews, C. P. (1996). Closed-form 2-D angle estimation with rectangular arrays in element space or beamspace via unitary ESPRIT. *Signal Processing. IEEE Transactions on*, *44*(2), 316–328. doi:10.1109/78.485927

Zonoozi, M., & Dassanayake, P. (1997). User Mobility Modeling and Characterization of Mobility Pattern. *IEEE Journal on Selected Areas in Communications*, *15*(7), 1239–1252. doi:10.1109/49.622908

About the Contributors

Ali Diab received the B.E. degree in Electronic Engineering from the Damascus University, Damascus, Syria, in 1999 and the Diploma degree in Computer Science and Automation from the same university in 2000. After that, he obtained his Dr.-Ing. Title from the Ilmenau University of Technology, Faculty of Computer Science in 2010. His dissertation focused on mobility management in IP-based networks. Currently, he is pursuing his postdoctoral degree at the Ilmenau University of Technology on the topic "Self-Organized Future Mobile Communication Networks."

* * *

Farhana Afroz obtained her MS degree and B.Sc. (Honours) degree from the department of Applied Physics, Electronics and Communication Engineering of University of Dhaka, Bangladesh. She is now pursuing Master of Engineering studies with Telecommunication Engineering major in University of Technology, Sydney, Australia. She worked as a Lecturer for the department of Electronics and Telecommunication Engineering in University of Liberal Arts Bangladesh, during 2007-2010 time period. Her research areas are 4G mobile networks and digital signal processing.

Furqan Ahmed received the B.E. and M.S. degrees in computer engineering and telecommunications engineering, respectively, from National University of Sciences & Technology, Pakistan. He is currently pursuing the Ph.D. degree at the Department of Communications & Networking, Aalto University, Finland. His research interests include resource allocation, self-organization, distributed algorithms, and related areas in wireless networking and signal processing.

Syed Muhammad Kashif Alam received MS – Information Technology degree from Graduate School of Engineering Sciences and Technology (GSESIT), Hamdard University Karachi in 2010, and Bachelor of Engineering from Hamdard Institute of Engineering and Technology (HIET), Hamdard University Karachi in 2004. His research Interests are Computer and Communication Networks, Computer Network Security and Cloud Computing.

Shouman Barua is a PhD research scholar at the University of Technology, Sydney. He received his BSc in Electrical and Electronic Engineering from Chittagong University of Engineering and Technology, Bangladesh and MSc in Information and Communication Engineering from Technische Universität Darmstadt (Technical University of Darmstadt), Germany in 2006 and 2014 respectively. He holds also more than five years extensive working experience in telecommunication sector in various roles including network planning and operation.

Anup Kumar Bhattacharjee completed B.E. (E&TC) in 1983,M.E.Tel.E. in1985 from Jadavpur University. He completed Ph.D.(Engg.) in 1989. He has more than 25 years of teaching and research experience. He has published more than 150 research paper in various international journals and conferences of repute. He guided six PhD students successfully. His areas of interests are Antenna, EM Theory, Radar Engg., Microwave Circuits, Microwave Devices, and Satellite Communication.

Angana Chakraborty is M-Tech(IT) from Indian Institute Of Engineering,Science and Technology(IIEST),Shibpur and B-Tech(CSE) from West Bengal University Of Technology(WBUT).Her research interests include Mobile WiMAX, Wireless Networks, Mobility Management. She published 3 papers in international conference proceedings of repute.

Surajit Deka is a Ph.D research scholar at the Department of Electronics and Communication Engineering, Gauhati University, India. He has completed his M-Tech in Electronics and Communication Technology in 2013 from Gauhati University.

Alexis Dowhuszko is a Post-Doctoral researcher at the Department of Communications and Networking, Aalto University, Finland, since April 2010. He received his Telecommunications Engineer degree from Blas Pascal University, Cordoba, Argentina, in 2002, and his Ph.D. degree in Engineering Sciences from the National University of Cordoba, Argentina, in 2010. Apart from his research activities, he has been responsible for practical project management tasks and international collaboration duties between Aalto and other national/international universities. He has also been involved in teaching activities of both under- and post-graduate courses. His current research interests lie in the area of radio resource management for future ultra-dense wireless networks, including multi-antenna communication schemes, time- and frequency-domain scheduling, power control techniques, and spectrum sharing approaches. He has authored or co-authored 10+ journal papers, 20+ conference articles, and 1 patent application.

Pantha Ghosal is a Graduate Research Assistant at University of Technology, Sydney. Prior to this, he completed B.Sc in Electrical and Electronic Engineering from Rajshahi University of Engineering & Technology, Bangladesh in 2007. He is an expert of Telecommunication network design and holds more than 7 years of working experience in this area.

K. Prathapchandran Kannimuthu received his Bachelor Degree in Computer Applications from Madurai Kamaraj University in the year 2005 and received his Masters Degree in Computer Applications from Gandhigram Rural Institute- Deemed University in the year 2008 and received Master of Philosophy in Computer Science from Bharathidasan University. At present he is a research scholar of the Department of Computer Science and Applications, Gandhigram Rural Institute- Deemed University. He has published papers both in conferences and journals and has a publication in a book chapter. He has two years of experience in teaching. His areas of interest are Computer Networks, Self-Organized Networks, Internet of Things, Mobile Ad Hoc Networks and Security.

Iqbal Uddin Khan received MS – Telecommunication degree from Graduate School of Engineering Sciences and Technology (GSESIT) Hamdard University Karachi in 2013, B-Tech Electronics from Indus Institute of higher Education Karachi Pakistan in 2011. Currently he is pursuing Ph.D. in Computer Science from Graduate School of Engineering Sciences and Technology (GSESIT) Hamdard University. His research Interests are Computer and Communication Networks, Computer Network Security and Cloud Computing.

Xiaoying Kong has broad interests in control engineering and software engineering.Her recent research work has been in GPS, inertial navigation systems, data fusion, robotics, sensor networks, Internet monitoring systems, agile software development methodologies, web technologies, web design methodologies, web architecture framework, analytical models of software development, and value-based software engineering. She has many years work experience in aeronautical industry, semiconductor industry, and software industry.

Nidhi Mathur has done post-graduation in Electronics Science and PhD from Lucknow University, Lucknow. She worked as a Lecturer in Electronics and Communication Engineering Department at Amity School of Engineering and Technology (Lucknow campus). She is a member of IETE (India.). Her current research interests are mobile communication and Image Processing.

Sumita Mishra received Ph.D degree in Electronics from DRML Awadh University, India. She received M. Tech. (Optical Communication) and M.Sc. (Electronics) degree from SGSITS, Indore and University of Lucknow in 2004 and 2000 respectively. She is currently Assistant professor at Amity School of Engineering and Technology, Amity University, Lucknow India. She is a member of IET (U.K.), IEEE, and IACSIT. Her current research interests include mobile communication, Optical interconnects and Machine Vision.

Andreas Mitschele-Thiel is a full professor at the Ilmenau University of Technology, Germany, and head of the Integrated Communication Systems lab. In addition, he is the head of the International Graduate School on Mobile Communications of the University. 2005 to 2009, he also served as Dean for the Faculty for Computer Science and Automation. In addition, he is co-founder and scientific director of two research spin-offs of the university, Cuculus and IDEO Labs. He received a Diploma in Computer Engineering from the Fachhochschule Esslingen in 1985, a M.S. in Computer and Information Science from The Ohio State University in 1989 and a Doctoral degree in Computer Science from the University of Erlangen in 1994. He completed his habilitation in Computer Science at the University of Erlangen in 2000. Andreas Mitschele-Thiel has held various positions in development, research and management at Alcatel and Lucent Bell Labs. His research focuses on the engineering of telecommunication systems. Special interests are in mobile communication networks, especially self-organization in next generation mobile networks, and IP-based mobile communication systems.

Asish K. Mukhopadhyay, Group Director, SR Group of Institutions, Jhansi is M.Tech (ECE) from IIT Kharagpur and Ph.D(Engg) in Computer Sc&Engg from Jadavpur University, Kolkata. He has 40 years of experience in teaching, research, administration, including 20 years at Indian Institute of Technology Kharagpur. He also served institutions like NERIST (Itanagar), CEMK (Kolaghat), NiT (Kolkata), BCREC (Durgapur), BCETW (Durgapur) and BITM (Shantiniketan), holding various positions such as Director, Principal, Dean, HoD, faculty etc. He served as Chairman, CSI, Durgapur Chapter, Honorary Secretary, Institution of Engineers (India), Kharagpur Centre, Member, Board of Studies, WB University of Technology, Kolkata and Vidyasagar University, Medinipur, Co-Chairman, Technical Committee, 14th Indian Engineering Congress (2001), Editor-in-Chief, IJRRNGN. He has over 85 research publications in national/ international journals and conference proceedings and delivered many key note/theme lectures in conferences. He is co-author of two book chapters published by IGI Global, USA and editor of one international book published by ACEEE. He is honorary Life Fellow of ACEEE, Life Fellow of the Institution of Engineers (I), Sr. Member, IEEE, Sr. Life Member, CSI; Life Member of ISTE, SSI and IETE.

Ahmar Murtaza completed his MS in Information Technology from Hamdard University, Karachi, Pakistan in 2013. He is working as Lab Instructor in Department of Computing, HIET, FEST, Hamdard University since last month. He is associated with the Hamdard University since 5 years as Lab Attendant. His research interests are Computer Networks, Mobile Computing, Electronic Engineering, etc.

Garimella Rama Murthy received Ph.D degree in Computer Engineering from Purdue University, West Lafayette, USA. He is currently a professor at the International Institute of Information Technology, Hyderabad, India. He is a fellow of IETE, Senior Member of ACM. He received Rashtriya Gaurav Award from IIFS, New Delhi, India. His research monograph entitled, "Multi-Dimensional Neural Networks: Unified Theory" is published by New Age International Publishers, New Delhi, India.

Wilson Ozuem teaches and researches communications issues in computer-mediated marketing environments (CMMEs). His research interests include the implications of information technology for decision making by marketers and consumers. The results of his research have been published in scholarly journals, books and international conference papers, including *European Journal of Marketing, Journal of Strategic Journal, Social Responsibility Journal, International Journal of Applied Behavioural Economics*. Dr Ozuem is a Visiting Senior Lecturer at the University of Hertfordshire (UK), and a Visiting Senior Lecturer at the University of Gloucestershire. Dr Ozuem is an Associate Professor in Digital Marketing at Regent's University, London (UK). Prof Wilson currently holds a visiting professorial fellowship in the Agribusines, Albania Agricultural University (Albania). He currently teaches MSc and MBA courses in marketing communications, digital marketing, research methodology, and marketing management. He is a Fellow of the Chartered Institute of Marketing (FCIM) and has worked as a marketing consultant for several companies and organizations. He received his BA in Business from the University of Portsmouth, MA Marketing from University of West London, MBA from London Metropolitan University, MEd (Educational Leadership and Management) from Open University (UK), PhD in Digital Marketing from Anglia Ruskin University and Postgraduate Diploma/MA in Educational Research from University of Cambridge.

Chandan Pradhan is currently pursuing MS by Research from IIIT Hyderabad, India. His area of interest are signal processing, wireless communication and cognitive radio.

Aqeel ur Rehman received MS in Information Technology from Hamdrad University, Karachi, Pakistan in 2001 and BS in Electronic Engineering from Sir Syed University of Engineering and Technology, Karachi, Pakistan in 1998. He completed his Ph.D. degree in Computer Science from National

University of Computer and Emerging Sciences, Karachi, Pakistan in 2012. He is associated with Faculty of Engineering Sciences and Technology, Hamdard University, Karachi, Pakistan as the Associate Professor and is the Chairman of Department of Computing. His current research interests include Sensor Networks, Ubiquitous Computing, Smart Agriculture, Context Modeling and Computer Networks.

Sajal Sahais currently working as Assistant professor in the department of computer application, Narula Institute of Technology, Kolkata, India since august 2004.He received his ME degree in InformationTechnology from West Bengal University of Technology (2007), MCA degree from Sikkim manipalInstitute of Technology, Sikkim (2004) and BSc (Physics) degree from Calcutta University(2001).His current research area includes Overlay based heterogeneous networks, Mobile IP route optimization,Remote Sensing and GIS. Shri saha is the recipient of Silver medal for securing 2nd position in ME examination. He has nine years of professional experience in educational and research field. He contributed significantly in the area of Mobile computing for the last five years. He published 20 papers in refereed international and national journals and conferences of repute. Total citation count of all his papers are 20 till now and h-index: 3(source Google Scholar). He authored a book titled "GIS and Remote Sensing:Applications in Flood Damage assessment" published by Lap Lambert Academic Publishing, Germany.

Kumbesan Sandrasegaran is an Associate Professor at UTS and Centre for Real-Time Information Networks (CRIN). He holds a PhD in Electrical Engineering from McGill University (Canada)(1994), a Master of Science Degree in Telecommunication Engineering from Essex University (1988) and a Bachelor of Science (Honours) Degree in Electrical Engineering (First Class) (1985). His current research work focuses on two main areas (a) radio resource management in mobile networks, (b) engineering of remote monitoring systems for novel applications with industry through the use of embedded systems, sensors and communications systems. He has published over 100 refereed publications and 20 consultancy reports spanning telecommunication and computing systems.

Kunal Sankhe is pursuing M.S. in ECE in IIIT Hyderabad. His research area is cognitive radio in cellular network.

Kandarpa Kumar Sarma serves as Associate Professor in the department of Electronics and Communication Technology at Gauhati University. His research interests include Mobile and Wireless Communication, Soft-Computation, Speech Processing, Antenna Design, Document Image Analysis, Digital Signal Processing.

S. Sivagurunathan Shanmugam is an Assistant Professor in the Department of Computer Science and Applications, Gandhigram Rural Institute- Deemed University. He received his B.sc degree in Physics from Madurai Kamaraj University in the year 1995 and M.C.A degree in Computer Applications and M.Phil degree in Computer Science from Madurai Kamaraj University in the year 1998 and 2004 respectively. He received his Ph.D degree in Network Security from Anna University in the year 2010. He has seventeen years of experience in teaching. He is a life member of Computer Society of India (CSI). He has more than fifteen publications in Journals and conferences and three publications as book chapters. His areas of interest include Mobile Ad Hoc Network, Network Security, Self-Organized Networks and Internet of Things.

Deepti Singhal is pursuing PhD in CSE from IIIT, Hyderabad. She has completed M. Tech in Information and Communication Technology (ICT) from Dhirubhai Ambani Institute of Information and Communication Technology (DAIICT), Gandhinagar, India, in 2007. She has 3+ years of industry research experience, and was part of IMT Advance India Evaluation Group. She is also a Technical Reviewer of - IEEE International Power and Energy Conference (PECON 2012) - IEEE Symposium on Wireless Telecommunications Applications (ISWTA2012) - IEEE Symposium on Business, Engineering & Industrial Applications (ISBEIA 2012) - IEEE Conference on Computer Applications and Industrial Electronics (ICCAIE 2011), http://www.i-cerg.net/iccaie2011/. - Journal of Engineering and Computer Innovations, http://www.academicjournals.org/JECI/.

Ramprasad Subramanian is an experienced telecom engineer in the field of 2G/3G and LTE/LTE-A. He holds M.S (By research) in Information and Communication from Institute of Remote Sensing, Anna University (India)(2007) and Bachelors of Engineering in Electronics and Communication engineering from Bharathidasan University (2001)(India). He has done many projects in the area of 2G/3G and LTE. He has done many consultative projects across Africs/Americas/Asia etc. He was the recipient of India's best invention award for the year 2004 from Indian Institute of Management Ahmadabad and Government of India. His current research focuses on 4G mobile networks and vehicular Ad hoc networks.

Olav Tirkkonen is associate professor in communication theory at the Department of Communications and Networking in Aalto University, Finland, where he has held a faculty position since August 2006. He received his M.Sc. and Ph.D. degrees in theoretical physics from Helsinki University of Technology in 1990 and 1994, respectively. Between 1994 and 1999 he held post-doctoral positions at the University of British Columbia, Vancouver, Canada, and the Nordic Institute for Theoretical Physics, Copenhagen, Denmark. From 1999 to 2010 he was with Nokia Research Center (NRC), Helsinki, Finland, most recently acting as Research Fellow. He has published some 190 papers, and is coauthor of the book "Multiantenna transceiver techniques for 3G and beyond". His current research interests are in coding theory, multiantenna techniques, as well as cognitive, heterogeneous and self-organizing wireless networks.

Index

J

L

M

N

O

P

Q

R

S

T

U

V

W

com/pod-product-compliance

06B/1328

8 1 5 2 2 5 0 2 3 9 5 *